T0211036

Lecture Notes in Computer Science 14123

Founding Editors

Gerhard Goos
Juris Hartmanis

The series Lecture Notes in Computer Science (LNCS), including its subseries Lecture Notes in Artificial Intelligence (LNAI) and Lecture Notes in Bioinformatics (LNBI), has established itself as a medium for the publication of new developments in computer science and information technology research, teaching, and education.

LNCS enjoys close cooperation with the computer science R & D community, the series counts many renowned academics among its volume editors and paper authors, and collaborates with prestigious societies. Its mission is to serve this international community by providing an invaluable service, mainly focused on the publication of conference and workshop proceedings and postproceedings. LNCS commenced publication in 1973.

Vladimir M. Vishnevskiy ·
Konstantin E. Samouylov · Dmitry V. Kozyrev
Editors

Distributed Computer and Communication Networks: Control, Computation, Communications

26th International Conference, DCCN 2023
Moscow, Russia, September 25–29, 2023
Revised Selected Papers

 Springer

Editors
Vladimir M. Vishnevskiy (ID)
V.A. Trapeznikov Institute of Control
Sciences of Russian Academy of Sciences
Moscow, Russia

Konstantin E. Samouylov (ID)
Peoples' Friendship University of Russia
Moscow, Russia

Dmitry V. Kozyrev (ID)
V.A. Trapeznikov Institute of Control
Sciences of Russian Academy of Sciences
Moscow, Russia

Peoples' Friendship University of Russia
(RUDN University)
Moscow, Russia

ISSN 0302-9743   ISSN 1611-3349 (electronic)
Lecture Notes in Computer Science
ISBN 978-3-031-50481-5   ISBN 978-3-031-50482-2 (eBook)
https://doi.org/10.1007/978-3-031-50482-2

This Springer imprint is published by the registered company Springer Nature Switzerland AG
The registered company address is: Gewerbestrasse 11, 6330 Cham, Switzerland

Paper in this product is recyclable.

Preface

This volume contains a collection of revised selected full-text papers presented at the 26th International Conference on Distributed Computer and Communication Networks (DCCN 2023), held in Moscow, Russia, during September 25–29, 2023. DCCN 2023 was jointly organized by the Russian Academy of Sciences (RAS), the V.A. Trapeznikov Institute of Control Sciences of RAS (ICS RAS), the Peoples' Friendship University of Russia (RUDN University), the National Research Tomsk State University, and the Institute of Information and Communication Technologies of Bulgarian Academy of Sciences (IICT BAS).

The conference was a continuation of the traditional international conferences of the DCCN series, which have taken place in Sofia, Bulgaria (1995, 2005, 2006, 2008, 2009, 2014); Tel Aviv, Israel (1996, 1997, 1999, 2001); and Moscow, Russia (1998, 2000, 2003, 2007, 2010, 2011, 2013, 2015–2022) in the last 26 years. The main idea of the conference was to provide a platform and forum for researchers and developers from academia and industry from various countries working in the area of theory and applications of distributed computer and communication networks, mathematical modeling, and methods of control and optimization of distributed systems, by offering them a unique opportunity to share their views, discuss prospective developments, and pursue collaboration in this area. The content of this volume is related to the following subjects:

- Communication networks, algorithms, and protocols
- Wireless and mobile networks
- Computer and telecommunication networks control and management
- Performance analysis, QoS/QoE evaluation, and network efficiency
- Analytical modeling and simulation of communication systems
- Evolution of wireless networks toward 5G
- Centimeter- and millimeter-wave radio technologies
- Internet of Things and fog computing
- Cloud computing, distributed and parallel systems
- Machine learning, big data, and artificial intelligence
- Probabilistic and statistical models in information systems
- Queuing theory and reliability theory applications
- High-altitude telecommunications platforms
- Security in infocommunication systems

The DCCN 2023 conference gathered 122 submissions from authors from 18 different countries. From these, 105 high-quality papers in English were accepted and presented during the conference. All submissions underwent a rigorous double-blind peer-review process with 3 reviews per submission. The current volume contains 41 extended papers which were recommended by session chairs and selected by the Program Committee for the Springer post-proceedings. Thus, the acceptance rate is 39%.

All the papers selected for the post-proceedings volume are given in the form presented by the authors. These papers are of interest to everyone working in the field of computer and communication networks.

We thank all the authors for their interest in DCCN, the members of the Program Committee for their contributions, and the reviewers for their peer-reviewing efforts.

September 2023

Vladimir M. Vishnevskiy
Konstantin E. Samouylov
Dmitry V. Kozyrev

Organization

Program Committee Chairs

V. M. Vishnevskiy (Chair)   ICS RAS, Russia
K. E. Samouylov (Co-chair)   RUDN University, Russia

Publication and Publicity Chair

D. V. Kozyrev   ICS RAS and RUDN University, Russia

International Program Committee

S. M. Abramov   Program Systems Institute of RAS, Russia
A. M. Andronov   Transport and Telecommunication Institute,
                               Latvia
T. Atanasova   IICT BAS, Bulgaria
S. E. Bankov   Kotelnikov Institute of Radio Engineering and
                               Electronics of RAS, Russia
A. S. Bugaev   Moscow Institute of Physics and Technology,
                               Russia
S. R. Chakravarthy   Kettering University, USA
D. Deng   National Changhua University of Education,
                               Taiwan
S. Dharmaraja   Indian Institute of Technology, Delhi, India
A. N. Dudin   Belarusian State University, Belarus
A. V. Dvorkovich   Moscow Institute of Physics and Technology,
                               Russia
D. V. Efrosinin   Johannes Kepler University Linz, Austria
Yu. V. Gaidamaka   RUDN University, Russia
Yu. V. Gulyaev   Kotelnikov Institute of Radio-Engineering and
                               Electronics of RAS, Russia
V. C. Joshua   CMS College Kottayam, India
H. Karatza   Aristotle University of Thessaloniki, Greece
N. Kolev   University of São Paulo, Brazil
G. Kotsis   Johannes Kepler University Linz, Austria

A. E. Koucheryavy	Bonch-Bruevich Saint-Petersburg State University of Telecommunications, Russia
A. Krishnamoorthy	Cochin University of Science and Technology, India
R. Kumar	Namibia University of Science and Technology, Namibia
N. A. Kuznetsov	Moscow Institute of Physics and Technology, Russia
L. Lakatos	Budapest University, Hungary
E. Levner	Holon Institute of Technology, Israel
S. D. Margenov	Institute of Information and Communication Technologies of Bulgarian Academy of Sciences, Bulgaria
N. Markovich	ICS RAS, Russia
A. Melikov	Institute of Cybernetics of the Azerbaijan National Academy of Sciences, Azerbaijan
E. V. Morozov	Institute of Applied Mathematical Research of the Karelian Research Centre RAS, Russia
A. A. Nazarov	Tomsk State University, Russia
I. V. Nikiforov	Université de Technologie de Troyes, France
S. A. Nikitov	Kotelnikov Institute of Radio Engineering and Electronics of RAS, Russia
D. A. Novikov	ICS RAS, Russia
M. Pagano	University of Pisa, Italy
V. V. Rykov	Gubkin Russian State University of Oil and Gas, Russia
R. L. Smeliansky	Lomonosov Moscow State University, Russia
M. A. Sneps-Sneppe	Ventspils University College, Latvia
A. N. Sobolevski	Institute for Information Transmission Problems of RAS, Russia
S. N. Stepanov	Moscow Technical University of Communication and Informatics, Russia
S. P. Suschenko	Tomsk State University, Russia
J. Sztrik	University of Debrecen, Hungary
S. N. Vasiliev	ICS RAS, Russia
M. Xie	City University of Hong Kong, China
A. Zaslavsky	Deakin University, Australia

Organizing Committee

| V. M. Vishnevskiy (Chair) | ICS RAS, Russia |
| K. E. Samouylov (Vice Chair) | RUDN University, Russia |

D. V. Kozyrev (Publication and Publicity Chair)	ICS RAS and RUDN University, Russia
A. A. Larionov	ICS RAS, Russia
Y. S. Aleksandrova	ICS RAS, Russia
S. P. Moiseeva	Tomsk State University, Russia
T. Atanasova	IIICT BAS, Bulgaria
I. A. Kochetkova	RUDN University, Russia

Organizers and Partners

Organizers

Russian Academy of Sciences (RAS), Russia
V.A. Trapeznikov Institute of Control Sciences of RAS, Russia
RUDN University, Russia
National Research Tomsk State University, Russia
Institute of Information and Communication Technologies of Bulgarian Academy of Sciences, Bulgaria
Research and Development Company "Information and Networking Technologies", Russia

Support

Information support was provided by the Russian Academy of Sciences. The conference was organized with the support of the IEEE Russia Section, Communications Society Chapter (COM19) and the RUDN University Strategic Academic Leadership Program.

Contents

Analytical Modeling of Distributed Systems

Computer and Communication Networks

Distributed Systems Applications

.

Influence of Access Points' Height and High Signal Relation in WLAN Fingerprinting-Based Indoor Positioning Systems' Accuracy

Mrindoko R. Nicholaus[1], Francis A. Ruambo[2,3,4], Elijah E. Masanga[4], Mohammed Saleh Ali Muthanna[5(✉)] [iD], and Andrei Lashchev[5]

[1] Computer Science and Engineering Department, College of Information and Communication Technology, Mbeya University of Science and Technology, Mbeya 131, Tanzania
[2] Information Systems and Technology Department, College of Information and Communication Technology, Mbeya University of Science and Technology, Mbeya 131, Tanzania
[3] School of Cyber Science and Engineering, Huazhong University of Science and Technology, Wuhan 430074, China
[4] Research and Development Department, Aifrruis Laboratories, Mbeya, Tanzania
[5] Institute of Computer Technologies and Information Security, Southern Federal University, 347922 Taganrog, Russia
muthanna@sfedu.ru

Abstract. Wireless Local Area Network (WLAN) Fingerprinting-based Indoor Positioning Systems (IPS) offer several key advantages over other indoor positioning technologies, including cost-effectiveness with high accuracy and passive tracking of entities' location without Global Positioning System (GPS) dependence. However, achieving a satisfactory accuracy performance of the WLAN fingerprinting-based IPS is still challenging, as it is affected by several factors, such as obstacles (e.g., walls and furniture) that can block or weaken wireless signals. Additionally, the IPS's accuracy can be affected by environmental changes, such as the height, addition, or removal of a WLAN Access Point (AP) or changes in radio frequency interference. This paper presents a novel algorithm incorporating a probabilistic analytical model employing only high signal relations that mitigate the effect of low signal relation on the WLAN fingerprinting-based IPS, thus improving the accuracy performance. Furthermore, it analyses the impact of AP heights on the accuracy performance of the WLAN IPS. Upon extensive experiments, the proposed algorithm improves the accuracy of the IPS by an average of 32.74% in terms of Root Mean Square Error (RMSE). These results imply that incorporating APs' height and high signal relations can significantly improve the accuracy performance of the WLAN fingerprinting-based IPS.

Keywords: Fingerprinting · Indoor Positioning · Probabilistic Analytical Model · WLAN

1   Introduction

Indoor Positioning Systems (IPS) are becoming essential and prevalent due to the wide range of wireless technologies. The IPS offers positioning services in various ways depending on the positioning techniques employed and can allow context-aware computing with position awareness [1,2]. For example, IPS is also crucial for disaster services, where Global Positioning System (GPS) has not been operating well. In this regard, position/location fingerprinting techniques based on the current Wireless Local Area (WLAN) planning infrastructure have been proposed for the indoor environment [3,4]. Such a system could provide more services to the existing WLAN settings [5,6]. There are several existing implementation techniques implanting WLAN IPS; however, upon comparison among these techniques like Time-of-Arrival (ToA) and Angle-of-Arrival (AoA), the fingerprinting technique is relatively simple to implement [7]. The TOA app-roach, in particular, is based on computing the variation in arrival time: the time it takes for a signal to travel from a radio transmitter to one or more signal receivers follows a straight line. After synchronizing the transmitter and each receiver, The traversing time difference will be determined by recording the sig-nal's reception time along each path [8]. In this regard, at least three reference points (encompassing relevant signals) are required for accurate TOA measure-ments to support the 2-D location, as shown in Fig. 1 [9]. The requirement for perfect time synchronization of all stations, particularly the mobile device, is a disadvantage of the ToA technique (which might be a frightening problem for some IEEE 802.11 client device implementations). Because propagation periods over a few meters' distance are in the order of nanoseconds, extremely slight differences in time synchronization can result in very substantial mistakes in position accuracy. A time measurement inaccuracy of 100 ns, for example, can result in a 30 m localization error.

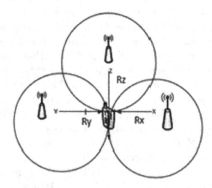

Fig. 1. Range-based Positioning in the ideal case [10]

The AoA approach, on the other hand, computes the transmitter's position by intersecting several pairs of angle track lines [11]. Figure 2 shows how the

AoA technique can use at least two well-known reference locations (A, B) and two approximated angles 1 and 2 to identify the 2-D position of the object P. AoA estimation (aka, direction finding), which can be accomplished using either directional antennae or a collection of the antenna [12]. Additionally, AoA can calculate a position with minimal measurements without requiring temporal synchronization among measuring devices. However, line-of-sight (LOS) communication is necessary to attain this ideal situation [13]. Non-line-of-sight (NLOS) mistakes are caused by the multi-path and shadowing effects, which can contribute to errors. Another disadvantage of AoA approaches is the considerable and sophisticated hardware requirements and location estimation degradation when the mobile target moves away from the measuring units [14].

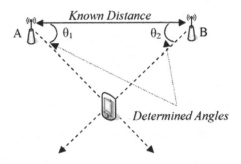

Fig. 2. Mechanism of AoA [Source: [15]]

In this regard, WLAN fingerprinting-based IPS has become a favorable option and is applicable in many applications, such as Indoor navigation: WLAN fingerprinting-based IPS can provide turn-by-turn directions for indoor navigation. Asset tracking: WLAN fingerprinting-based IPS can track the position of assets within a facility, such as equipment or inventory. Emergency response: In an emergency, such as a fire or natural disaster, WLAN fingerprinting-based IPS can help locate people quickly. Retail: WLAN fingerprinting-based IPS can track customer movement within a store and provide targeted advertising or offers. Overall, WLAN fingerprinting-based IPS is a powerful technology that can cost-effectively provide highly accurate indoor positioning, and it has a wide range of applications in various fields.

Generally, the role of WLAN fingerprinting-based indoor positioning systems (IPS) is to determine a device's position within a specific area using data from wireless signals. This technique is typically done by collecting data on the strength and characteristics of wireless signals in a given area and then using that data to create a "fingerprint" of the area. Then by matching its current wireless signal data to the fingerprint of the area, the determination of the device's position is possible. Typically, the development of WLAN-based fingerprinting positioning systems can be divided into two stages: offline training, during which the position fingerprints are collected by conducting a Received Signal Strength

Indicator (RSSI) site survey from various Access Points (APs). Then, the mobile user's position is determined while in the online phase by comparing the most recent RSSI measurements received with the fingerprint RSSI measurements that were previously gathered. As seen in Fig. 3, a rectangular network (grid) of points conceals the entire area under examination, experimentation, or implementation [16–18].

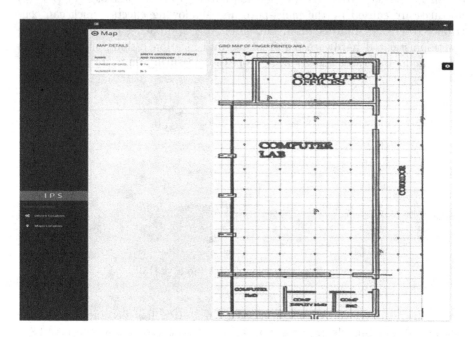

Fig. 3. Map with fingerprinted reference points and the grid.

Towards this research direction, several existing research works focus on improving accuracy performance using different approaches. Most recent work on indoor positioning systems focuses on studies that enhance positioning accuracy and estimation algorithms. In contrast, fewer studies consider the effect of transmitter height and RSSI signal relation on performance accuracy. In their research, Zeng et al. [19] revealed that adopting RSSI-based ranging technology in indoor three-dimensional spatial positioning systems was feasible and possible via the RSSI characteristics' research and analysis. Leveraging this idea for a 2.4 GHz wireless signal, Zhang et al. [20] examined the connection between the path loss and the height of the transmitting antenna. However, as the signal transmitter's height, the WLAN AP was at a fixed point without consideration of the effects of the height difference on the signal. In another research work, rather than dealing with 2.4 GHz radio frequency transmission, Hu et al. investigated the relationships between signal propagation properties and variables, including communication distance, AP height, and propagation path on a 5.8 GHz radio frequency transmission. Instead of measuring (monitoring) and examining the influencing factors in the indoor environment, research work by Li [21] concentrated on outdoor and indoor environments which showed that the difference in the antenna's height significantly changes the received signal intensity by the user's terminal. Another research work by Kaemarungsi and Krishnamurthy studied [22] user presence and orientation influence on RSSI. Furthermore, Bahl and Padmanabhan [23] observed that the user's orientation leads to a variation in RSSI level and recommended incorporating the RSSI data with the user's orientation-aware in computing the user's position, especially in human-related indoor positioning. However, these studies did not contemplate the effect of signal relation and APs' height.

Based on the reviewed literature in this study, the factors impacting the indoor RSSI characteristics, which affect positioning accuracy, have not been the subject of systematic quantitative research for the indoor positioning method based on RSSI studies. Moreover, no study in the existing literature review directly dwelt on the effect of the height of the AP on the WLAN fingerprinting-based IPS accuracy performance while considering high signal relations. Therefore, there is a need to analyze the impact of AP height on WLAN-based IPS accuracy while considering high signal relations. This work considers the height difference of the transmitter and high signal relation to quantitatively analyze the features of the 2.4 GHz RSSI gathered in indoor settings. It adopts a probabilistic analytical model to examine the effect of the accuracy performance of WLAN fingerprinting-based IPS. The primary contributions of this research paper are three-fold:

- We collect RSSI data for different APs' heights using a tailored-made approach for WLAN fingerprinting-based IPS.
- Propose an algorithm for selecting high-signal relations and computing position employing the position fingerprints and the detected RSS as two separate random variables using the probability density function (PDF) in the signal space.
- Analyze the performance accuracy of the proposed algorithm for the WLAN fingerprinting-based IPS.

The organization of this paper is as follows. First, Sect. 2 describes the WLAN fingerprinting-based IPS conceptual framework and its operation tailored to realize our idea. Then, Sect. 3 explains the experimental design and setup incorporating the proposed algorithm for improving the WLAN-based IPS accuracy performance. Next, the performance results, together with the discussion, are presented in Sect. 4. Finally, Sect. 5 concludes the paper and suggests future work.

2 WLAN Fingerprinting-Based Indoor Positioning Systems

This section describes the theoretical analysis and conceptual framework of a WLAN fingerprinting-based indoor positioning system, including data collection and processing, fingerprinting database construction, and location computation using signal relations.

2.1 Theoretical Analysis

Understanding the technical features of WLAN indoor fingerprinting and its limits is required for a theoretical study. The presence of multi-path propagation, in which signals reflected from walls and other obstacles generate interference, is one of the critical hurdles to indoor fingerprinting.

Furthermore, environmental changes, such as adding or removing obstructions, might alter the system's accuracy, changing the signal's behavior and resulting in inaccurate position estimates. In this situation, the radio signal is thought to go through four phases in a location fingerprinting system, either online or offline. Transmitting, traversing, receiving, and processing are the four steps. Figure 4 depicts an Ishikawa diagram (a fishbone diagram or a cause-and-effect diagram) with these four phases as the key components. Within the WLAN-based fingerprinting positioning system, from APs (offline or online), a radio signal is first transmitted, then traversed through the environment from transmitters to receivers (offline or online), and finally received at the unknown target location (online) or RPs (offline), before being processed to create a fingerprint database (offline) or position calculation (online). Every step might have an impact on performance. As a result, the critical aspects should be carefully addressed step by step.

In WLAN RSS position fingerprinting methods, all emitters are APs during the signal transmission stage. As previously indicated, this method is a cost-cutting solution that may complete the utilization of already existing APs. However, APs are prepared by several vendors and differ in device parameters like maximum transmit power. Meanwhile, APs can be placed with uneven distributions, densities, and heights. As a result, a receiver can access many APs in different places. In brief, differences in AP device models, distribution, density, and height can all impact the receiver's RSS values at this stage. The radio signal traverses from transmitters to receivers in the second stage. The RSSI can be influenced by the physical environment of the room as well as radio transmissions from other sources. For example, building materials, indoor structures, and furniture can influence radio signal reflection, diffraction, and attenuation [24]. Crowding in indoor spaces can impede and absorb radio signals, and other dynamic objective measures like temperature and humidity can influence radio signal propagation. WLAN signals from different APs or sources, such as a microwave oven, may interfere. In offline training, the WLAN fingerprinting RSS samples may be on RPs whose locations must be known before the signal-receiving stage. The complexity level and RSS values in fingerprinting radio databases will be affected by RP density, operators, height, and distribution variances. The critical effort in the online positioning phase is to seek the nearest RP using RSS collected at an unknown point, implying that sampling work is also required during the online phase. As a result, model differences between receivers in the online and offline phases can have an effect. For example, in the online positioning stage, one smartphone can create a fingerprint database, despite users' devices varying with different models and manufacturers.

In addition, unlike the number of samples and sampling rate, they can affect the samples in fingerprint databases and positioning findings. The final step is signal processing, which entails creating an RSS fingerprint database (offline) and computing position (online). During the database construction phase, the signal filtering method, feature extraction, and selection may alter the data and structure of the fingerprint database. As a result, the critical components of this step are signal filtering, feature extraction, feature selection, and the machine learning algorithm. These elements also influence the features used in online positioning. In online positioning processing, the positioning result is calculated from received RSS and a fingerprint database using machine learning techniques. As previously stated, indoor signal propagation is influenced by various factors and must be considered while picking regions for data gathering. Before establishing a conceptual framework in the following part, an Ishikawa diagram is utilized to bridge the probable effect elements influencing WLAN RSS location fingerprinting performance. In Fig. 4, the "fish head" represents WLAN RSS location fingerprinting positioning performance, and the "fish bones" represent the primary and potential variables. The considerations and assumptions made in this research are underlined by red rectangles in the Ishikawa diagram and are thoroughly detailed in the experiment design section.

3   Conceptual Framework of the Study

To establish the foundation for the scientific analysis in this research, we use the conceptual framework as an essential input for developing WLAN- based IPS.

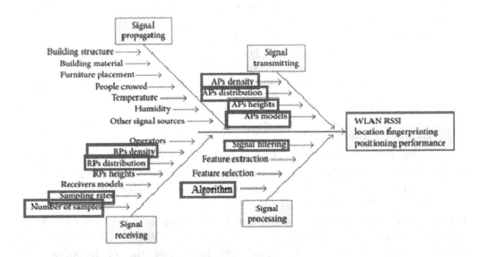

Fig. 4. The Ishikawa Diagram reflects the main and potential factors contributing to WLAN fingerprinting positioning performance

Utilizing theory to demonstrate the primary concepts to be scrutinized and their interrelationship, the conceptual framework aids in analyzing the research problem. Comprehending the appropriate conceptual framework is crucial, providing a general overview of the critical components and relationships that must be explored, thereby bridging the gap between theoretical concepts and empirical research.

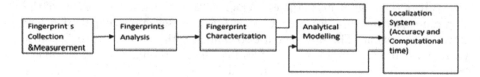

Fig. 5. The Conceptual Framework for Indoor Positioning System Analysis Model

The first stage of the process, as shown in Fig. 5, involves performing a site survey to measure and collect RSS data in a well-defined area, which is then used to build a radio map fingerprint. In the subsequent stage, the fingerprint structure is analyzed to comprehend how it is utilized to estimate the mobile user's position. Several fingerprint scheme characteristics are examined during this stage, such as the position fingerprint neighbor set and clusters. The distinct features of the RSS pattern in position fingerprints are also characterized. The analysis model is then developed to assess the system's performance before deployment. A reliable radio map is generated using a low signal relation elimination technique and a fingerprint mean filter. Ultimately, the positioning system can accurately determine the mobile user's position. The WLAN IPS model analysis's conceptual framework is significant in this regard. It provides this study's general view by showcasing the relationship between the major components of the WLAN-based fingerprinting indoor positioning model contribution.

Furthermore, as shown in Fig. 6, the indoor positioning analysis model flowchart with signal relation-aware holistically presents the WLAN-based IPS operations. Therefore, the detailed description of each step/process of variables presented by the flowchart for WLAN-based IPS with signal relation level awareness is given hereafter.

3.1   Extract Online Data

Online data extraction is vital in enhancing the IPS Model's performance. Thus, it is imperative to initialize and prepare all relevant data variables before initiating the extraction process. The primary data extracted during this stage include the WLAN AP MAC address and AP signal. The MAC address should be passed before the AP signal to ensure consistent data capturing. Two flags must work in unison to complete the task, with one flag for the WLAN AP MAC address and the other for the WLAN AP signal logic. In the last stage of the online data extraction process, verifying that the minimum required number of WLAN APs for positioning has been achieved accordingly is essential.

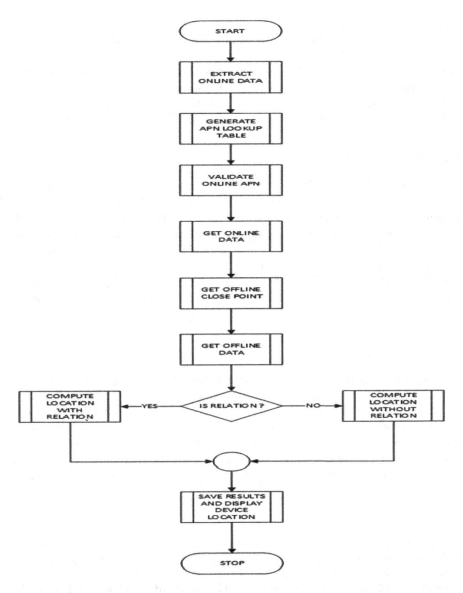

Fig. 6. The Flow Chart for Indoor Positioning Analysis Model with Signal Relation-Aware

3.2 Generate Access Point Lookup-Table

The AP lookup-table generation step is a crucial part of the IPS Model as it helps to reduce unnecessary data in the IPS database, leading to better performance. This process aids in removing redundant data from the IPS database, reducing computation time and improving accuracy. During this stage, the primary data to be collected is the AP MAC address, which is used to compare with the registered APs in the database. All relevant data variables should be initialized and prepared as in the previous stage to ensure that the necessary incoming data is handled appropriately. Every incoming WLAN AP MAC address must be included in the lookup table for comparison with the stored one within the database to ensure constant comparison and thus allows verification in the validation process stage before fetching the following WLAN AP MAC address

3.3   Get Online Data

The primary objective of the "get online data" stage is to collect verified WLAN APs data and pass it on to the subsequent sorting stage. This stage also involves assigning a tracking number to each AP to make it easier to keep and track records in the future. The essential data/information obtained/extracted during this phase includes the AP's MAC address and signal strength. AP validation should occur again before extraction to ensure accurate data capturing. In addition, the MAC address should be captured before the AP signal to maintain consistency. Finally, all data, including the AP track number, should be saved for usage in the next stage.

3.4   Get the Close Point

The main goal of the "get close point process" stage is to pinpoint the most appropriate reference points (RPs) by determining the one with the lowest combined RSS average value and then sending it to the position probability equation. The best points are computed by combining online signals and offline signals. The preliminary information gathered during this phase comprises the AP's MAC address and signal strength. In this stage, the MAC address must be obtained before capturing the AP signal to ensure consistent data collection.

3.5   Get Offline Data

The primary objective of the "get offline data" stage is to obtain new information/data from the three (3) most suitable RPs, which will be utilized along with online data to compute the signal correlation/relation. The key details to be gathered during this stage include the AP's MAC address and signal strength. The data collection process should begin with the highest-ranking close point and move downwards. Passing the MAC address before the AP signal is essential for consistent data capturing.

3.6   Compute Position with Enabled Signal Relation-Aware Functionality

The primary objective of the "compute position with the signal relation-aware functionality" stage is to identify signals with high relation based on the specific

AP. Therefore, signal relations are computed using only the three best-selected points. The mean and standard deviation of the high relation signal must be determined following signal relation calculation. Eventually, the position probability should be computed using the estimated/assessed mean and standard deviation from the high relation signal.

3.7 Compute Position Without Enabled Signal Relation-Aware Functionality

The "compute position without signal relation-aware functionality" phase ascertains a position without considering signals with a strong correlation/relation to a specific AP. As a result, the standard deviation and average/mean are calculated without considering the signal correlation/relation situation. Finally, the likelihood of the position is calculated using the standard deviation and mean (average) of the received signals.

3.8 Save and Display Device Position

The primary objective of the "save and display device position" stage is to store and exhibit the most accurate position of the user, which has the highest probability. The presented data includes the user's x and y coordinates and the probability of the user's position. The current and subsequent position information is compared to determine the optimal position.

Having comprehended and illustrated the principal theories to be examined and a graphical representation of the variables' interactions, which connects the gap between general concepts in theory and quantifiable indicators, the subsequent section outlines the experimental plan/design to implement/realize the proposed idea.

4 Experimental Design

This section describes the WLAN fingerprinting positioning system's experimental setup design and modeling. It details the requirements for establishing an experimental setup with its specifications and describes the modeling of the WLAN IPS.

4.1 Experiment Setup

The RSS samples were collected using a Dell Inspiron N5050 laptop computer with an Interactive Scanning Driver tool from five 2.4 radio frequency APs at each mapped Reference Point (RP). All APs set up in the testing environment with their power set to MEDIUM and at the predetermined height (1–3 m). During offline learning, the sampling RSS is gathered on grid-shaped RPs at one and a half (1–1.5) m. The in-service experiment area is estimated to be 350 m^2 with 74 training positions, as illustrated in Fig. 1, described earlier in the introduction.

4.2   WLAN Indoor Positioning System Modeling

Consider a connected IPS to a single floor, indoor WLAN in a classified facility/building. The presumption is that the environment contains N-IEEE 802.11b APs that are accessible and available in the investigated area. A well-defined four-sided grid completes the two-dimensional surface plan, and an approximative mobile position of choice is constrained to the points across this grid. Each database record includes a mapping from the grid point coordinates (x, y) to the corresponding vector of RSS values from all nearby APs.

4.3   Mathematical Model

The approximation of the mobile user's position is via using two vectors. The first vector consists of samples of the RSS gathered from N APs during the online sessions at the planned/intended area. It is referred to in this work as an RSS sample vector. The symbol for this vector is: $(x, y) = (s'AP_1, s'AP_2, s'AP_3, s'AP_4, ...s'AP_N)$. The indoor positioning algorithm calculates the mobile user's position using the RSS sample vector and other data (such as MAC address, RP, et cetera). With a few caveats, each component of this RSS vector is taken to be a random variable. The random variables $s'AP_i$ (in dBm) for all "i" are commonly independent. The random variables $s'AP_i$ are regularly distributed and expressed in dBm. The actual means of all the RSS random variables from N APs at the intended area, which are described in the position database, make up the other vector that builds the fingerprints of the place. For the sake of this work, it is referred to as the position fingerprints, position fingerprints, or RSS vector and denoted as $\mu_y = (\mu_{y1}, \mu_{y2}, \mu_3 ... \mu_{yn})$. Numerous studies support the claim that the RSS is a regularly distributed random variable [8,17]. Equation 1, which depicts and illustrates the idea of signal relations, highlights the notion of a signal relation. Scaling up the relationship between the signal and comparing the relationships between the database and user-send values in the algorithm's compassion controls the relation level. For example, if A = 80 dBm and B = 78 dBm, then Eq. 1 gives the signal relations value:

$$Z = A - B = 80 - 78 = 2\,dBm \tag{1}$$

Usually, the terminal device at a specific RP receives signal strength levels that are noticeably higher than elsewhere in the area under consideration. To establish a practical model algorithm, we leverage this information notion. Also, when the position is very close to the specific AP, the AP signal strength is high enough, boosting the likelihood of choosing the respective position. For example, assume that X and Y, two signal strengths, have a Gaussian distribution. The difference between the two signals is also distributed randomly, expressed as Z = X - Y, and serves as the random variable in the probabilistic relation-based approach. Equation 2 then establishes the random variable's mean denoted by Z.

$$\mu_z = \mu_x - \mu_y \tag{2}$$

And Eq. 3 defines the variance

$$\delta_z{}^2 = \delta_x{}^2 - \delta_y{}^2 \tag{3}$$

How the approach operates allows it to determine the probability density function values for each common relationship between user-retrieved and database beacons for each candidate position. In this context, "common" denotes that the relations share the same MAC addresses. The probability density function values are then added for each place and divided by the quantity (number) of relations. Choosing a position in the database as a candidate position requires determining the necessary number of common user signal relations. In this method, positions that should be determined by a sufficient number of signals in the right connection before using the probabilistic equation can be ignored. Accomplishing this approach is by introducing the iteration approach. For instance, if $k = 3$ in the computation of the first relation, then $k = k + 3$ is in the following step. At last, as indicated in Eqs. 4 and 5, the normal difference distributions density function is the final probability density function method employed.

$$f(z) = \frac{e^{\left(-\frac{(\mu_x - \mu_y)^2}{2(\delta_x{}^2 + \delta_y{}^2)}\right)}}{\sqrt{1 + (\delta_x{}^2 - \delta_y{}^2)}} \tag{4}$$

$$P(L) = \frac{\sum\limits_{i=1}^{n} f(z_i)}{n} \tag{5}$$

5   Results and Discussions

5.1   Effect of System Parameters on Performance

This subsection analyses the impact of height on the IPS model's accuracy performance using signal relation and non-signal relation mechanisms. This approach aims to confirm an intuition that introducing the signal relation technique can improve the accuracy of the IPS model as the height of access points changes. For a clear conclusion, the IPS model performance evaluation is conducted before comparison with other related research works.

5.2   Effect of WLAN Access Points Height on Accuracy Performance Based on Signal Relation Technique

Figure 7 demonstrates the position estimation performance aspect in RMSE over different Test Points (TPs). The study showed that the WLAN AP height caused the RMSE of 0.118 m, 1.172 m, and 1.198 m, respectively. It reveals that the RMSE observed when height is "one meter" is small compared to the "three-meter" setting. As the height increases, the RMSE also increases. This finding implies that the "low height" AP configuration can accurately deliver excellent interior position compared to no signal association.

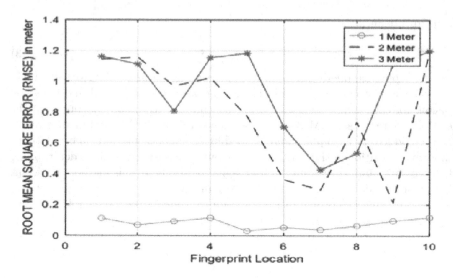

Fig. 7. Position estimation performance in meters over various Test Points (TPs) when the power level is "MEDIUM"

5.3   Effect of WLAN Access Points Height on Accuracy Performance Based on None-Signal Relation Technique

Figure 8 demonstrates the position estimation performance aspect's Root Mean Square Error (RMSE) over various Testing Points (TPs). The study showed that the AP height caused the RMSE of 1.217 m, 1.196 m, and 1.224 m, respectively. Upon height changes, the RMSE is not concisely determined if it increases or decreases due to a combination of low and high signal relation for position estimation. However, the RMSE observed is higher than the signal relation approach. This result suggests that the signal relation approach can provide better indoor position accuracy performance compared to the none signal relation approach. Figure 8 demonstrates the position estimation performance aspect in RMSE over various Test Points (TPs) when no signal relation approach is applied.

5.4   Comparison of Indoor Positioning Systems' Performance

As stated earlier in the introduction section of this paper, the study's main objective is to develop a novel algorithm to mitigate the effect of low signal relation caused by AP height on WLAN fingerprinting indoor positioning systems to improve accuracy. This study has instigated at the right time; although many studies have been conducted with considerable accuracy, many have not considered the potential of low signal relation caused by transmitter height on accuracy [18]. The empirical findings presented in this study provide a new theoretical understanding of the WLAN fingerprinting technique in positioning, particularly in addressing indoor positioning performance accuracy problems. Previous

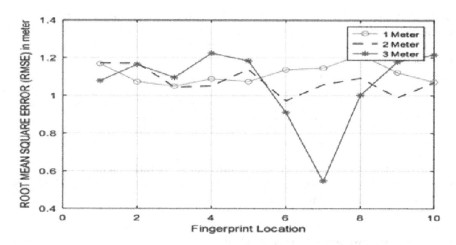

Fig. 8. Position estimation performance in meters over various Test Points (TPs) when the power level is "MEDIUM"

studies have described the effect of APs height and high signal fluctuation as barriers to the WLAN fingerprinting technique in improving indoor positioning accuracy. Despite these initiatives and many others, indoor positioning accuracy on WLAN- based fingerprinting technique solutions remains a challenge. The study reveals that transmitter height affects indoor positioning accuracy based on WLAN fingerprinting technique. Also, the study shows that introducing clustering techniques such as signal relation enhances the positioning accuracy of indoor positioning based on the WLAN fingerprinting technique. This finding corroborates the results of a study by [25] that introducing clustering has improved indoor positioning accuracy based on the WLAN fingerprinting technique. The findings are consistent with the results of other global studies investigating the effect of different parameters on indoor positioning models [26,27]. For example, [28] reported the impact of access point height on indoor positioning. The study by [29] concluded that there is no significant effect when the height variation is not considered high; however, the authors could not vary the transmitter height directly. Instead, the authors applied the method of changing the transmitter height of the mobile device used for RSS measurements. The authors' conclusion could not be proper because the height measurements of the mobile device are limited to the height of the person who carries the device. In contrast, in the case of transmitter height, the range of height measurements is limited to building height.

Furthermore, to understand the percentage accuracy improvement, Eq. 1.6 has been used to compute the percentile. Table 1 shows that the proposed model attains an average accuracy of 32.740% medium power setting upon introducing the concept of high signal relation. Again, the average accuracy was computed using Eq. 6.

$$H_p = [(H_n s - H_w s)/H_n s] * 100 \qquad (6)$$

Table 1. Comparison of Performance Accuracy between Signal and Non-Signal Relation Technique With Medium Power Level Setting

Height (in Meter)	1	2	3
Maximum RMSE with Signal Relation-Aware Algorithm (in Meter)	0.118	1.172	1.198
Maximum RMSE without Signal Relation-Aware Algorithm (in Meter)	1.285	1.196	1.224
Percentage accuracy improvement with Signal Relation Algorithm AME meter	90.817	2.00	2.12
Average Percentage accuracy improvement with Signal Relation Algorithm AME meter	32.740		

where H_p is the percentage accuracy improvement with the signal relation algorithm, $H_n s$ is the accuracy without the signal relation algorithm, and $H_w s$ is the accuracy with the signal relation algorithm.

6   Conclusion and Future Work

This paper presents a novel algorithm to mitigate the effect of low signal relation on the WLAN fingerprinting-based IPS, thus improving the accuracy performance. The proposed signal relation approach improves accuracy by eliminating the combination of RSS with high deviation for positioning determination. Furthermore, it analyses the AP heights' impact on the accuracy performance of the WLAN IPS using a probabilistic analytical model that employs only high signal relations. The results imply that incorporating APs' height and high signal relations can significantly improve the performance accuracy of the WLAN-based fingerprinting IPS. Additionally, the results provide a deep insight into the mechanism of handling the AP height effect behind the WLAN fingerprinting indoor positioning system based on the RSSI ranging model. The next step is to develop a model based on WLAN fingerprinting that considers the influence of AP power for IPS.

Acknowledgments. The research is supported by postdoc fellowship granted by the Institute of Computer Technologies and Information Security, Southern Federal University, project № P.D./22-01-KT.

References

1. Tinh, P.D., Bui, H.H., Nguyen, D.C.: A genetic based indoor positioning algorithm using Wi-Fi received signal strength and motion data. IAES Int. J. Artif. Intell. **12**(1), 328–346 (2023). https://doi.org/10.11591/ijai.v12.i1.pp328-346
2. Ding, G., Tan, Z., Wu, J., Zhang, J.: Efficient indoor fingerprinting localization technique using regional propagation model. IEICE Trans. Commun. **97**(8), 1728–1741 (2014). https://doi.org/10.1587/TRANSCOM.E97.B.1728
3. Adam, A.B.M., Muthanna, M.S.A., Muthanna, A., Nguyen, T.N., El-Latif, A.A.A.: Toward smart traffic management with 3D placement optimization in UAV-assisted NOMA IIoT networks. IEEE Trans. Intell. Transp, Syst. **24**, 15448–15458 (2022)

4. Wang, Q., et al.: Free-walking: pedestrian inertial navigation based on dual foot-mounted IMU. Elsevier. 25 Apr 2023. https://www.sciencedirect.com/science/article/pii/S2214914723000533

5. Basta, N., Kairo, A., Agypten.: A four-dimensional model for ITS destination prediction and mobility simulation (2023). Accessed 25 Apr 2023. https://oparu.uni-ulm.de/xmlui/handle/123456789/48328

6. Muthanna, M.S.A., et al.: Deep reinforcement learning based trans-mission policy enforcement and multi-hop routing in QoS aware LoRa IoT networks. Comput. Commun. **183**, 33–50 (2021)

7. Khoo, H.W., Ng, Y.H., Tan, C.K.: A fast and precise indoor positioning system based on deep embedded clustering (2022). https://books.google.com/. https://doi.org/10.2991/978-94-6463-082-4_6.]

8. Pirzada, N., Nayan, M.Y., Subhan, F., Abro, A., Hassan, M.F., Sakidin, H.: Location fingerprinting technique for WLAN device-free indoor localization system. Accessed 17 Apr 2023. http://eprints.utp.edu.my/id/eprint/12327/

9. Häb-Umbach, I., Höher, I.: WLAN fingerprinting based indoor positioning in the presence of censored and dropped data. Accessed 17 Apr 2023. https://d-nb.info/109096577X/34

10. Yuan, Y., Chao, D., Song, L.: Study of WLAN fingerprinting indoor positioning technology based on smart phone (2015). Accessed 17 Apr 2023. https://www.atlantis-press.com/proceedings/icismme-15/21137

11. Nicholaus, M.R., Nfuka, E.N., Greyson, K.A.: A novel probabilistic algorithm for indoor WLAN fingerprinting system. Accessed 17 Apr 2023. https://search.proquest.com/openview/cbe8ba134d939c78c073574f4222330c/1?pq-origsite=gscholar&cbl=2044551

12. Shata, A.M., Abd El-Hamid, S.S., Heiba, Y.A., Nasr, O.A.: Multi-site fusion for WLAN based indoor localization via maximum discrimination fingerprinting. Accessed 17 Apr 2023. https://ieeexplore.ieee.org/abstract/document/8316627/

13. Ko, D., Kim, M., Son, K., Han, D.: Passive fingerprinting reinforced by active radiomap for WLAN indoor positioning system. Accessed 17 Apr 2023. https://ieeexplore.ieee.org/abstract/document/9611210/

14. Alshami, I.H., Ahmad, N.A., Sahibuddin, S.: RSS certainty: an efficient solution for RSS variation due to device heterogeneity in WLAN fingerprinting-based indoor positioning system. Accessed 17 Apr 2023. https://ieeexplore.ieee.org/abstract/document/9636886/

15. Sangthong, J., Supanakoon, P., Promwong, S.: Indoor navigation and tracking using WLAN-fingerprinting technique and K-inverse harmonic means clustering on mobile device. Trans Tech Publ. Accessed 17 Apr 2023. https://www.scientific.net/AMM.781.77

16. Altaf Khattak, S.B., Fawad, N.M.M., Esmail, M.A., Mostafa, H., Jia, M.: WLAN RSS-based fingerprinting for indoor localization: a machine learning inspired bag-of-features approach. https://www.mdpi.com/. https://doi.org/10.3390/s17061339

17. Muthanna, M. S. A., Wang, P., Wei, M., Abuarqoub, A., Alzu'bi, A., Gull, H.: Cognitive control models of multiple access IoT networks using LoRa technology. Cogn. Syst. Res. **65**, 62–73 (2021). ISSN 1389–0417

18. Pirzada, N., Nayan, M.Y., Subhan, F., Abro, A., Hassan, M.F., Sakidin, H.: Location fingerprinting technique for WLAN device-free indoor localization system. Wirel. Pers. Commun. **95**(2), 445–455 (2017). https://doi.org/10.1007/S11277-016-3902-8

19. Zeng, C., Liu, H., Xu, K., Han, W., Zhang, H.: RSSI... - Google Scholar. https://scholar.google.com/scholar?hl=en&as_sdt=0
20. Lin, M., Chen, B., Zhang, W., Yang, J.: Characteristic analysis of wireless local area network's received signal strength indication in indoor positioning (2019). https://doi.org/10.1049/iet-com.2019.0681
21. Li, C.: Measurement and modeling of the influence... - Google Scholar. https://scholar.google.com/scholar?hl=en&as_sdt=0
22. Kaemarungsi, K.: Design of indoor positioning systems based on location finger-printing technique. In: Telecommunications (1994)
23. Bahl, P., Padmanabhan, V.N.: RADAR: an in-building RF-based user location and tracking system. Proc. - IEEE INFOCOM **2**, 775–784 (2000). https://doi.org/10.1109/INFCOM.2000.832252
24. Lin, M., Chen, B., Zhang, W., Yang, J.: Characteristic analysis of wireless local area network's received signal strength indication in indoor positioning. IET Commun. **14**(3), 497–504 (2020). https://doi.org/10.1049/IET-COM.2019.0681
25. Kokkinis, A., Raspopoulos, M., Kanaris, L., Liotta, A., Stavrou, S.: Multi-device Map-aided Fingerprint-based Indoor Positioning using Ray Tracing, no. 1 (2013)
26. Bisio, I., et al.: A trainingless WiFi fingerprint positioning approach over mobile devices, vol. 13 (2014). https://doi.org/10.1109/LAWP.2014.2316973
27. Wu, C., Yang, Z., Liu, Y.: Smartphones based crowdsourcing for indoor localization (2013). Accessed 11 June 2023. https://ieeexplore.ieee.org/abstract/document/6805641/
28. Feng, C., Au, W.S.A., Valaee, S., Tan, Z.: Received-signal-strength-based indoor positioning using compressive sensing. Accessed 11 June 2023. https://ieeexplore.ieee.org/abstract/document/6042868/
29. Wombacher, W., Dil, B.J., Havinga, P.J.M.: Practical indoor localization using Bluetooth (2012). Accessed 11 June 2023. https://essay.utwente.nl/61496/

Revolutionizing H2M Interaction: Telepresence System Enabling Sign Language Expansion for Individuals with Disabilities

Konstantin Kuznetsov[1], Ammar Muthanna[1(✉)], Abdelhamied A. Ateya[2,3],
and Andrey Koucheryavy[1]

[1] The Bonch-Bruevich Saint-Petersburg State University of Telecommunications,
St. Petersburg 193232, Russian Federation
`muthanna.asa@sut.ru`
[2] EIAS Data Science Lab, College of Computer and Information Sciences,
Prince Sultan University, Riyadh 11586, Saudi Arabia
[3] Department of Electronics and Communications Engineering, Zagazig University,
Zagazig 44519, Sharqia, Egypt

Abstract. The modern network technology ecosystem is expanding rapidly in multiple dimensions. Due to the demands of the Tactile Internet and fifth-generation (5G) networks, conventional approaches are no longer producing significant advancements. The imminent integration of immersive technologies holds the potential to revolutionize human existence, ushering in a new era of fully realized and enhanced human-to-machine interactions (H2M). This paradigm shift is poised to offer many benefits and adopt wireless technologies to enable real-time remote control. Concurrently, introducing the Tactile Internet necessitates a fundamental reevaluation of the foundational principles governing current and prospective communication networks. This article delves into an experimental endeavor that seeks to seamlessly integrate immersive technologies within 5G networks. A microservice application was developed for this purpose, simulating the nuanced movements of a person's hand in space and facilitating the expression of intricate gestures. The article also delves into the strategies employed to minimize latency, enhancing the overall experience.

Keywords: Immersive technologies · Internet of Things · 5G networks · telepresence services · robot avatar

1 Introduction

Immersive technologies are various methods and means that influence the spectrum of human sensations. Recently, such technologies have been gaining in popularity and are rapidly entering everyday life; for example, virtual reality (VR) and augmented reality (AR) are familiar to many active industries. Human vision is not very sensitive to changes, which allows us to easily implement the

© The Author(s), under exclusive license to Springer Nature Switzerland AG 2024
V. M. Vishnevskiy et al. (Eds.): DCCN 2023, LNCS 14123, pp. 21–33, 2024.
https://doi.org/10.1007/978-3-031-50482-2_2

above-described technology in the communication networks of previous generations. However, moving to other senses, the picture changes dramatically - the response to tactile sensations reaches 1 ms, which imposes weighty requirements to circular delay to transmit such sensations. Fifth-generation networks, namely the ultra-reliable low latency communication (URLLC) segment, are designed to achieve the lowest possible latency while maintaining high transmission reliability. Ultra-reliable low latency communications incorporate a revolutionary framework designed to guarantee that communication systems operate with unprecedented reliability and minimal delay. Despite the fact that traditional communication networks have evolved to provide remarkable performance, certain applications require a level of dependability and responsiveness that is currently unattainable. Industries such as healthcare, transportation, industrial automation, and emergency services require instantaneous data exchange, virtually zero delay, and consistent, real-time interactions; however, conventional communication systems struggle to deliver these characteristics consistently. Multiple factors drive the pursuit of URLLC technology in this environment of accelerated change. The Internet of Things (IoT) has enabled the proliferation of autonomously communicating devices, necessitating near-instantaneous responses to facilitate synchronized operations. In addition, the emergence of technologies such as autonomous vehicles, remote operations, and critical infrastructure management has heightened the importance of developing communication systems with high reliability and low latency. In order to implement immersive technologies, we proposed an H2M interaction system, which allows the creation of libraries of gestures for the language of deaf people. It is worth noting that previously created systems were aimed at reading and reproducing movements at a distance in real-time. Motion library creation will automate many technological and/or other processes and make it possible to train people in various professions or games, such as golf. It will also be possible to transform the counted movements into idealized ones based on such libraries. In this regard, WLAN fingerprinting-based IPS has become a favorable option and is applicable in many applications, such as Indoor navigation: WLAN fingerprinting-based IPS can provide turn-by-turn directions for indoor navigation. Asset tracking: WLAN fingerprinting-based IPS can track the position of assets within a facility, such as equipment or inventory. Emergency response: In an emergency, such as a fire or natural disaster, WLAN fingerprinting-based IPS can help locate people quickly. Retail: WLAN fingerprinting-based IPS can track customer movement within a store and provide targeted advertising or offers. Overall, WLAN fingerprinting-based IPS is a powerful technology that can cost-effectively provide highly accurate indoor positioning, and it has a wide range of applications in various fields.

Generally, the role of WLAN fingerprinting-based indoor positioning systems (IPS) is to determine a device's position within a specific area using data from wireless signals. This technique is typically done by collecting data on the strength and characteristics of wireless signals in a given area and then using that data to create a "fingerprint" of the area. Then by matching its current wireless signal data to the fingerprint of the area, the determination of the device's position is possible. Typically, the development of WLAN-based fingerprinting positioning systems can be divided into two stages: offline training, during which the position fingerprints are collected by conducting a Received Signal Strength Indicator (RSSI) site survey from various Access Points (APs). Then, the mobile user's position is determined while in the online phase by comparing the most recent RSSI measurements received with the fingerprint RSSI measurements that were previously gathered. As seen in Fig. 3, a rectangular network (grid) of points conceals the entire area under examination, experimentation, or implementation [16–18].

2 Relevance of Immersive Technology Transfer

Immersive technology is an ambiguous direction; not all people today are ready to understand how it will be possible to touch a person who is far away. But one way or another, the technology makes it possible, first of all, to secure an employee in hazardous production facilities. Large companies are already showing their interest in this direction. In addition, there are opportunities to improve production processes. For example, using a crane to install the engine in the car will not be necessary because the operator's muscles are a robot. There is also an opportunity to reduce the number of employees due to the possibility of using highly qualified specialists in different parts of the world. These points interest hot-headed engineers, marketers, and health and safety experts; this combination usually serves as an inevitable way to achieve the goal, which means that very soon, it will change our lives for the better. There are two major challenges in human sensory transmission: the limitations of circular delay to achieve real-time and the production of ultra-fast, accurate sensors and actuators. Haptic senses are the most demanding in terms of latency of the entire range of human senses. Their use in industrial environments imposes an obligation to provide the highest reliability, uptime, and safety. 5G mobile communication networks, according to the declared characteristics, will be able to provide the basis for the realization of the ideas of transferring the full range of human senses.

3 Problem Statement and Experiment

To study methods of implementation of immersive technologies in fifth-generation communication networks. To create a microservice application simulating the movement of a human hand in space, allowing to display letters of the alphabet of gesture language with movements. Analyze ways to reduce latency to achieve a level that allows realizing the full range of human sensations based on 5G networks. Immersive technology seeks to transport the fullest range of human senses into space. But what if the number of senses is limited in the person himself? Unfortunately, many people in the world have hearing, speech, and touch (tactile) impairments. Increasingly new technologies are being introduced to reduce the impact of such disabilities on a person's full life. The experiment aims to create a web application with a microservice architecture to control the robot arm and read the "glove" in real-time to reproduce and read the hand positions corresponding to a letter from the alphabet of the sign language. The hand should be able to show, and the glove should be able to read the letters in sign language, both in tactile mode and in avatar mode, repeating the movements when a certain event occurs. The sign language alphabet consists of 32 letters - 23 of these are represented by static gestures, and nine by dynamic gestures (Fig. 1).

Fig. 1. Alphabet of Sign Language

4   Booth Description

A visual representation of the entire stand is shown in Fig. 2. The main elements are the ESP-32 microcontrollers, the board with gyroscope and accelerometer MPU-6050, Processing, and MQTT Broker for communication between the robot arm and processing.

4.1   ESP-32

ESP-32 is a modern microcontroller manufactured by Espressif Systems, which is the successor of the worldwide popular ESP8266 board (Fig. 3). The controller is often used in the creation of IoT devices or other devices communicating via Wi-Fi, BLE, LoRa, etc.

- Processor: Tensilica Xtensa LX6 32-bit microprocessor, two cores
- Clock frequency: up to 240 MHz

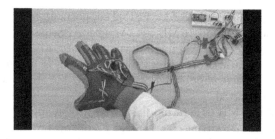

Fig. 2. Stand with glove horn

Fig. 3. ESP-32 microcontroller

- Performance: up to 600 DMIPS (ultra-low power co-processor: allows ADC conversions, calculations, and level thresholds in deep sleep state).
- Wireless connection:
  - Wi-Fi: 802.11 b/g/n/e/i (802.11n @ 2.4 GHz up to 150 Mbps)
  - Bluetooth: v4.2 BR/EDR and Bluetooth Low Energy (BLE).
- Memory:
  - Internal ROM memory: 448 Kbits
  - For download and basic functions: SRAM: 520 Kbits
    * For data and instructions, RTC fast SRAM: 8 Kbit
    * For data and main processor storage during RTC boot from deep sleep, RTC slow SRAM: 8 Kbit
    * For co-processor access during deep sleep eFuse: 1 kbit
  - Built-in flash memory:
    * Flash memory connected internally via IO16, IO17, SD_CMD, SD_CLK, SD_DATA_0 and SD_DATA_1 on ESP32-D2WD and ESP32-PICO-D4
      · 0 MB (ESP32-D0WDQ6, ESP32-D0WD and ESP32-S0WD chips)
      · 2 MB (ESP32-D2WD chip)
      · 4 MB (ESP32-PICO-D4 SiP module)

* External Flash and SRAM: ESP32 supports up to four external QSPI flash memories and 16 MB SRAM with AES-based hardware encryption to protect program and developer data.
  · Up to 16 MB of external flash memory is mapped to the CPU code space, supporting 8-, 16- and 32-bit access. Code execution is supported.
  · Up to 8 MB of external flash/SRAM is mapped to the CPU data space, supporting 8-, 16- and 32-bit access. Reading data from Flash and SRAM is supported. Writing data is supported in SRAM.
- Peripheral I/O: Rich peripheral interface with DMA including capacitive touch control, ADC (analog-to-digital converter), DAC (digital-to-analog converter), I²C (inter-network integrated interface), UART (universal asynchronous receiver/transmitter), CAN 2.0 (controller network), SPI (serial peripheral interface), I²S (integrated gateway), RMII (reduced media-independent interface), PWM (pulse width modulation), and more.
- Security: All IEEE 802.11 security features, including WFA, WPA/WPA2 and WAPI, are supported. Secure boot, flash encryption, 1024-bit OTP, up to 768-bit for clients. Cryptographic hardware acceleration: AES, SHA-2, RSA, Elliptic Curve Cryptography (ECC), Random Number Generator (RNG).

The controller has built-in processor core control functions, which allow isolating and significantly accelerating computational processes. Similar to its predecessor (ESP8266), the controller is primarily a wireless communication module that allows additional computation, making it a popular, universal SoC platform for IoT. The executable code is shown in Appendix 1.

4.2 MPU-6050

The MPU-6050 is a three-axis gyroscope and accelerometer for motion tracking designed to meet the low power, low cost, and high-performance requirements of smartphones, tablets, and wearable sensors. The module incorporates InvenSense MotionFusion software and runtime calibration, allowing manufacturers to eliminate the costly and complicated selection, qualification, and integration of discrete system-level devices into motion-enabled products, ensuring that sensor fusion algorithms and calibration procedures provide optimal performance for consumers. The MPU-6050 devices combine a 3-axis gyroscope and 3-axis accelerometer on a single silicon sensor with an integrated Digital Motion Processor (DMP) that processes complex 6-axis MotionFusion algorithms. The device can access external magnetometers or other sensors via an auxiliary I²C master bus, allowing the devices to collect a complete set of sensor data without system processor intervention. The devices are available in a 4 mm × 4 mm × 0.9 mm QFN package. For accurate tracking of both fast and slow movements, the module features a user-programmable gyroscope with a full range of ±250, ±500, ±1000 and ±2000 °/s (dps) and a user-programmable accelerometer with a full range of ±2g, ±4g, ±8g, and ±16g. Additional features include a built-in temperature sensor and on-chip oscillator with ±1% deviation over the operating temperature range. In order to simplify the interaction with the microcontroller, the project uses the NodeMCU debug board (Fig. 4).

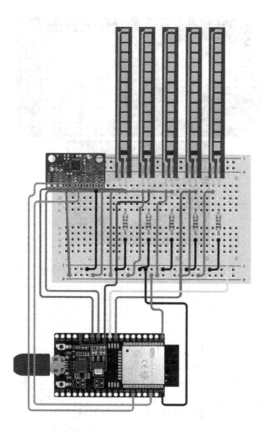

Fig. 4. Connection diagram

4.3   Processing

Processing is an open-source programming language based on Java that wants to program images, animations, and interfaces. Processing software is free, open-source, cross-platform software. The source archive includes a Java machine, the interpreter, a miniIDE, and several dozen examples. It includes tools for building graphical primitives, and 3D objects, working with light, text, and transformation tools, allowing you to import and export audio/video/sound files, processing mouse/keyboard events, and working with third-party libraries (openGL, PDF, DXF). As part of the experiment, processing is used to visualize the movements of a sketch of a mechanized robotic arm in the programming language Java. The developed program is versatile and allows one to conduct any experiments using three-dimensional visualization, including can be used to teach students to work with IoT devices and create new application cases of robotic gloves (Fig. 5).

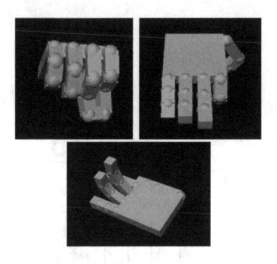

Fig. 5. Processing. 3D - visualization

4.4   MQTT Broker

EMQX is an open-source IoT MQTT message broker based on the Erlang/OTP platform (a Soft-Realtime, low-latency, distributed development platform). The broker is designed for mass customer access and implements fast message routing between massive physical network devices with minimal latency. It is one of the most scalable and reliable MQTT messaging platforms for connecting, moving, and processing data in business-critical IoT-era scenarios. Allows you to connect virtually any device using open standard MQTT, CoAP, and LwM2M IoT protocols. Easily scales to tens of millions of simultaneous MQTT connections using the EMQX Enterprise cluster [16].

5   Experiment with Delay Reduction

In order to analyze the level of delays and investigate methods of reducing them, it is proposed to consider the IoT system of interaction between the robot glove and its virtual avatar described above. In the experiment, it is proposed to perform a comparative analysis of three interaction schemes (Fig. 6):

1. All parts of the system are located in the local network. 2. The robot glove and the 3D visualization application are located locally, and the MQTT broker is on a remote Internet machine. The interaction is performed via Wi-Fi. 3. Robo-Glove is in the local network, and the application and the broker are migrated using Docker containers.

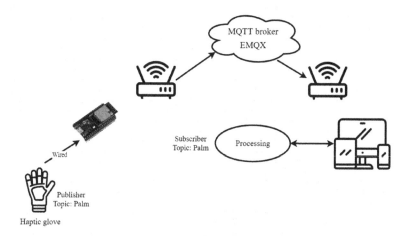

Fig. 6. General scheme of device interaction

5.1 Option №1

Consider the use of the bench in the local network. To make it more realistic, the experiment was conducted on the home Wi-Fi network, with connected clients not participating. The robot glove is connected to ESP-32, which communicates with the MQTT broker on Wi-Fi. The broker and the application rendering the avatar hand are located on the same local machine. Thus, increasing load on the MQTT broker has a negative impact as it fills up (Fig. 7). On average, the "threshold" value of the number of packets reaches 300 000 pcs. The work quality is restored after restarting the server (resetting the accumulated data volume).

5.2 Option №2

The experiment is conducted at a remote location of the Broker's MQTT on the Internet. The broker is deployed on virtual cloud servers (VPS) using a Docker container. The rest of the devices of the stand are located similarly to option 1. Considering the latency of ICMP packets to the described server of 50–70 ms, we can conclude about stabilizing the stand by deploying the broker on a separate, more productive computer (Fig. 8). Thus, the most significant impact has delays in the path of packets over the Internet. In order to reduce them, it is proposed to migrate the broker closer to the subscriber.

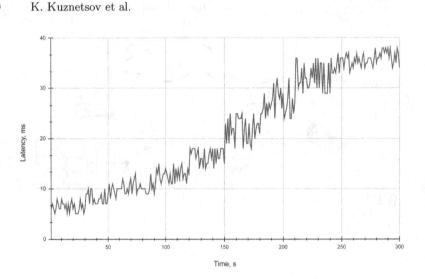

Fig. 7. Schedule of delay distribution for option 1

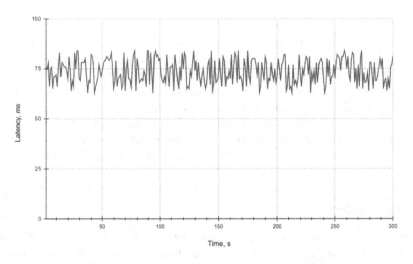

Fig. 8. Schedule of delay distribution for option 2

5.3 Option №3

Let's consider a way to reduce latency based on the data obtained in previous experiments. We propose placing an MQTT broker in a Docker container and migrating it to equipment near the terminal subscriber. We will migrate the broker, previously deployed on remote VPS servers, to a virtual machine located two hops from the terminal equipment for the experiment. This method involves using only the built-in functionality of Docker. Management will be done using Docker CLI, Docker Hub is the remote cloud storage, and Docker daemon is responsible for organizing the network, downloading the image, and deploying the container. The migration was performed ten times to establish the service migration time and stability. The average deployment time of the service was 1.56 if there was no pre-loaded image. Next, the analysis of delays during the work of the stand in the new configuration was carried out. It is worth noting that the equipment's performance significantly increased compared with option 1, as well as the throughput capacity of the broker (Fig. 9). The delay of ICMP packets passing between the end equipment of the test bench was reduced to 2–4 ms. Thus, the overall work of the test bench is within a 10–11 ms delay, which does not allow passing tactile sensations but is quite enough to perform the tasks. Compared with option 1, the stand began functioning in a stable state. Additionally, an experiment on the service operation for 2 h was performed, and the quality of the test bench remained unchanged.

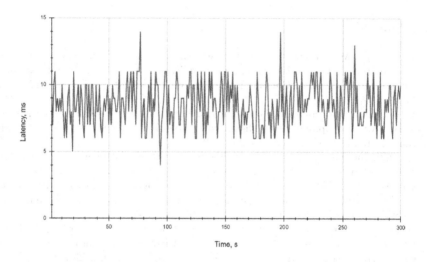

Fig. 9. Schedule of delay distribution for option 3

6   Conclusion

Today, the field of network technology is expanding in all directions. Increasing the channel width and improving the speed of the equipment no longer brings significant results because the requirements for the Tactile Internet and the fifth generation of mobile data networks surpass all of the known physical phenomena. A promising area of development is the development of predictive reasoning kernels, which are designed to anticipate an individual's actions in the future, thus significantly reducing the latency requirements while leaving the reaction time unchanged. Exceptions are applications that consider particularly precise positioning, such as surgery. On the way to fighting to achieve the requirements, in terms of the conceptual approach, there are even more difficulties associated with loss and attenuation in data paths, with packet processing time on active network equipment, with the creation of highly sensitive sensors and highly accurate actuators with minimal response time. The combined application of all the described methods will allow us to get as close as possible to the solution of the set tasks. That in the near future will make it possible to massively introduce immersive technologies into human life, secure it, accelerate learning processes, use narrowly-specialized specialists with greater efficiency, and much more. But getting close to the edges of physical research, building networks is no longer an evolutionary process.

Acknowledgments. The studies at St. Petersburg State University of Telecommunications. Prof. M.A. Bonch-Bruevich were supported by the Ministry of Science and High Education of the Russian Federation by the grant 075-15-2022-1137.

References

1. Ateya, A.A., Muthanna, A., Koucheryavy, A.: 5G framework based on multi-level edge computing with D2D enabled communication. In: 2018 20th International Conference on Advanced Communication Technology (ICACT), pp. 507–512. IEEE (2018)
2. Ateya, A.A., Muthanna, A., Gudkova, I., Abuarqoub, A., Vybornova, A., Koucheryavy, A.: Development of intelligent core network for tactile Internet and future smart systems. J. Sens. Actuator Netw. **7**(1), 1 (2018)
3. An introduction to immersive technologies. https://vistaequitypartners.com/insights/an-introduction-to-immersive-technologies
4. AR, MR, VR, XR - 4 abbreviations that we must be familiar with in the world of modern UX-UI. https://ergomania.eu/ar-mr-vr-xr-4-abbreviations-that-we-must-be-familiar-with-in-the-world-of-modern-ux-ui/
5. Ateya, A.A., Khayyat, M., Muthanna, A., Koucheryavy, A.: Toward tactile internet. In: 2019 11th International Congress on Ultra Modern Telecommunications and Control Systems and Workshops (ICUMT), pp. 1–7. IEEE (2019)
6. Pedersen, K.I., Berardinelli, G., Frederiksen, F., Mogensen, P., Szufarska, A.: A flexible 5G frame structure design. IEEE Commun. Mag. **54**, 53–59 (2016)
7. 5G new radio network. Nokia white paper (2018)
8. Kato, N., et al.: The deep learning vision for heterogeneous network traffic control: proposal, challenges, and future perspective. IEEE Wirel. Commun. **24**(3), 146–153 (2017)
9. Network Functions Virtualisation (NFV), Architectural Framework. Standard ETSI GS NFV 002 V1.2.1 (2014)
10. Faisal, V.A.: NFV MANO-Management & Orchestration. TelcoCloud Bridge (2020)
11. Muthanna, A., et al.: Framework of QoS management for time constraint services with requested network parameters based on SDN/NFV infrastructure. In: 2018 10th International Congress on Ultra-Modern Telecommunications and Control Systems and Workshops (ICUMT), pp. 1–6. IEEE (2018)
12. Multi-access Edge Computing (MEC), Framework and Reference Architecture. Standard ETSI GS MEC 003 V3.1.1 (2022)
13. Mobile Edge Computing (MEC); Deployment of Mobile Edge Computing in an NFV environment. Standard ETSI GR MEC 017 V1.1.1 (2018)

14. MQTT oasis standard. https://docs.oasis-open.org/mqtt/mqtt/v5.0/mqtt-v5.0.html. Accessed 24 Mar 2022

15. ESP32 Series Datasheet. Espressif systems (2022)

16. An Open-Source, Cloud-Native, Distributed MQTT Broker for IoT

17. Le, L.V., Sinh, D., Tung, L.P., Paul Lin, B.S.: A practical model for traffic forecasting based on big data, machine learning, and network KPIs. In: 15th IEEE Annual Consumer Communications Networking Conference, pp. 1–4. IEEE (2018)

18. Kumar, P. M., Devi, U., Manogaran, G., Sundarasekar, R., Ch, N.:

Efficient Transmission of Holographic Images: A Novel Approach Toward 6G Telepresence Services

Daniil Svechnikov[1], Bogdan Pankov[1], Yana Nesterova[1], Artem Volkov[1(✉)],
Abdelhamied A. Ateya[2,3], and Andrey Koucheryavy[1]

[1] The Bonch-Bruevich Saint-Petersburg State University of Telecommunications,
St. Petersburg 193232, Russian Federation
artemanv.work@gmail.com

[2] EIAS Data Science Lab, College of Computer and Information Sciences, Prince
Sultan University, Riyadh 11586, Saudi Arabia

[3] Department of Electronics and Communications Engineering, Zagazig University,
Zagazig, Sharqia 44519, Egypt

Abstract. In recent years, advancements in technology have brought
forth a new frontier in visual communication. Holography is a technique
that captures and reproduces three-dimensional (3D) images with an
unprecedented level of realism and depth, has emerged as a groundbreak-
ing method for conveying visual information. Unlike traditional images
and videos, holography recreates scenes with full parallax, enabling viewers to perceive objects from various angles. The transmission of holographic images presents both exciting possibilities and unique challenges.
To this end, this article investigates a novel method for transmitting holographic images efficiently across communication networks. This innovative approach involves transmitting data related to control point movements rather than the entire image. The technique focuses on capturing
the dynamic 3D movements of a person and transmitting information
about the shifts in their control points. To achieve this, a system comprising three distinct applications has been devised. These applications
are responsible for recording object motion data, converting it into key
point coordinates, transmitting it through communication networks, and
rendering the data as movements of a 3D object within an augmented
reality (AR) application.

Keywords: Immersive technologies · Internet of Things · 5G
networks · telepresence services · robot avatar

1 Introduction

In the realities of the modern world, there is a pressing need to obtain information
remotely. Various types of traffic, including voice and video traffic, have different
parameters, such as the volume of transmitted information, delay requirements,
and bandwidth. There is an ongoing "race" between people's and companies'
demands for content and services and the development of network hardware

V. M. Vishnevskiy et al. (Eds.): DCCN 2023, LNCS 14123, pp. 34–43, 2024.
https://doi.org/10.1007/978-3-031-50482-2_3

to meet these demands. Recently, university students and schoolchildren world-wide couldn't attend educational institutions in person. Web designers, testers, programmers, analysts, marketers, and representatives of many other professions were forced to work from home. Companies had to redirect their computing resources to support online conferencing services like "Zoom," "Google Meet," and others. These solutions provide fast access to video channels between the sender and the receiver. However, this video traffic consumes a significant portion of the Internet's resources, putting a heavy load on all the network hardware through which it passes. Another aspect of developing applications for remote work is the remote transmission of three-dimensional images, which allows for a more detailed understanding of human movements from all perspectives and opens up new possibilities for immersive communications [1,2]. Transmitting data about three-dimensional objects in their original form, such as streaming video from an array of cameras, requires significant bandwidth. Therefore, using technologies to transform the acquired object data into other formats becomes crucial [2,3]. Holographic communication refers to transmitting and receiving holographic images for visual communication. Unlike traditional images or videos, which are usually two-dimensional representations, holographic images capture and reproduce three-dimensional (3D) scenes with high realism and depth. This technology enables viewers to perceive objects and scenes from various angles, mimicking how objects appear in the physical world. In holographic communication, the goal is to transmit these complex 3D holographic images from one location to another, allowing remote users to experience and interact with lifelike visual content as if they were in the same physical space. This involves capturing the interference patterns created by the interaction of light waves and reconstructing those patterns at the receiving end to recreate the 3D image. Holographic communication can revolutionize various fields, including teleconferencing, entertainment, education, medical imaging, and more. It promises to enable remote collaboration and interaction in a much more immersive and natural way than traditional video conferencing or image-sharing methods. However, holographic communication also presents significant technical challenges. The transmission and rendering of holographic data require substantial bandwidth and processing power, making it necessary to develop efficient data compression, transmission, and reconstruction methods. Researchers actively explore various techniques and technologies to make holographic communication feasible and practical for real-world applications. This work discusses novel ways to reduce the network load by abandoning "heavy" video traffic and utilizing data transmission traffic instead. The video obtained from the camera will be converted into a set of numbers (coordinates) on the sender's local machine and transmitted over the network with minimal requirements. After receiving these data, the end client will perform the reverse transformation of numerical values into a 3D model and display it on the screen. This solution allows the transmission of three-dimensional images instead of conventional two-dimensional video, reduces the real-time traffic load, and enables lighter data transmission traffic.

2   Related Works

There are few works that consider holographic communications. In this article, we present recent studies in this context. In [4], the authors examined the problem of standardizing the requirements for data streams transmitted over the Internet from the Kinect device. As a result, insights were gained regarding the required network characteristics of communication lines and the performance of controlling, routing, switching, and user devices to enable the widespread implementation of holographic services. In [5], the authors conducted a thorough analysis of the potential of ultra-low latency communication networks in mitigating the digital gap between different areas of the Russian Federation. The study's findings can be applied in the implementation of the digital economy program by scientific and project organizations in the strategic development of communication networks, as well as by universities in the enhancement of the educational process. In [6], the authors presented an analysis of the development of fifth-generation communication networks, 5G/IMT-2020, and the fundamental changes expected in the evolution of communication networks by 2030. The article explains potential scenarios of applications that are expected to be supported by the networks in 2030. Additionally, key functions and technologies enabling the communication of 2030 are introduced. The article also discusses other crucial aspects beyond communication technologies that should be considered in developing networks for the year 2030. This article will explore an alternative approach to delivering telepresence services capable of operating on low-bandwidth communication networks and having low latency requirements. The application will have the potential for expansion and the addition of support for holographic services, which will impose significant demands on the networks of 2030. Information about the system described in this article is highly relevant and correlates with the issues discussed in the aforementioned articles.

3   Proposed System

The system, created for successful data transmission from the sender to the receiver, consists of three applications (Fig. 1):

- Sender application (Windows operating system).
- Server application.
- Receiver application (Windows or Android operating system).

The technologies used in the development allow the application to act as both a client and a server simultaneously. However, a decision was made to implement the server on a separate machine with a dedicated IP address. Such a decision ensures the security of user data and provides the convenience of connecting to a public address that will remain constant and not change over time.

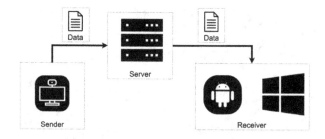

Fig. 1. Simplified Network Architecture

3.1 Booth Description

Data collection takes place within the sender application. For the application to function correctly, the following requirements must be met:

- Windows operating system.
- Stable internet access.
- Connected Kinect controller.

Data collection involves capturing body key points using the Kinect and further processing them with the Unity platform. Kinect is a touchless motion controller with two depth sensors, a color video camera, and a microphone array. The software enables complete 3D recognition of body movements, facial expressions, and voice [7]. The depth sensor includes an infrared projector and a monochrome CMOS matrix, allowing the Kinect sensor to obtain a three-dimensional image under natural lighting conditions. After successfully connecting to the server and capturing the human silhouette with the camera, a model will appear in the application that will fully replicate the person's movements on the sender side. This approach allows for faster development and testing of the application. It allows the user to see if the key points (Fig. 2) are being correctly detected or if they have gone out of the camera's view, causing data collection to be interrupted.

The application continuously sends information about the last position of the person to the server. If the client-sender goes out of the camera's frame, the model on the screen will disappear, and the network traffic volume will decrease since the network function for transmitting coordinates will not be invoked. Only packets necessary to maintain the connection to the server will occupy the bandwidth, ready to resume full data transmission as soon as the person reappears in the frame. This optimization helps reduce unnecessary data transmission and ensures efficient use of network resources.

3.2 Data Transmission

Data transmission occurs from the Windows application through the server to either the Android application or a similar Windows application, using the Mirror technology specifically designed for working with Unity. Mirror is a system

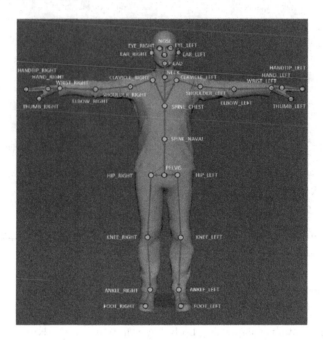

Fig. 2. Atlas of Body Key Points on a Human

for building multiplayer capabilities for Unity. It is built on top of the lower-level transport real-time communication layer and handles many of the common tasks required for multiplayer games. While the transport layer supports any kind of network topology, Mirror is a server-authoritative system. However, it allows one of the participants to be a client and the server simultaneously, so no dedicated server process is required. However, in the application being considered, a client-server-client system is implemented with the server running on a separate machine [8]. Figure 3 presents Mirror's layers.

The KCP network protocol is used for data transmission: KCP is a fast and reliable protocol that can achieve the transmission effect of a reduction of the average latency by 30% to 40% and a reduction of the maximum delay by a factor of three, at the cost of 10% to 20% more bandwidth wasted than TCP. It is implemented by using the pure algorithm and is not responsible for the sending and receiving of the underlying protocol (such as UDP), requiring the users to define their transmission mode for the underlying data packet and provide it to KCP in the way of callback. Even the clock needs to be passed in from the outside without any internal system calls [9]. The KCP Transport component in the Inspector window is shown in Fig. 4.

The network component in the client application allows sending requests to the server. The first request is to create a room to which the client-sender from the PC connects and the client-receiver from the smartphone or another PC. It is important to note that currently, only one data source can be in the room,

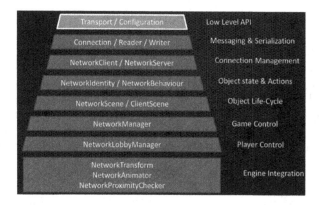

Fig. 3. Mirror's series of layers that add functionality [8]

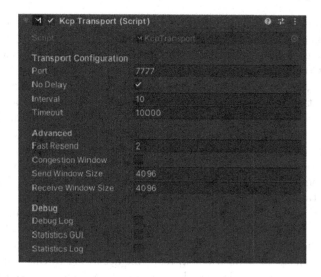

Fig. 4. The KCP Transport component in the Inspector window

while there can be any number of receivers. This room is hosted on the server, allowing all clients to connect and disconnect at any time without errors. On the server, a container is created where the sender application can store the data about the key points' coordinates and from which the receiver application can retrieve this data. Transmitting the body coordinates of the client-sender instead of their actual model reduces unnecessary network load and provides flexibility in displaying the model. It will be possible to write a custom shader that dynamically generates the human model in real-time based on the depth map provided by the RGB-D camera Kinect. However, at the moment, it requires pre-scanning the person, adding a mandatory skeletal structure, and loading it into the application as a separate model for reproduction.

4   Visualization on the Receiver Side

The key points can be visualized in the Android application using the ARCore
SDK. ARCore is a platform developed by Google for creating augmented reality
experiences. By utilizing various APIs, ARCore enables phones to gather data
about the real world and interact with it. Some API interfaces are available for
Android and iOS to facilitate cross-platform augmented reality (in this paper,
only the Android component is utilized) [10]. ARCore uses three key capabilities
to integrate virtual content with the real world as seen through your phone's
camera:

– Motion tracking.
– Environmental understanding.
– Light estimation.

ARCore's motion tracking technology uses the phone's camera to detect key
points, also known as features and tracks how they move over time. By combining the motion of these points with the information from the phone's inertial
sensors, ARCore determines the phone's position and orientation as it moves
in space [11]. ARCore's real-world data allows placing objects, annotations, or
other information in a way that seamlessly integrates them with the real world.
Users can move around and view the created objects from various angles, and if
they leave the room, the object will not disappear but will remain in its position.
The Android application connects to the server using an IP address and starts
scanning the surrounding planes using the phone's camera. After a short period,
the application recognizes a plane and draws an indicator to aid in object placement. Upon tapping the screen, an object with a network component appears,
instantly starting to receive data from the server. Suppose the server is aware
of the presence of one or more people in the frame of the client-sender. In that
case, the application creates a model for each client-receiver on the screen and
dynamically adjusts their positions in real-time according to the received data.
     The application has implemented the automatic removal of a specific model
from the screen and all its data on the server if the person exits the frame
or the removal of all models and their network components if the client-sender
disconnects from the server. This ensures efficient management of virtual objects,
ensuring that they are dynamically updated based on the presence or absence of
users. In addition to the Android application, there is a possibility to reproduce
the movements of a person on another PC. For this, during the first launch
of the application, the user will be prompted to choose their role: either as a
client-sender or a client-receiver. The display system on the PC-receiver has
similar functionality to the system on Android-receiver, except for placing the
object in real space. The display on the PC remains fixed at a single point, and
the movement of objects on the screen can only be triggered by the person's
movement in the frame on the sender's side.

5 Testing: Comparison with the Nearest Competitors

A comparative analysis was conducted between the described system with the working title MVP and the closest competitor applications: TrueConf, Zoom, and Yandex Telemost. All tests were performed within the same local network. During the investigation, the primary mode of video conferencing was tested: in the analog applications, web cameras were enabled for the communicating parties, while in the MVP system, the motion transmission to the model was achieved using developed technologies. The measurement results were entered in Table 1 and displayed in Fig. 5.

Table 1. Traffic analysis and Subjective assessment of the quality of information display organized using desktop applications "TrueConf," "Zoom," "Yandex Telemost," and "MVP."

Application	Average bandwidth (transmission side), Mbps		Average bandwidth (receiving side), Mbps		Subjective assessment of the quality of information display 4,7
	Load	No Load	Load	No Load	
TrueConf	11,3	7,1	11,1	7,0	4,2
Zoom	4,8	0,1	4,6	0,1	4,5
Yandex Telemost	4,5	3,4	4,4	3,3	4,7
MVP	0,5	0,4	0,5	0,4	

In Fig. 5: experiments "No Load" are represented from 1 to 6 - a person in the frame is sitting still, and nothing is moving in the background. Experiments "Load" are represented from 7 to 13 - the maximum number of changing pixels on the screen: the movement of a person in the frame and a changing background. The comparative analysis showed that all applications transmit less traffic when there are fewer movements in the frame. In the case of the MVP system, only packets for maintaining the connection are transmitted, similar to the Zoom application. However, when activity is detected in the frame, the Zoom application, like other competitors, requires higher bandwidth than MVP.

6 Conclusion

The article describes a system that synchronizes control points of the human body with corresponding points on a model. The system consists of the Sender Application (Windows operating system), the Server Application, and the Receiver Application (Windows or Android operating system). Such systems can

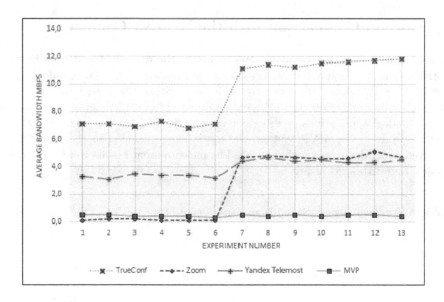

Fig. 5. Bandwidth comparison

facilitate interactions between people where the presence of two or more individuals is required, such as in teacher-student interactions, corporate meetings, gatherings, etc. They enable a significant reduction in network bandwidth requirements by transmitting only the data about the movement of objects instead of transmitting the actual image or 3D model. Further system development can include porting the receiving application to virtual reality headsets, eliminating the need to hold a phone. Additionally, functionality could be expanded for client applications, incorporating various 3D models and writing custom shaders to generate HTC traffic.

Acknowledgments. The studies at St. Petersburg State University of Telecommunications. prof. M.A. Bonch-Bruevich were supported by the Ministry of Science and High Education of the Russian Federation by the grant 075-15-2022-1137.

References

1. Clemm, A., et al.: Toward truly immersive holographic-type communication: challenges and solutions. IEEE Commun. Mag. **58**(1), 93–99 (2020)
2. Vybornova, A.: Immersive Telecommunications: a survey and future development. Telecom IT **9**(3), 1–10 (2021). https://doi.org/10.31854/2307-1303-2021-9-3-1-10
3. Cao, C., Preda, M., Zaharia, T.: 3D point cloud compression: a survey. In: The 24th International Conference on 3D Web Technology, pp. 1–9 (2019)
4. Pankov, B., Makolkina, M.: Research of network characteristics of holographic traffic. Telecom IT **10**(3), 20–31 (2022). https://doi.org/10.31854/2307-1303-2022-10-3-20-31

5. Chistova, N., Borodin, A., Koucheryavy, A.: Ultra-low latency networks and bridging the digital divide between Russian Federation regions. Telecom IT **9**(2), 1–20 (2021). https://doi.org/10.31854/2307-1303-2021-9-2-1-20
6. Volkov, A., Muthanna, A., Koucheryavy, A.: Fifth generation communication networks: on the way to networks 2030. Telecom IT **8**(2), 32–43 (2020). https://doi.org/10.31854/2307-1303-2020-8-2-32-43
7. Technologies of contactless computer control [Online resource]. https://clck.ru/32q5it. Accessed 04 Nov 2022
8. Mirror Documentation [Online resource]. https://mirror-networking.gitbook.io/docs. Accessed 07 Nov 2022
9. Description of the KCP protocol [Online resource]. https://clck.ru/32q5oW. Accessed 08 Nov 2022
10. ARCore Documentation [Online resource]. https://developers.google.com/ar/develop. Accessed 15 Nov 2022
11. How far has ARCore come [Online resource]. https://russianblogs.com/article/89641480790/. Accessed 16 Nov 2022

The Simulation of Finite-Source Retrial Queues with Two-Way Communication to the Orbit, Incorporating a Backup Server

Ádám Tóth$^{(\boxtimes)}$ [ID] and János Sztrik [ID]

University of Debrecen, University Square 1, Debrecen 4032, Hungary
{toth.adam,sztrik.janos}@inf.unideb.hu

Abstract. This paper investigates a two-way communication retrial queuing system with a server that may experience random breakdowns. The system is a finite-source M/M/1//N type, and the idle server can make calls to the customers in the orbit, also known as secondary customers. The service time of the primary and secondary customers follow independent exponential distributions with rates of μ_1 and μ_2, respectively. The novelty of this study is to analyze the impact of various distributions of failure time on the key performance measures using a backup server, such as the mean response time of an arbitrary customer. One could think of a backup server as a primary server that operates at a reduced rate during periods of repair. To ensure a valid comparison, a fitting process is conducted so that the mean and variance of every distribution are equal. The self-developed simulation program provides graphical illustrations of the results.

Keywords: Simulation · Queueing system · Finite-source model · Sensitivity analysis · Backup server · Unreliable operation · Outgoing calls

1 Introduction

Nowadays, due to the growth of traffic and the increasing number of users, analyzing communication systems or designing optimal patterns for these schemes is a challenging task. Information exchange is essential in every aspect of life, and it is crucial to develop mathematical and simulation models of telecommunication systems or modify the existing ones to keep pace with these changes. Retrial queues are effective and appropriate tools for modeling real-life problems that arise in telecommunication systems, networks, mobile networks, call centers, and similar systems. Numerous papers and books have been dedicated to studying a variety of retrial queuing systems with repeated calls like in [4,5].

We are investigating a retrial queuing system with two-way communication capabilities, which has become a popular research topic due to its resemblance to certain real-life systems. This is particularly relevant in call centers, where

© The Author(s), under exclusive license to Springer Nature Switzerland AG 2024
V. M. Vishnevskiy et al. (Eds.): DCCN 2023, LNCS 14123, pp. 44–55, 2024.
https://doi.org/10.1007/978-3-031-50482-2_4

service units may engage in additional activities such as sales, promotion, and product advertising while attending to incoming calls. In our study, the primary server calls in customers from the orbit, known as secondary customers, when it becomes idle after a random period of time. The utilization of the service unit is monitored and has been extensively studied in previous works, for example in [3, 10].

In some scenarios, it is assumed by researchers that service units are available continuously, but failures or sudden events may happen during operation resulting in the rejection of incoming customers. Devices used in different industries are subject to breakdowns, and considering their reliable operation is quite an optimistic and unrealistic approach. Similarly, in wireless communication, various elements can affect the transmission rate, and interruptions may occur during packet transmission. The unreliable nature of retrial queuing systems greatly affects the system's operation and performance measures. At the same time, completely stopping production is not feasible as it can lead to delays in fulfilling the orders. Hence, during such failures, machines or operators with lower processing rates can continue to work to ensure a smoother operation. Additionally, the authors examined the possibility of having a backup server available to provide service at a reduced rate in cases where the main server is unavailable. Many recent papers have extensively studied retrial queuing systems with unreliable servers, [7, 9] are just a few examples.

In service sectors, it is not uncommon for service providers to experience breakdowns due to various reasons, including the inability to access their database to address customer requests. When such breakdowns occur, service providers often resort to alternative measures such as accessing backup systems or gathering additional information from the customers to provide the required solutions. Here are some papers which thoroughly investigate the behaviour of systems that tries to enhance the service by adding a backup server like [1, 8, 11, 12] or [15].

The main objective of this study is to investigate the impact of the unreliable operation of a system by comparing various failure time distributions on performance measures such as the mean response time of a customer or the service unit utilization. This paper is a continuation of the previous work by the authors [13], where the system had an unreliable server, but now, if the server is unavailable a backup server takes its place to serve incoming requests. To obtain the desired performance measures, a simulation model was developed using SimPack [6], a set of C/C++ libraries and executable programs for computer simulation. Simulation is an excellent alternative to deriving exact formulas, particularly when it is problematic or almost impossible. The user can apply as many distributions as needed to approximate performance measures. In this paper, we present a sensitivity analysis of various failure time distributions on the main performance measures. We illustrate the results through graphical representations of interesting phenomena related to sensitivity problems.

2 System Model

The system under consideration (in Fig. 1) is a retrial queuing system with an unreliable server and a finite-source. The source contains N customers, each generating primary customer requests with a rate of λ, such that inter-arrival times are exponentially distributed with a parameter of λ. Note that our model does not include waiting queues, so incoming customers occupy the server only when it is available and not busy. The service time of primary customers follows an exponential distribution with a parameter of μ_1. Following a successful service, the customer returns to the source. However, if an arriving customer (either from the source or orbit) encounters the server in a busy or failed state, the request is forwarded to the orbit. In the orbit, the customer may make an attempt to get its service requirement after an exponentially distributed random time with a parameter of σ. The system is assumed to have an unreliable server that can break down according to different distributions such as gamma, hypo-exponential, hyper-exponential, Pareto, and lognormal, each with different parameters but the same mean value. The repair process begins immediately after the server fails, and the repair time is exponentially distributed with parameter γ_2. If the server is busy and fails, the customer is immediately transferred to the orbit. All customers in the source can generate requests even if the service unit is unavailable, but these requests are directed to the backup server, which serves at a reduced rate (this is also an exponentially distributed random variable with parameter μ_3) when the main server is unavailable. The backup server is assumed to be reliable and works only if the main server is down. In the case of a busy backup server, the incoming requests are placed in the orbit. However, when the server is idle, it can initiate an outgoing call to the customers in the orbit after a random time, which is exponentially distributed with rate τ. The service time of these secondary customers follows an exponential distribution with parameters μ_2. The assumption made during model creation is that all random variables are completely independent of one another.

3 Simulation Results

We utilized a statistical module class providing a statistical analysis tool that enables us to quantitatively estimate the mean and variance values of observed variables using the batch mean method. The method aggregates n successive observations of a steady-state simulation to generate a sequence of independent samples. The batch mean method is a common technique used to establish confidence intervals for the steady-state mean of a process. To ensure the sample averages are approximately independent, large batches are required. More information on the batch mean method can be found in [2]. We conducted simulations with a 99.9% confidence level, and the simulation run was halted when the relative half-width of the confidence interval reached 0.00001.

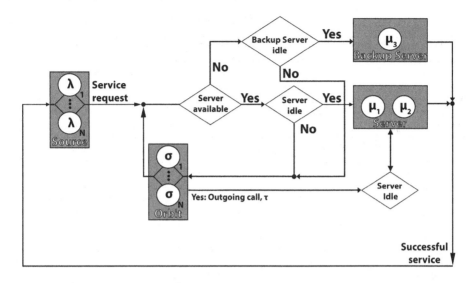

Fig. 1. The system model

3.1 First Scenario

In this section, we aimed to set the parameters of failure time for each distribution in such a way that the mean value and variance would be equal. The fitting process used for this purpose can be found in the following paper [14]. Four different distributions were considered in order to investigate their impact on performance measures. The hyper-exponential distribution was chosen to ensure that the squared coefficient of variation is greater than one. Table 2 presents the input parameters of the various distributions, while Table 1 shows the values of other applied parameters.

Table 1. Used numerical values of model parameters

N	λ	γ_2	σ	μ_1	μ_2	ν	μ_3
100	0.01	1	0.01	1	1.2	0.02	0.1; 0.6

The steady-state distribution for different failure time distributions is presented in Fig. 2. On the X-axes i represents the number of customers located in the system, and on the Y-axes $P(i)$ denotes the probability that exactly i customer is found in the system. Upon closer examination of the curves, it can be observed that all of them resemble the normal distribution. Although the Pareto distribution appears to have more customers in the system, there are no significant differences among the various distributions tested. Including a backup server results in a lower mean number of customers in the system in comparison with the paper of [13].

Table 2. Parameters of failure time

Distribution	Gamma	Hyper-exponential	Pareto	Lognormal
Parameters	$\alpha = 0.6$ $\beta = 0.5$	$p = 0.25$ $\lambda_1 = 0.41667$ $\lambda_2 = 1.25$	$\alpha = 2.2649$ $k = 0.67018$	$m = -0.3081$ $\sigma = 0.99037$
Mean	1.2			
Variance	2.4			
Squared coefficient of variation	1.6666666667			

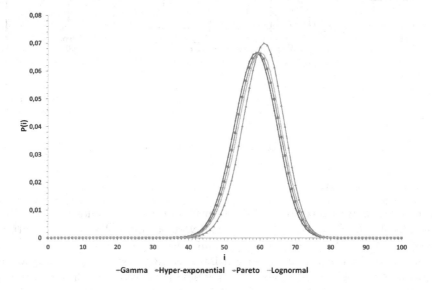

−Gamma ◦Hyper-exponential ⊸Pareto ⊸Lognormal

Fig. 2. Comparison of steady-state distributions

Figure 3 illustrates the relationship between the mean response time of customers and the arrival intensity. Consistent with the observations from Fig. 2, the highest mean response time is observed with the Pareto distribution. However, the differences among the other distributions are more noticeable. The gamma distribution yields the lowest mean response time. Interestingly, as the arrival intensity increases, the mean response time initially increases, but then starts to decrease after a certain point. This is a unique feature of retrial queuing systems with a finite source, and is a general characteristic when suitable parameter settings are used.

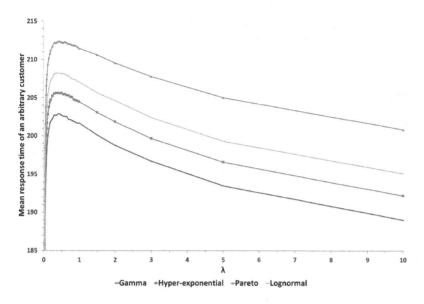

Fig. 3. Mean response time vs. arrival intensity

Figure 4 highlights the effect of using a backup server besides increasing arrival intensity. Under the "No backup server" it is meant when the normal server is either reliable or not but a backup server is not used at all. Actually, the expected behaviour occurs, the "No backup server, reliable normal server" case would be the ideal situation but in reality sudden acts or breakdowns can happen at any time. Comparing the obtained results when the service intensity of the backup is the lowest then the time spent in the system of the customers is the highest. From this figure we experience the advantage of using a backup server hence the customers spend less time in the system while they receive their service demand properly.

The final figure in this section depicts the utilization of the normal and backup service unit beside the arrival intensity using gamma distributed failure time. The red and orange curves represent the total time spent by the clients at the backup server. Upon careful examination of the figure, naturally, the utilization of the backup service unit decreases when the intensity of the service of the backup server increases. For the other distributions, the same tendency is observed, in this way, those ones are not depicted in this paper. As the arrival intensity increases, the utilization of the service units also increases. However, after reaching a certain arrival value (in this case it is around 5), utilization becomes basically stagnant.

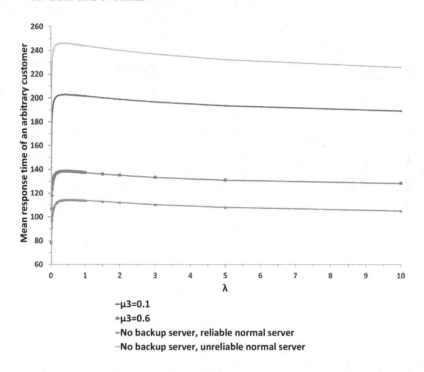

Fig. 4. Mean response time vs. arrival intensity

3.2   Second Scenario

We were curious about how the performance measurements are altered with the modification of the failure time parameters after observing the results of the previous section. The parameters were now selected to ensure that the squared coefficient of variation was below one. Because the squared coefficient of variation for a hypo-exponential distribution is always smaller than one, we replace the hyper-exponential distribution with hypo-exponential distribution. By utilizing the new failure time parameters, we will review the same figures as in the previous section to check the effect of newly chosen parameters, which is shown in Table 3. The other parameters remain unchanged (see Table 1) (Fig. 5).

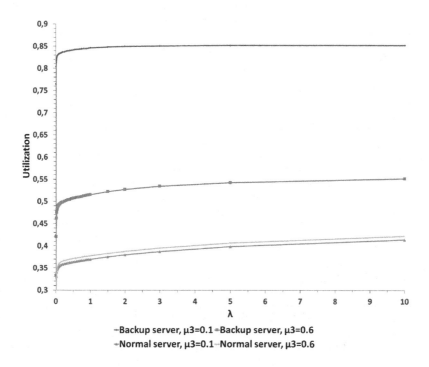

Fig. 5. Comparison of utilization

Table 3. Parameters of failure time

Distribution	Gamma	Hypo-exponential	Pareto	Lognormal
Parameters	$\alpha = 1.3846$ $\beta = 1.1538$	$\mu_1 = 1$ $\mu_2 = 5$	$\alpha = 2.5442$ $k = 0.7283$	$m = -0.0894$ $\sigma = 0.7373$
Mean	1.2			
Variance	1.04			
Squared coefficient of variation	0.722222			

Figure 6 displays the steady-state distributions with a squared coefficient variation of less than one. The curves overlap closely, indicating that regardless of the chosen distribution of failure time, the average number of customers in the system remains the same. In comparison to Fig. 2, the average values are a little bit greater and the curve of Pareto is closer to the other curves.

Figure 7 illustrates the development of the mean response time of an arbitrary customer as the arrival intensity increases. Upon closer inspection, it is evident that the curves are much closer to each other compared to Fig. 3, although there are minor differences among the chosen distributions. Similarly to Fig. 3, the highest values are observed in the case of Pareto distribution. When the squared coefficient of variation is less than one for each distribution, the mean waiting times are higher compared to the previous section (Fig. 7).

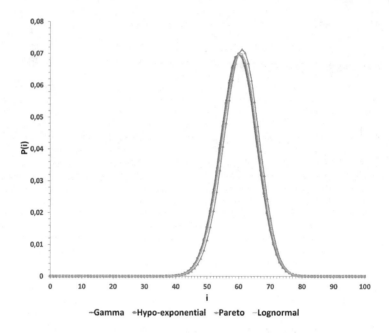

Fig. 6. Comparison of steady-state distributions

The final figure in this section illustrates the utilization of the primary and backup service units as a function of arrival intensity, using gamma distributed failure time. The utilization of the backup service unit is represented by the red and orange curves. Upon careful examination of the figure, it is observed that as the intensity of the service provided by the backup server increases, the utilization of the backup service unit decreases. This tendency is also observed for other distributions, although they are not depicted in this paper. With an increase in arrival intensity, the utilization of the service units also increases. However, after reaching a certain arrival value (in this case, approximately 5), the utilization reaches a plateau. In this figure, the same direction can be observed as in the previous section, the obtained values are basically equal to the previous scenario (Fig. 8).

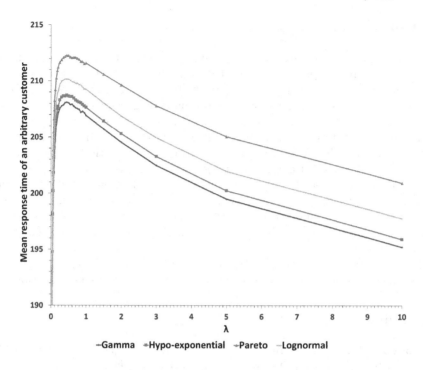

Fig. 7. Mean response time vs. arrival intensity

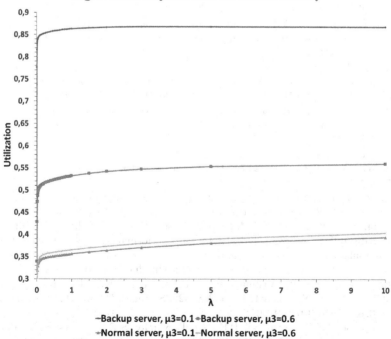

Fig. 8. Comparison of utilization

4   Conclusion

We present a retrial queuing system with finite source and two-way communication, where there is a primary server that is unreliable, and there is also a secondary service unit replacing it in the faulty periods. Moreover, we performed a sensitivity analysis using various random number generators to explore the effect of different distributions of failure time on the performance measures like the mean response time of an arbitrary customer or the utilization of the service units. We observe that when the squared coefficient of variation is greater than one, the mean response time of a customer exhibits some disparity among the values, but the influence is negligible when it is less than one. Using a backup service unit may significantly decrease the time spent in the system of the customers, especially in those scenarios where the primary service unit is under common breakdowns, or the channel is not reliable or the customers are moving rapidly.

In the future, we plan to make further modifications to the system, such as considering additional distributions and incorporating features like vacation.

References

1. Chakravarthy, S.R., Shruti, Kulshrestha, R.: A queueing model with server breakdowns, repairs, vacations, and backup server. Oper. Res. Perspect. **7**, 100131 (2020). https://doi.org/10.1016/j.orp.2019.100131. https://www.sciencedirect.com/science/article/pii/S2214716019302076
2. Chen, E.J., Kelton, W.D.: A procedure for generating batch-means confidence intervals for simulation: checking independence and normality. SIMULATION **83**(10), 683–694 (2007)
3. Dragieva, V., Phung-Duc, T.: Two-way communication M/M/1//N retrial queue. In: Thomas, N., Forshaw, M. (eds.) ASMTA 2017. LNCS, vol. 10378, pp. 81–94. Springer, Cham (2017). https://doi.org/10.1007/978-3-319-61428-1_6
4. Dragieva, V.I.: Number of retrials in a finite source retrial queue with unreliable server. Asia-Pac. J. Oper. Res. **31**(2), 23 (2014). https://doi.org/10.1142/S0217595914400053
5. Fiems, D., Phung-Duc, T.: Light-traffic analysis of random access systems without collisions. Ann. Oper. Res. **277**, 311–327 (2017). https://doi.org/10.1007/s10479-017-2636-7
6. Fishwick, P.A.: SimPack: getting started with simulation programming in C and C++. In: 1992 Winter Simulation Conference, pp. 154–162 (1992)
7. Gharbi, N., Nemmouchi, B., Mokdad, L., Ben-Othman, J.: The impact of breakdowns disciplines and repeated attempts on performances of small cell networks. J. Comput. Sci. **5**(4), 633–644 (2014)
8. Klimenok, V., Dudin, A., Semenova, O.: Unreliable retrial queueing system with a backup server. In: Vishnevskiy, V.M., Samouylov, K.E., Kozyrev, D.V. (eds.) DCCN 2021. LNCS, vol. 13144, pp. 308–322. Springer, Cham (2021). https://doi.org/10.1007/978-3-030-92507-9_25
9. Krishnamoorthy, A., Pramod, P.K., Chakravarthy, S.R.: Queues with interruptions: a survey. TOP **22**(1), 290–320 (2014). https://doi.org/10.1007/s11750-012-0256-6

10. Kuki, A., Sztrik, J., Tóth, Á., Bérczes, T.: A contribution to modeling two-way communication with retrial queueing systems. In: Dudin, A., Nazarov, A., Moiseev, A. (eds.) ITMM/WRQ -2018. CCIS, vol. 912, pp. 236–247. Springer, Cham (2018). https://doi.org/10.1007/978-3-319-97595-5_19
11. Liu, Y., Zhong, Q., Chang, L., Xia, Z., He, D., Cheng, C.: A secure data backup scheme using multi-factor authentication. IET Inf. Secur. 11(5), 250–255 (2017). https://doi.org/10.1049/iet-ifs.2016.0103. https://ietresearch. onlinelibrary.wiley.com/doi/abs/10.1049/iet-ifs.2016.0103
12. Satheesh, R.K., Praba, S.K.: A multi-server with backup system employs decision strategies to enhance its service. Research Square, pp. 1–31 (2023). https://doi. org/10.21203/rs.3.rs-2498761/v1
13. Sztrik, J., Tóth, Á., Pintér, Á., Bács, Z.: The effect of operation time of the server on the performance of finite-source retrial queues with two-way communications to the orbit. J. Math. Sci. 267, 196–204 (2022). https://doi.org/10.1007/s10958-022-06124-z
14. Toth, A., Sztrik, J., Kuki, A., Berczes, T., Effosinin, D.: Reliability analysis of finite-source retrial queues with outgoing calls using simulation. In: 2019 International Conference on Information and Digital Technologies (IDT), June 2019, pp. 504–511 (2019). https://doi.org/10.1109/DT.2019.8813419
15. Won, Y., Ban, J., Min, J., Hur, J., Oh, S., Lee, J.: Efficient index lookup for de-duplication backup system, pp. 383–384, September 2008. https://doi.org/10. 1109/MASCOT.2008.4770594

On Real-Time Model Inversion Attacks Detection

Junzhe Song(iD) and Dmitry Namiot[(✉)](iD)

Lomonosov Moscow State University, GSP-1, Leninskie Gory, Moscow 119991,
Russian Federation
`songjz@smbu.edu.cn`, `dnamiot@gmail.com`

Abstract. The article deals with the issues of detecting adversarial
attacks on machine learning models. In the most general case, adversarial
attacks are special data changes at one of the stages of the machine learning pipeline, which are designed to either prevent the operation of the
machine learning system, or, conversely, achieve the desired result for the
attacker. In addition to the well-known poisoning and evasion attacks,
there are also forms of attacks aimed at extracting sensitive information
from machine learning models. These include model inversion attacks.
These types of attacks pose a threat to machine learning as a service
(MLaaS). Machine learning models accumulate a lot of redundant information during training, and the possibility of revealing this data when
using the model can become a serious problem.

Keywords: machine learning · adversarial attacks · model inversion

1 Introduction

This article is an expanded version of a paper presented at the 2023 DCCN
conference [1].

Machine learning systems (and, at least now, it is a synonym for artificial intelligence systems) depend on data. This tautological statement leads, in
fact, to quite serious consequences. Changing the data then, generally speaking,
changes the performance of the model. Purposeful data changes are attacks on
machine learning models [2]. But the models themselves can be directly affected
during attacks. For example, weights can change on the fly, malicious code can be
loaded into weights, etc. Adversarial attacks, which are possible for any discrim-
inant machine learning models, pose a great threat to machine learning systems,
since they do not guarantee the results and quality of the system. And such guarantees are, for example, mandatory for the use of a machine learning (artificial
intelligence) system in critical areas such as avionics, automatic driving, special
applications, etc. [3,4]. An attack directly on the model also carries additional
risks of extracting private information stored in machine learning models [5].

Attacks aimed at stealing intellectual property from machine learning models must specifically form sequential requests to the model in order to extract

V. M. Vishnevskiy et al. (Eds.): DCCN 2023, LNCS 14123, pp. 56–67, 2024.
https://doi.org/10.1007/978-3-031-50482-2_5

the necessary information. Usually, when considering such attacks on machine learning models, first of all, the issues of countering such attacks are considered. This can be (most often) a special form of issuing the results of the model, which does not allow the attacker to correctly form subsequent requests. In the present work, we want to dwell on the detection (determining the fact of carrying out) such of attacks. Naturally, the definition of the fact of an attack does not negate the need for protection. But this, for example, is very important from the point of view of cybersecurity. Detecting an attack can help determine its source or links to other attackers. Attacks on the data contained in models are always related to polling those models. Accordingly, conceptually, the definition of the fact of an attack is similar to how it is done for web applications. First, we can weed out requests that look wrong. For example, incorrect headers, generated programmatically, etc. Further, you can filter out requests that come too often. And finally, use heuristics (rules) to identify attacks. This last point for machine learning systems is considered in this article.

The remainder of the article is structured as follows. In Sect. 2, we focus on attacks on intellectual property. Section 3 presents the developed algorithm for detecting extraction attacks. Section 4 presents the conclusion.

2   On Data Extraction from Machine Learning Models

In some classifiers, these types of attacks are also referred to as attacks aimed at stealing intellectual property. With the help of such attacks, one can, for example, restore the algorithm of the model or obtain various information about the training data. The American Standards Institute NIST in its glossary [6] defines 5 types of such attacks:

- Data Reconstruction
- Memorization
- Membership Inference
- Model Extraction
- Property Inference

The Model Extraction attack is perhaps the most understandable in its logic [15]. If we have the ability to interrogate the model, then we can accumulate a set of inputs and outputs <x, Y> and use this set as a training dataset to create a shadow model. This is one of the easiest ways to replicate the functionality of an existing model. In this regard, we can mention a multilayer perceptron, which just solves such problems (Fig. 1), by selecting hidden layers. We also note that all attacks of this class depend on the possibility of multiple polling of models. First of all, they are focused on attacks on MLaaS (machine learning as a service) systems [16].

Data reconstruction attacks should be recognized as the most serious in terms of access to private attributes. These attacks try to restore the input (training) data of the attacked model based on the results of its work [7]. Another name is model inversion attacks [8]. Recently, attention to this kind of attacks has

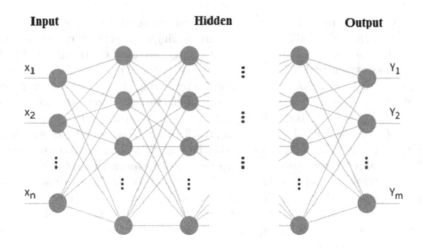

Fig. 1. Multilayer perceptron

grown significantly. Machine learning models accumulate a lot more data than traditional databases. In the latter (if we mean classical relational models), the data is structured and traditionally a lot of attention is paid to the efficiency of data storage. For training models, huge arrays of unstructured data are used (data markup for training, of course, is not structuring). This is where the idea of using this data comes from. A good overview of the work on this topic is presented, for example, in the paper [18].

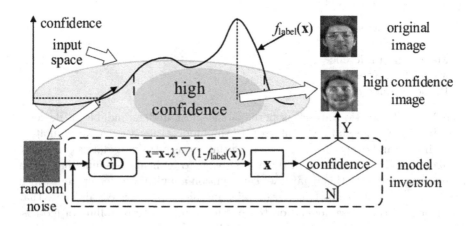

Fig. 2. Model inversion [9]

The idea is actually quite transparent - the data that was in the training set should be recognized better (the peak of the graph in Fig. 2) than those that were not in the set. If we have a model M with input x and output y, then the model inversion attack is the reconstruction of input x that is most likely classified as y by the model M. For example, if the attacked model is a face recognition model and some result of its work (the recognized image y), then the inversion attack is the reconstruction of the face image, which is most likely to be classified (recognized) as y. The point is that model M was trained on private dataset D. The attacker has access to the model (if it is MlaaS, then this is always the case), but, of course, there is no access to the private training dataset. Here the attacker is trying to restore it. In Fig. 2, the top image is a private image from the training dataset, the bottom image is recovered as a result of the attack [19]. In addition to the possible access to private information, such an attack allows you to restore the training dataset, which can be used to train the shadow model and then organize adversarial attacks. The accuracy of such attacks can be quite large. For example, the attacks described in [19] showed accuracy greater than 90% (Fig. 3 - original and restored images)

Fig. 3. Model Inversion on CelebA dataset [19]

Memorization attacks are a class of techniques that allow an attacker to extract training data from generative machine learning models such as language models [10]. Generalization and memorization in machine learning models are coupled, and neural networks can remember randomly selected datasets: deep learning models (in particular, generative models) often remember rare details about the training data that are completely unrelated to the task at hand. This "extra" data becomes the target of the attack.

Membership Inference attacks are aimed at determining whether a particular record or data sample was part of the training data set [11]. As a rule, such attacks are adapted to be executed in the black box mode (Fig. 4).

Unlike membership inference attacks that aim to expose the privacy of individual training set samples, property inference attacks aim to infer global properties of the entire training set of a machine learning model, such as inferring the proportion of certain sensitive properties of the training set [20]. In property inference attacks, the attacker tries to learn global information about the distribution of the training data. The goal is to disclose confidential information

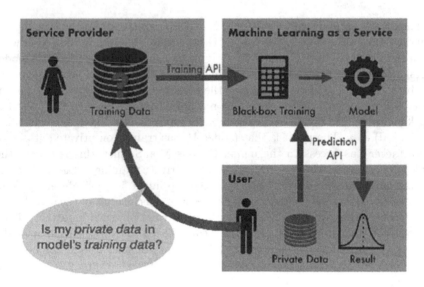

Fig. 4. Membership inference [12]

about the training sample (for example, dependence on some attributes, etc.)
[12].

3   On the Proposed Model Inversion Attack Detector

Consider a model M that performs image classification, i.e. $M: X \rightarrow Y$, where M is the set of answers such that $M \in \{0, ..., k\}$. The model maps X features to Y, and Y is a finite set. In image classification, Y represents the set of categories that the images in the dataset belong to. Typically, as an image classification model, M takes an image tensor X as input and produces a tuple of labels and corresponding probabilities P, that is,

$$M(X) = [(0, P_0), (1, P_1), ..., (k, P_k)], \text{ where}$$
$$\sum_{i=1}^{k} P_i = 1$$

The action of a model inversion attack can be formalized as a mathematical sequence [13]. To illustrate, consider a sequence of input values represented as

$$X^k = \{X_0^k, X_1^k, ..., X_n^k\}$$

where X_0^k is the initial input, and X_n^k is the result closest to the specified label k. The elements in a given sequence are related to each other. The connection between them can be described by the function f (this is the model inversion attack itself):

$$X_{n+1}^k = f(X_n^k)$$

The goal of the attack is to use model inversion techniques to gradually approximate the original pattern in the dataset corresponding to the selected target label. In other words, the attack is aimed at gradually changing the original data until it becomes similar to the original sample of the desired label:

$$\lim_{x \to \infty} X_n^k = X_{origin}^k$$

The Fig. 5 shows a generic form of the model inversion attack, where the attacker wishes to infer sensitive information related to the class represented by *label*. Let $\tilde{f}_{label}(x)$ represents the output confidence that the input vector x matches the target class, represented by *label*. E.g., in the case of facial recognition, *label* presents the name (ID) of a person whose facial features are the goal of an attack.

Algorithm 1 General Model Inversion Attack

Require: Initial vector x with non-sensitive features and random value for sensitive features
Ensure: Output vector x with inferred sensitive features
   while $\tilde{f}_{label}(x)$ < THRESHOLD **do**
      $x \leftarrow x + \alpha \cdot \nabla \tilde{f}_{label}(x)$

Fig. 5. A typical model inversion attack [14]

The algorithm receives a vector $x = (x_0, x_1, .., x_n)$, where some of the values may be non-sensitive data that the attacker could obtain about a target user and unknown sensitive data (they are initialized with any values). The goal of the attack is to maximize confidence (or minimize error) in the prediction according to the given threshold. This is achieved in a standard way for machine learning, using gradient $\nabla \tilde{f}_{label}(x)$ movement with some step α.

Our Model Inversion attack detection technique is to find the difference between the results of the (n)-th and (n + 1)-th iterations. Initially, we define a window in which ten image requests can be placed. When the user submits the second request, we start calculating the L2 norms. If the L2-norm value is less than 100 (this threshold may differ in different datasets), we can identify this query as a possible attack behavior. We then calculate the cosine similarity for each of these suspicious queries. If the cosine similarity consistently exceeds 0.9, this can be classified as a model inversion attack given that the average

---

Algorithm   Detector for Model Inversion Attack

```
 1: function Detector(inputs:list, window:int, L2_max, cosine_max)
 2:     L2_value = []
 3:     cosine_value = []
 4:     if len(inputs)>=2 then
 5:         L2_value.append(L2_norm(inputs[-1],inputs[-2]))
 6:         cosine_value.append(cosine_similarity(inputs[-1],inputs[-2]))
 7:     end if
 8:     if in L2_value has continuous window items >= L2_max then
 9:         if in cosine_value has continuous window items >= cosine_max then
10:             print('This is Model Inversion Attack')
11:         end if
12:     end if
13: end function
```

Fig. 6. The proposed model inversion attack detector

user rarely repeats the same query for the same image more than ten times. The pseudo-code is presented in Fig. 6.

The experimental parameters of our study are described as follows. We performed a model inversion attack on two different datasets: a 1-channel dataset, namely the MNIST handwritten digits set, and a 3-channel dataset, namely CIFAR10.

To implement the attack, we used the classic MI-Face attack [21]. The attack methodology uses a cost function, denoted as c, depending on the face recognition model $\tilde{f}$ and a special helper function $AuxTerm$. The $AuxTerm$ function allows you to include any additional information that can improve the cost function. The MI-Face function performs gradient descent over a maximum of α iterations using steps of size λ. The $Process$ function can manipulate the results of gradient descent, including despreading and sharpening. Gradient descent will terminate if the value of the candidate cost function does not improve after β iterations or if the value of the cost function exceeds a given threshold γ, returning the optimal result (Fig. 7).

The classification accuracy of the MNIST model was 98.7300% and that of CIFAR10 was 84.9700%. Figures 8 and 9 illustrate the reconstructed images after 300 and 10000 iterations.

In this case, they are not very different. But the first recognizable images appeared no earlier than the 40th iteration. This gives an estimate of the number of lookups that an attacker needs to create.

To test our idea, on 300 iterations for 10 labels (0–9) in the MNIST handwritten digit dataset, we evaluated the performance of the L metrics and cosine similarity (Figs. 10 and 11).

Algorithm   Inversion attack
1: **function** MI-Face(label,α, β, γ, λ)
2:     $c(x) \overset{def}{=} 1 - \tilde{f}_{label}(x) + AuxTerm(x)$
3:     $x_0 \leftarrow 0$
4:     **for** $i \leftarrow 1...\alpha$ **do**
5:         $x_i \leftarrow Process(x_{i-1} - \lambda \cdot \nabla c(x_{i-1}))$
6:         **if** $c(x_i) \geqslant max(c(x_{i-1}), ..., c(x_{i-\beta}))$ **then**
7:             break
8:         **end if**
9:         **if** $c(x_i) \leqslant \gamma$ **then**
10:             break
11:         **end if**
12:     **end for**
13:     **return** $[\arg\min_{x_i}(c(x_i)), \min_{x_i}(c(x_i))]$
14: **end function**

Fig. 7. MI-Face attack

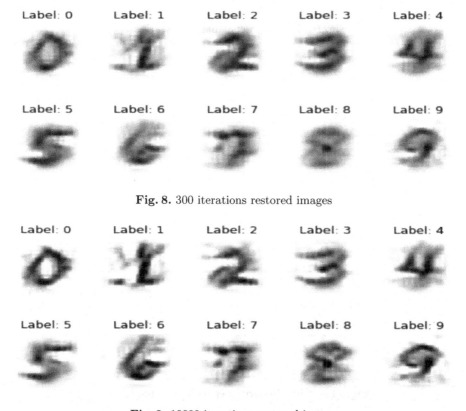

Fig. 8. 300 iterations restored images

Fig. 9. 10000 iterations restored images

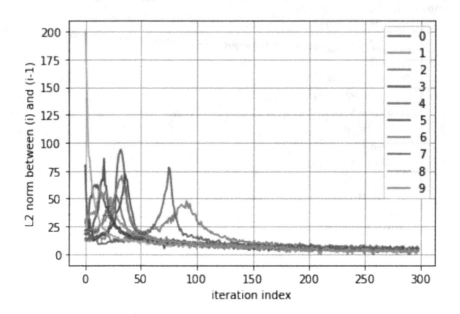

Fig. 10. L2 norm depending on iterations

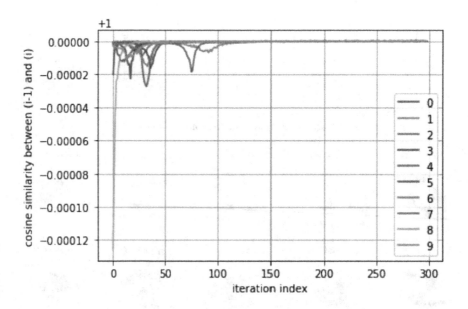

Fig. 11. Cosine similarity depending on iterations

The Fig. 11 shows the fast saturation of the cosine similarity parameter. Differences in subsequent queries become insignificant as the number of iterations (successive queries) increases. This is the main signal of the ongoing attack. And the study of the L2 norm allows you to select candidate queries for verification. The limit of 100 was chosen based on the results of experiments (there are peaks at these points on the graph).

A necessary condition for the target (attacked) model in this study is the presence of the Softmax layer as its output for the purposes of image classification. In addition, it is necessary that the model be highly accurate, with an accuracy of at least 85%, since the success of the model inversion attack is highly dependent on the classification accuracy. Accordingly, a higher level of target model accuracy will yield better results.

The source code for the proposed detection method is made publicly available and can be used to develop this direction [17].

When will this attack detection method not work? Obviously - with distributed attacks. If several attackers send requests in a coordinated manner, then successive lookup requests will be distributed among the participants in the attack. But then the attack will be more expensive and will be performed much more slowly. In general, it should be noted that attacks related to access to hidden information of machine learning models pose a serious threat to the MLaaS model. In such systems, we cannot avoid user requests - models are designed to serve such requests. Therefore, the definition of attacks must, of course, be combined with defenses against them. For example, regularization, adding controlled noise to the output of a confidence function, minimizing the variance of confidence estimates, etc. One of the main goals of precisely determining the fact of attacks is the ability to conduct investigations and look for attackers.

4 Conclusion

The article presents a new algorithm for detecting inversion attacks of machine learning models. We focused specifically on detecting the fact of an attack, and not on protecting against it. Model inversion attacks are always associated with the need for the attacker to make many related queries to the model. The idea of attack detection is that during an inversion attack, the attacker will sequentially refine the solution, and, accordingly, successive requests will be similar to each other. It is the detection of the fact that successive requests contain small regular changes relative to the previous ones that serves as a signal for an attack. The proposed algorithm has been tested on real attacks, and the implementation code has been made publicly available.

Acknowledgement. The work was carried out as part of the development of the program of the Faculty of Computational Mathematics and Cybernetics of the Lomonosov Moscow State University "Artificial intelligence in cybersecurity". The results of this work were used in the preparation of this article and are mentioned in the bibliography section.

The authors are grateful to the staff of the Department of Information Security of the Faculty of Computational Mathematics and Cybernetics of Lomonosov Moscow State University for the discussion of our work, criticism, and valuable additions.

This research has been supported by the Interdisciplinary Scientific and Educational School of Moscow University "Brain, Cognitive Systems, Artificial Intelligence".

References

1. DCCN 2023. https://dccn.ru. Accessed Jul 2023
2. Ilyushin, E., Namiot, D., Chizhov, I.: Attacks on machine learning systems-common problems and methods. Int. J. Open Inf. Technol. **10**(3), 17–22 (2022). (in Russian)
3. Namiot, D., Ilyushin, E., Chizhov, I.: The rationale for working on robust machine learning. Int. J. Open Inf. Technol. **9**(11), 68–74 (2021). (in Russian)
4. Namiot, D., Ilyushin, E.: On the robustness and security of artificial intelligence systems. Int. J. Open Inf. Technol. **10**(9), 126–134 (2022). (in Russians)
5. Namiot, D.: Schemes of attacks on machine learning models. Int. J. Open Inf. Technol. **11**(5), 68–86 (2023). (in Russian)
6. White Paper NIST AI 100–2e2023 (Draft) Adversarial Machine Learning: A Taxonomy and Terminology of Attacks and Mitigations. https://csrc.nist. gov/publications/detail/white-paper/2023/03/08/adversarial-machine-learning-taxonomy-and-terminology/draft. Accessed Jul 2023
7. Malekzadeh, M., Gunduz, D.: Vicious classifiers: data reconstruction attack at inference time. arXiv preprint arXiv:2212.04223 (2022)
8. Song, J., Namiot, D.: A survey of the implementations of model inversion attacks. In: Vishnevskiy, V.M., Samouylov, K.E., Kozyrev, D.V. (eds.) Distributed Computer and Communication Networks, DCCN 2022. CCIS, vol. 1748, pp. 3–16 Springer, Cham (2023). https://doi.org/10.1007/978-3-031-30648-8_1
9. Zhang, J., et al.: Privacy threats and protection in machine learning. In: Proceedings of the 2020 on Great Lakes Symposium on VLSI (2020)
10. Carlini, N., et al.: The secret sharer: evaluating and testing unintended memorization in neural networks. In: USENIX Security Symposium, vol. 267 (2019)
11. Hisamoto, S., Post, M., Duh, K.: Membership inference attacks on sequence-to-sequence models: is my data in your machine translation system? Trans. Assoc. Comput. Linguis. **8**, 49–63 (2020)
12. De Cristofaro, E.: An overview of privacy in machine learning. arXiv preprint arXiv:2005.08679 (2020)
13. Junzhe, S.: A survey of model inversion attacks and countermeasures. In: Junzhe, S., Namiot, D.E. (eds.) Proceedings of the Institute for Systems Analysis Russian Academy of Sciences, vol. 73, no. 1, pp. 82–93 (2023). EDN MQODQW. https:// doi.org/10.14357/20790279230110
14. Alves, T.A.O., França, F.M.G., Kundu, S.: MLPrivacyGuard: defeating confidence information based model inversion attacks on machine learning systems. In: Proceedings of the 2019 on Great Lakes Symposium on VLSI (2019)
15. Gong, X., et al.: Model extraction attacks and defenses on cloud-based machine learning models. IEEE Commun. Mag. **58**(12), 83–89 (2020)
16. Miao, Y., et al.: Machine learning-based cyber attacks targeting on controlled information: a survey. ACM Comput. Surv. (CSUR) **54**(7), 1–36 (2021)
17. MI detector. https://github.com/UnReAlKiNg/detection-of-model-inversion-attack. Accessed Jul 2023

18. Dibbo, S.V.: SoK: model inversion attack landscape: taxonomy, challenges, and future roadmap In: 2023 IEEE 36th Computer Security Foundations Symposium (CSF). IEEE Computer Society (2023)
19. Nguyen, N.-B., et al.: Re-thinking model inversion attacks against deep neural networks. arXiv preprint arXiv:2304.01669 (2023)
20. Hu, H., Pang, J.: PriSampler: mitigating property inference of diffusion models. arXiv preprint arXiv:2306.05208 (2023)
21. Fredrikson, M., Jha, S., Ristenpart, T.: Model inversion attacks that exploit confidence information and basic countermeasures. In: Proceedings of the 22nd ACM SIGSAC Conference on Computer and Communications Security (2015)

Distributed System for Scientific and Engineering Computations with Problem Containerization and Prioritization

Aleksander Sokolov$^{(\boxtimes)}$ [ID], Andrey Larionov [ID], and Amir Mukhtarov [ID]

V.A. Trapeznikov Institute of Control Sciences Russian Academy of Sciences,
Profsoyuznaya Str. 65, 117997 Moscow, Russia
`aleksandr.sokolov@phystech.edu`

Abstract. A key challenge in computer modeling is obtaining numerical results on large input datasets. For instance, this problem arises when researchers need to produce a visualization of an economic or physical process, as well as during the computation of characteristics of complex mathematical models using the Monte Carlo method. Each problem requires repeating the same program on a large set of inputs, consuming much time, from several hours to days and weeks. Computational algorithms may be implemented in different programming languages and may require various specific tools like GNU Octave, NS-3, or OMNeT++. This article describes a distributed system architecture, which can speed up the process of obtaining results for such problems. The system comprises a backend server, a control service (supervisor), worker nodes, and a database. These algorithms are executed in Docker containers to abstract from particular languages and tools required for computational algorithms. The system supports several strategies for problem prioritization to operate efficiently under heavy loads introduced by multiple users. To use the system, the user only needs to build a Docker image with an encapsulated algorithm, describe the input dataset in a JSON file, and upload them via the web interface. The system can be deployed in any public cloud. In this article, we describe the system architecture and numerical results obtained from computations on various clouds and local platforms. We demonstrate that the computation time for CPU-bound problems in different public clouds varies greatly. Finally, we show the influence of different prioritization strategies on the duration of computations under a moderate workload.

Keywords: distributed computing · container virtualization · cloud computing

The research was funded by the Russian Science Foundation, project no. 22-49-02023.

V. M. Vishnevskiy et al. (Eds.): DCCN 2023, LNCS 14123, pp. 68–82, 2024.
https://doi.org/10.1007/978-3-031-50482-2_6

1 Introduction

It is often necessary to obtain numerical results using mathematical models or algorithms on a large amount of data during applied research. In this case, models and algorithms are implemented in different programming languages or software tools such as GNU Octave, NS-3, or OMNeT++. The peculiarity of such problems is that results for each set of inputs can be obtained independently. Therefore, it is possible to reduce the time for obtaining results using horizontal scaling, doing calculations simultaneously on multiple work nodes. It is proposed to use a distributed computing system to automate such calculations. The system uses Docker containerization technology. In order to efficiently handle task flow, which arrives from many users, various prioritization strategies are used. The containerization allows abstracting from used programming languages and scientific computing tools. The prioritization strategy allows for allocating time among many users and tasks effectively. The proposed system is a web app with a microservice architecture, and it can be deployed in any public cloud.

Authors have already used the distributed computing system to conduct several numerical experiments in wireless networks and queueing theory modeling. For instance, in [18], the characteristics approximation of a complex queuing system was conducted using artificial neural networks, which were trained with a large set of synthetic data samples. Each element of this set included the queuing system parameters and their characteristic values, which were calculated using a simulation model. The total size of the data sample was 500'000 records. Each record was solved using a simulation model of around 3 s on average.

Using the proposed distributed computing system made it possible to obtain numerical results in less than a day without requiring changes in the code of a single-threaded simulation model or writing any special scripts for multithreaded calculations.

There are many systems for organizing distributed computing. The systems closest to the proposed in the article system are systems for voluntary computing [2–5,8,11,13,14,17], which provide computing power to many users. Also, the proposed system close to locally distributed computing systems [1,6,7,9,10,12,16]. Note many such systems are highly specialized and are developed to solve problems in physics, biology, chemistry, or astronomy (e.g. [15,19]).

In several voluntary computing systems, users should use special libraries to work [2–4,8,13,17]. So, for the BOINC (Berkeley Open Infrastructure for Network Computing) system [3,8,13], a user installs the SDK on the computer, defines the resources provided, and chooses the project he is ready to take into work. Since calculations can be carried out on different platforms, developers have to maintain different versions of programs on various operating systems and architectures. The specific SDK is also used in other systems, e.g., Golem [17]. Note Golem, like some other voluntary computing systems (for example, VFuse [4]), allows you to organize mutual settlements between consumers and providers of computing power using blockchain technology. The Golem system had 11628 cores available on 620 servers at the time of writing. A distinctive feature of the project [2] is the programming language JavaScript for calculations and Web Workers technology to run multiple threads and perform calculations in browsers on computers connected to the system.

A large number of systems used for scientific calculations locally need to provide capacities for public use. The Everest [16], and Nimrod [1] systems can be cited as an example. A distinctive feature of the Nimrod system is the presence of a declarative language for describing problems and the ability to establish links between problems and use the results of some problems as input for others. Articles [6,7,9,10] consider a system for calculating time-consuming JINR problems in which OpenNebula technology is used to manage resources. The main advantage of using OpenNebula technology is the integration of various virtual machines working with different virtualization technologies (VMware, KVM, LXD, Firecracker) into a single cluster for computing. Performing calculations on virtual machines is also implemented in the Everest system [16]. So, notice many systems for distributed computing in local networks rely on the MPI library [12,15,19].

The distributed calculation system proposed in this article differs from existing systems. Firstly, it is not necessary to use any special libraries to implement computational algorithms. Moreover, containerization technology allows effectively isolating the environment in which tasks are performed from the operating system of the worker node. It leads to a reduction in resources compared to the traditional virtualization technique. Secondly, in the proposed system, tasks are not connected and are not synchronized in any way. Moreover, problems queue within a single problem is not defined in advance. It is not required to use libraries for interprocess communication like MPI accordingly. Finally, thirdly, the application is designed for minor research problems and simulations of relatively small systems.

The article is organized as follows. The section 'System Architecture' provides a detailed description of the components and algorithms, a description of the life cycle of the task, and implemented prioritization schemes. This section also provides the analysis of overhead costs arising from asynchronous interaction between supervisor and workers and containerization using. The section 'Numerical results' provides an example of a task description, analysis of the tasks execution time in a Docker container for different programming languages, comparison of calculation duration on various cloud platforms and a local workstation, and calculation duration using different prioritization strategies. The section 'Conclusion' contains general conclusions and a description of plans for further system development.

2   System Architecture

In its structure, the distributed calculations system is a micro-service web application consisting of several services shown in Fig. 1. The user interacts with the system using the web interface or REST API. The backend is the server-side application for controlling and monitoring tasks, authorizing users, and collecting task performance statistics. The supervisor service is used to manage workers and monitor the task queue. Workers perform subtasks by creating and launching containers from user-submitted Docker images. Each subtask is executed in a separate container.

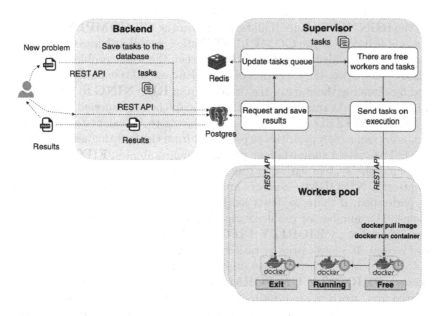

Fig. 1. The distributed computing system architecture.

The system uses two databases: relational databases Postgres and Redis. Unlike Postgres, Redis data is stored in RAM, which provides quick access, but creates problems for long-term storage. For this reason, Redis is used to store data that is relevant only during the execution of the service. Redis is also used for communication between components. Postgres is used to store task parameters, calculation results, user profiles, and other data that need access all the time.

Let us define the terms 'problem' and 'task,' which will be used later in the article. The problem is characterized by an algorithm for solving it, a set of parameters, and an array of input data. The software implementation of the algorithm is contained in the Docker image. The system automatically decomposes the task into small parts, which are called subtasks. A separate set of input data characterizes each subtask.

2.1 Problems: Definition, Life Cycle, Prioritization

Let us consider the life cycle of a problem and tasks. After the user uploads the file with the problem, it is decomposed into tasks. The problem and all its tasks are currently in the **INITIALIZED** state. After the user has submitted the problem for execution, it goes into the **RUNNING** state. Tasks are added to the queue according to their priority, passing into the **ST-QUEUED** state. When the released worker takes a task for execution, it goes into the **ST-RUNNING** state, from where it goes either to **ST-COMPLETED** on successful completion or **ST-ERROR** on error. When all the tasks have been completed, the problem leaves the **RUNNING** state. If all tasks were in the **ST-COMPLETED** state, then a problem goes to the **COMPLETED** state, and if at least one task failed with an error and ended up in **ST-ERROR**, then the whole problem goes to the **ERROR** state. Also, the execution of a problem and all remaining tasks can be paused by changing the state of the problem from **RUNNING** to **STOPPED**. In this case, all tasks running or waiting in the queue are canceled and returned to the **INITIALIZED** state. For some problems to be calculated faster than others, users can assign priorities to problems, from the first (highest) to the fifth (lowest). The system supports four prioritization strategies: **FIFO** – a problem that entered the system earlier is executed earlier. In this mode, priority does not matter; **UNIFORM** – workers are equally divided between the problems being performed. If there are fewer workers than problems, workers take earlier problems. The number of problems running simultaneously never exceeds the number of workers; **PRIORITY FIFO** – problems are sorted by priority and time of receipt. If a lower priority task was being performed in the system at the time of receipt of a higher priority problem, then the execution of the latter is suspended; **PRIORITY UNIFORM** – problems with the highest priority are calculated first, workers are distributed equally among problems with the same priority. Suppose a lower-priority problem was being performed in the system at the time of receipt of a higher-priority task. In that case, the new subtasks of the latter are not put on execution until the calculation of all priority tasks is completed.

2.2   Supervisor

The supervisor manages the execution of problems by distributing tasks among workers. The service monitors the status of workers, and the degree of filling of the task buffer, transfers tasks from a buffer to workers for execution as they are released, writes the calculation results to the Postgres database, and stops execution when receiving a command from the backend. The service is implemented in Python.

The supervisor uses Redis to interact with the backend and storage of the problem buffer. The backend informs the supervisor about the appearance of a new problem or requests the suspension of an already running problem in case of receiving the **STOP** command or the appearance of a higher priority problem. Redis also stores a buffer that stores tasks assigned to workers when released. The tasks in the buffer are sorted according to the selected prioritization strategy. When the next worker is released, the supervisor takes the first problem from the buffer, assigns it to the free worker, and puts a new task from the Postgres database in the vacant place. If a new, higher-priority problem has come from the backend, the supervisor can rewrite all or part of the buffer. The tasks in the buffer have the **ST-QUEUED** state, and the tasks still waiting to get into it are **INITIALIZED**. An example of transferring tasks to the buffer and assigning them to workers is shown in Fig. 2.

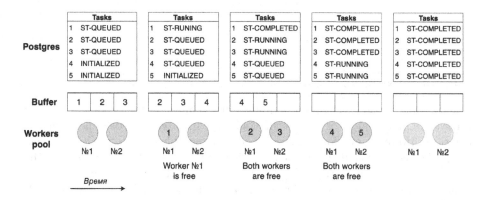

Fig. 2. An example of buffer state in the system.

The supervisor processes commands and polls workers in a loop. In each cycle, the supervisor performs four actions:

- collects the results of workers who have completed the execution of tasks, processes backend commands;
- assigns new tasks to the released workers;
- fills the buffer with new tasks.

By default, the interval between polls is one second. The REST API is used to interact with workers, which will be described later, and a Redis queue is used to receive commands from the backend.

2.3 Worker

The worker service performs tasks by creating and launching Docker containers from a Docker image provided by the user. The worker contains two execution threads: an HTTP server runs in one thread, and a Docker container with a task runs in the other. The HTTP server provides a REST API for the supervisor to submit a problem for execution, check the worker's status, and finish his work. The service is written in Python using the FastAPI framework.

Figure 3 shows a diagram of worker states. At the start, the service is in the **FREE** state. After a request from the supervisor for execution arrives, a Docker container with a subtask is launched, and the worker enters the **RUNNING** state. If successful, the worker switches to the **EXCITED** state. After that, the service supervisor can get the execution results and delete the container in which the task was executed using a POST-request */api/complete*, after which the working service returns to the **FREE** state and waits for new tasks to be executed.

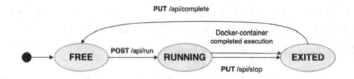

Fig. 3. Worker state machine.

2.4 Task Execution Time

The containerization technology simplifies the creation of new problems, making the system independent of what technologies the researcher needs to perform calculations. However, any form of virtualization inevitably leads to additional overhead. The asynchronous interaction of microservices also leads to their growth.

In order to estimate the overhead, let us look at the task execution cycle in more detail. Figure 4 shows a time diagram of the supervisor's interaction with working services. For simplicity, consider the case when only one worker is in the system. As noted above, to find out the worker's status and check whether he is available to perform new tasks, the supervisor implements a periodic polling. At time t_1, the supervisor begins to poll all workers, sending GET-requests for status. After receiving a response that the service is free (in the **EXITED** state) at time t_3, the supervisor accesses Redis in order to take a problem from the buffer for execution and sends it to the worker (time t_5), passing a POST-request /*api/run*. When receiving the request (t_6), the worker runs an image with a task in a separate process, executing the Docker run command. The task starts its execution at time t_6. After its launch, the worker reports this to the supervisor (t_7). If there are still workers in the system, the supervisor completes the survey at time t_8.

During the following survey, the working service is still performing the task; in response to the GET-request /*api/status*, the supervisor receives a **RUNNING** response. In the next iteration of the survey (t_{12}), the supervisor receives a response to the status request from the worker that it has finished executing and is in the **EXITED** state. The supervisor, having learned that the worker has finished its execution, sends a PUT-request /*api/complete* to complete the container execution. In the time interval (t_{14}, t_{15}), the worker completes the execution of the Docker container and deletes it from the system, sending the result and the execution log to the supervisor. The supervisor receives the worker's response and writes the problem result in the database. After that, the supervisor can send a new problem to the worker for execution.

3 Numerical Experiments

We carried out several numerical experiments to investigate the system's performance. At the beginning of this section, we show the impact of containerization on problem computation time. Then we offer the results of evaluating the effect

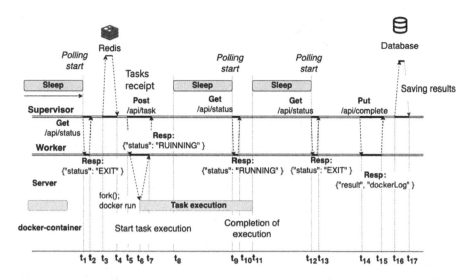

Fig. 4. Timing diagram of tasks execution in the system.

of the number of workers on problem computation time for a fixed number of cores. In the experiments, the choice of computing platform also significantly impacts task computation time. We conclude the section by showing how the prioritization strategy affects task execution time and the time to obtain the results of the first task.

3.1 The Impact of Containerization on Computations

To measure the computation overhead of containerization, we developed a test task and compared its execution time in a server operating system without containerization (Ubuntu 20.10) and inside a Docker container. We used the computation of the Fibonacci number recursively without using dynamic programming as a test task. This algorithm (do not use it for calculating Fibonacci numbers!) is convenient because it is guaranteed to load the CPU sufficiently long. It does not require switching the execution context to kernel space; thus, its different runs on the same hardware take approximately the same time. The source code of the program is in C++ is shown in Fig. 5:

The fib() function takes as input a Fibonacci number. We set the number of iterations in the function main(). Note that it is unnecessary to specify optimization flags during compilation. Otherwise, the compiler may delete the main loop in lines 18–20 because the program does not use the calculation result in the right parts of expressions.

We show the execution time of programs that calculate Fibonacci numbers in Python, JavaScript, C++, and Java programming languages in Table 1. To execute the program in JavaScript, we used Node JS. For Java, we used the standard OpenJDK. For the C++ program, we used the Clang compiler without

```
#include <chrono>
#include <iostream>
#include <iomanip>
#include <string>
#include <cstdlib>

using namespace std::chrono;

long fib(unsigned n) {
  if (n <= 1) return n;
  return fib(n-1) + fib(n-2);
}

int main(int argc, char **argv) {
  int n_iter = std::stoi(argv[1]);
  auto t_start = system_clock::now();
  long x{0};
  for (int i = 0; i < n_iter; i++) {
    x = fib(40);
  }
  auto t_end = system_clock::now();
  std::cout << std::fixed
    << duration_cast<milliseconds>(t_end - t_start).count()
    << std::endl;
}
```

Fig. 5. Implementation of Fibonacci numbers on C++.

optimizations. We selected the Fibonacci and iterations numbers so that the computation took about 0.5 s for each implementation. Notably, the code written in Python ran tens of times slower than code written in other languages.

Table 1. Execution time of Fibonacci number via recursive algorithm on a couple of programming languages

Programming language	Fibonacci number	Execution time, ms
Python	30	482
Node JS	38	431
C++	40	600
Java	41	634

Figure 6 shows the dependence of the speed of the Fibonacci number calculation on the number of iterations. As can be seen, the execution times in the Docker container and without containerization are different. The most significant difference in execution time is observed in Python implementation. It reaches 15% at 50 iterations. The deviation for all iterations for C++, Java, and Node JS implementations does not exceed 5%. Thus, depending on the type of task and the programming language used, the computation time in a Docker container can be much longer than the computation time on a computer.

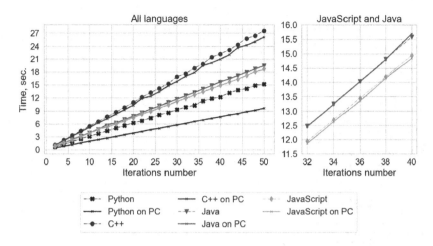

Fig. 6. Comparison of task execution time in Docker container and in a system without containerization for various programming languages.

3.2 Dependence of Problem Execution Time on the Size of the Worker's Pool

The supervisor polls the workers periodically, assigns new tasks, and stores the results in the database. The duration of this polling depends on the number of workers and affects how long a worker has to wait after the computation is complete before processing the next task. In addition, due to the limited number of vCPUs on the server, there is competition for CPU time between workers. The number of workers affects the problem computation speed.

We conducted a few experiments to evaluate the dependence of problem completion time on the number of workers. The problem in each experiment consisted of 100 tasks, calculating the Fibonacci number. For experiments, we used cloud platforms such as:

- Yandex Cloud - Intel Xeon Gold 6338 processor, 2.00 GHz;
- Selectel - Intel Xeon Gold 6140 processor, 2.3 GHz;
- VK Cloud - Intel Xeon Gold 6230, 2.1 GHz.

Also, in experiments, we used a workstation. The cloud servers used in the experiments had eight vCPUs. The workstation had an Intel Core i7-9700K processor with eight cores and did not support hyper-threading. In addition, we considered two modes of allocating cores to workers: without binding, when cores are assigned to processes by the system scheduler, and with binding, when cores are assigned manually to each computational process.

The execution time for each subtask was 10 s, i.e., it would take $10 \times 100 = 1000$ s or 16 min and 40 s to compute the entire task on a single processor. The jobs were executed in Docker containers. We used the Docker-compose program to run workers. Workers accessed the host system's Docker server from separate processes.

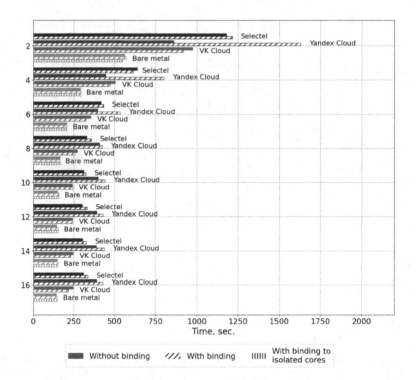

Fig. 7. Problem execution time for different execution modes: without binding workers to cores, with binding of workers to cores and with binding of workers to dedicated cores.

3.3 Theoretical Estimation of Computation Time

Let us denote the number of CPUs as m and n as the number of workers, p as the number of tasks, and t_0 as the computation duration of one task by one worker on a free CPU. Then the theoretical dependence of the problem execution time on the number of workers and the number of CPUs in the system is defined as follows:

$$T(n, m, p) = \begin{cases} \frac{pt_0}{n}, n \leq m \\ \frac{pt_0}{m}, n > m \end{cases} \tag{1}$$

The average computation time per task, taking into account that CPU time may be divided among several workers, is defined as

$$\bar{t} = t_0 max(1, \frac{n}{m}) \tag{2}$$

3.4 Estimating Overhead. Impact of Explicitly Linking Workers to Cores

The above theoretical formulas do not take into account two factors: the time spent on switching the context of running processes if the scheduler decides to change the running task (e.g., allow some system service to execute) and the time spent by the worker process that provides APIs to the supervisor. Figure 7 shows the dependence of task execution time on different platforms on the number of workers.

We experimented with different container startup modes: without binding the worker to the kernel, with binding the worker to the kernel. For the workstation, we experimented with binding workers to isolated cores. We used the Docker variable cpuset_cpus to bind Dockercontainer to CPU. In the experiment with isolated cores, only six were used for computation. The operation system used two of eight cores for system processes during experiments.

The results demonstrate a poor sense of explicitly binding workers to processor cores. Moreover, explicitly binding processors to workers gives a small gain only when the number of workers is less or equal to the number of reserved cores.

When a worker executes a task, time is spent not only on the task execution itself but also on waiting for the container to load, start and delete, for the supervisor to start polling, for the processor to release a system service, and other purposes. Figure 8 shows the difference between the actual time to execute a single task and the time to execute the container directly. It can be seen that as the number of workers increases, the time cost increases. This can be explained by the increased time required for the system to switch processors from computational tasks to service and system processes. If there are more workers than cores, the average execution time for each task grows linearly as CPU time is divided among several workers.

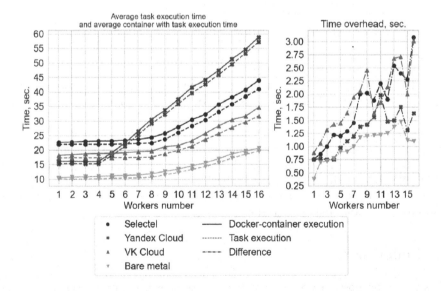

Fig. 8. Average task execution time depends on workers' count.

3.5 Problems and Tasks Prioritization

To investigate the application of prioritization, we conducted a numerical experiment in which two types of problems came to the system: high-priority (HP) and low-priority (LP). We investigated the effectiveness of different prioritization strategies in the experiment. Each problem contained 100 tasks. In the experiment, we used a workstation with Intel Core i7-9700K, which had 8 CPUs and did not support hyper-threading. As a task, we used the Fibonacci number calculation as before. The number of iterations and the calculated number are chosen so that the calculation takes about 10 s (C++ implementation). Running 50 tasks for eight workers took seven iterations number to run the workers. The total time is equal to the product of the number of iterations of launching workers by the time of executing one task: 7 * 10 = 70 s. Priority problems came at an intensity of one every two minutes, and non-priority problems came at one every minute and a half.

Table 2. Computation metrics for different prioritization modes

Prioritization mode	Execution time from problem creation to complete execution, s		Time until the start of execution first task, s	
	HP	LP	HP	LP
FIFO	526	589	446	501
PRIORITY FIFO	98	844	12	752
UNIFORM	1166	1159	8	12
PRIORITY UNIFORM	182	1542	5	84

Table 2 shows the metrics for the case when each task has a constant execution time. The most effective mode is PRIORITY FIFO. The time a problem is in the system in this prioritization mode is minimal and equals 98 s. PRIORITY UNIFORM is slightly inferior to PRIORITY FIFO. The time a problem is in the system equals 182 s. But this mode shows the best time to start the first task - 5 s. In not-prioritized modes, FIFO is the most efficient. Compared to UNIFORM, the time in the system is shorter for prioritized and non-prioritized problems. In FIFO mode, the problem is usually in the execution queue - i.e., the time before the first task starts execution.

4 Conclusion

This paper proposes a system architecture for scientific and engineering computing using Docker containerization technology and task prioritization. We described the architecture and interaction of services of this system. We proposed several modes for prioritization: FIFO is sorting problems by arrival time;

PRIORITY FIFO is sorting by priority and arrival time; UNIFORM is the distribution of problems among workers, and PRIORITY UNIFORM is priority distribution of workers when all workers deal with priority tasks first. We conducted a numerical experiment in this paper to study each priority modes effectiveness. The PRIORITY UNIFORM mode was the most effective for task prioritization. Service time for priority problems is ten times less than for non-priority problems using this mode. Numerical experiments on different cloud platforms showed that the efficiency significantly depends on the mode of assigning virtual processors to physical cores. In the future, we plan to add support for multiple supervisors to the system for simultaneous work on different clouds. Also, we plan to improve the supervisor and worker communication protocol, and instead of REST API, use a message broker (Rabbit MQ or Redis). We also plan to explore the possibility of using alternative containerization technologies (primarily LXC) and investigate various performance optimization options.

References

1. Abramson, D., Giddy, J., Kotler, L.: High performance parametric modeling with Nimrod/G: killer application for the global grid? In: Proceedings 14th International Parallel and Distributed Processing Symposium, IPDPS 2000, pp. 520–528. IEEE Computer Society (2000). https://doi.org/10.1109/IPDPS.2000.846030. http://ieeexplore.ieee.org/document/846030/

2. Agliamzanov, R., Sit, M., Demir, I.: Hydrology@Home: a distributed volunteer computing framework for hydrological research and applications. J. Hydroinformatics **22**(2), 235–248 (2020). https://doi.org/10.2166/hydro.2019.170. https://iwaponline.com/jh/article/22/2/235/71586/HydrologyHome-a-distributed-volunteer-computing

3. Anderson, D.P.: BOINC: a platform for volunteer computing. J. Grid Comput. **18**(1), 99–122 (2020). https://doi.org/10.1007/s10723-019-09497-9. http://link.springer.com/10.1007/s10723-019-09497-9

4. Antelmi, A., D'Ambrosio, G., Petta, A., Serra, L., Spagnuolo, C.: A volunteer computing architecture for computational workflows on decentralized web. IEEE Access **10**, 98993–99010 (2022). https://doi.org/10.1109/ACCESS.2022.3207167. https://ieeexplore.ieee.org/document/9893800/

5. Arora, R., Redondo, C., Joshua, G.: Scalable software infrastructure for integrating supercomputing with volunteer computing and cloud computing. Commun. Comput. Inf. Sci. **964**, 105–119 (2019)

6. Balashov, N., et al.: Service for parallel applications based on JINR cloud and HybriLIT resources. EPJ Web Conf. **214**, 07012 (2019). https://doi.org/10.1051/epjconf/201921407012

7. Baranov, A.V., Balashov, N.A., Makhalkin, A.N., Mazhitova, Y.M., Kutovskiy, N.A., Semenov, R.N.: New features of the JINR cloud. In: CEUR Workshop Proceedings, vol. 2267, pp. 257–261 (2018)

8. Barranco, J., et al.: LHC@Home: a BOINC-based volunteer computing infrastructure for physics studies at CERN. Open Eng. **7**(1), 379–393 (2017). https://doi.org/10.1515/eng-2017-0042. https://www.degruyter.com/document/doi/10.1515/eng-2017-0042/html

9. Kadochnikov, I., Korenkov, V., Mitsyn, V., Pelevanyuk, I., Strizh, T.: Service monitoring system for JINR Tier-1. EPJ Web Conf. **214**, 08016 (2019). https://doi.org/10.1051/epjconf/201921408016.   https://www.epj-conferences.org/10.1051/epjconf/201921408016

10. Korenkov, V., et al.: The JINR distributed computing environment. EPJ Web Conf. **214**, 03009 (2019). https://doi.org/10.1051/epjconf/201921403009. https://www.epj-conferences.org/10.1051/epjconf/201921403009

11. Mengistu, T.M., Che, D.: Survey and taxonomy of volunteer computing. ACM Comput. Surv. **52**(3) (2019). https://doi.org/10.1145/3320073

12. Nguyen, N., Bein, D.: Distributed MPI cluster with Docker Swarm mode. In: 2017 IEEE 7th Annual Computing and Communication Workshop and Conference (CCWC), January 2017, pp. 1–7. IEEE (2017). https://doi.org/10.1109/CCWC.2017.7868429. http://ieeexplore.ieee.org/document/7868429/

13. Nikitina, N., Manzyuk, M., Podlipnik, Č, Jukić, M.: Volunteer computing project SiDock@home for virtual drug screening against SARS-CoV-2. In: Byrski, A., Czachórski, T., Gelenbe, E., Grochla, K., Murayama, Y. (eds.) ANTICOVID 2021. IAICT, vol. 616, pp. 23–34. Springer, Cham (2021). https://doi.org/10.1007/978-3-030-86582-5_3

14. Nouman Durrani, M., Shamsi, J.A.: Volunteer computing: requirements, challenges, and solutions. J. Netw. Comput. Appl. **39**, 369–380 (2014). https://doi.org/10.1016/j.jnca.2013.07.006.   https://linkinghub.elsevier.com/retrieve/pii/S1084804513001665

15. Stelly, C., Roussev, V.: SCARF: a container-based approach to cloud-scale digital forensic processing. In: Proceedings of the 17th Annual DFRWS USA, DFRWS 2017 USA, pp. S39–S47 (2017). https://doi.org/10.1016/j.diin.2017.06.008

16. Sukhoroslov, O., Putilina, E.: Cloud services for automation of scientific and engineering computations science. Bus. Soc. **1**(2), 6–9 (2018)

17. Uriarte, R.B., DeNicola, R.: Blockchain-based decentralized cloud/fog solutions: challenges, opportunities, and standards. IEEE Commun. Stan. Mag. **2**(3), 22–28 (2018). https://doi.org/10.1109/MCOMSTD.2018.1800020. https://ieeexplore.ieee.org/document/8515145/

18. Vishnevsky, V., Klimenok, V., Sokolov, A., Larionov, A.: Performance evaluation of the priority multi-server system MMAP/PH/M/N using machine learning methods. Mathematics **9**(24), 3236 (2021). https://doi.org/10.3390/math9243236. https://www.mdpi.com/2227-7390/9/24/3236

19. Zhou, J., Bie, S.W., Miao, L., Zhang, Y., Jiang, J.: Docker-enabled scalable parallel MLFMA system for RCS evaluation. Prog. Electromagn. Res. M **67**, 169–176 (2018). https://doi.org/10.2528/PIERM18021907. http://www.jpier.org/PIERM/pier.php?paper=18021907

Overview of Research Works on Applications of UHF RFID on Vehicles for Data Transmission

Vilmen Abramian$^{(\boxtimes)}$ (ID), Andrey Larionov, and Ivan Fedotov (ID)

V. A. Trapeznikov Institute of Control Sciences of RAS, Moscow, Russia
{abramian.vl,fedotov.ia15}@physics.msu.ru

Abstract. RFID (Radio Frequency Identification) technology has found wide application in many areas of science and technology and people's everyday lives (in libraries, shops, when registering passengers on the subway and buses, etc.). Every year the number of RFID applications is growing, and at the same time the number of scientific papers exploring certain issues related to this topic is increasing. One of these rapidly developing areas is the use of RFID to identify fast-moving transport objects (cars, trains), which is currently poorly reflected in existing reviews. This article fills this gap.

The paper provides an overview of publications in the field of RFID technologies and standards. A description of theoretical and experimental results is given, as well as the architecture and hardware and software for the practical implementation of land vehicle identification systems used in various fields.

Keywords: RFID · transport · overview

1 Introduction

Radio frequency identification technology has found wide application in many areas of science and technology and people's everyday lives (in libraries, shops, when registering passengers on the subway and buses, etc.). Every year the number of RFID applications is growing, and at the same time the number of scientific papers exploring certain issues related to this topic is increasing.

The review by [34] highlights the following areas of RFID application: transport; agriculture and livestock farming; health and social welfare; environmental applications; and protection and safety. These areas can be expanded and clarified by the use of radio frequency identification in construction, transportation of materials and cargo, when tracking and controlling the speed of special equipment, as well as organizing its parking [52].

Logistics can be identified as a separate large area of application of RFID. Dozens of articles are devoted to this issue, as can be seen from the review of [22]. The effective use of radio frequency identification for determining the location of objects, navigation, as well as marking and searching for various

V. M. Vishnevskiy et al. (Eds.): DCCN 2023, LNCS 14123, pp. 83–94, 2024.
https://doi.org/10.1007/978-3-031-50482-2_7

objects indoors, where the use of GPS technology is difficult, is described in the review [37]. Tagging is carried out by placing cheap passive RFID tags containing information about their exact location. Such systems are often used as assistants for visually impaired people in public spaces, for example, in the metro [37].

One of the intensively developing areas of RFID application in the last decade is radio frequency identification of fast moving vehicles (cars, trains). However, at present, theoretical and practical research in this area has been poorly reflected in reviews published in the world literature. This article fills this gap.

The structure of the article includes the following sections. The second section provides a brief description of RFID technologies and standards. The terms used in the following are given. The third section is devoted to a review of publications on the use of RFID in land transport (roads and railways). A description of theoretical and experimental results is given, as well as the architecture and hardware and software for the practical implementation of land vehicle (VV) identification systems used in various fields.

2   RFID Standard

Radio frequency identification is a technology for marking and automatically identifying objects, in which data stored in RFID tags is read or written using radio signals. The simplest RFID system consists of at least two parts: tags - radio-electronic devices located on the identified object and readers - active devices that initiate the data exchange procedure. An important feature of RFID systems is that tags are usually passive devices that do not have a built-in power source and await initiation of data exchange from the reader. Thanks to this, tags become so cheap and simple that it becomes possible to equip them with most household products in retail stores, for example, clothing tags, book covers in libraries, etc.

An RFID system may contain other devices to expand its functionality or be part of a larger system. For example, the data received by the reader is sent via other communication protocols to a remote database. To increase the number of tasks solved using RFID, semi-passive and active tags were created. Semi-passive tags have their own energy source (lithium or solar battery), which provides power to the built-in microcircuit. Accordingly, they are capable of receiving and transmitting signals over significantly greater distances compared to passive tags. At the same time, semi-passive tags cannot generate their own high-frequency signal, thereby initiating a communication session. They can only modulate the reader's field in the same way that passive tags do. Active tags have a different operating principle, as they include an active transmitter and sometimes a receiver. To transmit data to the reader, they emit a high-frequency electromagnetic field instead of modulating the reader's field.

RFID uses 3 main information transmission ranges, each of which has its own standards. Low frequency range: 30–300 kHz (LF) is used for close communication between the tag and the reader at a distance of up to 1 cm [25]. High frequency range: 3–30 MHz (HF, HF) is used for communication over a distance of 0–1 m [21]. Ultrahigh frequency range: 300 MHz–3 GHz (microwave, UHF) is used for communication and identification at distances of more than 1 m [26].

RFID technology has been widely used in various fields since the end of the 20th century. In addition to the classic use of RFID for identifying objects in stores and warehouses, creating electronic access cards, etc., this technology can be used for positioning mobile objects [16, 23, 46] or to help find people with disabilities lost in the city capabilities [30]. More information about RFID technology in general can be read in the book RFID handbook by Klaus Finkenzeller [28]. Further in the review, only UHF RFID will be considered, since it is this technology that allows to identify objects at distances that satisfy the height of the mounting on road poles along the highway.

The EPC Class 1 Generation 2 standard [26] describes the physical (PHY) and channel (MAC) levels of the radio frequency identification system for passive and semi-passive tags. At the physical level, the standard describes methods for modulating and encoding signals, and at the data link level, it describes the data exchange protocol between the reader and tags. The protocol is based on the Slotted ALOHA standard described in [10]. It allows you to deal with collisions when there are several tags in the reader area, and also contains tools for reading and writing data to tags.

To connect RFID readers to control systems, standard protocols are used that are supported by most manufacturers. The Low Level Reader Protocol (LLRP) [8] defines a low-level interface between the reader and the controller. This protocol allows you to perform random access and inventory operations, configure the reader's radio interface parameters, obtain data on the state of the radio channel and diagnostic data on the operation of the reader. The LLRP protocol is supported by most existing readers. The standard defines the ability of the protocol to operate over TLS (transport layer security) channels. In addition to the LLRP, EPCglobal has published several standards [4–7] that can be used in reader and control system development. The Reader Management 1.0.1 (RM) [4] standard defines a system model and MIB (Management Information Base) for collecting data about the reader's operating status via the SNMP protocol. The Discovery, Configuration, and Initialization (DCI) for Reader Operations [7] standard allows reader, controllers, and LLRP clients to find each other on the network, authenticate between controllers and readers, manage reader operation, download new software images, and perform other service functions. The Application Level Events (ALE) Standard [5, 6] describes guidelines for middleware development.

RFID standards are constantly being improved, with new additions and specifications appearing. In 2020, the RAIN RFID Alliance proposed an interface specification between the reader and the controller [9]. This specification is higher level than LLRP. It describes basic actions, including configuration and obtaining reader status, radio protocol settings, and access to tags. The specification defines messages sent between the reader and controller in JSON format, as well as the use of [45] transport identification to solve the problem.

3   Application of RFID in Transport

Radio frequency identification is used in many areas of industry and everyday life, including ground transportation. Historically, one of the first applications of RFID technology was its use on toll roads for vehicle identification and toll collection. For the first time such systems began to be used in the 80 s of the 20th century in the USA. In the 90 s, the use of RFID on toll roads became widespread not only in the USA, but also in some European countries [19,38]. Some areas of application of RFID in transport are given in the RAIN RFID report [15].

One of the most common uses of RFID in ground transportation is vehicle recognition. A large number of works are devoted to this issue. Thus, the work of [33] draws attention to the fact that the movement of the RFID reader and tag relative to each other complicates the reading process. A mathematical model is presented to calculate the probability of successful tag reading. Information about identified vehicles can be effectively used, for example, in traffic flow studies, as well as for vehicle control. The work [48] describes a hardware and software complex for creating a control system for controlling vehicle access to a certain territory. Two experiments were carried out: one on a university campus, in which 20 cars with RFID tags took part, and the second in a small city, in which the system was tested on 6 cars over several weeks. Vehicle control is also studied in the work [31], where a method was proposed for estimating the travel time of vehicles using data from RFID readers installed above highways. Travel time statistics were collected using data obtained from built-in RFID tags in cars. The results of the developed theoretical model are compared with experimental data.

The works [58,62,63] describe the principles of constructing and implementing an automated system for monitoring traffic violations using RFID technologies. A large-scale experiment was carried out with the participation of 800 cars, the license plates of which had passive RFID tags installed, and the readers were located above the road surface. Three-month tests conducted in winter showed that the use of RFID technology provides a vehicle detection probability of about 95% [60]. The results of the experiment coincided with theoretical studies assessing the probability of radio frequency identification of vehicles [39,40,61]

The works [18,51,55] discuss the use of RFID for controlling barriers, paying tolls for various types of vehicles, as well as for adjusting the phases of traffic lights and searching for stolen cars. Using RFID, you can transmit not only the car ID (license plate), but also other data (for example, about the presence of fines [43]). The work of [35] also proposes the use of RFID technology for vehicle identification and payment for toll roads. This involves using active tags in cars to increase the range of their interaction with the reader and the ability to send more information. Radio frequency identification can also be used to ensure law and order. For example, the work of [17] raises the problem of car theft in the Philippines and proposes a method to combat it by installing RFID tags inside the car and entering information about it into a special database. The article [41] discusses a system for automating the issuing of fines, which, according to the authors, can help in the fight against corruption on the road. Researchers from Bangladesh presented a paper [11], which examined a toll collection system on toll roads using RFID technology. In the article [12] the authors developed a concept and simulation model of a system that identifies passing vehicle traffic on highways and intersections.

Using RFID technology, the speed of vehicles can be measured [24,32,68]. This is realized by changing the power of the tag signal received by the reader. The works [32,68] propose a method for measuring vehicle speed by placing RFID tags in the road surface and RFID readers on cars. The article [24] proposes using RFID to track traffic and determine the speed of cars by placing an RFID tag on a vehicle license plate and a reader next to the road. The article also describes experiments in which the speed of an electric bicycle was measured, the maximum value of which was 35 km/h.

Recently, automotive self-organizing networks (Vehicular ad hoc networks, VANETs) [13,59] have become a hot topic, which provide high-speed communication from various devices on the roads, including RFID readers, with control center databases. VANET networks are one of the main components of the Intelligent Transportation System (ITS) [27,59,65]. Such systems involve the use of a large number of technologies and various communication systems, with RFID occupying a special place in them [39,40,47,53,58,64,66,69,70]. The work [70] presents a system for monitoring traffic flow and detecting road accidents using radio frequency identification. The problem of finding optimal locations for installing readers is being solved in order to minimize the cost of deploying the system.

The works [49,57,65] developed a method for optimal placement of wireless network base stations along long transport highways to collect information from video recording systems and promptly transmit data on traffic violations to the control center. This method was used to design the implementation of an efficiently operating broadband high-speed wireless network along the Kazan ring road (M7 Volga).

The paper [69] presented a method for detecting traffic congestion in urban environments that uses RFID-based vehicle positioning. This implies that all vehicles will be able to exchange data with each other and form clusters connected by a VANET network. The work [53] presents a prototype of a vehicle traffic monitoring system. It consists of piezo sensors installed under the road surface to assess the flow density, as well as RFID readers that receive information from tags located on cars to provide priority traffic at intersections to special vehicles (ambulances, police, etc.) In the article [47] a system is being investigated that identifies a large number of cars per unit of time, and the zones in which RFID tags are read and their dependence on the speed of transport are also considered. A mathematical model is built that takes into account the features of the RFID protocol, movement speed and other parameters. Analytical results are compared with laboratory experimental data.

The performance assessment of vehicle identification systems based on UHF RFID is considered in the work [56]. The possibility of identifying tags located on cars and trains is being considered. Models have been developed to determine the optimal location of the tag on various vehicles.

A significant number of works are devoted to the study of determining the location of a vehicle using RFID technologies. Despite the fact that satellite technologies (GLONASS and GPS) have now become widespread, the issue of increasing the accuracy and fault tolerance of such navigation systems remains relevant. One way to clarify vehicle location data is to use radio frequency identification. By placing the reader on a moving vehicle, and the tags on various road infrastructure objects. At the same time, the tags store information about their exact location. The paper [73] presents a vehicle positioning method using GPS data refined using RFID technology. The article develops mathematical models for calculating the location error. In the work [42], where the RFID protocol is used to determine the location of the vehicle and subsequent transmission of this data through the VANET network to the server for planning and estimating the route in real time. Comparison results are presented for two cases: using only GPS technology and using RFID in conjunction with VANET. The work [29] discusses a system for determining the location and identification of obstacles in low visibility conditions. The article describes the analytical model of the communication channel, including for the case of two-beam propagation, carries out mathematical modeling of the proposed system, as well as laboratory and field experiments.

Tags can be placed not only on road infrastructure, but also directly on the road surface, as proposed in the work [50]. Thanks to this, machines equipped with readers can determine their location. The article examines algorithms for dealing with collisions that arise when several readers try to read one tag at the same time. The effectiveness of the proposed algorithms is studied using simulation.

Location determination is relevant not only on public roads, but also on railway transport for the precise positioning of trains. The works [71, 72] explore the identification of ultra-high-speed trains accelerating up to 500 km/h. The readers are supposed to be fixed between the sleepers, and the trains will be equipped with a set of tags attached along the entire length of the train. The article [20] proposes another way to determine the location of railway transport by attaching a reader to the train, which will receive location information from tags distributed along its route. A similar method is considered in the work [36], where it is proposed to use RFID in the metro system to clarify location data obtained by the odometry method, traditional for underground railways.

In 2022, the work [64] was published, in which, for the first time in world practice, a description was given of the developed hybrid vehicle identification complex based on the joint use of existing vehicle video recording systems and RFID technologies. The hybrid complex, designed to implement pilot zones of a new system for improving road safety in the cities of Moscow, St. Petersburg and Kazan, was successfully tested at the traffic police test site of the Republic of Tatarstan.

RFID technology in the field of transport can also be used to solve various problems not directly related to object identification. Many traffic accidents happen due to driver fatigue and falling asleep while driving. Currently, systems that prevent the driver from falling asleep are beginning to be actively used in public transport [1]. The work of [67] proposes a similar system that uses RFID technology in its work. Another example of the use of RFID is presented in the work of [54], where a motion algorithm and laboratory bench are being developed for a small-sized robotic machine equipped with an RFID reader, thanks to which the robot moves along a trajectory marked by RFID tags (Table 1).

Table 1. Application areas of RFID in transport

Application of RFID	Articles
Transport identification	$[11, 12, 14, 17, 18, 31, 35, 41, 43, 44, 47, 48, 51, 55]$
Speed measurement	$[24, 32, 68]$
Smart roads	$[13, 27, 47, 53, 56, 69, 70]$
Location determination	$[29, 42, 50, 73]$
Railway	$[20, 36, 71, 72]$
Non-Standard Applications	$[54, 67]$

4   Conclusion

This paper provides a review of articles in the field of the rapidly developing area of using RFID for identifying fast-moving vehicles, which is currently poorly reflected in existing global reviews. A description of the theoretical and experimental results, as well as the architecture and hardware and software for the practical implementation of ground vehicle identification systems used in various fields is also given.

Acknowledgement. This research was supported by grants from the Russian Science Foundation No. 23-29-00795 [3], https://rscf.ru/project/23-29-00795/ and No. 22-49-02023 [2] https://rscf.ru/project/22-49-02023/.

References

1. The antison system helps to control the attention of drivers. https://mosmetro.ru/news/details/1700
2. Development and research of methods for increasing the reliability of new generation tethered high-altitude unmanned telecommunication platforms. https://rscf.ru/project/22-49-02023/
3. Modeling of a new generation optical packet switching network with non-stationary data flow. https://rscf.ru/project/23-29-00795/

4. Reader Management 1.0.1. EPCGlobal (2007)
5. The Application Level Events (ALE) Specification, Version 1.1.1. Part I: Core Specification. EPCGlobal (2009)
6. The Application Level Events (ALE) Specification, Version 1.1.1. Part II: XML and SOAP Bindings. EPCGlobal (2009)
7. Discovery, Configuration, and Initialization (DCI) for Reader Operations. Version 1.0 Ratified Standard. EPCGlobal (2009)
8. Low Level Reader Protocol (LLRP). Version 1.1 Ratified Standard. EPCGlobal (2010)
9. RAIN Reader Communication Interface Guideline. vol 4.0. RAIN RFID Alliance (2020)
10. Abramson, N.: THE ALOHA SYSTEM. In: Proceedings of the Fall Joint Computer Conference on - AFIPS 1970 (Fall), 17–19 November 1970, p. 281. ACM Press, New York (1970). https://doi.org/10.1145/1478462.1478502
11. Ahmed, S., Tan, T.M., Mondol, A.M., Alam, Z., Nawal, N., Uddin, J.: Automated toll collection system based on RFID sensor. In: Proceedings - International Carnahan Conference on Security Technology, October 2019 (2019). https://doi.org/10.1109/CCST.2019.8888429
12. Al-Naima, F.M., Al-Any, H.: Vehicle location system based on RFID. In: Proceedings - 4th International Conference on Developments in eSystems Engineering, DeSE 2011, pp. 473–478 (2011). https://doi.org/10.1109/DeSE.2011.11
13. Al-Shareeda, M.A., Manickam, S.: A systematic literature review on security of vehicular ad-hoc network (VANET) based on VEINS framework. IEEE Access (2023). https://doi.org/10.1109/ACCESS.2023.3274774
14. Albin Libi Madana, L.S.: Improved contactless RFID detections in transport system. In: Proceedings of International Conference on Intelligent Engineering and Management, ICIEM 2020, pp. 465–470 (2020). https://doi.org/10.1109/ICIEM48762.2020.9160053
15. Alliance, R.R.: Electronic vehicle identification (evi) (2018)
16. Errington, A.F.C., Brian, L.F., Daku, A.F.P.: Initial position estimation using RFID tags: a least-squares approach. IEEE Trans. Instrument. Meas. **59**(11), 2863–2869 (2010). https://doi.org/10.1109/TIM.2010.2046366
17. Balbin, J.R., et al.: Vehicle identification system through the interoperability of an ultra high frequency radio frequency identification system and its database. In: HNICEM 2017–9th International Conference on Humanoid, Nanotechnology, Information Technology, Communication and Control, Environment and Management, 2018-January, pp. 1–5 (2017). https://doi.org/10.1109/HNICEM.2017.8269457
18. Bhavke, A., Pai, S.: Smart weight based toll collection & vehicle detection during collision using RFID. In: 2017 International Conference on Microelectronic Devices, Circuits and Systems, ICMDCS 2017, January 2017, pp. 1–6 (2017). https://doi.org/10.1109/ICMDCS.2017.8211569
19. Blythe, P.: RFID for road tolling, road-use pricing and vehicle access control. IEE Colloquium (Digest) **123**, 67–82 (1999). https://doi.org/10.1049/ic:19990681
20. Buffi, A., Nepa, P.: An RFID-based technique for train localization with passive tags. In: 2017 IEEE International Conference on RFID, RFID 2017, pp. 155–160 (2017). https://doi.org/10.1109/RFID.2017.7945602
21. Cards, I.J.S., security devices for personal identification (eds.): ISO/IEC TS 24192-2:2021. Cards and security devices for personal identification - Communication between contactless readers and fare media used in public transport. ISO/IEC (2021)

22. Casella, G., Bigliardi, B., Bottani, E.: The evolution of RFID technology in the logistics field: a review. Procedia Comput. Sci. **200**, 1582–1592 (2022). https:// doi.org/10.1016/j.procs.2022.01.359

23. Cho, J.H., Cho, M.W.: Effective position tracking using b-spline surface equation based on wireless sensor networks and passive UHF-RFID. IEEE Trans. Instrum. Meas. **62**(9), 2456–2464 (2013). https://doi.org/10.1109/TIM.2013.2259099

24. Choy, J.L.C., Wu, J., Long, C., Lin, Y.B.: Ubiquitous and low power vehicles speed monitoring for intelligent transport systems. IEEE Sens. J. **20**(11), 5656–5665 (2020). https://doi.org/10.1109/JSEN.2020.2974829

25. Electronics, I.S.A. (ed.): ISO 14223-1:2011. Radiofrequency identification of animals. ISO/IEC (2011)

26. EPCGlobal: EPC^{TM} Radio-Frequency Identity Protocols Generation-2 UHF RFID Standard. Specification for RFID Air Interface Protocol for Communications at 860 MHz–960 MHz. Release 2.1. EPCGlobal (2018)

27. Zhu, F., Lv, Y., Chen, Y., Wang, X., Xiong, G., Wang, F.-Y.: Parallel transportation systems: toward IoT-enabled smart urban traffic control and management. IEEE Trans. Intell. Transport. Syst. **21**(10), 4063–4071 (2020). https://ieeexplore. ieee.org/document/8824092

28. Finkenzeller, K.: RFID Handbook (2003). https://doi.org/10.1002/0470868023

29. Garcia Oya, J.R., Martin Clemente, R., Hidalgo Fort, E., Carvajal, G.: Passive rfid-based inventory of traffic signs on roads and urban environments. Sensors **18**, 2385 (2018). https://doi.org/10.3390/S18072385. https://www.mdpi.com/ 1424-8220/18/7/2385/htm

30. Griggs, W.M., Verago, R., Naoum-Sawaya, J., Ordonez-Hurtado, R.H., Gilmore, R., Shorten, R.N.: Localizing missing entities using parked vehicles: an RFID-based system. IEEE Internet Things J. **5**(5), 4018–4030 (2018). https://doi.org/10.1109/ JIOT.2018.2864590

31. Gu, J., Li, M., Yu, L., Li, S., Long, K.: Analysis on link travel time estimation considering time headway based on urban road RFID data. J. Adv. Transport. **2021** (2021). https://doi.org/10.1155/2021/8876626

32. Jing, T., Li, X., Cheng, W., Huo, Y., Xing, X.: Speeding detection in RFID systems on roads. In: 2013 International Conference on Connected Vehicles and Expo, ICCVE 2013 - Proceedings, pp. 953–954 (2013). https://doi.org/10.1109/ICCVE. 2013.6799939

33. Jo, M., Youn, H.Y., Cha, S.H., Choo, H.: Mobile RFID tag detection influence factors and prediction of tag detectability. IEEE Sens. J. **9**(2), 112–119 (2009). https://doi.org/10.1109/JSEN.2008.2011076

34. Jung, K., Lee, S.: A systematic review of RFID applications and diffusion: key areas and public policy issues. J. Open Innov. Technol. Mark. Complexity **1**(1) (2015). https://doi.org/10.1186/s40852-015-0010-z

35. Khan, A.A., Yakzan, A.I., Ali, M.: Radio frequency identification (RFID) based toll collection system. In: Proceedings - 3rd International Conference on Computational Intelligence, Communication Systems and Networks, CICSyN 2011, pp. 103–107 (2011). https://doi.org/10.1109/CICSyN.2011.33

36. Kostrominov, A.M., Tyulyandin, O.N., Nikitin, A.B., Vasilenko, M.N., Osminin, A.T.: RFID-based navigation of subway trains. In: 2020 IEEE East-West Design and Test Symposium, EWDTS 2020 - Proceedings (2020). https://doi.org/10. 1109/EWDTS50664.2020.9225125

37. Kunhoth, J., Karkar, A.G., Al-Maadeed, S., Al-Ali, A.: Indoor positioning and wayfinding systems: a survey. Human-Centric Comput. Inf. Sci. **10**(1) (2020). https://doi.org/10.1186/s13673-020-00222-0

38. Landt, J.: The history of RFID. IEEE Pot. **24**, 8–11 (2005)
39. Larionov, A., Ivanov, R., Vishnevsky, V.: A stochastic model for the analysis of session and power switching effects on the performance of UHF RFID system with mobile tags. In: 12th Annual IEEE International Conference on RFID, RFID 2018, pp. 1–8 (2018). https://doi.org/10.1109/RFID.2018.8376204
40. Larionov, A.A., Ivanov, R.E., Vishnevsky, V.M.: UHF RFID in automatic vehicle identification: analysis and simulation. IEEE J. Radio Freq. Identificat. **1**(1), 3–12 (2017). https://doi.org/10.1109/JRFID.2017.2751592
41. Lonkar, B.B., Sayankar, M.R., Charde, P.D.: Design and monitor smart automatic challan generation based on RFID using GPS and GSM. SSRN Electron. J. (2018). https://doi.org/10.2139/ssrn.3164882
42. Lu, Y., Wang, M.: RFID assisted vehicle navigation based on VANETs, pp. 541–553 (2021)
43. Meneses González, R., Orosco Vega, R., Linares Y Miranda, R.: Some considerations about RFID system performance applied to the vehicular identification. In: 2011 IEEE International Conference on RFID-Technologies and Applications, RFID-TA 2011, pp. 123–127 (2011). https://doi.org/10.1109/RFID-TA.2011.6068626
44. Pandit, A.A., Talreja, J., Mundra, A.K.: RFID tracking system for vehicles (RTSV). In: 2009 1st International Conference on Computational Intelligence, Communication Systems and Networks, CICSYN 2009, pp. 160–165 (2009). https://doi.org/10.1109/CICSYN.2009.50
45. Paper, W.: Electronic Vehicle Identification (EVI) RAIN RFID White Paper Electronic Vehicle Identification (EVI). Technical Report (2018)
46. Park, S., Lee, H.: Self-recognition of vehicle position using UHF passive RFID tags. IEEE Trans. Ind. Electron. **60**(1), 226–234 (2013). https://doi.org/10.1109/TIE.2012.2185018
47. Pawłowicz, B., Trybus, B., Salach, M., Jankowski-Mihułowicz, P.: Dynamic RFID identification in urban traffic management systems. Sensors (Switzerland) **20**(15), 1–26 (2020). https://doi.org/10.3390/s20154225
48. Pedraza, C., Vega, F., Mañana, G.: PCIV, an RFID-based platform for intelligent vehicle monitoring. IEEE Intell. Transp. Syst. Mag. **10**(2), 28–35 (2018). https://doi.org/10.1109/MITS.2018.2806641
49. Pershin, O., Vishnevsky, V.M., Mukhtarov, A., Larionov, A.A.: Optimal placement of base stations as part of a comprehensive wireless network design. In: Information Technology and Computing Systems, pp. 12–25 (2022)
50. Qin, H., Chen, W., Chen, W., Li, N., Zeng, M., Peng, Y.: A collision-aware mobile tag reading algorithm for RFID-based vehicle localization. Comput. Netw. **199** (2021). https://doi.org/10.1016/j.comnet.2021.108422
51. Rajeshwari, S., Santhoshs, H., Varaprasad, G.: Implementing intelligent traffic control system for congestion control, ambulance clearance, and stolen vehicle detection. IEEE Sensors J. **15**(2) (2015)
52. Sharma, D.K., Alqattan, S., Mahto, R.V., Harper, C.: Role of RFID technologies in transportation projects: a review. Int. J. Technol. Intell. Plan. **12**(4), 349 (2020). https://doi.org/10.1504/ijtip.2020.10032144
53. Shirabur, S., Hunagund, S., Murgd, S.: VANET based embedded traffic control system. In: Proceedings - 5th IEEE International Conference on Recent Trends in Electronics, Information and Communication Technology, RTEICT 2020, pp. 189–192 (2020). https://doi.org/10.1109/RTEICT49044.2020.9315602

54. Teng, J.H., Hsiao, K.Y., Luan, S.W., Leou, R.C., Chan, S.Y.: RFID-based autonomous mobile car. In: IEEE International Conference on Industrial Informatics (INDIN), pp. 417–422 (2010). https://doi.org/10.1109/INDIN.2010.5549708

55. Tseng, J.D., Wang, W.D., Ko, R.J.: An UHF band RFID vehicle management system. In: 2007 IEEE International Workshop on Anti-counterfeiting, Security, Identification, ASID, pp. 390–393 (2007). https://doi.org/10.1109/IWASID.2007.373662

56. Unterhuber, A.R., Iliev, S., Biebl, E.M.: Estimation method for high-speed vehicle identification with UHF RFID systems. IEEE J. Radio Freq. Identificat. **4**(4), 343–352 (2020). https://doi.org/10.1109/JRFID.2020.2989900

57. Vishnevsky, V.M., Semenova, O.V., Larionov, A.A.: Performance evaluation of a high speed wireless tandem network using centimeter wave channels in road safety management systems. In: Control Problems (2013)

58. Vishnevsky, V.M., Minnikhanov, R.N., Dudin, A.N., Klimenok, V.I., Larionov, A.A., Semenova, O.V.: new generation of road safety systems and their application in intelligent transport systems. In: ITiVS, pp. 80–89 (2013)

59. Vishnevsky, V., Krishnamoorthy, A., Kozyrev, D., Larionov, A.: Review of methodology and design of broadband wireless networks with linear topology. Indian J. Pure Appl. Math. **47**, 329–342 (2016)

60. Vishnevsky, V.M., A, L.A., Celikin, U., Ivanov, R., Kozirev, D.: Experience in implementing a road safety system using uhf radio frequency identification. In: Proceedings of the 20th International Conference, Distributed Computer and Communication Networks, pp. 152–163 (2017)

61. Vishnevsky., V.M., Larionov, A.A., Mikhaylov, E., Fedotov, I., Abramian, V.: Methods for assessing the effectiveness of radio frequency identification systems for vehicles. In: Information Technology and Computing Systems, pp. 59–70 (2023)

62. Vishnevsky., V.M., Minnikhanov, R.N.: A new, innovative hardware and software complex for a remote monitoring system for traffic violations using RFID technologies. In: Proceedings of the 10th International Scientific and Practical Conference "Organization and Traffic Safety in Large Cities, Innovations: Resources and Opportunities", pp. 297–305 (2012)

63. Vishnevsky., V.M., Minnikhanov, R.N.: Automated system for monitoring traffic violations using RFID technologies and the latest wireless technology. In: Proceedings of the 2nd International Scientific and Practical Conference "Modern Security Problems: Theory and Practice", pp. 52–62 (2012)

64. Vishnevsky, V.M., Minnikhanov, R.N., Barsky, I.V., Larionov, A.A.: Development of a hybrid vehicle identification system based on video recognition and RFID. In: Proceedings of the 2022 International Conference on Information, Control, and Communication Technologies, ICCT 2022 (2022). https://doi.org/10.1109/ICCT56057.2022.9976609

65. Vishnevsky, V.M., Larionov, A., Smolnikov, R.V.: Optimization of topological structure of broadband wireless networks along the long traffic routes. In: Vishnevsky, V., Kozyrev, D. (eds.) DCCN 2015. CCIS, vol. 601, pp. 30–39. Springer, Cham (2016). https://doi.org/10.1007/978-3-319-30843-2_4

66. Vishnevsky, V.M., Larionov, A.: Design concepts of an application platform for traffic law enforcement and vehicles registration comprising RFID technology. In: 2012 IEEE International Conference on RFID-Technologies and Applications, RFID-TA 2012, pp. 148–153 (2012). https://doi.org/10.1109/RFID-TA.2012.6404501

67. Yang, C., Wang, X., Mao, S.: Unsupervised drowsy driving detection with RFID. IEEE Trans. Veh. Technol. **69**(8), 8151–8163 (2020). https://doi.org/10.1109/TVT.2020.2995835

68. Zhai, Y., Guo, Q., Min, H.: An effective velocity detection method for moving UHF-RFID tags. In: RFID-TA 2018–2018 IEEE International Conference on RFID Technology and Applications (2018). https://doi.org/10.1109/RFID-TA.2018.8552825

69. Zhang, E.Z., Zhang, X.: Road traffic congestion detecting by VANETs (2019). https://doi.org/10.2991/eee-19.2019.39

70. Zhang, W., Lin, B., Gao, C., Yan, Q., Li, S., Li, W.: Optimal placement in RFID-Integrated VANETs for intelligent transportation system. In: RFID-TA 2018–2018 IEEE International Conference on RFID Technology and Applications (2018). https://doi.org/10.1109/RFID-TA.2018.8552765

71. Zhang, X., Lakafosis, V., Traille, A., Tentzeris, M.M.: Performance analysis of "fast-moving" RFID tags in state-of-the-art high-speed railway systems. In: Proceedings of 2010 IEEE International Conference on RFID-Technology and Applications, RFID-TA 2010, pp. 281–285 (2010). https://doi.org/10.1109/RFID-TA.2010.5529918

72. Zhang, X., Tentzeris, M.: Applications of fast-moving RFID tags in high-speed railway systems. Int. J. Eng. Bus. Manag. **3**(1), 27–31 (2011). https://doi.org/10.5772/45676

73. Zheng, K., Yang, Q.: Vehicle positioning method based on RFID in VANETs. In: ACM International Conference Proceeding Series (2018). https://doi.org/10.1145/3207677.3277932

On the Identification of a Finite Automaton by Its Input and Output Sequences in Case of Distortions

S. Yu. Melnikov[1]([⊠])[iD], K. E. Samouylov[1,2][iD], and A. V. Zyazin[3]

[1] RUDN University, 6 Miklukho-Maklaya Street, Moscow 117198, Russian Federation
{melnikov-syu,samuylov-ke}@rudn.ru
[2] Federal Research Center "Computer Science and Control" of the Russian Academy
of Sciences (FRC CSC RAS), 44-2 Vavilov Street, Moscow 119333, Russian Federation
[3] MIREA – Russian Technological University, 78 Vernadsky Avenue, Moscow
119454, Russian Federation
ziazin@mirea.ru

Abstract. A method is proposed for checking whether a pair of observed distorted sequences can correspond to the true input and output sequences of a reference automaton, provided that the initial state of the automaton is unknown, and the proportion of deletions, insertions, and replacements of symbols is small. An example of applying the method for the case of a generalized shift register is given.

Keywords: finite state machine · distorted sequence · random number generator

1 Introduction

Pseudo-random sequence generators (PRGs) are one of the basic elements of information security systems. When assessing the quality of PRG, the question arises about the possibility of its identification [1]. If full or partial identification of the PRG itself or its individual nodes by the available output and input (or intermediate) data is possible, then this indicates the poor quality of this PRG and the inappropriateness of its use in security systems.

The ideas of geometric interpretation of the PRG properties have a long history. One of the classic statistical tests – checking the uniformity of the distribution of m-grams – when aggregating the results, it usually uses the union of points according to the principle of proximity in the sense of Euclidean distance. In the well-known article by Marsaglia [2], which at one time led to a radical revision of the principles for choosing the parameters of a linear congruential generator (LCG), geometric terminology is used, the distances between hyperplanes, on which the m-grams (points) formed by the LCG are located,

This paper has been supported by the RUDN University Strategic Academic Leadership Program.

are studied. This point of view, apparently, was based on studies of the correctness of using various PRG generators for Monte Carlo calculations. Later, similar approaches were proposed for other algorithms, for example, "xorshift128+" [3]. In addition, as part of the search for "successful" generators and their parameters for solving problems by stochastic modeling methods in parallel computing, the ideas of I.M. Sobol [4]. A group of researchers led by H. Niederreiter and their followers [5,6] moved from "sequences" to "networks" of pseudo-random points, i.e. to finding ways to construct a pseudo-random subset of points in an m-dimensional cube.

The listed ideas of using geometric constructions are also used to describe the statistical properties of the output sequence of the automaton in the analysis of prohibitions and semi-prohibitions of Boolean functions in the shift register scheme ([7] and others). Since the mid-1990s, an approach has been developed [8] related to the construction of geometric images of automata, in which the behavior of an automaton is displayed in geometric figures, in particular, in curves on a plane. According to this approach, the geometric image of the initial automaton $A_{q_0} = (X, Y, Q, h, f, q_0)$ is the set $\rho_{q_0} = \bigcup_{p \in X^*} \{(p, f(q_0, p))\}$, twice linearly ordered by introducing a linear order $\bar{\omega}_1$ on the set X^* and a linear order $\bar{\omega}_2$ on Y. The geometric image of the automaton is represented by points in the coordinate system with the abscissa axis $(X^*, \bar{\omega}_1)$ and the ordinate axis $(Y, \bar{\omega}_2)$. Using this approach allows a number of properties of automata to be represented by the properties of geometric curves. In [9], to characterize the automaton, the construction of a set of points in the real plane is used, the coordinates of which are given by input and output sequences.

In [10], a geometric description of the joint properties of the relative frequencies of occurrence of multigrams in the input and output sequences of the automaton was proposed using convex polyhedra.

2   Formulating the Problem

Let $A = (X, Y, Q, h, f)$ be a finite strongly connected Mealy automaton, where $X = Y = \{0, 1\}$ are the input and output alphabets, Q is the set of states; $h : Q \times X \to Q$ is a transition function; $f : Q \times X \to Y$ is the output function.

Let the automaton A, starting from the initial state $q_0 \in Q$, process the sequence $\chi^{(N)} = (x^{(1)}, x^{(2)}, \dots x^{(N)})$ into the sequence $\gamma^{(N)} = (y^{(1)}, y^{(2)}, \dots y^{(N)})$, where $x^{(i)}, y^{(i)} \in \{0, 1\}$, $1 \leq i \leq N$, $N \geq 1$. The result for the case when only the character frequencies in the input and output sequences are considered is as follows. The automaton A corresponds to a convex polygon R_A located in the square $[0, 1] \times [0, 1]$,

$$R_A = Conv \left\{ \left(\frac{c_x}{l(c)}, \frac{c_y}{l(c)} \right), c \in C_A \right\}, \tag{1}$$

where C_A is the set of elementary cycles in the transition graph of the automaton, c_x, c_y – sums of the elements of the input and output labeling of the cycle c, $l(c)$

– cycle length, c, $c \in C_A$, $Conv$ – is the convex hull of the set of points on the plane.

For the point $z^{(N)} = \frac{1}{N} \left(\sum_{i=1}^{N} x^{(i)}, \sum_{i=1}^{N} y^{(i)} \right)$, the inequality

$$\rho(z, R_A) \leq \frac{D}{D + N}, \tag{2}$$

is true, where D is the diameter of the automaton A, ρ – Chebyshev distance on the plane, $\rho((x, y), (x', y')) = \max\{|x - x'|, |y - y'|\}$.

Let us assume that it is not the true sequences $\chi^{(N)}$ and $\gamma^{(N)}$ that are observed, but the results of the transformations that have distorted them. The purpose of our work is to propose a way to check whether the observed distorted sequences can correspond to the true ones for the reference automaton, provided that the initial state of the automaton is unknown, and the proportion of deletions, insertions, and replacements of characters is small.

3   Sequence Transformations

3.1   Removing Characters from Sequences

Let $\Omega_N = \{1, 2, ..., N\}$ be the set of the first N natural numbers. Let two subsets $D_\chi, D_\gamma \subset \Omega_N$ be given, consisting of d_χ and d_γ elements, respectively, $0 \leq d_\chi$, $d_\gamma \leq N - 1$. From the sequence $\chi^{(N)} = \left(x^{(1)}, x^{(2)}, ...x^{(N)} \right)$ we delete the characters located in those places whose numbers belong to D_χ, and from the sequence $\gamma^{(N)} = \left(y^{(1)}, y^{(2)}, ...y^{(N)} \right)$ we remove the characters whose numbers belong to D_γ. We denote the sequences obtained after deletion, respectively, χ_D and γ_D. The lengths of these sequences are equal to $N - d_\chi$ and $N - d_\gamma$. Obviously, for $\chi^{(N)}$ and $\gamma^{(N)}$ there are $\binom{N}{d_\chi}$ and $\binom{N}{d_\gamma}$ variants of new sequences, respectively.

3.2   Inserting Characters in a Sequences

Let in the sequences $\chi^{(N)}$ and $\gamma^{(N)}$ we insert i_χ and i_γ binary characters, respectively, $i_\chi, i_\gamma = 0, 1, ...$ The sets of indexes of the inserted elements will be denoted as $I_\chi \subseteq \Omega_{N+i_\chi}$ and $I_\gamma \subseteq \Omega_{N+i_\gamma}$. Denote sequences with inserted elements by χ_I and γ_I respectively. The lengths of these sequences are equal $N + i_\chi$ and $N + i_\gamma$. Obviously, for $\chi^{(N)}$ and $\gamma^{(N)}$, $\binom{N + i_\chi}{i_\chi} \cdot 2^{i_\chi}$ and $\binom{N + i_\gamma}{i_\gamma} \cdot 2^{i_\gamma}$ variants of sequences are possible.

3.3   Replacing Characters in Sequences

Let two subsets $R_\chi, R_\gamma \subseteq \Omega_N$ be given, consisting of r_χ and r_γ elements, respectively, $0 \leq r_\chi, r_\gamma \leq N$. In the sequence $\chi^{(N)}$, we select the characters located in those places whose numbers belong to R_χ, and in the sequence $\gamma^{(N)}$ we select the characters whose numbers belong to R_γ. Let's replace the selected characters with other binary characters. The sequences obtained after the substitutions will be denoted by χ_R and γ_R, respectively. The lengths of these sequences are equal N. Obviously, for $\chi^{(N)}$ and $\gamma^{(N)}$, $\binom{N}{r_\chi}$ and $\binom{N}{r_\gamma}$ variants of new sequences are possible.

4   Main Result

For a finite binary sequence ε, we will denote by $\|\varepsilon\|$ the sum of its elements.

Lemma 1. *The inequalities*

$$\left| \frac{\|\chi^{(N)}\|}{N} - \frac{\|\chi_D\|}{N - d_\chi} \right| \le \frac{d_\chi}{N - d_\chi}, \quad \left| \frac{\|\gamma^{(N)}\|}{N} - \frac{\|\gamma_D\|}{N - d_\gamma} \right| \le \frac{d_\gamma}{N - d_\gamma}, \quad (3)$$

$$\left| \frac{\|\chi^{(N)}\|}{N} - \frac{\|\chi_I\|}{N + i_\chi} \right| \le \frac{i_\chi}{N + i_\chi}, \quad \left| \frac{\|\gamma^{(N)}\|}{N} - \frac{\|\gamma_I\|}{N + i_\gamma} \right| \le \frac{i_\gamma}{N + i_\gamma}, \quad (4)$$

$$\left| \frac{\|\chi^{(N)}\|}{N} - \frac{\|\chi_R\|}{N} \right| \le \frac{r_\chi}{N}, \quad \left| \frac{\|\gamma^{(N)}\|}{N} - \frac{\|\gamma_R\|}{N} \right| \le \frac{r_\gamma}{N}, \quad (5)$$

are valid.

Proof. Let us prove the inequality $\left| \frac{\|\chi^{(N)}\|}{N} - \frac{\|\chi_D\|}{N - d_\chi} \right| \le \frac{d_\chi}{N - d_\chi}$. Assume for simplicity that the last $d = d_\chi$ elements are removed from the sequence $\chi^{(N)} = \left(x^{(1)}, x^{(2)}, \dots x^{(N)} \right)$, while the first $N - d$ elements remain unchanged. Denote $S_N = \sum_{j=1}^{N} x^{(j)}$, $S_{N-d} = \sum_{j=1}^{N-d} x^{(j)}$, $S_d = \sum_{j=N-d+1}^{N} x^{(j)}$. Then

$$\left| \frac{\|\chi^{(N)}\|}{N} - \frac{\|\chi_D\|}{N - d_\chi} \right| = \left| \frac{S_N}{N} - \frac{S_{N-d}}{N - d} \right| =$$

$$= \left| \frac{(S_d + S_{N-d})(N - d) - N S_{N-d}}{N(N - d)} \right| = \left| \frac{N S_d - d S_N}{N(N - d)} \right|.$$

Expanding the modulus in the last expression, and taking into account that $0 \le S_N \le N$ and $0 \le S_d \le d$, $d < N$, we obtain $\left| \frac{N S_d - d S_N}{N(N-d)} \right| \le \frac{d}{N-d}$.

Let us prove the inequality $\left| \frac{\|\chi^{(N)}\|}{N} - \frac{\|\chi_I\|}{N + i_\chi} \right| \le \frac{i_\chi}{N + i_\chi}$. Suppose that $i = i_\chi$ new terms $\left(x^{(N+1)}, x^{(N+2)}, \dots x^{(N+i)} \right)$ are assigned to the sequence $\chi^{(N)} = \left(x^{(1)}, x^{(2)}, \dots x^{(N)} \right)$. Denote $S_N = \sum_{j=1}^{N} x^{(j)}$, $S_I = \sum_{j=N+1}^{N+i} x^{(j)}$. Then

$$\left| \frac{\|\chi^{(N)}\|}{N} - \frac{\|\chi_I\|}{N - i_\chi} \right| = \left| \frac{S_N}{N} - \frac{S_N + S_I}{N + i} \right| =$$

$$= \left| \frac{S_N(N + i) - N(S_N + S_I)}{N(N + i)} \right| = \left| \frac{N S_I - i S_N}{N(N + i)} \right|.$$

Expanding the modulus in the last expression, and taking into account that $0 \le S_N \le N$ and $0 \le S_I \le i$, we obtain $\left| \frac{NS_I - iS_N}{N(N+i)} \right| \le \frac{i}{N+i}$.

Let's prove the inequality $\left| \frac{\|\chi^{(N)}\|}{N} - \frac{\|\chi_R\|}{N} \right| \le \frac{r_\chi}{N}$. Let us assume that in the sequence $\chi^{(N)} = \left(x^{(1)}, x^{(2)}, ... x^{(N)} \right)$ the last $r = r_\chi$ elements of $\left(x^{(N-r+1)}, x^{(N-r+2)}, ..., x^{(N)} \right)$ are inverted, while the first $N - r$ elements remain unchanged. Let's denote $S_{N-r} = \sum_{j=1}^{N-r} x^{(j)}$, $S_r = \sum_{j=N-r+1}^{N} x^{(j)}$.

Then

$$\left| \frac{\|\chi^{(N)}\|}{N} - \frac{\|\chi_R\|}{N} \right| = \left| \frac{1}{N} \sum_{j=1}^{N} x^{(j)} - \frac{1}{N} \left(\sum_{j=1}^{N-r} x^{(j)} + \left(r - \sum_{j=N-r+1}^{N} x^{(j)} \right) \right) \right| =$$

$$= \left| \frac{S_{N-r} + S_r}{N} - \frac{S_{N-r} + r - S_r}{N} \right| = \left| \frac{2S_r - r}{N} \right|.$$

Considering that $0 \le S_r \le r$, it is easy to see that $\left| \frac{2S_r - r}{N} \right| \le \frac{r}{N}$.

Let be: $z_D = \left(\frac{\|\chi_D\|}{N - d_\chi}, \frac{\|\gamma_D\|}{N - d_\gamma} \right)$, $z_I = \left(\frac{\|\chi_I\|}{N + i_\chi}, \frac{\|\gamma_I\|}{N + i_\gamma} \right)$, $z_R = \left(\frac{\|\chi_R\|}{N}, \frac{\|\gamma_R\|}{N} \right)$ − points on the plane constructed from the observed distorted sequences, ρ − Chebyshev distance on the plane, defined as the maximum modules of the difference in the coordinates of two points. The distance between a point z and a set R is defined as $\rho(z, R) = \inf\{\rho(z, z'), z' \in R\}$, the automaton polygon is defined by (1).

Theorem 1. *Let R_A be the polygon of the automaton A corresponding to the frequencies of ones in the output and input sequences, and D be the diameter of the automaton A. The inequalities*

$$\rho(z_D, R_A) \le \frac{\max\{d_\chi, d_\gamma\}}{N - \max\{d_\chi, d_\gamma\}} + \frac{D}{N + D}, \tag{6}$$

$$\rho(z_I, R_A) \le \frac{\max\{i_\chi, i_\gamma\}}{N + \max\{i_\chi, i_\gamma\}} + \frac{D}{N + D}, \tag{7}$$

$$\rho(z_R, R_A) \le \frac{\max\{r_\chi, r_\gamma\}}{N} + \frac{D}{N + D}. \tag{8}$$

are satisfied.

Proof. Let $z^{(N)} = \left(\frac{\|\chi^{(N)}\|}{N}, \frac{\|\gamma^{(N)}\|}{N} \right)$ be a point on the plane corresponding to the relative frequencies of the symbols "1" in the sequences $\chi^{(N)}$ and $\gamma^{(N)}$. Then, according to (2), the inequality $\rho(z^{(N)}, R_A) \le \frac{D}{N+D}$ is satisfied. This means that there is a point $\xi \in R_A$ for which $\rho(z^{(N)}, \xi) \le \frac{D}{N+D}$ is valid. Applying the triangle inequality for points $\xi, z^{(N)}, z_D$, and Lemma, we get

$$\rho(z_D, \xi) \le \rho(z_D, z^{(N)}) + \rho(z^{(N)}, \xi) \le \max \left\{ \frac{d_\chi}{N - d_\chi}, \frac{d_\gamma}{N - d_\gamma} \right\} + \frac{D}{N + D}.$$

Using the fact that for $0 < x < N$, the function $\frac{x}{N-x}$ monotonically increases in x, we have

$$\max\left\{\frac{d_\chi}{N-d_\chi}, \frac{d_\gamma}{N-d_\gamma}\right\} = \frac{\max\{d_\chi, d_\gamma\}}{N - \max\{d_\chi, d_\gamma\}}.$$

Since $\rho(z_D, R_A) \leq \rho(z_D, \xi)$, the first inequality of the theorem is proved. The proof of the two remaining inequalities is carried out in a similar way.

The proved theorem provides a method to check whether the observed distorted sequences can correspond to the true ones for the reference automaton, assuming that the initial state of the automaton is unknown. The method consists of two stages. At the first stage, a polygon R_A of the reference automaton is constructed. At the second stage, one of the points $z_D = \left(\frac{\|\chi_D\|}{N-d_\chi}, \frac{\|\gamma_D\|}{N-d_\gamma}\right)$, $z_I = \left(\frac{\|\chi_I\|}{N+i_\chi}, \frac{\|\gamma_I\|}{N+i_\gamma}\right)$ or $z_R = \left(\frac{\|\chi_R\|}{N}, \frac{\|\gamma_R\|}{N}\right)$ is built, in accordance with the expected type of distortion. Further, the fulfillment of the corresponding inequality, (6), (7) or (8) is checked. If the inequality is violated, then the observed sequences do not correspond to any possible variant of the true sequences of the reference automaton for any of its initial states. If the inequality is true, then conclusions cannot be drawn.

Let us analyze the computational complexity of the proposed method if the number $|Q|$ of states of the automaton A is large. It is determined by the contribution of two terms. Firstly, the preliminary polygon construction complexity, and secondly, the complexity of checking inequality (6), (7) or (8). It follows from the results of [10] that the complexity of the preliminary construction of a polygon is limited by the value of $O(Q * 2^Q)$. Note that analytical methods for construction of polygon are possible for some automaton classes.

If the polygon R_A is already constructed, then the inequalities (6–8) checking complexity, as is easy to see, is not more than 4 times the complexity of checking whether a given point belongs to a convex polygon R_A. The computational complexity of the last problem can be estimated [11] by the value $O(Log\,v)$, where v is the number of vertices of the polygon. To estimate v, we use the fact that all the vertices of the polygon R_A have the form $\left(\frac{p_1}{q_1}, \frac{p_2}{q_2}\right)$, $0 \leq p_i \leq q_i \leq |Q|$, $i = 1, 2$. Estimating the possible number of different abscissas of the polygon vertices, due to the polygon convexity, we obtain $v \leq |Q|^2$. Therefore, the complexity of checking inequalities (6–8) in the case of a previously constructed polygon is estimated as $O(Log\,|Q|)$.

5   Example. Generalized Shift Register

In [12], generalized shift registers are defined, whose transition graphs are generalized in the sense of Imase and Ito [13] de Bruijn graphs. A binary generalized shift register of the order m, $m = 1, 2, ...$, is a Moore automaton $A_f^{(m)} = (X, Y, Q, h, f)$, where the input and output alphabets are $X = Y = \{0, 1\}$, the set of states is $Q = \{0, 1, ..., m-1\}$, the transition function is defined by the rule $h(q, \varepsilon) = (2q + \varepsilon) \mod m$, $q \in Q$, $\varepsilon = 0, 1$, the output function is some mapping $f : Q \rightarrow \{0, 1\}$. At $m = 2^t$ binary generalized shift register is a binary pass-through shift register with a accumulator of the capacity t.

Denote the transition graph of the automaton under consideration by $G(2, m)$.

The graph $G(2, m)$ is a directed graph with m vertices, and arcs from the vertex i to the vertex $2i + \varepsilon \pmod m$, $0 \leq i \leq m - 1$, $0 \leq \varepsilon \leq 1$. Such a graph has $2m$ arcs, is strongly connected, and is regular of degree 2. For $m = 2^t$, $t = 1, 2, ...$, the graph under consideration is a classical binary de Bruijn graph of degree t.

We consider using the Theorem for an automaton $A_f^{(6)}$, which is a binary generalized shift register of order 6, with an output function

$$f(q) = \begin{cases} 0, & \text{if} \quad q < 3, \\ 1, & \text{if} \quad q \geq 3. \end{cases} \tag{9}$$

5.1 Construction of the Polygon of the Automaton $A_f^{(6)}$

The graph $G(2, 6)$ is shown in Fig. 1.

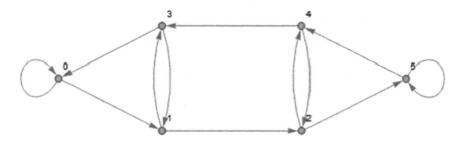

Fig. 1. The graph $G(2, 6)$.

It is easy to see that the graph contains 10 elementary cycles, $c_1 = \{0\}$, $c_2 = \{5\}$ are of length 1, $c_3 = \{2, 4\}$, $c_4 = \{1, 3\}$ are of length 2, $c_5 = \{2, 5, 4\}$, $c_6 = \{0, 1, 3\}$ are of length 3, $c_7 = \{1, 2, 3, 4\}$ is of length 4, $c_8 = \{1, 2, 5, 4, 3\}$, $c_9 = \{0, 1, 2, 4, 3\}$ are of length 5, $c_{10} = \{0, 1, 2, 5, 4, 3\}$ is a complete cycle of length 6.

The sequences of input symbols, under the action of which the automaton goes through each of these cycles, starting from the first state of the cycle (the input markings of these cycles), have the form: $\{0\}$, $\{1\}$, $\{0, 0\}$, $\{1, 1\}$, $\{1, 0, 0\}$, $\{1, 1, 0\}$, $\{0, 0, 1, 1\}$, $\{0, 1, 0, 1, 1\}$, $\{1, 0, 0, 1, 0\}$, $\{1, 0, 1, 0, 1, 0\}$. The output sequences of the automaton when moving through these cycles (the output markings of these cycles) have the form: $\{0\}$, $\{1\}$, $\{0, 1\}$, $\{0, 1\}$, $\{0, 1, 1\}$, $\{0, 0, 1\}$, $\{0, 0, 1, 1\}$, $\{0, 0, 1, 1, 1\}$, $\{0, 0, 0, 1, 1\}$, $\{0, 0, 0, 1, 1, 1\}$.

Let us calculate the vectors $\left(\frac{c_x}{l(c)}, \frac{c_y}{l(c)}\right)$ of the relative frequencies of occurrence of "1" in the input and output markings of each of the cycles:

$$z(c_1) = (0/1, 0/1) = (0, 0),$$
$$z(c_2) = (1/1, 1/1) = (1, 1),$$
$$z(c_3) = (0/2, 1/2) = (0, 1/2),$$
$$z(c_4) = (2/2, 1/2) = (1, 1/2),$$
$$z(c_5) = (1/3, 2/3),$$
$$z(c_6) = (2/3, 1/3),$$
$$z(c_7) = (2/4, 2/4) = (1/2, 1/2),$$
$$z(c_8) = (3/5, 3/5),$$
$$z(c_9) = (2/5, 2/5),$$
$$z(c_{10}) = (3/6, 3/6) = (1/2, 1/2).$$

Let us construct the convex hull of the found set of points according to (1):
$Conv\{z(c_i), i = 1, 2, ..., 10\} = Conv\{(0, 0), (1, 1), (0, 1/2), (1, 1/2)\}$ (See Fig. 2).

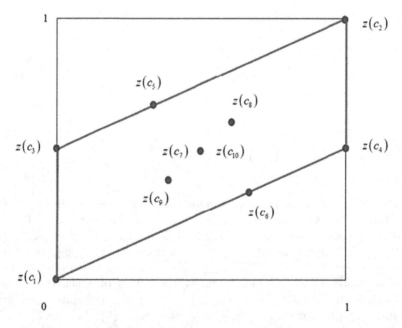

Fig. 2. Construction of the automaton $A_f^{(6)}$ polygon.

Thus, in the plane (x, y), the polygon of the automaton $A_f^{(6)}$ is the set of points of the square $\{(x, y), 0 \le x, y \le 1\}$ for which the double inequality $\frac{1}{2}x \le$

$y \leq \frac{1}{2}x + \frac{1}{2}$ is valid. By direct calculations, one can make sure that four automata $A_f^{(6)}$ have the same polygon, the output functions of which are given by the vectors of their table assignment: $(0,0,0,1,1,1)$, $(0,0,1,1,0,1)$, $(0,1,0,0,1,1)$, $(0,1,1,0,0,1)$.

5.2   An Example of Using the Theorem

Denote by ρ_0 the boundary indicated in the Theorem for distortions of the replacement type,

$$\rho_0 = \frac{\max\{r_\chi, r_\gamma\}}{N} + \frac{D}{N+D}. \tag{10}$$

Simple geometric calculations show that a point z_r cannot be located above the line $y = \frac{x}{2} + \frac{1}{2} + \frac{3}{2}\rho_0$ and below the line $y = \frac{x}{2} - \frac{3}{2}\rho_0$ (see Fig. 3).

If the point $z_R = \left(\frac{\|\chi_R\|}{N}, \frac{\|\gamma_R\|}{N} \right)$, built according to the sequences χ_R and γ_R, fell into the forbidden zones shown as shaded, then we can conclude that either the sequences $\chi^{(N)}$ and $\gamma^{(N)}$ are not input and output for the automaton $A_f^{(6)}$, or their distortions exceed the level corresponding to the value ρ_0.

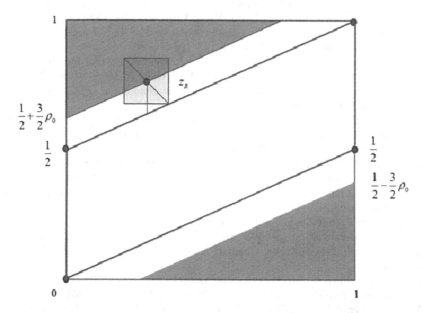

Fig. 3. Automaton $A_f^{(6)}$ polygon. Forbidden areas, where the point z_R cannot be located, are shaded.

6   Conclusion

A generalization of the automaton identification method based on polygons by the frequency characteristics of the input and output sequences is proposed. The generalization is carried out for the case when these sequences are subjected to three types of distortions: insertion, removal, replacement of symbols. In the case of distortions, the automaton identification method based on polygons remains operable. However, with an increase in the proportion of distortions, the information content of the method decreases. The computational complexity of the method is practically independent of the level of distortion.

References

1. Bhattacharjee, K., Maity, K., Das, S.: A search for good pseudo-random number generators: survey and empirical studies. arXiv:1811.04035 (2018)
2. Marsaglia, G.: Random numbers fall mainly in the planes. Proc. Natl. Acad. Sci. **61**, 5–28 (1968)
3. Haramoto, H., Matsumoto, M.: Again, random numbers fall mainly in the planes: xorshift128+. arxiv.org/abs/1908.10020 (2020)
4. Sobol', I.M.: The distribution of points in a cube and the approximate evaluation of integrals. U.S.S.R. Comput. Math. Math. Phys. **7**(4), 86–112 (1967)
5. Niederreiter, H.: Low-discrepancy point sets obtained by digital constructions over finite fields. Czechoslovak Math. J. **42**, 143–166 (1992)
6. L'Ecuyer, P., Marion, P., Godin, M., Puchhammer, F.: A tool for custom construction of QMC and RQMC point sets. In: Keller, A. (ed.) Monte Carlo and Quasi-Monte Carlo Methods, pp. 51–70. Springer, Heidelberg (2022). https://doi.org/10.1007/978-3-030-98319-2_3
7. Nikonov, V.G., Nikonov, N.V.: Zaprety k-znachnyh funkcij i ih svyaz' s problemoj razreshimosti sistem uravnenij special'nogo vida. Vestnik RUDN. Prikladnaya i komp'yuternaya matematika. **2**(1), 79–93 (2003)
8. Epifanov, A.: Recognition methods of geometrical images of automata models of systems in control problem. J. Mech. Mater. Mech. Res. **1**(1) (2018)
9. Anashin, V.S., Khrennikov, A.U.: Applied Algebraic Dynamics. Berlin, de Gruyter Expositions in Mathematics, p. 558 (2009)
10. Melnikov, S.Y., Samouylov, K.E.: Polyhedra of finite state machines and their use in the identification problem. In: NEW2AN/ruSMART -2020. LNCS, vol. 12526, pp. 110–121. Springer, Cham (2020). https://doi.org/10.1007/978-3-030-65729-1_10
11. Preparata, F.P., Shamos, M.I.: Computational Geometry - An Introduction, p. 398. Springer, Heidelberg (1985). https://doi.org/10.1007/978-1-4612-1098-6
12. Maksimovskiy, A.Y., Melnikov, S.Y.: Spectral and combinatorial characteristics of the reduced de brijn graphs. Voprosy kiberbezopasnosti **4**(28), 70–76 (2018)
13. Imase, M., Itoh, M.: Design to minimize diameter on building-block network. IEEE Trans. Comput. **30**, 439–442 (1981)

Analysis of Tethered Unmanned High-Altitude Platform Reliability

V. M. Vishnevsky[1], E. A. Barabanova[1(✉)], K. A. Vytovtov[1],
and G. K. Vytovtov[2]

[1] V. A. Trapeznikov Institute of Control Sciences of RAS, Profsoyuznaya Street 65,
08544 Moscow, Russia
elizavetaalexb@yandex.ru

[2] Astrakhan State Technical University, Tatishchev Street. 16, 414056 Astrakhan,
Russia

Abstract. The reliability of the tethered unmanned high-altitude platform as a single complex system is investigated in this paper. The reliability block diagram of the platform has been developed. It consists of the main subsystems and its elements connected in series. The method of calculating the reliability metrics of the tethered unmanned high-altitude platform such as the reliability function, and the mean time to failure is presented. Reliability metrics of communication system elements are calculated. The approach for calculation of tethered unmanned platform service life is presented.

Keywords: Tethered drone · Reliability function · Mean time to failure · Failure rate · Service life

1 Introduction

Currently, tethered high-altitude unmanned telecommunication platforms of long-term operation are widely used for solving a class of very important problems, including long-term communication and remote video surveillance for critical infrastructure sites for a long time [1]. Thus the main parameter of such drones is reliability, which is primarily determined by the reliability indicators of its main elements.

In some of previous works k out-of-n type models were used to study the reliability of engines of tethered drones in the case when k out of n engines failed [2,3]. The so-called "$(2,3) - out - of - 6 : F$" system model on the basis of a multidimensional Markov process described and applied for evaluation of the reliability characteristics of a tethered multirotor high-altitude platform based on a hexacopter is considered in [2]. The model allows to takes into account the increase in the functional load after the failure of an element on the remaining

The reported study was funded by Russian Science Foundation, project number 22-49-02023.

operating ones, and the location of the failed elements. The "4 − out − of − 8" system model is presented in [3]. The influencing of reliability function of such factors as the coefficient of variation of the lifetime of elements and various (non-exponential) distributions of their lifetime are considered. The sensitivity issues of k out-of-n type models is considered in [4].

The other important problem is to calculate the reliability of the local hybrid navigation system consists of optical and radio subsystems in the cases of bad weather conditions or lifting and landing drone time moments. The reliability analysis of such navigation systems is performed using Markov models considered in transient mode [4].

In a number of works, the investigating the autonomous quadrocopters reliability was carried out using the fault tree analysis method [5–8]. The application of the analytical method for calculating the reliability of the power batteries, drone engines and the electrical board was considered in [5]. The probability approach is based on the structural tree structure reliability diagrams. Authors of [6] have proposed the classification of failures according to severity, which includes the following categories such as catastrophic, critical, marginal, minor.

The new logistic approach based on reliability and maintenance assessment have been developed in [7]. The authors have proposed the method of determination a more efficient interval for the maintenance activities for unmanned aerial vehicles (UAV). In [8] a combination of fault-tree analysis and failure mode and effects analysis models to identify, analyse, and evaluate the most critical failure modes and mechanisms in a drone prototype have been developed.

Presented reliability methods may be used for almost any kind of reliability analysis for different type of UAV subsystems. Unlike autonomous drones tethered drones consist of more complex power, communication and control subsystems so the methods of autonomous drone reliability calculation cannot be applied to tethered ones.

The aim of this work is investigating the reliability of tethered drone which has been presented as a one complex system, and calculating its reliability metrics such as the reliability function, and the mean time to failure.

The paper is organized as follows. In Sect. 2 the reliability block diagram of the tethered unmanned high-altitude platform is given. Section 3 presents the algorithm of communication system reliability calculation. Calculation of the service life of a tethered unmanned platform is presented in Sect. 4. The paper is concluded in Sect. 5.

2   The Statement of the Problem

A tethered unmanned platform consisting of two main modules is being considered (Fig. 1). The first module is a ground station which is the spatial car with the power block and the second main module is a multi-rotor type unmanned aerial vehicle. The power line between the ground station and drone provide the long-term operation of the system.

Fig. 1. The tethered unmanned high-altitude platform

It is necessary to analyse the tethered unmanned high-altitude platform entire system reliability of taking into account possible failures of its elements.

3   The Reliability Block Diagram of the Tethered Unmanned High-Altitude Platform

The approach of constructing the reliability block diagram of the tethered unmanned high-altitude platform is used in this work to investigate the reliability of tethered drone (Fig. 1).

The reliability block diagram shows the function connection of the tethered unmanned high-altitude platform elements. It is not predictably the schematic diagram of the tethered unmanned high-altitude platform, but the functional components of the system [10].

The main tethered drone subsystems are the unmanned aerial vehicle subsystem, the Terrestrial module, and the control subsystem. As shown in Fig. 2, the subsystem of the unmanned aerial vehicle includes the following main components such as the flight controller, the power system, the electric engine system, the navigation and landing device, and the communication system. The terrestrial module includes the intelligent lifting mechanism subsystem, a power subsystem, and a communication system. The tethered unmanned aerial platform control subsystem includes the unmanned aerial vehicle (UAV) control unit, and the landing platform control unit.

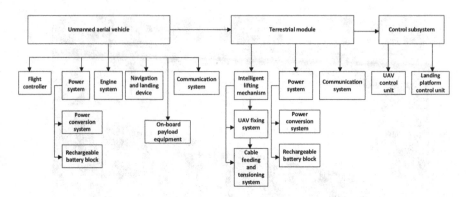

Fig. 2. The reliability block diagram of the tethered unmanned high-altitude platform

The ground-to-aircraft transmission of high-power energy system (power system) consists of the power conversion system and the rechargeable battery block. The intelligent lifting mechanism subsystem consists of the intelligent fixing and the cable feeding and tensioning systems.

As the main problem of building tethered drone is increasing weight of payload the presented reliability scheme doesn't contain duplicate elements, since additional elements increase the weight of the drone. Thus in proposed approach all elements of the reliability block diagram are connected in series. Taking into account the exponential distribution of failure rates of tethered unmanned aerial platform components, it is possible to calculate the failure rate of the entire system using the formula [10]:

$$\Lambda = \sum_{i=1}^{n} \lambda_i, \tag{1}$$

where n is the number of system elements.

As the initial data for calculating the reliability of individual elements of the tethered drone, the time between failures specified in the technical data sheet of the corresponding equipment or its analogues can be used. The failure rates of system elements also can be calculated by knowing the mean time to failure of their radio electronic components and the methodology for calculating their reliability [10]. The failure rates of some subsystems such as engine system and navigation system can be found using the algorithms proposed in [2] and [12].

The failure rates of the system elements calculated in accordance with [11] are presented in Table 1.

After substituting λ_i from the Table 1 into the expression (1) we obtain $\Lambda = 7.1275$ 1/h which is the value of the entire tethered platform failure rate.

The reliability function of the tethered unmanned high-altitude platform can be calculated using well-known formula:

Table 1. Reliability metrics of tethered high-altitude unmanned platform subsystems

Subsystems of the tethered unmanned platform	Failure rates λ_i, 1/h
Flight controller	$0.8963 \cdot 10^{-6}$
Power conversion system	$7,5276 \cdot 10^{-5}$
Rechargeable battery block	$6.8325 \cdot 10^{-5}$
Engine system	$2.5734 \cdot 10^{-6}$
Navigation and landing device	$1.4576 \cdot 10^{-5}$
Communication system	$9,5935 \cdot 10^{-6}$
UAV fixing system	$5,7834 \cdot 10^{-5}$
Cable feeding and tensioning system	$4.4529 \cdot 10^{-4}$
UAV control unit	$2.7623 \cdot 10^{-6}$
Landing platform control unit	$3.5623 \cdot 10^{-5}$

$$R(t) = e^{-\Lambda t} \tag{2}$$

Mean time to failure $MTTF$ of the tethered drone can be found using the following expression:

$$\int\limits_{0}^{\infty} R(t)\mathrm{d}t = \frac{1}{\Lambda} \tag{3}$$

4 Communication System Reliability Calculation

As communication system of the tethered drone is a complex system the approach of calculating the reliability of a communication system for an unmanned aerial platform has been considered.

Let us assume that the equipment presented in Table 2 is used to build the communication system of the tethered drone.

Table 2. Reliability metrics of communication system elements

Communication elements	Type of equipment	$MTTF$, h	λ, 1/h
Ethernet switch	SW-70202	144890	$6,9018 \cdot 10^{-6}$
Fiber transceiver	SFP-S1SC13	445890	$2,2427 \cdot 10^{-6}$

To calculate the failure rate of an optical cable λ_e using for providing communication from the terrestrial module to the unmanned aerial vehicle the following mathematical model obtained as a result of empirical research is used [10]:

$$\lambda_e = [\lambda_{b.1} \cdot m \cdot K_{T1} + \lambda_{b.2} \cdot K_{T2}] \cdot L_K \cdot K_O + \lambda_{b.3} \cdot m \cdot K_{T1} \cdot K_{RG1}, \tag{4}$$

where $m = 2$ is the number of optical fibers in the cable; $\lambda_{b.1} = 2.33 \cdot 10^{-15}$ 1/h·m is the basic failure rate of optical fibers in the process of their operating time referred to one meter of cable type length; $\lambda_{b.2} = 6.86 \cdot 10^{-15}$ 1/h·m is the basic rate of sudden failures of the cable structure in the process of their operating time referred to one meter of the cable type length; $\lambda_{b.3} = 1.21 \cdot 10^{-11}$ 1/h·m is the basic rate of gradual failures of fiber-optical cables in the process of their operating time; $L_K = 110$ m is the length of the fiber-optical cable; $K_O = 4$ is the coefficient of rigidity of operating conditions for a given group of equipment [11].

The value of the suitability criterion KG_1 is calculated by the formula:

$$KG_1 = \frac{d}{K_{T4}} \tag{5}$$

Here $d = 13$ dBm is the maximum allowable value of the attenuation coefficient in the fibre-optic cable. The value of d is determined by the sensitivity of the receiving optical module SFP. $K_{T4} = 1,3$ is the temperature coefficient characterizing the maximum reversible changes of the fiber-optical cable attenuation coefficient in the range of negative operating temperatures [11]. Therefore $KG_1 = \frac{13}{1,3} = 10$.

The values of temperature coefficients K_{T1} and K_{T2} are determined by the formulas:

$$K_{T1} = exp[-K_{E1}(\frac{1}{T_{eq}} - \frac{1}{298})], \tag{6}$$

$$K_{T2} = exp[-K_{E2}(\frac{1}{T_{eq}} - \frac{1}{298})], \tag{7}$$

where $K_{E1} = 18,06 \cdot 10^3$ is the coefficient that depends on the activation energy of degradation processes [10] for claddings of optical fibers; $K_{E2} = 8,05 \cdot 10^3$ is the coefficient depending on the activation energy of degradation processes [11] for fiber- optical sheaths; T_{eq} is the equivalent operating temperature of components which can be calculated by the formula:

$$T_{eq} = (\frac{1}{T_{max}} + \frac{1}{K_E}ln\frac{(\sum_{i=1}^{n} t_i}{\sum_{i=1}^{n} t_i^* \cdot t_{T_{max}}})^{-1}, \tag{8}$$

where

$$t_i^* = t_i\{exp[-K_E(\frac{1}{T_i} - \frac{1}{T_{max}})]\}. \tag{9}$$

Here t_i is the total time interval of the communication system element at a temperature T_i; $T_{max} = 313$ K is the maximum operating temperature; $t_{T_{max}}$ is the total time interval of the component operation at the maximum operating temperature.

The example of the distribution of a tethered platform operating time at a certain temperature during the year is presented in Table 3. Here it is assumed that during the year the tethered unmanned aerial platform operates 3456 h and 18 h of them at the maximum operating temperature.

Table 3. Operating time of a tethered unmanned aerial platform at a given temperature T_i during the month t_i, h

T_i,K	Jan	Feb	Mar	Apr	May	Jun	Jul	Aug	Sep	Oct	Nov	Dec
233	–	18	–	–	–	–	–	–	–	–	–	–
238	18	36	–	–	–	–	–	–	–	–	–	–
243	36	72	–	–	–	–	–	–	–	–	–	18
248	72	72	–	–	–	–	–	–	–	–	–	36
253	72	36	–	–	–	–	–	–	–	–	–	72
258	36	36	18	–	–	–	–	–	–	–	18	72
263	36	18	36	18	–	–	–	–	–	18	36	36
268	18	–	72	36	–	–	–	–	18	36	72	36
273	–	–	72	72	18	–	–	–	36	72	72	18
278	–	–	36	72	36	18	–	18	72	72	36	–
293	–	–	36	36	72	36	18	36	72	36	36	–
288	–	–	18	36	72	72	36	72	36	36	18	–
293	–	–	–	18	36	72	72	72	36	18	–	–
298	–	–	–	–	36	36	72	36	18	–	–	–
303	–	–	–	–	18	36	36	36	–	–	–	–
308	–	–	–	–	–	18	36	18	–	–	–	–
313	–	–	–	–	–	–	18	–	–	–	–	–

Taking into account the valuers of operating time presented in Table 3 and expressions (6)–(9), we have obtained the following coefficients: $K_{T1} = 1821,4$; $K_{T2} = 1393,7$. After substituting K_{T1} and K_{T2} into (4), we get $\lambda_e = 4,49 \cdot 10^{-7}$ 1/h. Then, according to (1)–(3) and taking into account the series connection of elements in the reliability block diagram (Fig. 2), we obtain the reliability metrics of the communication subsystem for one year of operation: $\Lambda = 95.935 \cdot 10^{-7}$ 1/h; $MTTF = 104237$ h; $R(t) = 0.92$.

5 Numerical Calculation of the Tethered Unmanned Platform Service Life

The tethered unmanned platform service life is one of important metrics that determines the time during which the equipment can function effectively without the need for repair or replacement. The service life is the calendar duration of operation of an object from the start of operation or its resumption after major repairs until the object reaches its limit state.

According to [13] one of the criteria for the limit state may be the economic inexpediency of further operation of the facility. This criterion is determined based on a comparison of the cost of electronic equipment taking into account its depreciation at the end of its service life and the cost of its restoration within one year of operation and can be described by the formula:

$$\frac{C_p}{Z_T} \geq 1 \tag{10}$$

where C_p is the cost of restoration of tethered unmanned high-altitude platform per year of maintenance; Z_T is the cost of tethered unmanned high-altitude platform taking into account its depreciation for the year of calculating the cost of restoration.

When calculating the service life of the designed tethered unmanned high-altitude platform, it is first necessary to determine the cost of annual repairs of the tethered unmanned high-altitude platform C_p in connection with the technical resource of its elements then the cost of production of the tethered unmanned high-altitude platform Z_T taking into account depreciation under the influence of the technical process. By comparing these values, the optimal value of service life (durability) can be determined using the relationship:

$$C_p \geq Z_T \tag{11}$$

Optimal durability (service life) T_{sl} is found as the intersection point of the curves $C_p = f_1 (T)$ and $Z_T = f_2 (T)$. It is not advisable to increase T_{sl} beyond this point, since the costs of repairs exceed the costs of manufacturing a new tethered unmanned high-altitude platform.

Let's consider a numerical example of calculating the service life of a tethered platform. When determining the annual cost of tethered unmanned high-altitude platform repair, the cost of the tethered unmanned high-altitude platform component requiring replacement was taken into account, as well as the cost of wages for maintenance personnel. The replacement frequency (number of replacements per year N) was calculated taking into account the failure rate of tethered unmanned high-altitude platform components. The number of changing each components of tethered unmanned high-altitude platform and the total number of changing by year N is presented in Table 4. Table 5 shows the cost of each component of the tethered unmanned high-altitude platform C_e.

Tables 4, 5 use the following notations: UAV is the Unmanned aerial vehicle; PCS is the Power conversion system; RBB is the Recharge able battery block; PS is the power switch; CS is the Control subsystem; TM is the Terrestrial module; CS TM is the control system of Terrestrial module; CF TS is the Cable feedingand tensioning system; FS is the UAV fixing system; LM TM is the Lifting mechanism of Terrestrial module; CP TM is the Control Panel of Terrestrial module; RCU is the Remote control unit.

When calculating the salaries of engineers E who replace faulty tethered unmanned high-altitude platform components, the number of replacements per year N, the time required to replace one component ($t = 1$ h), the cost of a standard hour $C = 625$ rubles were taken into account. The basic and additional wages are taken into account (12.85% of the basic salary), as well as the percentage of deductions for social needs (30.2%).

Table 4. Number of changing components of tethered unmanned high-altitude platform by year

Year	UAV	PCS	RBB	PS	CS	TM	CS TM	CF and TS	FS	LM TM	CP TM	RCU	N
1	1	0	0	0	0	4	0	0	0	0	0	0	5
2	1	0	0	0	3	0	1	0	1	0	0	0	6
3	1	1	1	0	0	4	1	0	0	0	0	1	9
4	2	0	0	0	0	3	0	1	1	1	0	0	8
5	1	1	1	0	0	4	0	0	0	0	0	0	7
6	1	0	0	0	1	3	1	1	0	1	0	1	9
7	1	1	1	0	0	4	0	0	0	0	0	0	7
8	2	0	0	0	0	3	0	1	1	1	0	0	8
9	1	1	1	0	0	4	1	0	0	0	0	1	9
10	1	0	0	0	0	3	0	1	0	1	0	0	6
11	1	1	1	0	0	4	0	0	0	0	0	0	7
12	2	0	0	0	1	3	1	1	1	1	1	1	12
13	1	1	1	0	0	4	0	0	0	0	0	0	7
14	1	0	0	0	0	3	0	1	0	1	0	0	6
15	1	1	1	0	0	4	1	0	0	0	0	1	9

Table 5. The cost of each component of the tethered unmanned high-altitude platform C_e in thousand rubles

The name	UAV	PCS	RBB	PS	CS	TM	CS TM	CF TS	FS	LM TM	CP TM	RCU
C_e	566	379	103	4.5	278	536	150	59	49	21	96	21

$$E = N \cdot t \cdot C \cdot 0.1285 + (N \cdot t \cdot C + 0.1285 \cdot N \cdot t \cdot C) \cdot 0.302 \qquad (12)$$

The cost of the entire set of equipment C_z that requires replacement every year will be different and is determined by the formula:

$$C_z = \sum_{i=1}^{12} n_i C_{ei} \qquad (13)$$

where n_i is the number of number of system element replacements from the Table 4; C_{ei} is the cost of a system element from the Table 5. The total cost of the costs C_p is determined by the sum of the costs for the salaries of the engineers carrying out the replacement of equipment E and the cost of the entire set of equipment C_z.

$$C_p = E + C_z \qquad (14)$$

C_p values are recalculated taking into account the reduction in cost of repairs under the influence of the technical process according to the formula:

$$C'_p = C_p(1 - \alpha_p)^T \qquad (15)$$

where C_p' is repair costs in the $T - th$ year of the tethered unmanned high-altitude platform operation recalculated taking into account their cost reduction under the influence of technical progress; α_p is the average annual reduction in machine repair costs (in fractions of a unit) usually take $\alpha_p = 0.01 \div 0.015$. Here in numerical calculations $\alpha_p = 0.01$.

The results of calculating the cost of repairing the tethered unmanned high-altitude platform by year are presented in Table 6.

Table 6. The cost of the tethered unmanned high-altitude platform repair by year

$Year$	$C_z, rubles$	N	$E, rubles$	$C_p, rubles$	$C_p', rubles$
1	2707000	5	4591.00	2711591.0	2684475.09
2	2252000	6	5509.90	2257509.9	2212585.45
3	3360000	9	8264.85	3368264.85	3268224.02
4	2312000	8	7346.54	2319346.54	2227955.03
5	3191000	7	6428.22	3765346.54	3580807.09
6	2701000	9	8264.85	2709264.85	2550719.08
7	3191000	7	6428.22	3197428.22	2980212.05
8	2312000	8	6428.22	2318428.22	2139317.34
9	3360000	9	8264.85	3368264.85	3076968.04
10	2252000	6	5509.9	2257509.9	2041651.49
11	3191000	7	6428.22	3197428.22	2862779.80
12	2856000	12	11019.8	2867019.8	2541282.98
13	3191000	7	6428.22	3197428.22	2805810.48
14	2252000	6	5509.90	2257509.9	1961202.27
15	3360000	9	8264.85	3368264.85	2896904.32

The value of Z_T is determined by the formula:

$$Z_T = \frac{Z_0}{(1 + I_n)^T} \tag{16}$$

where Z_T is the cost of the tethered unmanned platform T years after its creation, taking into account its depreciation under the influence of technical progress; Z_0 is initial cost of the tethered unmanned platform; I_n is average annual increase in labor productivity in the country in fractions of a unit; T is the number of years that expires from the date of creation of the tethered unmanned platform.

The value of I_n usually varies from 0.07 to 0.1. Let's take $I_n = 0.08$. Here in numerical calculations the initial cost of the tethered unmanned platform is taken equal to 5800000 rubles.

Based on the results of calculating the cost of repairs and the cost of repro-duction of the tethered unmanned platform Z_T, graphs $C'p(t)$ and $Z_T(t)$ are constructed. The intersection point of the graphs determines the service life of the tethered unmanned platform, which is 11 years. This service life is quite acceptable for this type of equipment (Fig. 3).

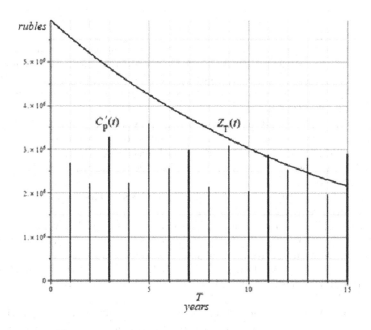

Fig. 3. Determination of the tethered unmanned high-altitude platform service life

6 Conclusion

This paper proposes the methodology and the mathematical models for calculating and analysis the tethered high-altitude unmanned platform reliability as a complex single system. The approach is based on the constructing of the tethered high-altitude unmanned platform reliability block diagram, which is presented for the first time in this paper. The example of calculating the reliability metrics of the tethered unmanned platform communication subsystem such as reliability function and mean time to failure is considered. The method and numerical calculation of the tethered unmanned platform service life are described.

References

1. Vishnevsky, V.M., Efrosinin, D.V., Krishnamoorthy, A.: Principles of construction of mobile and stationary tethered high-altitude unmanned telecommunication platforms of long-term operation. In: Vishnevskiy, V.M., Kozyrev, D.V. (eds.) DCCN 2018. CCIS, vol. 919, pp. 561–569. Springer, Cham (2018). https://doi.org/10.1007/978-3-319-99447-5_48
2. Kozyrev, D.V., Phuong, N.D., Houankpo, H.G.K., Sokolov, A.: Reliability evaluation of a hexacopter-based flight module of a tethered unmanned high-altitude platform. In: Vishnevskiy, V.M., Samouylov, K.E., Kozyrev, D.V. (eds.) DCCN 2019. CCIS, vol. 1141, pp. 646–656. Springer, Cham (2019). https://doi.org/10.1007/978-3-030-36625-4_52
3. Vishnevsky, V., Selvamuthu, D., Rykov, V., Kozyrev, D., Ivanova, N.: Reliability modeling of a flight module of a tethered high-altitude telecommunication platform. In: 2022 International Conference on Information, Control, and Communication Technologies (ICCT). IEEE, New York (2022). https://ieeexplore.ieee.org/document/9976764
4. Ivanova, N.M.: On importance of sensitivity analysis on an example of a k-out-of-n system. Mathematics 11, 1100 (2023)
5. Vishnevsky, V.M., Vytovtov, K.A., Barabanova, E.A., Buzdin, V.E.: Mathematical model for reliability indicators calculation of tethered UAV hybrid navigation system. In: Journal of Physics: Conference Series, vol. 2091 (2021)
6. Imani, K., Gholami, A., Dehaghi, M.B.: Reliability calculation with error tree analysis and breakdown effect analysis for a quadcopter power distribution system. Maint. Reliabil. Cond. Monit. 2, 45–57 (2022)
7. de Oliveira Martins Franco, B.J., Góes, L.C.S: Failure analysis methods in unmanned aerial vehicle (UAV) applications. In: 19th International Congress of Mechanical Engineering (2007)
8. Petritoli, E., Leccese, F., Ciani, L.: Reliability and maintenance analysis of unmanned aerial vehicles. Sensors 18, 3171 (2018). https://doi.org/10.3390/s18093171
9. Shafiee, M., Zhou, Z., Mei, L., Dinmohammadi, F., Karama, J., Flynn, D.: Unmanned aerial drones for inspection of offshore wind turbines: a mission-critical failure analysis. Robotics 10, 26 (2021). https://doi.org/10.3390/robotics10010026
10. Birolini, A.: Reliability Engineering: Theory and Practice. Springer, Berlin (1999). https://doi.org/10.1007/978-3-662-05409-3
11. Reliability of electrical devices. Manual. - M.: MDRF (2006). (in Russian)
12. Vishnevsky, V., Vytovtov, K., Barabanova, E., Buzdin, V.E., Frolov, S.A.: Local hybrid navigation system of tethered high-altitude platform. In: Vishnevskiy, V.M., Samouylov, K.E., Kozyrev, D.V. (eds.) DCCN 2021. LNCS, vol. 13144, pp. 67–79. Springer, Cham (2021). https://doi.org/10.1007/978-3-030-92507-9_7
13. Blank, L., Targuin, A.: Engineering Economy, 7th edn. The McGraw-Hill Companies, New York (2012)

Analytical Modeling of Distributed Systems

Information Spreading
in Non-homogeneous Evolving Networks
with Node and Edge Deletion

Natalia M. Markovich$^{(\boxtimes)}$ and Maksim S. Ryzhov

V.A. Trapeznikov Institute of Control Sciences, Russian Academy of Sciences,
Profsoyuznaya Street 65, 117997 Moscow, Russia
markovic@ipu.rssi.ru, nat.markovich@gmail.com, maksim.ryzhov@frtk.ru

Abstract. A preferential attachment (PA) has been suggested to model
network evolution and to explain conjectured power-law node degree
distributions in real-world networks. In Markovich, Ryzhov (2022a,b),
the schemes of the linear PA proposed in Wan et al. (2020) for the
network evolution were suggested for information spreading. The PA and
the well-known algorithm SPREAD proposed in Mosk-Aoyama, Shah
(2006) were compared regarding the minimum number of evolution steps
K^* required to spread a single message among a fixed number of nodes
in non-homogeneous directed networks. This comparison was done in
Markovich, Ryzhov (2022a,b) without node and edge deletion during
the evolution. The objective of the current study is to investigate the
impact of the PA parameters on spreading of a single message to a fixed
number of nodes in the graph when an existing node or edge is uniformly
deleted at each step of the PA evolution. The results are provided for
simulated and real graphs.

Keywords: non-homogeneous evolving network · information
spreading · linear preferential attachment · SPREAD algorithm · node
and edge deletion

1 Introduction

Let $G_t = (V_t, E_t)$ be the graph with sets of nodes V_t and edges E_t generated by
the attachment of a new node and a new edge at time moment t. The preferential
attachment (PA) has been suggested to model network evolution in such a way
to explain conjectured power-law degree distributions in real-world networks,
[1]. Such networks contain rare giant nodes with a large number of links (node
degrees). The node degrees grow fast at each evolution step since new nodes
prefer to be attached to valuable nodes. The evolving networks are generally
non-homogeneous, i.e. node degrees and other influence measures like PageRanks
may be non-stationary distributed. In the paper, we focus on the rate of the
spreading of one message among a fixed number of nodes in the non-homogeneous

The authors were supported by the Russian Science Foundation RSF, project number
22-21-00177.

direct networks. The necessity of the information spreading arises particularly for distributed computation [2].

In [3,4] the linear PA $\alpha-$, $\beta-$, $\gamma-$ schemes proposed in [5] to model the network evolution were suggested for information spreading. Namely, a message to be spread among a fixed number of nodes is assumed to be in the disposal of some node set $S_t \subset V_t$. A new node v_t that is appended to the network at a time t may get the message from one of the nodes in S_t (let say, node i) if v_t follows the node i, i.e. a new edge is created such that it is directed from v_t to $i \in S_t$. This case corresponds to the $\alpha-$scheme where the directed edge is created from a "new" node to an "old" node. For example, if a customer follows a Web page, then he(she) can upload the content of the page. Peer-to-peer networks provide another example when a follower tends to attach to a node with a large number of peers to upload a required peer. Another way to spread an information is to create a new edge between a pair of existing nodes i and j. If the node i has the message, i.e. $i \in S_t$ and the node j has not for the time being, and if there is an edge $j \to i$, then node i shares the message with node j. We assume that the message cannot be spread to the node j if the edge $i \to j$ is created at the evolution step. Both existing nodes may have not the message and then no sharing is happening. The $\beta-$scheme corresponds to the creation of the edge with any direction between two "old" nodes.

The PA spreading has been compared with a well-known algorithm SPREAD proposed in [2] by the minimum number of evolution steps K^* required to spread a message among a fixed number of nodes in the non-homogeneous directed networks. The idea of the original SPREAD is to share all messages nodes have among all nodes in an undirected network. In [4] the modified SPREAD has been applied to spread a single message among a fixed number of nodes in directed networks. The latter comparison was done in [3,4] without nodes and edges deletion at each step of the evolution.

Less attention in the literature related to the PA is devoted to the node and edge deletion, see for instance [5–7]. In [6] the node deletion is interpreted in the context of agent systems. Then the collapsing agent that loses its incoming connections (consumption links) leads to the breaking of some production links of its neighbors, and thus also leading to their collapse. In [5] the node deletion is provided in the evolving graph at the final time of the evolution but not permanently at each step when a new node is appended. In [7] the tail index of the PageRank and the Max-linear model used as node influence characteristics is estimated nonparametrically for graphs generated by the PA $\alpha-$, $\beta-$, $\gamma-$ schemes where the existing node or edge is deleted.

Our objective is to extend the results in [3,4] by investigating the impact of the PA parameters α, β, γ on the spreading of one message to a fixed number of nodes in the graph, when a node or an edge is uniformly deleted at each step of the evolution. To this end, we aim to compare the spreading rate of the linear PA schemes and of the SPREAD algorithm for different values of α, β, γ.

The message may be lost in the network due to the deletion of nodes with the message. Such situation is impossible for classical SPREAD where nodes share all messages what they have with each other. If an edge is deleted, then the message cannot be lost as in the case of node deletion, but it may not be transmitted further.

The paper is organized as follows. In Sect. 2, related works regarding the linear PA schemes (Sect. 2.1) and the spreading information (Sect. 2.2) are described. In Sect. 3, our main results concerning the spreading with the node or edge deletion for homogeneous simulated and non-homogeneous real temporal graphs are presented. The exposition is finalized with conclusions.

2   Related Works

2.1   Preferential Attachment

The linear PA schemes [1,5] start with an initial directed graph $G(k_0)$ with at least one node and k_0 edges and construct a growing sequence of directed random graphs $G(k) = (V(k), E(k))$ for evolution steps $k \geq 1$. A graph $G(k)$ is produced from $G(k-1)$ by adding a directed edge. Let us denote the number of nodes at step k as $N(k)$, and in- and out-degree of node w in the graph $G(k)$ with k edges as $I_k(w)$ and $O_k(w)$. The edge creation proposed in [1,5] is activated by flipping a 3-sided coin with probabilities α, β and γ such that $\alpha + \beta + \gamma = 1$. The independent identically distributed (i.i.d.) trinomial r.v.s with values 1, 2 and 3 and the corresponding probabilities α, β and γ are generated to select schemes. Let $\Delta_{in}, \Delta_{out}$ be other PA parameters.

- By the α-scheme, one appends to $G(k-1)$ a new node $v \in V(k) \setminus V(k-1)$ and a new edge $(v \rightarrow w)$ to an existing node $w \in V(k-1)$ with probability α. The node w is chosen with probability depending on its in-degree in $G(k-1)$

$$P(choose\ w \in V(k-1)) = \frac{I_{k-1}(w) + \Delta_{in}}{k - 1 + \Delta_{in} N(k-1)}.$$

- By the β-scheme, one adds a new edge $(v \rightarrow w)$ to $E(k-1)$ with probability β, where the existing nodes $v, w \in V(k-1)$ are chosen independently from the nodes of $G(k-1)$ with probabilities

$$P(choose\ (v, w)) = \frac{O_{k-1}(v) + \Delta_{out}}{k - 1 + \Delta_{out} N(k-1)} \cdot \frac{I_{k-1}(w) + \Delta_{in}}{k - 1 + \Delta_{in} N(k-1)}.$$

- By the γ-scheme, one adds a new node $v \in V(k) \setminus V(k-1)$ and an edge $(w \rightarrow v)$ with probability γ. The existing node $w \in V(k-1)$ is chosen with probability

$$P(choose\ w \in V(k-1)) = \frac{O_{k-1}(w) + \Delta_{out}}{k - 1 + \Delta_{out} N(k-1)}.$$

Note that $N(k) = N(k-1)$ for the β-schema and $N(k) = N(k-1) + 1$ for the others. These scenarios realize a 'rich-get-richer' mechanism, when a node with a large degree can likely increase it with a high probability. As mentioned in [5], the PA schemes allow creating multiple edges between two nodes and self loops.

2.2 Information Spreading

Let us describe the idea of the information spreading first by the SPREAD method and then by the PA. We assume that all nodes in the network have asynchronized clocks. Let $k \geq 0$ denote the index of a tick, on which at most one node can receive messages by communicating with another node. To this end, on a clock tick one of n nodes (let say a node i) of the graph is chosen uniformly. Then the node i chooses a node j uniformly among its neighbors with probability $P_{ij} = 1/d_{\max}$, where $d_{\max} = \max_{i \in V} d_i$, d_i is the degree of node i. The usage $P_{ij} = 1/d_i$ as in [8] allows us to avoid the necessity to know the maximal node degree in the network.

In Algorithm 1 by [4], the SPREAD algorithm proposed in [2] for undirected graphs has been modified for directed graphs assuming that a single message of an initial graph G_0 can be spread to a part of the rest nodes. The node i may share the message with the node j if there is a directed edge from j to i. We use $P_{ij} = 1/I_i$, where I_i is a node in-degree in contrast to [4] where out-degree O_i was used instead of I_i. Such a mechanism simulates the message spreading when users search and collect data from pages in the Internet and can share it further with other users through their web-pages. The next node j is proposed to select uniformly among nodes $V \setminus S(k)$ without the message at the clock tick k. $S(k)$ denotes the set of nodes that have the message at the end of the clock tick k.

The linear PA can also be used for spreading using the directed edge $(j \rightarrow i)$ from the new node j to the old node i or between two existing nodes i and j [4]. The edge $(j \rightarrow i)$ can be created by the $\alpha-$ or $\beta-$ schemes.

As in [4] we compare the SPREAD algorithm and the PA schemes by the minimal number of clock ticks or evolution steps required to disseminate the message from G_0 to n nodes with probability not less than $1 - \delta$, namely,

$$K^* = K^*(n, \delta) = inf\{0 < k \leq K' : Pr(\|S(k)\| = n) > 1 - \delta\}, \quad \delta \in (0, 1), (1)$$

where $\|S(k)\|$ is a cardinality of the set. The number of steps is limited by K'. If $K^* \leq K'$ holds, then $S(K^*) = n$ is likely hold for a sufficiently small δ. If $S(K^*) < n$ holds, then K' steps of the evolution are likely not enough to disseminate the message to n nodes.

3   Comparison of the PA Schemes and the SPREAD Algorithm

3.1   Spreading in Simulated Directed Homogeneous Graphs

We compare a spreading ability of the PA and the SPREAD for simulated directed graphs and three deletion strategies. The first strategy 'without node and edge deletion' has been studied in [3,4]. Here, we focus on two strategies of uniform node or edge deletion at each evolution step when a new node is appended.

The conditions of our experiment are the following. We generate 100 graphs starting with an initial graph G_0 by the PA schemes up to a step $K^*(n, \delta) \leq K'$ with $K' = 2.5 \cdot 10^5$ and $\delta = 0.01$ in Eq. (1). We aim to spread a message from G_0 to $n = 100$ other nodes. The probability in Eq. (1) is approximated by a proportion of the event $\{\|S(k)\| = n\}$ for a given k over 100 graphs. A triangle of connected nodes has been used as G_0 for the PA evolution without node and edge deletion in [4]. If we use a triangle as G_0 in the case of the evolution with node deletion at each step, then the evolved graph degenerates and always contains 3 nodes and a random number of edges. This depends on a frequency of using the β−scheme since the node is deleted with its edges. If an edge is deleted during evolution, then the graph may contain several nodes with only 3 edges. This does not allow us to investigate the message propagation properly.

To give more opportunities for the message spreading in case of node or edge deletion, we take G_0 sufficiently large. Namely, G_0 is generated by 10^3 steps of the evolution by the PA schemes with parameters $(\alpha, \beta, \gamma, \Delta_{in}, \Delta_{out}) = (0.4, 0.2, 0.4, 1, 1)$. Starting with G_0, 100 graphs are evolved by the PA for each set of parameters α, β, γ which all are taken in the interval $[0.01, 0.99]$ with step 0.01 as far as $\Delta_{in} = \Delta_{out} = 1$. The comparison of the PA and the SPREAD algorithms is shown in Fig. 1.

The PA schemes spread the information faster for larger values of α, that corresponds to spreading among newly appended nodes mostly, and smaller β; see Fig. 1(top line). In contrast, the SPREAD is faster for smaller α's and smaller β's apart from the case with node deletion; Fig. 1(third line). In the latter case, the SPREAD is faster if there are a few edges directed from the existing nodes to newly appended. Then the risk of deleting an existing node with the message is smaller.

The impact of α and β on the proportion of events $\{S(K') < n\}$, i.e. when the message cannot be delivered to n nodes by K' steps, is shown in Fig. 1(second and fourth lines). The PA and the SPREAD without node and edge deletion and with edge deletion lead to the full spreading of the message among n nodes independently of α and β. The case of the node deletion is different. The PA may spread to all n nodes for any α if β is large enough. For the small β, the message will be delivered to a part of n nodes. Similar conclusions can be done for the SPREAD.

Figure 2 shows the impact of α and β on K^*_{PA}/K^*_{SPREAD}, where K^*_{PA} and K^*_{SPREAD} are the minimum number of steps required for the PA and SPREAD algorithms, respectively, to propagate a single message to $n = 100$ new nodes. The options $\alpha + \beta > 1$ were not considered because $\alpha + \beta + \gamma = 1$ holds. One can see areas where the PA is faster spreader than the SPREAD, i.e. $K^*_{PA} < K^*_{SPREAD}$, and vice versa. Figure 2a and c look similar that is in agreement with Fig. 1.

3.2   Spreading in Non-homogeneous Temporal Graphs

Let us investigate the SPREAD and the PA spreading methods for real temporal graphs. The temporal graphs gathered in [9] have edges with timestamps.

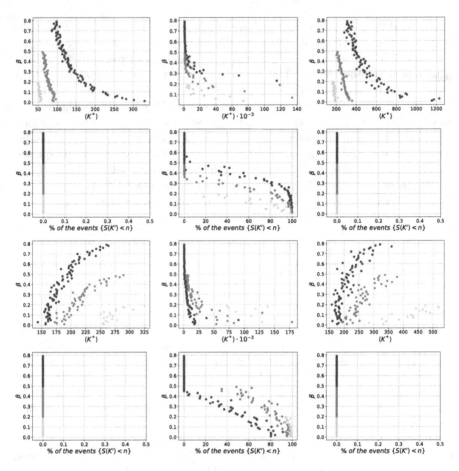

Fig. 1. The PA parameter β against the average number of steps $\langle K^* \rangle$ over 100 graphs and against the proportion of the events $\{S(K') < n\}$ for the PA schemes without node and edge deletion (left column); with node deletion (middle column); with edge deletion (right column) for spreading by the PA schemes (two top lines) and by the SPREAD algorithm (two bottom lines). The dark, light dark and grey points correspond to $\alpha \in \{0.2, 0.5, 0.8\}$, respectively.

A directed edge (u, v, t) implies that a node u sends a message to a node v at the time t. Graphs of messages and comments from websites (sx-mathoverflow, sx-askubuntu, CollegeMsg), graphs of bitcoin transactions (soc-sign-bitcoin-otc, soc-sign-bitcoin-alpha) and graphs of e-mail communication (email-Eu) provide examples of the graphs. Their description can be found in Table 1.

The size of some graphs is large. It makes computationally difficult to model the information spreading by the PA and SPREAD. To overcome the problem, we estimate the parameters of the PA model by the methodology described in [5] and find the corresponding points (α, β) at Fig. 2 to compare the rate of the spreading algorithms by means of simulation results of Sect. 3.1.

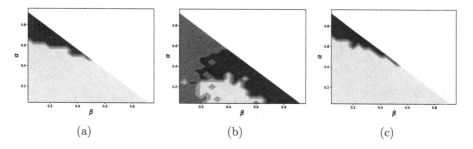

Fig. 2. The dependence of K^*_{PA}/K^*_{SPREAD} averaged over 100 graphs for the PA parameters α and β and the evolution without node and edge deletion (Fig. a), with uniform node deletion (Fig. b) and with uniform edge deletion (Fig. c). The dark area indicates $K^*_{PA} < K^*_{SPREAD}$, the grey area - $K^*_{PA} > K^*_{SPREAD}$. In Fig. b, the light dark area indicates that the message is missed for both PA and SPREAD algorithms.

Table 1. Temporal graphs given in [9] with their description.

Name	Number of nodes, temporal edges	Description
sx-mathoverflow (M-Overflow)	24818, 506550	Comments, questions, and answers on Math Overflow
sx-askubuntu (AskUb)	159316, 964437	Comments, questions, and answers on Ask Ubuntu
email-Eu (EU)	986, 332334	E-mails between users at a research institution
CollegeMsg (ColMsg)	1899, 20296	Messages on a Facebook -like platform at UC-Irvine
soc-sign-bitcoin-otc (Bit-otc)	5881, 35592	Bitcoin OTC web of trust network
soc-sign-bitcoin-alpha (Bit-alpha)	3783, 24186	Bitcoin Alpha web of trust network

The PA was naturally provided to explain conjectured power-law degree distributions in real networks, see [1]. In [1] and [5], the power-law model of their asymptotic in- and out-degree distributions was obtained depending on parameters of the PA model without any node and edge deletion. Assuming as in [5] that the removal of nodes and edges will leave heavy tails of the degree distributions, we study the temporal graphs. To this end, we verify the presence of heavy tails by means of the extreme value index γ (EVI). The positive EVI indicates the heavy tail.

Let $X_{i,n}, 1 \leq i \leq n$ be the associated ascending order statistics of i.i.d. r.v.'s $\{X_n\}_{n>1}$. We apply the mixed moment estimator [10] of the EVI that can be used for real-valued EVIs. The estimator is a bias-reduced in comparison with the Hill's estimator and it may have better performance in terms of the mean squared error (MSE) than a basic moment estimator. The mixed moment estimator is determined by

$$\widehat{\gamma}_n^{MM}(k) = \frac{\widehat{\varphi}_n(k) - 1}{1 + 2\min\left(\widehat{\varphi}_n(k) - 1, 0\right)}, \tag{2}$$

where

$$\widehat{\varphi}_n(k) = \frac{M_n^{(1)}(k) - L_n^{(1)}(k)}{\left(L_n^{(1)}(k)\right)^2},$$

$$L_n^{(1)}(k) = \frac{1}{k}\sum_{i=1}^{k}\left(1 - \frac{X_{n-k,n}}{X_{n-i+1,n}}\right), \quad M_n^{(1)}(k) = \frac{1}{k}\sum_{i=1}^{k}\left(\ln\frac{X_{n-i+1,n}}{X_{n-k,n}}\right).$$

The number of k largest order statistics in Eq. (2) can be found by the bootstrap procedure [11] in the same way as for the Hill's estimator $M_n^{(1)}(k)$.

Table 2. The mixed moment estimates Eq. (2) of the EVI and their bootstrap confidence intervals in brackets for the in-degree $\widehat{\gamma}_n^{ind}(k)$ and out-degree $\widehat{\gamma}_n^{out}(k)$ for the temporal graphs from Table 1.

Name	$\widehat{\gamma}_n^{ind}(k)$	$\widehat{\gamma}_n^{out}(k)$
M-Overflow	0.5021	0.1481
	(0.1341, 0.8098)	(0.0238, 0.744)
AskUb	0.7503	0.9096
	(0.5802, 1.0394)	(0.7395, 1.1942)
EU	0.1455	0.1178
	(0.024, 0.5271)	(0.0428, 0.1408)
ClMsg	0.1613	0.3086
	(0.0251, 0.5034)	(0.1773, 0.6919)
Bit-otc	0.0463	0.5364
	(0.0135, 0.4598)	(0.1993, 0.8355)
Bit-alpha	0.2842	0.198
	(0.0153, 0.3396)	(0.0243, 0.2085)

The mixed moment estimates of the in- and out-degrees of each temporal graph are shown in Table 2. One can see that the EVI estimates have positive values for all graphs both for the in- and out-degrees. It means that degree distributions are heavy-tailed and the PA model may describe the evolutionary behavior of the temporal graphs.

For each temporal graph, the PA parameters are evaluated by means of the Snapshot method [5]. The estimates are denoted as $(\widehat{\alpha}^{SN}, \widehat{\beta}^{SN}, \widehat{\gamma}^{SN}, \widehat{\Delta}_{in}^{SN}, \widehat{\Delta}_{out}^{SN})$. The spreading rate of both the SPREAD and PA algorithms is compared for different values of the latter parameters. The results are presented in Fig. 3, where $\widehat{\Delta}_{in}^{SN}$ and $\widehat{\Delta}_{out}^{SN}$ were found to be close to 1 for all graphs. Figure 3 depicts Fig. 2 by uncolored areas: note I indicates an area where the SPREAD delivers message faster, note II - the PA delivers message faster, note III signs area where the message was missed with both algorithms. Each graph is depicted by

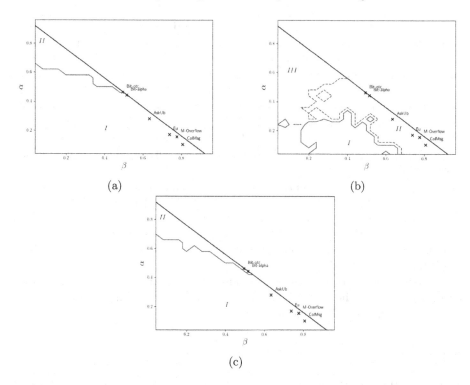

Fig. 3. K_{PA}^*/K_{SPREAD}^* for real graphs from Table 1 corresponding to the Snapshot estimates $\widehat{\alpha}^{SN}$ and $\widehat{\beta}^{SN}$. The evolution without node and edge deletion, with the uniform node deletion and the uniform edge deletion are shown in Fig. a, b and c, respectively. The areas denoted as *I, II, III* where $K_{PA}^*/K_{SPREAD}^* > 1$, $K_{PA}^*/K_{SPREAD}^* < 1$ hold or the message is lost, respectively, correspond to the colored areas in Fig. 2.

the point at some area corresponding to the values $(\widehat{\alpha}^{SN}, \widehat{\beta}^{SN})$. The condition $\widehat{\alpha}^{SN} + \widehat{\beta}^{SN} + \widehat{\gamma}^{SN} = 1$ is fulfilled for the area under the diagonal line.

One may conclude the following from Fig. 3. For the evolution without node and edge deletion in Fig. 3a all graphs apart of the Bit-otc fall in the area *I* that means that the SPREAD delivers the message faster than the PA. All considered graphs fall in the area *II* for the evolution with the uniform node deletion and the PA is faster, Fig. 3b. For the evolution with the uniform edde deletion, all graphs fall in the area *I* apart of the bitcoin ones that fall in the area *II* that implies mostly the superiority of the SPREAD, Fig. 3c. No one graph falls in the area *III* that means that the message cannot be lost.

4   Conclusions

We study the linear PA schemes to spread one message from an initial graph to n rest nodes of the network. The spreading is investigated both for homogeneous simulated (Sect. 3.1) and non-homogeneous temporal (Sect. 3.2) graphs with uniform node or edge deletion and without any deletion. We compare the PA and the well-known SPREAD algorithm on directed graphs with possible cycles and multiple edges generated by the PA with different sets of its parameters. We assume that the message may be transmitted only from an existing node with the message to a new appending node or between two existing nodes if one of them has the message but another one not. These two cases correspond to the PA $\alpha-$ and $\beta-$ schemes. The PA $\gamma-$ scheme cannot lead to the spreading since it corresponds to the edge directed from a new node without the message to the existing node.

Considering the simulated graphs, one may conclude that the PA $\alpha-$ and $\beta-$ schemes may be the faster spreader than the SPREAD algorithm for large values of α and for the evolution without node and edge deletion or edge deletion. The PA may be faster than the SPREAD for smaller values of α in the case of the node deletion. For node deletion, the message may be lost, and both spreading algorithms will not be effective.

For non-homogeneous graphs and parameters $(\Delta_{in}, \Delta_{out})$ taken equal to one, the PA may spread the message faster in all investigated temporal graphs only for the evolution with the uniform node deletion.

References

1. Bollobás B., Borgs C., Chayes J., Riordan O.: Directed scale-free graphs. In: Society for Industrial and Applied Mathematics, SODA 2003, USA, pp. 132–139 (2003)
2. Mosk-Aoyama, D., Shah, D.: Computing separable functions via gossip. In: Proceedings of the 25th ACM Symposium on Principles of Distributed Computing, PODC 2006, pp. 113–122. ACM, New York (2006)
3. Markovich, N.M., Ryzhov, M.S.: Information spreading with application to non-homogeneous evolving networks. In: CCIS, vol. 1552, pp. 284–292 (2022)
4. Markovich, N.M., Ryzhov, M.S.: Information spreading and evolution of non-homogeneous networks. Adv. Syst. Sci. Appl. **2**, 21–33 (2022)
5. Wan, P., Wang, T., Davis, R.A., Resnick, S.I.: Are extreme value estimation methods useful for network data? Extremes **23**, 171–195 (2020)
6. da Cruz, J.P., Lind, P.G.: The bounds of heavy-tailed return distributions in evolving complex networks. Phys. Lett. A **377**, 189–194 (2013)
7. Markovich, N.M., Ryzhov, M.S., Vaičiulis, M.: Tail index estimation of PageRanks in evolving random graphs. Mathematics **10**(16), 3026 (2022)
8. Censor-Hillel, K., Shachnai, H.: Partial information spreading with application to distributed maximum coverage. In: Proceedings of the 29th ACM Symposium on Principles of Distributed Computing, PODC 2010, pp. 161–170. ACM, New York (2010)
9. Leskovec J., Krevl A., SNAP Datasets: Stanford Large Network Dataset Collection (2014). http://snap.stanford.edu/data
10. Fraga Alves, M.I., Gomes, M.I., de Haan, L., et al.: Mixed moment estimator and location invariant alternatives. Extremes **12**, 149–185 (2009)
11. Markovich, N.: Nonparametric Analysis of Univariate Heavy-Tailed Data: Research and Practice. Wiley, Hoboken (2007)

Comparative Analysis of a Resource Loss System with the Finite Buffer and Different Service Disciplines

V. A. Stepanov[1] , A. V. Daraseliya[1]([⊠]) , E. S. Sopin[1,2] ,
and S. Ya. Shorgin[1,2]

[1] Peoples' Friendship University of Russia (RUDN University), Moscow, Russian
Federation
{daraselia-av,sopin-es}@rudn.ru
[2] Institute of Informatics Problems, Federal Research Center Computer Science and
Control of Russian Academy of Sciences, Moscow, Russian Federation
sshorgin@ipiran.ru

Abstract. Fifth-generation (5G) networks are expected to revolutionize wireless communication by enabling faster and more reliable data
transfer rates. However, the use of millimeter-wave frequencies in 5G
networks introduces unique challenges in data transmission due to phenomena such as dynamic signal blockage, which can result in sudden and
intense resource demands for customers. This paper presents a mathematical model of a 5G base station using a resource allocation system
with waiting and non-homogeneous resource requirements for customers.
The study aims to determine the most efficient way to select customers
from the waiting buffer to increase system performance and minimize the
probability of blocking data customers and reduce waiting time in the
waiting buffer. As a result of the research, it was found, that the service
discipline that prioritized customers with the highest resource requirements and took them from the waiting buffer until the system reached its
limit, demonstrates the best system performance under certain parameters. In particular, in this paper, an analytical model was obtained for one
of the service disciplines with the calculation of a system of equilibrium
equations and performance measures.

Keywords: 5G networks · Millimeter-wave frequencies · Dynamic
signal blockage · Data session continuity · Resource allocation ·
Queuing theory

1 Introduction

The emergence of 5G technology has brought about significant advancements
in the field of data transmission. With its high speed, low latency, and massive

The research was funded by the Russian Science Foundation, project no.22-79-10128,
https://rscf.ru/en/project/22-79-10128/.

capacity, 5G promises to revolutionize the way we communicate and consume data. However, the implementation of 5G also poses unique challenges that must be addressed for the technology to reach its full potential. One of these challenges is the dynamic blocking of signals, which causes a sudden increase in resource demands for a particular session, leading to resource congestion and blocking of other sessions.

To address this challenge, several techniques have been proposed, including the use of Channel Quality Indicator (CQI) and Modulation and Coding Scheme (MCS) in 5G networks [1]. CQI and MCS are used to optimize data transmission by selecting the appropriate modulation and coding scheme based on the channel conditions. This optimization can significantly improve the efficiency of data transmission and reduce the likelihood of blocking.

In several studies [2–5], resource loss systems (LS) with finite resources and random requirements have been used as one of the tools for modeling 5G cellular networks. The main advantage of using these systems in 5G network modeling is due to their ability to describe important features of session serving process. Integration of resource LS with stochastic geometry models is common practice to effectively capture the stochastic characteristics arising from user locations relative to base stations (BS) and the dynamic traffic service at BS. Recent studies have leveraged these models to explore diverse session continuity mechanisms in 5G NR systems, encompassing multiconnectivity [6,7], resource reservation [5], and their combined functionality [8]. However, the authors did not take into account the possibility of storing customers in the waiting buffer. Our previous research [9] has been conducted on the analysis of resource LSs with waiting buffer, for which the equilibrium equations for the stationary probabilities of the system, as well as, the loss probability of the system, average waiting time, average number of customers, and average resource requirements of blocked customers were analytically derived.

In the current paper, we extend this work [9] by considering the specific case of 5G networks and analyzing different methods for selecting customers from the waiting buffer to minimize the probability of blocking and waiting time. The simulation model is implemented in Python using methods from [10], and the parameters of the system are defined by the number of servers, the number of available resource units, and the waiting buffer size. We compare six different methods of selecting customers from the waiting buffer and evaluate their performance based on several metrics.

2   Model and Service disciplines Description

We consider a resource LS with finite resources R, limited number of servers N and waiting buffer size V, see Fig. 1. Customers arrive according to the Poisson law with rate λ. Serving process of each customer requires a server and a random number r_i of resources, $1 \leq i \leq N$, $0 < r_i \leq R$. If there are not available servers in the system, the customers are placed into a waiting buffer. In case there is space available in the waiting buffer, the customers are held until the system has sufficient available resources and available servers to handle them. Upon entering service, customers are serviced with a service rate μ, and once their service is complete, they exit the system. Customers that cannot be accepted into the waiting buffer due to lack of buffer space are lost from the system.

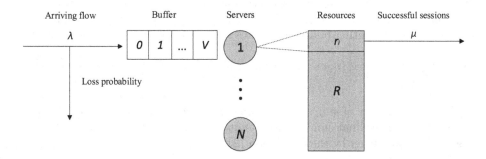

Fig. 1. Resource LS with the waiting buffer

The performance of the system is evaluated in terms of various metrics, including the blocking probability, the average number of customers in service, the average number of occupied resources, the average number of customers in the buffer, and the average waiting time in the buffer. These metrics of interest are calculated based on the chosen method of selecting customers from the buffer for service, i.e. service discipline.

Additionally, six different service disciplines for selecting customers from the buffer for service are studied:

- *Discipline 1*: service is provided to customers in the buffer until the system runs out of resources or the buffer is full. The number of customers selected from the buffer for service can vary from one to all, and the arrival of customers follows the First-Come-First-Served (FCFS) principle.;
- *Discipline 2*: customers are sorted in ascending order by resource requirement, and the ones with the lowest requirement are serviced first until the system runs out of resources or the buffer is full;
- *Discipline 3*: customers are sorted in descending order by resource requirement, and the ones with the highest requirement are serviced first until the system runs out of resources or the buffer is full;
- *Discipline 4*: one random customer is selected from the buffer for service. However, if there is enough space in the system for multiple customers, only one randomly chosen customer enters the system [9];
- *Discipline 5*: several random customers are selected from the buffer until the system runs out of resources or the buffer is full;
- *Discipline 6*: one customer that is first in the buffer is selected from the buffer. The discipline follows FCFS principle as well.

During the system analysis, we store the size of the number of occupied resources for each customer in the system in a vector of size N. However, only the aggregate value for the entire system is used in the analysis. Moreover, the time a customer spends in service is recorded when it is accepted into the system.

The behavior of the system can be described as a random process $X(t) = (\xi(t), \delta(t), \theta(t))$, where $\xi(t)$ represents the number of devices in the system, $\delta(t)$ represents the number of occupied resources and $\theta(t)$ represents the number of customers in the waiting buffer. The state space is given by the following form:

$$S = \bigcup_{0 \leq n \leq N} S_n, S_n = \left\{ (n, m, r) : 0 \leq m \leq M, 0 \leq n \leq N, 0 \leq r \leq R, p_m^{(n)} > 0 \right\}. \tag{1}$$

Let's sort the set of states by the number of occupied resources and denote $I(n, m, r)$ as the ordinal number of the state (n, m). The steady-state probabilities of $X(t)$ are given by the expression (2).

$$q_n(m) = \lim_{t \to \infty} P\{\xi(t) = n, \delta(t) = m, \theta(t) = r, (n, m) \in S_n\}. \tag{2}$$

3   The System of Equilibrium Equations and Metrics of Interest for One of the Service Disciplines

In [9] we have performed analytical calculations for service discipline with the choice of one random customer from the waiting buffer.

In this mathematical model, we assume that the distribution of the resource requirements of customers in the buffer is equal to the original one, although this is not so. The fact is that the first customers entering the buffer that was empty before, additional conditions are imposed on it, that cu did not fit into the system, because of this, the distribution is different. These assumptions do not lead to significant measures of inaccuracy in Sect. 4.2.

In this section we have complicated the mathematical model by considering the case for the 1st service discipline in which one or several customers are selected from the buffer.

3.1   System Analysis

Let $\phi_{m,j}(k, s, l), l \geq 0$ be the probability that exactly m customers are accepted for service with a total requirement of j resource units having k free servers, s free resource units in the system and l customers in the waiting buffer, $m \leq l, m \leq k, j \leq s$.

Lemma 1. $\phi_{m,j}(k, s, l)$ is given by

$$\phi_{m,j}(k, s, l) = p_j^{(m)} \left(1 - \sum_{i=0}^{s-j} p_i \right), \tag{3}$$

where $l \geq 0, m \leq l, m \leq k, j \leq s$.

Let us denote the stationary probability that there are n customers on the servers that totally occupy r resources and k customers in the waiting buffer by $P_{n,k}(r)$. Then the system of equilibrium equations for the simplified model can be written as follows:

$$\lambda P_{0,0}(0) = \mu \sum_{j=1}^{R} P_{1,0}(j), \tag{4}$$

$$(\lambda + n\mu)\, P_{n,0}(r) = \lambda \sum_{j=0}^{r-1} P_{n-1,0}(j) p_{r-j} \tag{5}$$

$$+ (n+1)\mu \sum_{j=r+1}^{R} P_{n+1,0}(j) \frac{p_{j-r} p_r^{(n)}}{p_j^{(n+1)}}$$

$$+ \sum_{m=1}^{\min(V,n)} (n-m+1)\mu \sum_{j=n-m}^{R} P_{n-m+1,m}(j) \sum_{i=1}^{j-n+1} \frac{p_i p_{j-i}^{(n-m)}}{p_j^{(n-m+1)}}$$

$$\times \phi_{m,r-j+i}(N-n+m, R-j+i, m), \quad n \in [1, N),$$

$$(\lambda + N\mu) P_{N,0}(r) = \lambda \sum_{j=N-1}^{r-1} P_{N-1,0}(j) p_{r-j} \tag{6}$$

$$+ \sum_{m=1}^{\min(V,N)} (N-m+1)\mu \sum_{j=N-m}^{R} P_{N-m+1,m}(j) \sum_{i=1}^{j-N+1} \frac{p_i p_{j-i}^{(N-m)}}{p_j^{(N-m+1)}}$$

$$\times \phi_{m,r-j+i}(m, R-j+i, m),$$

$$(\lambda + n\mu)\, P_{n,1}(r) = \lambda \sum_{j=R-r+1}^{R} P_{n,0}(r) p_j \tag{7}$$

$$+ (n+1)\mu \sum_{j=r+1}^{R} P_{n+1,1}(j) \frac{p_{j-r} p_r^{(n)}}{p_j^{(n+1)}} \phi_{0,0}(N-n, R-r, 1)$$

$$+ \sum_{m=1}^{\min(n,V-1)} (n-m+1)\mu \sum_{j=n}^{R} P_{n-m+1,m+1}(j) \sum_{i=1}^{j-n+1} \frac{p_i p_{j-i}^{(n-m)}}{p_j^{(n-m+1)}}$$

$$\times \phi_{m,r-j+i}(N-n+m, R-j+i, m+1), \quad n \in [1, N),$$

$$(\lambda + n\mu) P_{n,k}(r) = \lambda P_{n,k-1}(r) \tag{8}$$

$$+ (n+1)\mu \sum_{j=r+1}^{R} P_{n+1,k}(j) \frac{p_{j-r} p_r^{(n)}}{p_j^{(n+1)}} \phi_{0,0}(N-n, R-r, k)$$

$$+ \sum_{m=1}^{\min(n,V-k)} (n-m+1)\mu \sum_{j=n}^{R} P_{n-m+1,k+m}(j) \sum_{i=1}^{j-n+1} \frac{p_i p_{j-i}^{(n-m)}}{p_j^{(n-m+1)}}$$

$$\times \phi_{m,r-j+i}(N-n+m, R-j+i, m+k), \quad n \in [1, N], k \in [2, V],$$

$$\left(\lambda \sum_{j=1}^{R-r} p_j + n\mu\right) P_{n,V}(r) = \lambda P_{n,V-1}(r) \tag{9}$$

$$+ (n+1)\mu \sum_{j=r+1}^{R} P_{n+1,V}(j) \frac{p_{j-r} p_r^{(n)}}{p_j^{(n+1)}} \phi_{0,0}(N-n, R-r, V), \quad n \in [1, N],$$

$$(\lambda + N\mu) P_{N,k}(r) = \lambda P_{N,k-1}(r) \tag{10}$$

$$+ \sum_{m=1}^{\min(N,V-k)} (N-m+1)\mu \sum_{j=N}^{R} P_{N-m+1,k+m}(j) \sum_{i=1}^{j-N+1} \frac{p_i p_{j-i}^{(N-m)}}{p_j^{(N-m+1)}}$$

$$\times \phi_{m,r-j+i}(m, R-j+i, m+k), \quad k \in [1, V],$$

$$N\mu P_{N,V}(r) = \lambda P_{N,V-1}(r), \tag{11}$$

where $r \in [1, R]$.

3.2   Performance Metrics of Interest

Now, we proceed with performance metrics of interest. With the stationary distribution, one can obtain the blocking probability π.

$$\pi = \sum_{r=N}^{R} P_{N,V}(r) + \sum_{n=1}^{N-1} \sum_{r=n}^{R} P_{n,V}(r) \sum_{i=R-r+1}^{R} p_i. \tag{12}$$

The average number $\bar{N}$ of customers in the waiting buffer is

$$\bar{N} = \sum_{n=1}^{N} \sum_{r=n}^{R} \sum_{k=1}^{V} k P_{n,k}(r). \tag{13}$$

The average waiting time W takes the following form:

$$W = \frac{\bar{N}}{\lambda(1-\pi)}. \tag{14}$$

4   Numerical Results

4.1   Comparison of Service Disciplines

For numerical analysis we consider a system with $N = 100$ servers, and $R = 100$ resource units, $V = 50$ waiting buffer size. The service time for each customer is exponentially distributed with $\mu = 1$. The system is modeled using two different resource requirement distributions: the geometric distribution with parameter values of $p = 0.7$ and the Poisson distribution with parameter 4.

The results of the simulation are presented in tabular form for ease of comparison between the different selection methods. Overall, the performance of the system is heavily dependent on the resource requirement distribution and the method used to select customers for service.

We start with Fig. 2a and 2b illustrating the effect of the average number of occupied resourses on the arrival rate for the geometric and Poisson distributions, respectively. It is evident that under low load conditions, the methods exhibit approximately similar values. However, as the arrival rate increases, it becomes apparent that the method that prioritizes the most resource-demanding customers from the waiting buffer demonstrates the highest average number of occupied resources in the system. This dependency is observed for both the geometric and Poisson distributions, with the only distinction being that the graphs for the Poisson distribution start to noticeably increase earlier.

Figure 3a and 3b present the dependence of the average waiting time for customers on the arrival rate. At low loads, a significant gain in time is observed for the discipline 2, and discipline 3, on the contrary, shows the worst values. In the case of the geometric distribution, at $\lambda = 28$, the waiting time is 69% less compared to other waiting buffer scheduling methods. At the incoming flow rate of $\lambda = 31$, the time difference is slightly less and amounts to about 67%. With further increase in the load, the situation is completely changing and the average waiting time of discipline 3 relative to other disciplines decreases. For the Poisson distribution, the system behaves similarly, however with a smaller time advantage of discipline 3.

Figure 4 shows the blocking probability for the geometric distribution. At low loads, the probability is zero because the number of occupied servers in the system has not yet reached its maximum, as can be seen in Fig. 3a and 3b. With an increase in λ, the blocking probability also increases, which is due to the fact that a larger number of customers enter the system, occupying resources and space in the waiting buffer.

4.2   Comparison of the Average Resource Requirement in the Waiting Buffer

To check the assumption that the distribution of resource requirements is close to the original distribution, we plotted Fig. 5 which illustrates the relationship between the average resource requirements in buffer and the arrival rate for five different service disciplines. It is evident that the discipline 2 exhibits the highest

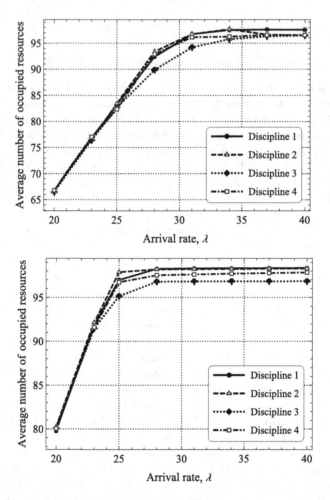

Fig. 2. Average number of occupied resources. a) Geometric distribution b) Poisson distribution

value, where customers are sorted in ascending order and accepted into the system as long as there is available resources and devices. This can be attributed to the fact that the system initially receives customers with the lowest resource requirements, while more demanding customers remain in the buffer.

Discipline 1, which takes customers from the buffer without sorting as long as there is available resources, and discipline 3 in which customers are sorted in descending order demonstrate the lowest resource requirement in the buffer,

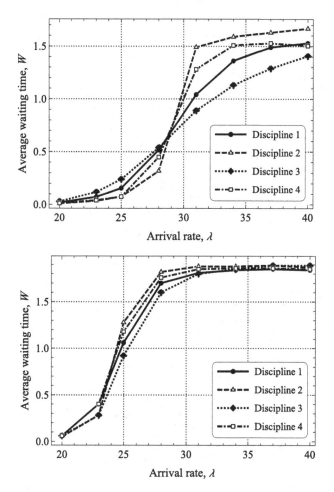

Fig. 3. Average waiting time. a) Geometric distribution b) Poisson distribution

moreover, unlike other disciplines, it decreases with increasing load. This can be explained by the fact that in discipline 1, only those customers that enter the buffer at times when it is empty will have a distribution of resource requirements different from the original one. Because if the buffer is empty, then the customer is buffered only if it has too high a demand, due to which it cannot immediately go for service. And if a customer arrives into the system when there is already something in the buffer, then regardless of its requirement, it gets into the buffer.

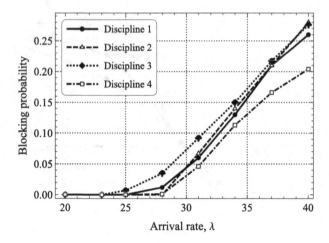

Fig. 4. Blocking probability. Geometric distribution

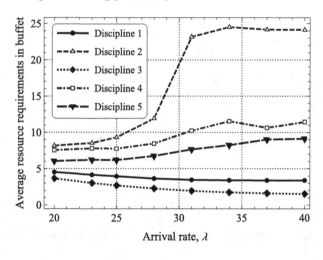

Fig. 5. The average resource requirements in the buffer. Geometric distribution

For such customers, the resource requirement has exactly the same distribution as the original one. Therefore, at "light" loads, when the buffer is often empty, the proportion of first customers is higher, so it increases the average resource requirement of customers from the buffer. And under heavy loads, the buffer is almost always non-empty, so there are almost no such customers. Thus, as the load increases, the average requirement for the resource of customers from the buffer asymptotically tends to the original one. Note that the average resource requirement for discipline 3 showed the results closest to the original distribution.

Disciplines 4 and 5 which accepts one or several random customers from the buffer, showed closer values. In discipline 5, customers follow the same principle as in discipline 1, but due to a different way of selecting customers from the buffer, the behavior of the system changes greatly. In discipline 1, for example, there is a buffer for servicing a "heavy" customer. It may not get there right away, it may have to wait until several customers leave the service, but with each end of the service of the next customer, the free resource in the system becomes more and more, and, finally, it will also get serviced. Thus, she can wait for some time until enough resources are freed, but all other, "light" customers will also wait with her. And in discipline 5, such a "heavy" customer with some probability will be ahead of other, "lighter" customers. As a result, the number of free resource of the system may not decrease for a very long time to the values at which this "heavy" customer can be serviced. It turns out that discipline 5, implicitly, is still more likely to miss "light" customers for servicing and delay heavy ones. And this effect increases with increasing load, as the choice of customers in the buffer becomes larger.

5   Conclusion

In our paper we consider a model of a multi-server loss system with the multi-type of resources and waiting buffer and investigated the system performance indicators for different service disciplines. The research revealed that, under specific parameters, the method that prioritized customers with the highest resource requirements and served them until the system reached its capacity, demonstrated the most optimal system performance. The results of this study can be used in the design of 5G networks.

A   Appendix

In this Appendix, we provided the proof of our Lemma introduced in Subsect. 3.1.

Proof of Lemma 1.

Let ξ_i be the requirement of the i-th customer to the resource, then

$$\phi_{m,j}(k, s, l) = P\{\xi_1 + \xi_2 + ... + \xi_m = j, \xi_1 + ... + \xi_m + \xi_{m+1} > s\} \qquad (15)$$

$$= P\{\xi_1 + ... + \xi_m = j\}P\{\xi_{m+1} > s - j\} = p_j^{(m)}\left(1 - \sum_{i=0}^{s-j} p_i\right),$$

where $l \geq 0, m \leq l, m \leq k, j \leq s$. Consider 2 cases of intervals, $k \geq l > 0$ and $l > k > 0$, respectively.

1. If $k \geq l > 0$, then
   (a) $m = 0, j = 0$

$$\phi_{0,0}(k, s, l) = \left(1 - \sum_{i=0}^{s} p_i\right); \qquad (16)$$

(b) $m < l \leq k$

$$\phi_{m,j}(k,s,l) = p_j^{(m)}\left(1 - \sum_{i=0}^{s-j} p_i\right);$$
(17)

(c) $m = l \leq k$

$$\phi_{m,j}(k,s,l) = p_j^{(l)}.$$
(18)

To prove the veracity of the expression (15), we sum the obtained probabilities (16)–(18) as

$$S = p_0^{(0)}\left(1 - \sum_{i=0}^{s} p_i\right) + \sum_{m=1}^{l-1}\sum_{j=0}^{s} p_j^{(m)}\left(1 - \sum_{i=0}^{s-j} p_i\right) + \sum_{j=0}^{s} p_j^{(l)}.$$
(19)

To further simplify the sum (19), we divide it into 3 parts, which we will separately transform:

$$S_1 = p_0^{(0)}\left(1 - \sum_{i=0}^{s} p_i\right) = \left(1 - \sum_{i=0}^{s} p_i\right);$$
(20)

$$S_2 = \sum_{m=1}^{l-1}\sum_{j=0}^{s} p_j^{(m)}\left(1 - \sum_{i=0}^{s-j} p_i\right)$$
(21)

$$= \sum_{m=1}^{l-1}\sum_{j=0}^{s} p_j^{(m)} - \sum_{m=1}^{l-1}\sum_{j=0}^{s} p_j^{(m)}\sum_{i=0}^{s-j} p_i$$

$$= \sum_{m=1}^{l-1}\sum_{j=0}^{s} p_j^{(m)} - \sum_{m=1}^{l-1}\sum_{j=0}^{s} p_j^{(m+1)}.$$

Then finally substituting the simplified sums (20) and (21) in (19) we obtained that the normalizing condition for the probabilities is satisfied.

$$S = \sum_{m=0}^{l}\sum_{j=0}^{s} p_j^{(m)} - \sum_{m=0}^{l-1}\sum_{j=0}^{s} p_j^{(m+1)}$$
(22)

$$= \sum_{m=0}^{l}\sum_{j=0}^{s} p_j^{(m)} - \sum_{m=1}^{l}\sum_{j=0}^{s} p_j^{(m+1)} = \sum_{j=0}^{s} p_j^{(0)} = 1.$$

2. If $l > k > 0$, then similar for the calculation for the interval above

(a) $l > k > 0$

$$\phi_{0,0}(k,s,l) = \left(1 - \sum_{i=0}^{s} p_i\right);\qquad(23)$$

(b) $m < k \le l$

$$\phi_{m,j}(k,s,l) = p_j^{(m)}\left(1 - \sum_{i=0}^{s-j} p_i\right);\qquad(24)$$

(c) $m = k \le l$

$$\phi_{m,j}(k,s,l) = p_j^{(l)}.\qquad(25)$$

By analogy with formulas (16)–(18), proved in the previous paragraph, the sum of the probabilities (23)–(25) is also 1, which concludes the proof.

References

1. 3GPP, TR 38.901, vol 16.1.0. Evolved Universal Terrestrial Radio Access (E-UTRA); Carrier Aggregation; Base Station (BS) radio transmission and reception (2020)
2. Naumov, V., Beschastnyi, V., Ostrikova, D., Gaidamaka, Y.: 5G new radio system performance analysis using limited resource queuing systems with varying requirements. In: Vishnevskiy, V.M., Samouylov, K.E., Kozyrev, D.V. (eds.) DCCN 2019. LNCS, vol. 11965, pp. 3–14. Springer, Cham (2019). https://doi.org/10.1007/978-3-030-36614-8_1
3. Lu, X., et al.: Integrated use of licensed- and unlicensed-band mmWave radio technology in 5G and beyond. IEEE Access 7, 24376–24391 (2019)
4. Daraseliya, A.V., et al.: Analysis of 5G NR base stations offloading by means of NR-U technology. Inf. Appl. **15**(3), 98–111 (2021)
5. Begishev, V., Moltchanov, D., et al.: Quantifying the impact of guard capacity on session continuity in 3GPP new radio systems. IEEE Trans. Veh. Technol. **68**(12), 12345–12359 (2019)
6. Petrov, V., et al.: Achieving end-to-end reliability of mission-critical traffic in softwarized 5G networks. IEEE J. Sel. Areas Commun. **36**(3), 485–501 (2018)
7. Petrov, V., et al.: Dynamic multi-connectivity performance in ultra-dense urban mmWave deployments. IEEE J. Sel. Areas Commun. **35**(9), 2038–2055 (2017)
8. Kovalchukov, R., Moltchanov, D., et al.: Improved session continuity in 5G NR with joint use of multi-connectivity and guard bandwidth. In: 2018 IEEE Global Communications Conference (GLOBECOM), pp. 1–7. IEEE (2018)
9. Daraseliya, A., Sopin, E.S., Shorgin, S.Y., On approximation of the time-probabilistic measures of a resource loss system with the waiting buffer. In: Vishnevskiy, V.M., Samouylov, K.E., Kozyrev, D.V. (eds.) DCCN 2022, CCIS, vol. 1748, pp. 282–295. Springer, Cham (2023). https://doi.org/10.1007/978-3-031-30648-8_23
10. Buslenko N.P.: Simulation of complex systems [Modelirovaniye slojnih sistem], 2 edn. Revised (1978)

Batch Service Polling System: Mathematical Analysis and Simulation Modeling

Vladimir Vishnevsky[1] , Olga Semenova[1] , Van Hieu Nguyen[2(✉)] ,
and Minh Cong Dang[2,3]

[1] Institute of Control Sciences of Russian Academy of Sciences, Profsoyuznaya Street
65, Moscow 117997, Russia
[2] Moscow Institute of Physics and Technology, Institutskiy per. 9, Dolgoprudny,
Moscow Region 141701, Russia
{hieu.nguyen,dang.mk}@phystech.edu
[3] University of Engineering and Technology, Vietnam National University Hanoi,
Xuan Thuy Street 144 E3, Hanoi, Vietnam

Abstract. In this paper, we study the mathematical and simulation
models of the polling system with cyclic polling order and batch service.
The service discipline is considered to be either exhaustive or gated. We
study mathematical analysis of system with limited queue capacities and
exhaustive discipline to measure the performance parameters. We discuss
the numerical simulation results of systems with different batch size scenarios to examine the correlation between the system characteristics and
the batch sizes of the queues.

Keywords: Polling system · Batch service · Exhaustive service
discipline · Gated service discipline · Cyclic polling order

1 Introduction

Over the years, the polling system has been widely used to model and evaluate
the performance of broadband wireless networks, such as Wi-Fi and Wi-Max.
In queuing theory, a polling system is a system consisting of N queues with a
common server for all of them [7].

The server chooses a queue to serve based on the *polling order*, with the
most common one being cyclic order: the server serves from Q_1 to Q_N and then
returns to the first queue. The time the server takes to service from the first
queue to the last one is called a *cycle*. At any given time, only one queue is
connected to the server for service.

Usually, in a polling system, the server can only know the information about
the number of customers in a queue at the time the server finishes connecting to

The reported study was funded by the Russian Science Foundation within scientific
project No. 22-49-02023 "Development and study of methods for reliability enhancement of tethered high-altitude unmanned telecommunication platforms".

the queue (*polling moment*). After connecting to a queue, the server will serve that queue based on the *service discipline*, which is the number of customers served in the queue by the server in one polling.

Although the polling system has been studied extensively [1,6] in the last decades, these papers only considered the case where the server serves customers individually, while the polling system with batch service is much less explored. Boxma et al. [2] and Wal et al. [9] considered the case where the server serves the queue as a single batch, and the batch size is the number of customers present in the queue at the polling moment. Vlasiou and Yechieli [8] studied a system in which, instead of one server serving a queue, there is an infinite number of servers simultaneously connecting in groups to the queue for service. The time that the group of servers visits the queue is independently distributed and varies between queues. Van Oyen [5] studied optimal scheduling for cyclic systems, specifically whether the server waits for more customers to serve in a batch or not at the polling moment. Xia et al. [10] considered the problem of dynamic allocation of a single server with finite batch processing capability to a set of parallel queues. The customer arrival flows are independent Poisson processes with equal arrival rates. The batch service times are exponentially distributed and identical. The server has zero switchover time from one queue to another. Also, for the system with homogeneous arrival processes and zero switchover time, Liu and Nain [4] considered optimal dynamic routing policies when the server has complete freedom of visits. The service discipline is either exhaustive or locally gated. Dorsman et al. [3] studied a system including the inner part and outer part. Upon entering the system, customers need to wait in the outer part until there are enough customers for one batch, and then they will be transferred to the inner part which functions as an ordinary polling system.

Compared with previous models, in this paper, we consider a system in which each queue has an individual number k_i that characterizes the batch size of that queue. The purpose of this paper is to provide a mathematical model of this system and a mathematical analysis for the case with limited waiting space in queues and exhaustive service discipline. Our purpose is also to build a simulation model to study the main system's performance characteristics. Based on the simulation results, we will explore the correlation between the system characteristics and the batch size of the queue.

The structure of this paper is as follows: Sect. 1 presents the problem statement. In Sect. 2, we describe the details of the mathematical model. In Sect. 3, we analyze the system with limited queue waiting space and exhaustive service discipline, calculate its final stationary probabilities and performance characteristics. In Sect. 4, we discuss the structure of the system's simulation model. Section 5 presents the numerical results of the simulation model and provides the analysis comparison.

2   Mathematical Model

The system consists of a single server and $N(N \geq 1)$ queues with the queue waiting space (*queue capacity*) $N_i = \overline{1, \infty}$, $i = \overline{1, N}$. The arrival process of the queue Q_i is an independent Poisson process with parameter λ_i. The system's polling order is *cyclic*, which means that in each working cycle, the server visits each queue from Q_1 to Q_N to serve the customers. The switchover times to the queues are independent and identically distributed with the distribution function $S_i(t)$ and the mean $s_i = \int_0^\infty t dS_i(t)$ for the queue Q_i. Each queue has a parameter k_i that characterizes its batch size, which is the maximum number of customers the server can take simultaneously for service when it stays in that queue. The service times of batches in the queue Q_i are independent and identically distributed with the distribution function $B_i(t)$ and the mean $b_i = \int_0^\infty t dB_i(t)$. The service discipline, in this case, can be *exhaustive*, meaning the server leaves the queue only when there are no more customers there, or *gated*, where the server only serves customers present in the queue at the polling moment, and customers arriving later will be served in the next cycle.

Let's look at an example of how exhaustive batch service works. Suppose the server has finished connecting to the queue Q_i with parameter $k_i = 3$. Assume that there are 5 customers in the queue. The server will serve the first batch of 3 customers. If, during that time, no more customers enter this queue, after serving the first batch, the server will serve the second batch of only 2 customers. After that, the server checks if there are any customers in the queue. Suppose during the service time of the second batch, a customer enters the queue; then the server will continue to serve the third batch of only 1 customer and check if the queue is empty or not. Note that the service times of all 3 batches are independent, sharing the same distribution function $B_i(t)$ and the mean b_i, independent of the batch size. If the queue is empty, the server switches to the next queue.

The gated service case is similar, with the only difference being that customers who arrive after the polling moment will be served in the next cycle.

It is easy to see that if $k_i = 1$, this batch service polling model becomes a regular polling system with cyclic order and exhaustive or gated service, which has been studied intensively in [7].

3   Mathematical Analysis of System with Limited Queue Capacities and Exhaustive Service Discipline

In this section, we consider the case of an exhaustive service discipline and derive the balance equations for the steady-state probabilities of the system. To ensure computability, we assume that each queue has a finite capacity, denoted by N_i, corresponding to queue Q_i. We assume that the system is in a steady state, and the probabilities of the system parameters are independent of time. A steady state Y of the system is described by the following parameters: $Y = (\phi, n_1, ..., n_N, k_c)$, where:

- ϕ: represents the phases of the server, which can take the following values:
  C_i - server is connecting to the queue Q_i
  S_i - server is serving customers in the queue Q_i
- $n_1, ..., n_N$: denote the current length of each queue.
- k_c: denotes the current batch size. If the server is in phase C, $k_c = 0$. If the server is in phase S, k_c takes a value from 1 to k_i, which is the maximum batch size of queue Q_i. Note that k_c cannot exceed the current queue length n_i.

The server switches between phases according to the process illustrated in Fig. 1.

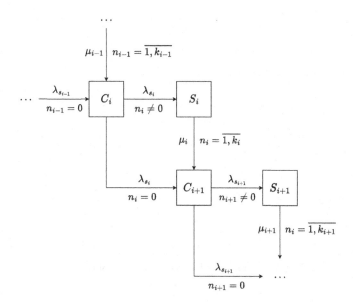

Fig. 1. Transitions between server phases

The server switches from phase C_i to S_i only if the queue Q_i is not empty ($n_i \neq 0$). After completing service at queue Q_i, the server switches to the next one (phase C_{i+1}). If queue Q_i is empty, the server skips phase S_i and proceeds directly to phase C_{i+1}.

3.1 Global Balance Equations

Phase C. Assuming the system is in the state $(C_i, n_1, ..., n_i, ..., n_N, 0)$, the server is currently connecting to queue Q_i. The current number of customers in the queues is $n_1, ..., n_N$, respectively. During this phase, the server is not serving any customers, so the current batch size k_c is 0.

The customer arrival flow for each queue follows a Poisson distribution with parameter λ_i corresponding to queue Q_i. The service and switchover flows follow exponential distributions with parameters b_i and s_i, respectively. Therefore, the service and switchover rates are $\mu_i = 1/b_i$ and $\lambda_{s_i} = 1/s_i$.

In the phase C_i, the system state parameters must satisfy the following conditions:

$$\begin{cases} 0 \le n_i \le N_i \\ k_c = 0 \end{cases} \tag{1}$$

Figure 2 illustrates the transitions between states in phase C.

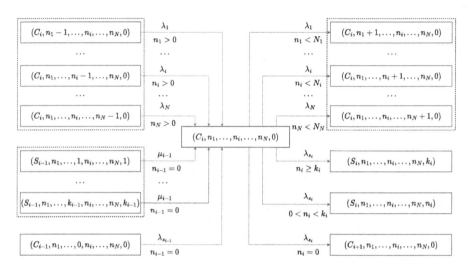

Fig. 2. Transitions between states in phase C

Let $I_{\{A\}}$ denote the indicator function of event A. $I_{\{A\}} = 1$ when event A occurs; otherwise, $I_{\{A\}} = 0$.

Based on Fig. 2, we can derive the following balance equation for the state $(C_i, n_1, ..., n_i, ..., n_N, 0)$, where $p(C_i, n_1, ..., n_i, ..., n_N, 0)$ represents the steady state probability of the system being in this state:

$$p(C_i, n_1, ..., n_i, ..., n_N, 0) \left(\sum_{j=1}^{N} \lambda_j I_{\{n_j < N_j\}} + \lambda_{s_i} \right)$$

$$= \sum_{j=1}^{N} p(C_i, n_1, ..., n_j - 1, ..., n_N, 0) \lambda_j I_{\{n_j > 0\}}$$

$$+ \left(\sum_{j=1}^{k_{i-1}} p(S_{i-1}, n_1, ..., j, n_i, ..., n_N, j) \mu_{i-1} \right.$$

$$\left. + \quad p(C_{i-1}, n_1, ..., 0, n_i, ..., n_N, 0) \lambda_{s_{i-1}} \right) I_{\{n_{i-1} = 0\}}$$

$$\tag{2}$$

In phase C, the length n_i of any queue Q_i can range from 0 to N_i. The number of states in the different C phases is the same and can be calculated using the following equation:

$$num_C = \prod_{i=1}^{N}(N_i + 1) \tag{3}$$

Phase S. Assuming the system is in the state $(S_i, n_1, ..., n_i, ..., n_N, k_c)$, the server is currently serving customers in queue Q_i. The number of customers in each queue is $n_1, ..., n_N$ respectively. It's important to note that for the server to transition from phase C_i to S_i, there must be at least one customer in queue Q_i ($n_i \geq 1$). The current batch size is denoted by k_c, where k_c must satisfy the conditions $k_c \leq k_i$ and $k_c \leq n_i$.

In the phase S_i, the system state parameters must satisfy the following conditions:

$$\begin{cases} 0 \leq n_j \leq N_j, \ j = \overline{1, N}, j \neq i \\ 1 \leq n_i \leq N_i \\ 1 \leq k_c \leq k_i \\ k_c = k_i \ if \ n_i \geq k_i, \ otherwise \ k_c = n_i \end{cases} \tag{4}$$

Figure 3 illustrates the transitions between states in phase S.

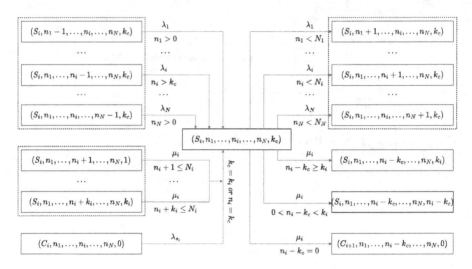

Fig. 3. Transitions between states in phase S

Similar to phase C, based on Fig. 3, we can derive the following balance equation for the state $(S_i, n_1, ..., n_i, ..., n_N, k_c)$:

$$p(S_i, n_1, ..., n_i, ..., n_N, k_c) \left(\sum_{j=1}^{N} \lambda_j I_{\{n_j < N_j\}} + \mu_i \right)$$

$$= \sum_{\substack{j=1 \\ j \neq i}}^{N} p(S_i, n_1, ..., n_j - 1, ..., n_N, k_c) \lambda_j I_{\{n_j > 0\}}$$

$$+ \quad p(S_i, n_1, ..., n_i - 1, ..., n_N, k_c) \lambda_i I_{\{n_i > k_c\}} \tag{5}$$

$$+ \quad \left(\sum_{j=1}^{k_i} p(S_i, n_1, ..., n_i + j, ..., n_N, j) \mu_i I_{\{n_i + j \leq N_i\}} \right.$$

$$\left. + \quad p(C_i, n_1, ..., n_i, ..., n_N, 0) \, \lambda_{s_i} \right) I_{\{k_c = k_i \, \cup \, n_i = k_c\}}$$

The number of states in phase S_i can be calculated using the following equation:

$$num_{S_i} = \prod_{\substack{j=1 \\ j \neq i}}^{N} (N_j + 1) \cdot \sum_{j=1}^{k_i} (N_i + 1 - j) \tag{6}$$

We have obtained a system of equations in the form of (2) and (5). The solution to this system of equations gives the probabilities of the steady states.

The total number of equations in the system can be calculated using the following equation:

$$num_{eq} = N.num_C + \sum_{i=1}^{N} num_{S_i} \qquad (7)$$

To find the solution to this system of equations, we need to use the following normalization condition:

$$\sum_{i=1}^{N} \sum_{n_1=0}^{N_1} \cdots \sum_{n_N=0}^{N_N} p(C_i, n_1, ..., n_N, 0)$$

$$+ \sum_{i=1}^{N} \sum_{n_1=0}^{N_1} \cdots \sum_{n_i=1}^{N_i} \cdots \sum_{n_N=0}^{N_N} \sum_{\substack{k_c=1 \\ k_c \leq n_i}}^{k_i} p(S_i, n_1, ..., n_i, ..., n_N, k_c) = 1 \qquad (8)$$

or

$$\sum_{Y \in E} p_Y = 1 \qquad (9)$$

where E - system state space.

3.2   Performance Characteristics Evaluation

Based on the final stationary probabilities mentioned above, let's evaluate the system's performance characteristics.

In our model, a customer only leaves the queue after being served. Therefore, the number of customers in the queue at any given time is the sum of the number of customers waiting for service and the number of customers being served at that time. Similarly, the time a customer spends in the queue (*sojourn time*) is the sum of their waiting time and service time.

The probability that the queue Q_i has j customers at any given time is the sum of the probabilities of all states with $n_i = j$. The number of customers in queue Q_i can take a value from 0 to the queue capacity N_i. Therefore, the mean number of customers (*mean queue length L_i*) in queue Q_i is given by:

$$L_i = \sum_{j=0}^{N_i} \sum_{\{Y \in E | n_i = j\}} j.p_Y, \qquad i = \overline{1, N} \qquad (10)$$

When the queue is full, customers will not be able to enter the system. The probability that a customer is rejected at queue Q_i (*loss rate P_{loss_i}*) is given by:

$$P_{loss_i} = \sum_{\{Y \in E | n_i = N_i\}} p_Y, \qquad i = \overline{1, N} \qquad (11)$$

The effective arrival rate λ_{e_i} of queue Q_i is calculated as follows::

$$\lambda_{e_i} = P_{loss_i}.\lambda_i, \qquad i = \overline{1, N} \qquad (12)$$

Applying Little's law, we obtain the mean sojourn time of customers V_i in queue Q_i:

$$V_i = \frac{L_i}{\lambda_{e_i}}, \qquad i = \overline{1, N} \tag{13}$$

The traffic intensity ρ_i to queue Q_i is the probability that the server is serving this queue:

$$\rho_i = \sum_{\{Y \in E | \phi = S_i\}} p_Y, \qquad i = \overline{1, N} \tag{14}$$

Mean sojourn time V of all the customers in the system:

$$V = \sum_{i=1}^{N} \rho_i V_i \tag{15}$$

The mean cycle time C is the sum of the time the server takes to service the queues and the time the server switchover to those queues: $C = \rho C + s$, where $\rho = \sum_{i=1}^{N} \rho_i$ - total traffic intensity of the system, and $s = \sum_{i=1}^{N} s_i$ - the mean server's total switchover time in the cycle.

So,

$$C = \frac{s}{1 - \rho} \tag{16}$$

In this work, the solution to the balance equation system and the evaluation of performance characteristics are implemented using the Python language with the Scipy and PyPardiso libraries. An example of their numerical results is provided in Sect. 5.

4   Simulation Model

The simulation model of the batch service polling system in the paper is built on the OMNeT++ 6.0.1 platform in the C++ language. OMNeT++ is an object-oriented discrete event network simulation framework that allows the building of programs to simulate complex wired and wireless networks. Some of the main advantages of OMNeT++ over other frameworks are its very fast simulation speed, its easy-to-get-used-to C++ syntax, its support for the component model, which makes maintenance and extension easy, and its large academic user community.

The structure of the simulation model is depicted in Fig. 4. The program reads input data containing information about the system configuration, such as the random number generator class, queue number, customer arrival rates, service discipline, service times, switchover times, and batch sizes. This information is sent to the Time Interval Generator to generate the customer arrival flow, the queue switchover flow, as well as the service flow. After being served, the customers are transferred to the Sink, where the simulation termination condition is checked. Then, the program exports files containing output data, including

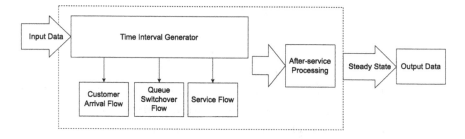

Fig. 4. Structure of the simulation model

mean queue lengths, mean loss rate, mean effective arrival rate, mean customer waiting time and sojourn time, mean cycle time in each queue and the server, mean intervisit time, and the number of cycles.

5   Numerical Results

5.1   Systems with Limited Queue Capacities

In this section, we will review some of the numerical results of the mathematical model and simulation presented above.

First, let's consider the systems described in Sect. 3, which have an exhaustive service discipline and limited queue capacities. The service and switchover flows follow exponential distributions with mean values of $b_i = 0.01$ s and $s_i = 0.001$ s, respectively. The corresponding batch size of Q_i is $k_i = 2$. Numerical results for systems with 2 queues are shown in Table 1, with mean arrival rates $\lambda_1 = 30$ and $\lambda_2 = 60$. The results for systems with 3 queues are shown in Table 2, with $\lambda_1 = 20$, $\lambda_2 = 30$, and $\lambda_3 = 40$. The column labeled "S" represents the results obtained from the simulation model in Sect. 4, while the column labeled "A" shows the analytical results obtained using the formulas from Sect. 3.

It can be observed that regardless of the values of queue capacities, the system performance characteristics obtained by solving the system of balance equations to find the final stationary probabilities closely match the simulation results, with an error of no more than 2% (column "Δ"). As N_i increases, the systems behave similarly to systems with infinite queue capacities. However, due to the exponential growth in the number of balance equations, which requires significant computing power, we will only consider simulation results for cases where $N_i = \infty$ and $k_i > 1$, where $i = \overline{1, N}$.

5.2   Systems with Unlimited Queue Capacities

Let's assume that the number of queues $N = 2$, queue capacities $N_i = \infty$, the mean service time of a batch at each queue $b_i = 0.01$ s, and the mean switchover time to each queue $s_i = 0.001$ s.

Table 1. $N = 2$

Parameters	$N_i = 3$			$N_i = 20$		
	A	S	Δ, %	A	S	Δ, %
L_1	0.6416	0.6404	0.19	0.8037	0.8032	0.06
L_2	0.9946	0.9942	0.04	1.3096	1.3101	0.04
V_1	0.0227	0.0226	0.44	0.0268	0.0268	0.0
V_2	0.0190	0.0190	0.0	0.0218	0.0218	0.0
λ_{e1}	28.318	28.303	0.05	29.995	30.000	0.02
λ_{e2}	52.317	52.352	0.07	60.013	60.000	0.02
P_{loss1}	0.0572	0.0566	1.06	0.0	0.0	0.0
P_{loss2}	0.1276	0.1275	0.08	0.0	0.0	0.0

Table 2. $N = 3$

Parameters	$N_i = 3$			$N_i = 20$		
	S	A	Δ, %	S	A	Δ, %
L_1	0.5007	0.5013	0.12	0.5687	0.5705	0.32
L_2	0.6835	0.6844	0.13	0.8066	0.8069	0.04
L_3	0.8275	0.8277	0.02	1.0009	1.0025	0.16
V_1	0.0259	0.0259	0.0	0.0284	0.0285	0.35
V_2	0.0244	0.0244	0.0	0.0269	0.0269	0.0
V_3	0.0228	0.0228	0.0	0.0251	0.0251	0.0
λ_{e1}	19.328	19.328	0.0	20.003	20.000	0.02
λ_{e2}	28.049	28.071	0.08	30.016	30.000	0.05
λ_{e3}	36.322	36.321	0.0	39.936	40.000	0.16
P_{loss1}	0.0336	0.0336	0.0	0.0	0.0	0.0
P_{loss2}	0.0640	0.0643	0.47	0.0	0.0	0.0
P_{loss3}	0.0924	0.0920	0.43	0.0	0.0	0.0

Table 3 presents the simulation results (column "S") for the customers' mean sojourn time in the system V in seconds in the case when the batch size $k_i = 1$ and the arrival rate λ_i gradually increases from 2 to 40. When comparing these results with the analytical results according to the formula presented in [7] (column "A"), we observe that the error of the simulation model is very small, less than 1% (column "Δ").

Simulation results for the cases when $k_i = 2$ and $k_i = 3$ are presented in Table 4. They are also shown in Fig. 5 along with the case when $k_i = 1$.

Figure 6 illustrates the relationship between the mean sojourn time of customers and the mean arrival rate, similar to Fig. 5, but for the case of 3 queues. The arrival rate, λ_i, ranges from 2 to 30.

Table 3. Mean sojourn time V when $k_i = 1$

Exhaustive Service				Gated Service			
λ_i	S	A	Δ, %	λ_i	S	A	Δ, %
2	0.0119	0.0119	0.0	2	0.0120	0.0120	0.0
5	0.0127	0.0127	0.0	5	0.0128	0.0128	0.0
10	0.0141	0.0141	0.0	10	0.0144	0.0144	0.0
15	0.0160	0.0160	0.0	15	0.0164	0.0164	0.0
20	0.0185	0.0185	0.0	20	0.0191	0.0192	0.01
25	0.0218	0.0220	0.02	25	0.0230	0.0230	0.0
30	0.0272	0.0273	0.01	30	0.0288	0.0288	0.0
35	0.0361	0.0360	0.01	35	0.0384	0.0383	0.01
40	0.0538	0.0535	0.03	40	0.0574	0.0575	0.01

Table 4. Mean sojourn time V when $k_i = 2$ and $k_i = 3$

Parameter	Exhaustive Service		Gated Service	
λ_i	$k_i = 2$	$k_i = 3$	$k_i = 2$	$k_i = 3$
2	0.0119	0.0119	0.0119	0.0119
5	0.0125	0.0125	0.0126	0.0126
10	0.0136	0.0136	0.0138	0.0138
15	0.0148	0.0146	0.0150	0.0149
20	0.0160	0.0157	0.0164	0.0161
25	0.0173	0.0168	0.0178	0.0172
30	0.0187	0.0179	0.0193	0.0183
35	0.0202	0.0190	0.0210	0.0194
40	0.0219	0.0201	0.0229	0.0205

From the graphs, we can conclude that, for cases with low customer arrival rates or low system loads, batch service does not significantly affect system performance. However, as the arrival rate increases, the efficiency of the batch service improves. It is important to note that this increase is not linear. With the system parameters described above, the optimal batch size is 2 or 3. For cases with larger batch sizes, their graphs almost coincide with the graph when $k_i = 3$.

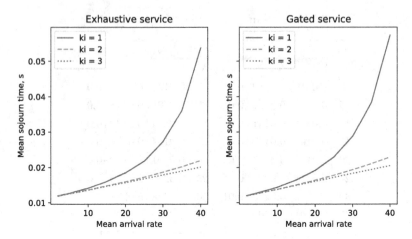

Fig. 5. Dependence of the mean sojourn time V on the mean arrival rate with different batch sizes. $N = 2$

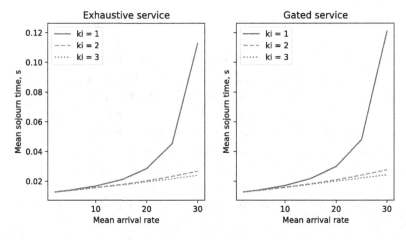

Fig. 6. Dependence of the mean sojourn time V on the mean arrival rate with different batch sizes. $N = 3$

6 Conclusion

In the paper, we have presented the mathematical model and analysis, developed the simulation model of the batch service polling system with exhaustive and gated services. By studying the simulation results, we have analyzed the correlation between the system's performance characteristics and the batch size of the queues.

References

1. Borst, S., Boxma, O.: Polling: past, present, and perspective. TOP **26**, 335–369 (2018). https://doi.org/10.1007/s11750-018-0484-5
2. Boxma, O., Wal, J., Yechiali, U.: Polling with batch service. Stochast. Models **24**, 604–625 (2008). https://doi.org/10.1080/15326340802427497
3. Dorsman, J.P., Mei, R., Winands, E.: Polling systems with batch service. OR Spectrum **34**, 743–761 (2011). https://doi.org/10.1007/s00291-011-0275-y
4. Liu, Z., Nain, P.: Optimal scheduling in some multiqueue single-server systems. IEEE Trans. Autom. Control **37**(2), 247–252 (1992). https://doi.org/10.1109/9.121629
5. Oyen, M.P.V., Teneketzis, D.: Optimal batch service of a polling system under partial information. Math. Methods Oper. Res. **44**(3), 401–419 (1996). https://doi.org/10.1007/bf01193939
6. Takagi, H., Kleinrock, L.: Analysis of polling systems. Perform. Eval. **5**(3), 206 (1985). https://doi.org/10.1016/0166-5316(85)90016-1
7. Vishnevsky, V., Semenova, O.: Polling systems and their application to telecommunication networks. Mathematics **9**(2), 117 (2021). https://doi.org/10.3390/math9020117
8. Vlasiou, M., Yechiali, U.: M/g/∞ polling systems with random visit times. Probab. Eng. Inf. Sci. **22**(1), 81–106 (2008). https://doi.org/10.1017/S0269964808000065
9. van der Wal, J., Yechiali, U.: Dynamic visit-order rules for batch-service polling. Probab. Eng. Inf. Sci. **17**(3), 351–367 (2003). https://doi.org/10.1017/S0269964803173044
10. Xia, C.H., Michailidis, G., Bambos, N., Glynn, P.W.: Optimal control of parallel queues with batch service. Probab. Eng. Inf. Sci. **16**(3), 289–307 (2002). https://doi.org/10.1017/S0269964802163029

Analysis of the Queueing System Describing a Mobile Network Subscriber's Processing Under Varying Modulation Schemes and Correlated Batch Arrivals

Alexander Dudin[(✉)] and Olga Dudina

Department of Applied Mathematics and Informatics, Belarusian State University,
220030 Minsk, Belarus
dudin@bsu.by, dudina@bsu.by

Abstract. The model of subscriber's processing in a mobile communication network consisting of a finite number of zones with the subscribers' service rate depending on a zone where the subscriber is currently residing, is analysed. The subscribers arrivals are defined by a batch marked Markovian process. The total number of subscribers that can receive service in a cell at the same time is restricted. The operation of the cell is described by a multi-server queueing system whose dynamics is defined by a multi-dimensional Markov chain. The generator of the chain having the upper-block-Hessenbergian structure is derived. The stationary distribution of the chain and the primary indicators of subscriber quality of service are found. Dependencies of the variety of important performance characteristics of the cell on the average rate of subscribers arrival and limiting number of subscribers that can be processed simultaneously are numerically highlighted.

Keywords: adaptive modulation · batch correlated arrival process · stationary distribution

1 Introduction

The so-called multiplexing is widely used for information transmission in telecommunication networks. It assumes creation of many logical channels for subscribers service in one physical channel. Traditionally used for this goal are Frequency Division Multiplexing (FDM), Time Division Miplexing (TDM)) and Wavelength Division Multiplexing (WDM). The transmission of information over a physical channel in the literature and engineering practice was usually described in terms of a multi-server queueing system in which the server device corresponds to a logical channel. The service (transmission) time of any subscriber is random. It has a known distribution, the parameters of which are determined by the channel speed and the subscriber information length (in bits). Such systems are fairly well studied in the literature. However, progress in the development of transmission equipment has led to the possibility of using several multiplexing schemes simultaneously. This, in turn, has led to possibility

V. M. Vishnevskiy et al. (Eds.): DCCN 2023, LNCS 14123, pp. 156–170, 2024.
https://doi.org/10.1007/978-3-031-50482-2_13

of changing the subscriber service rate over time. Multi-server queueing systems with dynamically changing service rates are more or less well studied in the literature, see, e.g., [1, 2] only in context of so called systems operating in the random environment in which service rate varies in all servers synchronously. Systems where the subscribers change service rate independently of each other are quite poorly studied in the existing literature. In this paper, we investigate the system where of the current service rate of an subscriber depends on his/her distance from the base station and various obstacles to the radio signal propagation. A smaller distance entails an amplification of the received signal and, accordingly, an increase in the speed of servicing the subscriber.

The more detailed description of the considered here problem and a brief survey of the related literature can be found in paper [3]. The model considered in the present paper is a practically important extension of the model recently considered in [3] to the case of possibility of batch generation of subscriber requests. This generalization implies that the multi-dimensional Markov chain (MC) under study does not fall into the class of Quasi-Birth-and-Death processes (QBD) exhaustively studied by M. Neuts, see [4]. This essentially complicates the algorithmic analysis of the model.

The remainder of the text is as follows. Queueing model is completely defined in Sect. 2. In Sect. 3, behaviour of the considered system is described by the multidimensional MC. Required denotations are introduced. The infinitesimal generator of this MC is derived. Section 4 contains formulas for computation of values of various performance characteristics of the queueing system. Numerical examples demonstrating dependencies of the system performance measures on its parameters are given in Sect. 5. Brief conclusion is given in Sect. 6.

2   Mathematical Model

We consider the subscriber's processing as service in an N-server queueing system having no buffer. This means that no more than N subscribers can receive service simultaneously.

The cell consists of R, $1 < R < \infty$, non-overlapping zones. The strength of a base station signal, which mainly defines the subscriber's processing rate, depends on the number of the zone where the mobile subscriber is currently residing. To reflect this fact, we formally classify arriving subscribers into R types corresponding to the zone where the subscriber tries to start service. The subscribers' arrivals are assumed to be defined by the $BMMAP$ (batch marked Markov arrival process).

This process is the essential extension of the more well-known marked Markov arrival process $(MMAP)$, see [5], accounting a possibility of batch arrivals, or the batch Markov arrival process $(BMAP)$, see [6–8], accounting the possibility of different types of arriving subscribers. The potential moments of subscriber's arrivals in the $BMMAP$ are defined as the epochs of the jumps of an underlying process that is an irreducible continuous-time MC ν_t, $t \geq 0$, having a state space $\{1, 2, ..., W\}$ where W is a finite integer number.

The $BMMAP$ is defined by the collection of the matrices D_0, $D_r^{(l)}$, $r = \overline{1, R}$, $l = \overline{0, L}$, of size W. Here L is the maximum batch size. The elements of the matrix $D_r^{(l)}$ define the transition rates of the MC ν_t which cause the arrival of l type-r subscribers. We denote $D = \sum_{r=1}^{R} \sum_{l=1}^{L} D_r^{(l)}$. The matrix's D_0 non-diagonal elements define the rates of the transitions of the MC ν_t at which subscribers do not arrive. The diagonal entries of the matrix D_0 are negative and specify the rate of exit of the chain ν_t from the associated states up to the sign. The matrix $D(1) = D_0 + D$ is the MC's ν_t generator.

The mean rate λ_r of type-r subscribers generation is defined as

$$\lambda_r = \boldsymbol{\theta} \sum_{l=1}^{L} l D_r^{(l)} \mathbf{e}, r = \overline{1, R},$$

where $\boldsymbol{\theta}$ is row vector defining the invariant probability distribution of the MC ν_t and $\mathbf{e}$ is column vector with all components equal to 1. The mean total rate λ of subscribers arrival is given by formula

$$\lambda = \sum_{r=1}^{R} \lambda_r.$$

For more detailed information about the $BMMAP$, see, e.g., book [9].

Due to the possibility of batch arrival of subscribers and a availability of a finite number of servers, we have to fix the subscriber's acceptance discipline. Here, we suggest the partial admission discipline. If an arbitrary group that consists of l subscribers of any type arrives when there are k, $k \le l$, free servers, then $l - k$ subscribers start service and the rest of subscribers leave the system without service (are lost), $l = \overline{1, L}$. The known disciplines of complete subscribers rejection and complete subscribers admission can be analysed by analogy with the analysis presented below.

Let us analyse the fate of a subscriber who establishes a connection in the rth zone or arrives into this zone from another one during the service. We suggest that the subscriber's residence time in the rth zone is exponentially distributed with the known parameter $\mu_r, r = \overline{1, R}$.

There are various scenarios of a subscriber departure from zone number r :

- the processing of the subscriber is successfully completed. The probability of realization of this scenario is denoted by $p_{r,serv}$;
- the subscriber enters the kth zone (becomes a type-k subscriber). The probability of realization of this scenario is denoted $p_{r,k}$;
- the service of the subscriber is interrupted due to a loss of connection. The probability of realization of this scenario is denoted $p_{r,loss}$;
- the service of the subscriber is terminated due to handover to another cell. The probability of realization of this scenario is denoted $p_{r,hand}$.

Note, that

$$p_{r,serv} + p_{r,loss} + p_{r,hand} + \sum_{k=1, k \ne r}^{R} p_{r,k} = 1$$

for each r. Via multiplication of the rate μ_r by the probabilities of these listed scenarios, we obtain the rates $\mu_{r,serv}$, $\mu_{r,k}$, $\mu_{r,loss}$ and $\mu_{r,hand}$ of successful finish of the service, transfer of the subscriber to the kth zone, service interruption and handover, respectively. It is obvious that the following relation is valid:

$$\mu_r = \mu_{r,serv} + \mu_{r,loss} + \mu_{r,hand} \sum_{k=1,k\neq r}^{R} \mu_{r,k}.$$

We will interpret the total service time of a subscriber in the cell as the time during which an irreducible MC η_t, $t \geq 0$, with the collection of R transient states $\{1, 2, \ldots, R\}$ reaches one of the three absorbing states. The dynamics of this MC is described as follows. The state of the MC η_t at the service beginning epoch is selected among the existing transient states corresponding to the type of a subscriber. If this is a type-r subscriber, the state of the MC η_t is chosen as r, $r = \overline{1, R}$.

The transitions rates of the process η_t among the transient states form the sub-generator of form

$$S = \begin{pmatrix} -\mu_1 & \mu_{1,2} & \mu_{1,3} & \cdots & \mu_{1,R} \\ \mu_{2,1} & -\mu_2 & \mu_{2,3} & \cdots & \mu_{2,R} \\ \vdots & \vdots & \vdots & \ddots & \vdots \\ \mu_{R,1} & \mu_{R,2} & \mu_{R,3} & \cdots & -\mu_R \end{pmatrix}.$$

When the process η_t resides in the transient state rth, the subscriber receives service of type-r. The transition of the process η_t from one transient state to another one occurs during the moment when the subscriber transits between the corresponding zones.

The transition of the process η_t to the first absorbing state occurs at the moment of successful service completion. The transition rates to this absorbing state form the the column vector $\mathbf{S}_{serv} = (\mu_{1,serv}, \mu_{2,serv}, \ldots, \mu_{R,serv})^T$.

The transition of the process η_t to the second absorbing state occurs at the moment of service interruption (due to subscriber impatience or his/her dissatisfaction by the quality of provided service). The transition rates to this absorbing state constitute the column vector $\mathbf{S}_{loss} = (\mu_{1,loss}, \mu_{2,loss}, \ldots, \mu_{R,loss})^T$.

The transition of MC to the third absorbing state occurs at the moment of subscriber's departure the cell due to handover. The transition rates to this absorbing state constitute the column vector $\mathbf{S}_{hand} = (\mu_{1,hand}, \mu_{2,hand}, \ldots, \mu_{R,hand})^T$.

After the formal definition of subscribers generation and processing, we have an opportunity can start the analysis of the system behaviour.

3 Markov Process Defining the Dynamics of the Cell Operation

The dynamics of the cell under consideration can be described by the irreducible continuous-time MC

$$\xi_t = \{n_t, \nu_t, \mathbf{m}_t\}, \ t \geq 0,$$

where at the moment t the component n_t defines the number of subscribers in the cell, ν_t is the state of the underlying process of the $BMMAP$, and $\mathbf{m}_t$ is the vector $\mathbf{m}_t = (m_t^{(1)}, \ldots, m_t^{(R)})$, where $m_t^{(r)}$ is the number of servers providing the rth phase of the service (the current number of subscribers residing in the rth zone), $m_t^{(r)} = \overline{0, n_t}$, $\sum_{r=1}^{R} m_t^{(r)} = n_t$, $r = \overline{1, R}$.

The MC ξ_t, $t \geq 0$, is regular, irreducible and has a finite state space. Thus, the stationary invariant probabilities of its states

$$\varphi(n, \nu, m^{(1)}, \ldots, m^{(R)}) = \lim_{t \to \infty} P\{n_t = n, \nu_t = \nu, m_t^{(1)} = m^{(1)}, \ldots, m_t^{(R)} = m^{(R)}\}$$

exist for any choice of the parameters of the cell operation.

Let us enumerate the states of MC ξ_t in the reverse lexicographic order of the components $(m^{(1)}, \ldots, m^{(R)})$ and the direct lexicographic order of the component ν and build the row vectors $\boldsymbol{\varphi}_n$, $n = \overline{0, N}$, of the stationary invariant probabilities.

It is well known that the probability vectors $\boldsymbol{\varphi}_n$, $n = \overline{0, N}$, represent an unique solution to the following system of linear algebraic equations:

$$(\boldsymbol{\varphi}_0, \boldsymbol{\varphi}_1, \ldots, \boldsymbol{\varphi}_N)\mathcal{F} = \mathbf{0}, \ (\boldsymbol{\varphi}_0, \boldsymbol{\varphi}_1, \ldots, \boldsymbol{\varphi}_N)\mathbf{e} = 1 \tag{1}$$

where $\mathcal{F}$ is the infinitesimal generator of the MC ξ_t, $t \geq 0$, and $\mathbf{0}$ denote the row vector with all components equal to 0. Therefore, to calculate these vectors it is firstly necessary to derive explicit expression for generator Q. To do this, we need the following notation.

Given that exactly n subscribers reside in the cell at a transition moment, let us introduce the following matrices:

- $P_n(\beta_r)$ defines the transition probabilities of the vector process $\mathbf{m}_t$, $t \geq 0$, at the epoch when a subscriber of type-r starts service;
- $L_n(\mathbf{S}_{serv})$ defines the transition rates of the process $\mathbf{m}_t$ when service of an arbitrary subscriber ends successfully;
- $L_n(\mathbf{S}_{loss})$ defines the transition rates of the process $\mathbf{m}_t$ when one of the subscribers departs from the cell due to the loss of connection;
- $L_n(\mathbf{S}_{hand})$ defines the transition rates of the process $\mathbf{m}_t$ when one of the subscribers departs from the cell to another cell due to handover;
- $A_n(S)$ defines the transitions rates of the process $\mathbf{m}_t$ when one of subscribers transits to another zone;
- the $\Delta^{(n)}$ is the diagonal matrix the moduli of the diagonal entries of which are equal to the total rate of departure the of the process $\mathbf{m}_t$ from the associated state.

The following statement is valid:

Theorem 1. *The infinitesimal generator $\mathcal{F}$ of the MC ξ_t, $t \geq 0$, has the following block structure*

$$\mathcal{F} = \begin{pmatrix} \mathcal{F}_{0,0} & \mathcal{F}_{0,1} & \mathcal{F}_{0,2} & \mathcal{F}_{0,3} & \cdots & \mathcal{F}_{0,N-1} & \mathcal{F}_{0,N} \\ \mathcal{F}_{1,0} & \mathcal{F}_{1,1} & \mathcal{F}_{1,2} & \mathcal{F}_{1,3} & \cdots & \mathcal{F}_{1,N-1} & \mathcal{F}_{1,N} \\ O & \mathcal{F}_{2,1} & \mathcal{F}_{2,2} & \mathcal{F}_{2,3} & \cdots & \mathcal{F}_{2,N-1} & \mathcal{F}_{2,N} \\ \vdots & \vdots & \vdots & \vdots & \ddots & \vdots & \vdots \\ O & O & O & O & \cdots & \mathcal{F}_{N,N-1} & \mathcal{F}_{N,N} \end{pmatrix}. \tag{2}$$

The blocks $\mathcal{F}_{i,l}$, $i = \overline{0, N}$, $l = \overline{\max\{0, i-1\}, N}$, are defined as follows:

$$\mathcal{F}_{0,0} = D_0,$$

$$\mathcal{F}_{n,n} = D_0 \oplus (A_n(S) + \Delta^{(n)}), \, n = \overline{1, N-1},$$

$$\mathcal{F}_{N,N} = D_0 \oplus (A_N(S) + \Delta^{(N)}) + D \otimes I_{T_N},$$

$$\mathcal{F}_{n,n+l} = \sum_{r=1}^{R} D_r^{(l)} \otimes \prod_{i=n}^{n+l-1} P_i(\boldsymbol{\beta}_r), \, n = \overline{0, N-2}, \, l = \overline{1, \min\{N-1-n, L\}},$$

$$\mathcal{F}_{n,n+l} = O_{WT_n \times WT_{n+l}}, \, n = \overline{0, N-2}, \, l = \overline{L+1, N-1-n},$$

$$\mathcal{F}_{n,N} = \sum_{r=1}^{R} \sum_{l=N-n}^{L} D_r^{(l)} \otimes \prod_{i=n}^{N-1} P_i(\boldsymbol{\beta}_r), \, n = \overline{\max\{0, N-L\}, N-1},$$

$$\mathcal{F}_{n,N} = O_{WT_n \times WT_N}, \, n = \overline{0, N-L-1},$$

$$\mathcal{F}_{n,n-1} = I_W \otimes (L_n(\mathbf{S}_{serv}) + L_n(\mathbf{S}_{loss}) + L_n(\mathbf{S}_{hand})), \, n = \overline{1, N},$$

where

I is the identity matrix and O is a zero matrix of a suitable size. If the size is not clear from context it is indicated by a suffix. E.g., $O_{WT_n \times WT_{n+l}}$ means the non-square zero matrix of size $WT_n \times WT_{n+l}$;

the symbols $\otimes$ and $\oplus$ represent the Kronecker product and sum of matrices, correspondingly, see, e.g., [10];

$$T_n = \frac{(n+R-1)!}{n!(R-1)!}, \, n = \overline{0, N};$$

$\boldsymbol{\beta}_r$ is the row vector of size R the rth component of which is equal to 1 while the other components are equal to 0, $r = \overline{1, R}$.

The detailed descriptions of the matrices $P_n(\boldsymbol{\beta}_r)$, $L_n(\mathbf{S}_{serv})$, $L_n(\mathbf{S}_{loss})$, $L_n(\mathbf{S}_{hand})$, $A_n(S)$, $\Delta^{(n)}$, and the recursive algorithms available for their calculation are presented in [3].

Mention that the idea of the use of the process $\mathbf{m}_t$ for description of service in many parallel servers goes back to the work [11].

Proof. The theorem is proved by analyzing the intensities of various transitions of the MC ξ_t, $t \geq 0$, across an infinitesimal length interval and rewriting them into the matrix form. Since not more than one subscriber can leave arrive to the cell or depart from the cell during such an interval, $\mathcal{F}_{i,j} = O$ for all pairs of indices i, j such that $i > j + 1$.

The diagonal entries of the matrices $\mathcal{F}_{n,n}$, $n = \overline{0, N}$, are strictly less than 0. The modulus of these entries are equal to rate of the exit from the corresponding state.

The reasons for MC ξ_t departing from its current state are as follows:

a) The departure of the underlying process of subscribers arrival from the current state. The rates of such departures are defined by the modulus of the respective diagonal elements of the matrix D_0 if $n = 0$ (the cell is empty), and $D_0 \otimes I_{T_n}$ if n, $n = \overline{1, N}$, subscribers reside in the cell.
   Mention, that when N subscribers already stay in the cell at an epoch of the underlying process of arrivals transition with generation of a batch of subscribers, all the arriving subscribers are lost and exit from the corresponding state does not happen. The corresponding diagonal elements of the matrix $D \otimes I_{T_N}$ provide the intensities of these transitions.
b) Transition of the state of the process $\mathbf{m}_t$, $t \geq 0$, into another state. When the number of subscribers in the cell is equal to n, the rates of such transitions are given by the modulus of the associated diagonal elements of the matrix $I_W \otimes \Delta^{(n)}$, $n = \overline{1, N}$.

The non-diagonal entries of the matrices $\mathcal{F}_{n,n}$ are equal to the transition intensities of the components of the MC ξ_t that do not cause the change of the subscribers number in the cell, $n = \overline{0, N}$. The respective transitions are the following ones:

a) The change of the state of the arrivals underlying process without the arrival of a new subscriber (the intensities of these transitions are the non-diagonal entries of the matrix D_0 if $n = 0$, and $D_0 \otimes I_{T_n}$ for $n = \overline{1, N}$).
b) There N subscribers residing in the cell. The process ν_t implements a transition from some state to another one with an arrival of a group of arbitrary type subscribers. This group of subscribers is lost and the corresponding transition rates are given by the entries of the matrix $D \otimes I_{T_N}$.
c) The jump of the process $\mathbf{m}_t$, $t \geq 0$, corresponding to a subscriber transfer to another zone. The respective transition rates are the non-diagonal entries of the matrices $I_W \otimes A_n(S)$, $n = \overline{1, N}$.

Thus, we obtained the formulas for the matrices $\mathcal{F}_{n,n}$, $n = \overline{0, N}$.

Now let us derive the blocks $\mathcal{F}_{n,n-1}$, $n = \overline{1, N}$. Their entries define the transitions of the MC ξ_t that imply the decrease in the number of subscribers n in the cell by one. The transition rates are given as the entries of the matrix: $I_W \otimes L_n(\mathbf{S}_{serv})$ when a subscriber successful finishes the service; matrix $I_W \otimes L_n(\mathbf{S}_{loss})$ when a subscriber leaves the system due to loss of connection; matrix $I_W \otimes L_n(\mathbf{S}_{hand})$ when a subscriber leaves the system due to handover.

Taking in mind these scenarios, we obtain the presented above form of the blocks $\mathcal{F}_{n,n-1}$.

The elements of the matrices $\mathcal{F}_{n,n+l}$, $n = \overline{0, N-1}, l = \overline{1, N-n}$, define the rates of transitions of the components of the MC ξ_t that cause the increase in the number of subscribers n in the cell by l. This can happen only if l arriving subscribers are admitted to the cell and start service. The associated rates are given by the entries of the matrices $\sum\limits_{r=1}^{R} D_r^{(l)} \otimes \prod\limits_{i=n}^{n+l-1} P_i(\boldsymbol{\beta}_r)$, for $n = \overline{0, N-2}$, $L + 1 \leq l \leq N-1-n$, (if the whole arriving group of subscribers is admitted) and of the matrices $\sum\limits_{r=1}^{R} \sum\limits_{l=N-n}^{L} D_r^{(l)} \otimes \prod\limits_{i=n}^{N-1} P_i(\boldsymbol{\beta}_r)$ for $n = \overline{\max\{0, N-L\}, N-1}$ (if the size of the arriving group of subscribers exceeds the number of available slots for service in the cell).

The number of equations of the finite system (1) with the matrix having the upper-Hessenberg form (2) equals $W \sum\limits_{n=0}^{N} T_n + 1$ and may be quite large. To solve system (1), we recommend the numerically stable algorithm from [12] which efficiently exploits the block matrix structure of the generator $\mathcal{F}$.

4   Computation of the Main Performance Indicators of the Cell

The mean number of subscribers in the cell at an arbitrary point in time is computed by

$$N^{sys} = \sum_{n=1}^{N} n \boldsymbol{\varphi}_n \mathbf{e}.$$

The probability of an arbitrary subscriber loss upon arrival to the r-th zone caused by the overflow of the cell is

$$P_r^{ent-loss} = \frac{1}{\lambda} \sum_{l=1}^{L} \sum_{n=N-l+1}^{N} \boldsymbol{\varphi}_n [(l - (N-n)) D_r^{(l)} \otimes I_{T_n}] \mathbf{e}, \ r = \overline{1, R}.$$

The probability of an arbitrary subscriber loss due to the cell overflow is computed by

$$P^{ent-loss} = \frac{1}{\lambda} \sum_{r=1}^{R} \sum_{l=1}^{L} \sum_{n=N-l+1}^{N} \boldsymbol{\varphi}_n [(l - (N-n)) D_r^{(l)} \otimes I_{T_n}] \mathbf{e}.$$

The output rate from the r-th zone is computed by

$$\lambda_r^{out} = \sum_{n=1}^{N} \boldsymbol{\varphi}_n (I_W \otimes L_n(\mathbf{S}_r^{serv})) \mathbf{e}, \ r = \overline{1, R}.$$

The rate of the output flow of the subscribers received complete service is computed by

$$\lambda^{out} = \sum_{n=1}^{N} \varphi_n (I_W \otimes L_n(\mathbf{S}_{serv}))\mathbf{e}.$$

The probability of an arbitrary subscriber loss due to a poor connection in the r-th zone is

$$P_r^{loss-connection} = \frac{1}{\lambda} \sum_{n=1}^{N} \varphi_n (I_W \otimes L_n(\mathbf{S}_r^{loss}))\mathbf{e}, \; r = \overline{1,R}.$$

The probability of an arbitrary subscriber loss due to a poor connection is

$$P^{loss-connection} = \frac{1}{\lambda} \sum_{n=1}^{N} \varphi_n (I_W \otimes L_n(\mathbf{S}_{loss}))\mathbf{e}.$$

The probability of an arbitrary subscriber loss due to his/her handover to another cell is

$$P^{loss-handover} = \frac{1}{\lambda} \sum_{n=1}^{N} \varphi_n (I_W \otimes L_n(\mathbf{S}_{hand}))\mathbf{e}.$$

The probability of an arbitrary subscriber loss due to handover to another cell from the r-th zone is given by

$$P_r^{loss-handover} = \frac{1}{\lambda} \sum_{n=1}^{N} \varphi_n (I_W \otimes L_n(\mathbf{S}_r^{hand}))\mathbf{e}.$$

Here, the column vectors $\mathbf{S}_r^{serv}$, $\mathbf{S}_r^{loss}$ and $\mathbf{S}_r^{hand}$ are obtained from zero column vector by replacement of the rth entry by $\mu_{r,serv}$, $\mu_{r,loss}$, $\mu_{r,hand}$, $r = \overline{1,R}$, respectively.

The loss probability of an arbitrary subscriber is computed by

$$P^{loss} = 1 - \frac{\lambda^{out}}{\lambda} = P^{ent-loss} + P^{loss-connection} + P^{loss-handover}.$$

The loss probability of an arbitrary subscriber from the r-th zone is computed by

$$P_r^{loss} = P_r^{ent-loss} + P_r^{loss-connection} + P_r^{loss-handover}, \; r = \overline{1,R}.$$

The probability that the system is empty at an arbitrary time is computed by

$$P_{idle} = \varphi_0 \mathbf{e}.$$

5   Numerical Examples

In the numerical experiment, we study the system's operation under different values of maximal number of simultaneously serviced subscribers N and the average arrival intensity of subscribers λ. To do this, we suppose that the cell consists of three zones $R = 3$ (there are three types of subscribers) and consider the $BMMAP$ arrival process that has the mean arrival rate $\lambda = 1$ and is determined by the following set of matrices:

$$D_0 = \begin{pmatrix} -0.15822 & 0.000312175 \\ 0.000580329 & -0.562582 \end{pmatrix},$$

$$D_1^{(1)} = \begin{pmatrix} 0.00369311 & 8.0197019 \times 10^{-6} \\ 0.0000301976 & 0.104678 \end{pmatrix},$$

$$D_1^{(2)} = \begin{pmatrix} 0.00624764 & 9.049392 \times 10^{-6} \\ 0.0000203463 & 0.0695853 \end{pmatrix},$$

$$D_1^{(3)} = \begin{pmatrix} 0.00950012 & 0.0000170641 \\ 0.0000336134 & 0.0644012 \end{pmatrix}, D_1^{(4)} = \begin{pmatrix} 0.0102377 & 0.000425044 \\ 0.0000128711 & 0.0653913 \end{pmatrix},$$

$$D_2^{(1)} = \begin{pmatrix} 0.0265734 & 0.000039554 \\ 0.0000291085 & 0.0529324 \end{pmatrix},$$

$$D_2^{(2)} = \begin{pmatrix} 0.0440595 & 1.683147 \times 10^{-6} \\ 0.000034059 & 0.0987686 \end{pmatrix}, D_2^{(3)} = D_2^{(4)} = O,$$

$$D_3^{(1)} = \begin{pmatrix} 0.0181688 & 0.00386178 \\ 8.366232 \times 10^{-6} & 0.0687508 \end{pmatrix},$$

$$D_3^{(2)} = \begin{pmatrix} 0.0176244 & 0.000176239 \\ 4.4553899 \times 10^{-7} & 0.024539 \end{pmatrix},$$

$$D_3^{(3)} = \begin{pmatrix} 0.0172278 & 0.0000371282 \\ 4.4553899 \times 10^{-7} & 0.012786 \end{pmatrix}, D_3^{(4)} = O.$$

The average arrival intensities of the first, second and third type of subscribers are equal to $\lambda_1 = 0.617522$, $\lambda_2 = 0.232501$ and $\lambda_3 = 0.149977$, respectively.

The subscriber's sojourn time in the first, second and third zone has an exponential distribution with the rate $\mu_1 = 0.5$, $\mu_2 = 0.3$ and $\mu_3 = 0.1$, respectively. A subscriber of the first type leaves the first zone and transits to the second or third zone with probability $p_{1,2} = 0.1$ and $p_{1,3} = 0.002$ respectively. With probabilities $p_{2,1} = \frac{1}{12}$ and $p_{2,3} = 0.1$ a subscriber of the second type transits from the second zone to the first and third zone respectively. A third type subscriber with probabilities $p_{3,1} = 0.01$ and $p_{3,2} = 0.1$ transits from the third zone to the first and second zone, respectively.

We suggest that the stay of the subscriber in the r-th zone ends with successful service completion with the probability $p_{r,serv}$, $r = \overline{1,3}$, where $p_{1,serv} = 0.892$, $p_{2,serv} = 0.78$ and $p_{3,serv} = 0.64$. If service of the subscriber is terminated due to a loss of connection, then the subscriber leaves the cell without complete service with probabilities $p_{1,loss} = 0.006$, $p_{2,loss} = 0.02$ and $p_{3,loss} = 0.1$

depending on its type. With the probability $p_{1,hand} = 0$, $p_{2,hand} = \frac{1}{60}$ and $p_{3,hand} = 0.15$, service of the subscriber of the first, second and third type, respectively, is terminated due to the handover to another cell.

To highlight behavior of the cell performance indicators, we vary the value of the maximal number of subscribers N, which can be serviced in the cell at the same time, in the interval $[5; 60]$ with step 5 and the mean arrival rate λ in the interval $[1; 7]$ with step 1. Variation of the value of λ is implemented via the corresponding scaling of of the matrices $D_k^{(r)}$ elements.

Figure 1 illustrates the dependence of the output rate λ^{out} of completely serviced subscribers and the respective output rates λ_r^{out} from the r-th zone, $r = 1, 2, 3$, on values of N and λ.

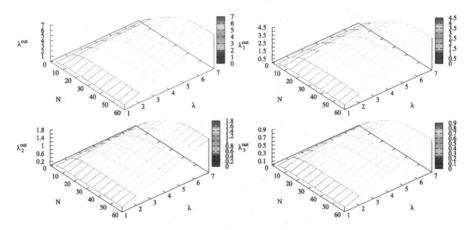

Fig. 1. Relationship of the output flow rate λ^{out} of serviced subscribers and the rates λ_r^{out}, $r = 1, 2, 3$, of the output flow from the r-th zone with the parameters N and λ

The dependence of the probability $P^{ent-loss}$ of an arbitrary subscriber loss upon arrival due to the system overflow and the loss probabilities $P_r^{ent-loss}$, $r = 1, 2, 3$, of an arbitrary subscriber upon arrival to the r-th zone due to the cell overflow on parameters N and λ is presented in Fig. 2.

Figures 3, 5 and 5 demonstrate the relationship between the probability of an arbitrary subscriber loss P^{loss} and the probabilities P_r^{loss}, $r = 1, 2, 3$, of an arbitrary subscriber loss from the r-th zone, the loss probability $P^{loss-connection}$ of an arbitrary subscriber due to a loss of connection and the loss probabilities $P_r^{loss-connection}$, $r = 1, 2, 3$, of an arbitrary subscriber due to a loss of connection in the r-th zone and the probability $P^{loss-handover}$ of an arbitrary subscriber loss due to handover to another cell and the loss probabilities $P_r^{loss-handover}$, $r = 1, 2, 3$, of an arbitrary subscriber due to handover to another cell from the r-th zone with the parameters N and λ.

The behaviour of the mean number of subscribers in the cell at an arbitrary epoch N^{sys} and the probability that the system is empty P_{idle} for distinct values of the parameters N and λ is demonstrated in Fig. 6.

Presented here dependencies match well to their intuitively anticipating shape. In particular, the loss probabilities of subscribers at the entry to the cell pretty sharply increase with the growth of the subscribers arrival rate and reduction of the number N of subscribers that can be serviced simultaneously. Oppositely, the loss probabilities of subscribers due to the loss of connection

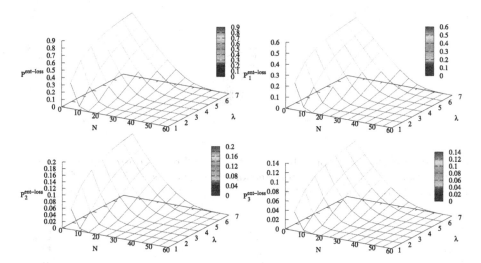

Fig. 2. Relationship of the probability $P^{ent-loss}$ of a subscriber loss upon arrival due to the cell overflow and the respective loss probabilities $P_r^{ent-loss}$, $r = 1, 2, 3$, with the parameters N and λ

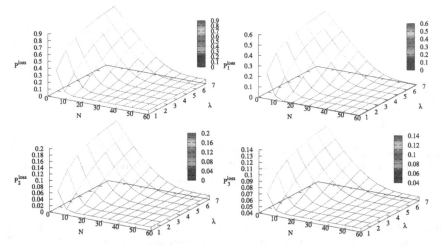

Fig. 3. Relationship of the probability of an arbitrary subscriber loss P^{loss} and the probabilities P_r^{loss}, $r = 1, 2, 3$, of an arbitrary subscriber loss from the r-th zone with the parameters N and λ

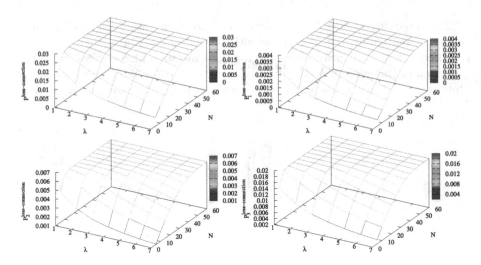

Fig. 4. Relationship of the probability $P^{loss-connection}$ of an arbitrary subscriber loss due to a poor connection and the corresponding loss probabilities $P_r^{loss-connection}$, $r = 1, 2, 3$, and the parameters N and λ

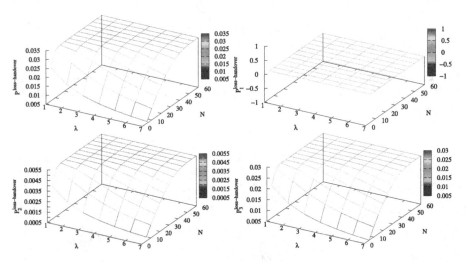

Fig. 5. Relationship of the loss probability $P^{loss-handover}$ of an arbitrary subscriber due to handover to another cell and the corresponding loss probabilities $P_r^{loss-handover}$, $r = 1, 2, 3$, and the parameters N and λ

increase then N increases. Existence of such opposite trends when the parameter N increases explains the non-monotonic behaviour of the output rate λ^{out} of serviced subscribers and the rates λ_r^{out}, $r = 1, 2, 3$ earlier observed on Fig. 1.

The value and importance of the presented algorithmic and numerical results consists of the fact that they allow estimate the dependencies not only quali-

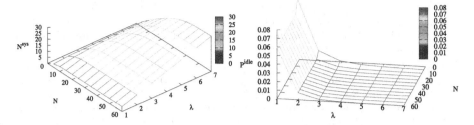

Fig. 6. Relationship of the mean number of subscribers in the cell N^{sys} and the probability P_{idle} that the cell is empty with the parameters N and λ

tatively, but quantitatively. This is important for making managerial decisions, e.g., decisions concerning the identification of the maximal values of arrival rate λ, required bandwidth of base station and the number N for which the arbitrarily fixed set of constraints imposed on quality of subscribers service is fulfilled.

6 Conclusion

Obtained algorithmic results can be used for evaluation of the performance of the cell and subscriber's quality of service indicators under the fixed pattern of arriving flow, form of zones, configuration of the equipment and available modulation schemes as well as for optimization of the limiting number N of simultaneously serviced subscribers under various criteria of the quality of the system's operation and possibly imposed by the SLA (Service Level Agreement) constraints on the selected performance characteristics.

Considered queueing model can be generalized into direction of existence of limitation not only on the number of subscribers simultaneously residing in the cell, but also on the total bandwidth required by these subscribers, see, e.g. [13,14], with possibility of temporal use of the reduced transmission rate. Also possibility of differentiation of arriving subscribers to having the strict and the flexible requirements to the transmission rate can be incorporated into the model following to [15].

References

1. Kim, C.S., et al.: The $BMAP/PH/N$ retrial queueing system operating in Markovian random environment. Comput. Oper. Res. **37**(7), 1228–1237 (2010)
2. Kim, C., et al.: Analysis of an queueing system operating in a random environment. Int. J. Appl. Math. Comput. Sci. **24**(3), 485–501 (2014)
3. Kim, C., et al.: Mathematical model of operation of a cell of a mobile communication network with adaptive modulation schemes and handover of mobile users. IEEE Access **9**, 106933–106946 (2021)
4. Neuts, M.F.: Matrix-Geometric Solutions in Stochastic Models -An Algorithmic Approach. Johns Hopkins University Press, Charles Village, Baltimore (1981)

5. He, Q.M.: Queues with marked customers. Adv. Appl. Probab. **28**, 567–587 (1996)
6. Chakravarthy, S.R.: Introduction to Matrix-Analytic Methods in Queues 1: Analytical and Simulation Approach - Basics. ISTE Ltd., London and John Wiley and Sons, New York (2022)
7. Chakravarthy, S.R.: Introduction to Matrix-Analytic Methods in Queues 2: Analytical and Simulation Approach - Queues and Simulation. ISTE Ltd., London and John Wiley and Sons, New York (2022)
8. Lucantoni, D.: New results on the single server queue with a batch Markovian arrival process. Commun.-Stat.-Stoch. Model. **7**, 1–46 (1991)
9. Dudin, A.N., Klimenok, V.I., Vishnevsky, V.M.: The Theory of Queuing Systems with Correlated Flows. Springer, Cham (2020). https://doi.org/10.1007/978-3-030-32072-0
10. Graham A.: Kronecker Products and Matrix Calculus with Applications. Courier Dover Publications (2018)
11. Ramaswami, V.: Independent Markov processes in parallel. Comm. Statist.-Stoch. Models **1**, 419–432 (1985)
12. Dudin, A., et al.: Analysis of single-server multi-class queue with unreliable service, batch correlated arrivals, customers impatience, and dynamical change of priorities. Mathematics **9**(11), 1257 (2021)
13. Kim, C.S., et al.: Mathematical models for the operation of a cell with bandwidth sharing and moving users. IEEE Trans. Wireless Commun. **19**(2), 744–755 (2020)
14. Sas, B., et al.: Modelling the time-varying cell capacity in LTE networks. Telecommun. Syst. **55**(2), 299–313 (2014)
15. Dudin, A., Dudin, S., Dudina, O.: Analysis of a queueing system with mixed service discipline. Methodol. Comput. Appl. Probab. **25**(2), 57 (2023)

Analysis of Queuing Systems under N Policy with Different Server Activation Strategies

Greeshma Joseph[1]([✉]), Varghese Jacob[2][iD], and Achyutha Krishnamoorthy[1][iD]

[1] CMS College, Kottayam, India
elizebethgreeshma@gmail.com
[2] Government Arts and Science College, Nadapuram, India
http://www.cmscollege.ac.in

Abstract. In this paper, we consider single server queuing systems under N policy with different activation strategies for the server. Customers arrive according to a Markovian Arrival Process and the service time follows a phase type distribution. The activation time of the server is exponentially distributed. We obtain stationary distribution of the queuing process using the Matrix Geometric Method. Using these distribution we calculate performance measures of the system. We analyse these models with numerical examples in order to have a comparison of the performance measures associated with it. A cost function is developed for each model and the optimum value of N is investigated numerically. Finally, a comparison study between the models is presented.

Keywords: queuing systems under N policy · activation time · forced activation · Matrix Geometric Method · cost analysis

1 Introduction

Many real-world situations involve queuing systems in which the server may be unavailable occasionally when the system becomes empty. In such systems, the server's idle times or service facilities can be utilised for other important purposes in the system to enhance its efficiency. These queuing systems have been investigated and applied extensively in various engineering systems, like production units, inventory systems, computers networks, flexible manufacturing domains, and telecommunication systems.

In literature there are a large number of policies regulating vacations in queuing systems like single vacation policy, working vacation policy, multiple vacation policy, N policy, T policy, D policy Krishnamoorthy, Ushakumari [1], and some of their combinations Chakravarthy et al. [2]. The concept of N policy was first introduced by Yadin and Naor [3] in an $M/G/1$ queueing system. In N policy models, the server is turned on when the total number of customers in the system reaches a desired level $N \geq 1$, and is turned off whenever the system

---

The original version of the chapter has been revised. The chapter authors' affiliations have been updated. A correction to this chapter can be found at
https://doi.org/10.1007/978-3-031-50482-2_42

becomes empty. Among these operating policies, N policy is more acceptable in the area of optimal design and control of queues due to its straightforward nature. An extensive research has been done in N policy queuing models. Baker [4] examined operating policies in the $M/M/1$ queue with exponential start-up. Tian et al. [5] worked on an $M/M/1$ queue with single working vacation where the server gives service at a lower rate during it's vacation. Recently Sreenivasan et al. [6] extended it to $MAP/PH/1$ queue with working vacations, vacation interruptions and N policy. The main objective behind control mechanisms in queuing systems is to find effective strategies to turn the server on and off in order to maintain a minimum long-run cost. It can be seen that in many production systems, when all jobs are finished, the machine remains idle until the arrival of the next job. If there is a cost for operating the machine, the most efficient way to operate the system is to shut down the machine when no jobs are left and bring it up again as the number of jobs in queue reaches a predetermined level N. Also there are many real life situations in which the server needs time to do preparatory works in the system before starting its service.

Recently, Greeshma et al. [7] have conducted a comparative study of queuing systems under N policy with variant of activation times and customer impatience. In this model, the authors have introduced the notion of forced activation into the queuing systems, where the server requires positive amount of time to activate before service. The work of [7] is extended in the following manner in this paper. Firstly we use a more general point process for modelling the arrivals, known as Markovian Arrival Process (MAP). Then for services we use the phase type distribution, which is the generalised version of some of the prominent distributions like exponential, hyperexponential and Erlang.

In stochastic modelling, Poisson processes are widely used because of their simple and easily tractable nature. MAP generalizes the Poisson process by retaining its tractability for modelling purposes. The MAP, developed by Neuts [8] provides a way to model correlated arrivals. Suppose that an irreducible Continuous Time Markov Chain(CTMC) has m transient states. The sojourn time in state i is exponential distributed with parameter λ_i. After completing sojourn time in state i, with probability $p_{ij}(0)$ the CTMC can move to any of the $m-1$ transient states (except i) without an arrival and with probability $p_{ij}(1)$ the CTMC can move to any of the m transient states including the state i with an arrival. Only with an arrival, the CTMC can go from a state to itself. A MAP is completely described by a pair of matrices (D_0, D_1), where $D_0 = (d_{ij}^{(0)})$ and $D_1 = (d_{ij}^{(1)})$ such that $d_{ii}^{(0)} = -\lambda_i$, $1 \leq i \leq m$, $d_{ij}^{(0)} = \lambda_i p_{ij}(0)$, for $j \neq i$, $1 \leq i, j \leq m$ and $d_{ij}^{(1)} = \lambda_i p_{ij}(1)$, $1 \leq i, j \leq m$, with $\sum_{j} p_{ij}(0) + \sum_{j \neq i} p_{ij}(1) = 1$, for $1 \leq i \leq m$. The underlying Markov chain has the generator matrix Q^* given by $Q^* = D_0 + D_1$. Let ξ be the steady state probability vector of the Markov process with generator Q^*. Hence, ξ is the unique solution to the equation,

$$\xi Q^* = \mathbf{0}; \ \xi \mathbf{e} = 1.$$

The expected number of arrivals per unit time will be $\lambda = \xi D_1 \mathbf{e}$. For more details see Chakravarthy [9].

Exponential distribution plays a prominent role in modelling queuing systems because of its tractable nature and memoryless property. However, the assumptions of exponential distribution are highly constrained. Neuts [10] developed phase type distributions to overcome the limitations of the exponential distribution. A phase type distribution is defined as the distribution of the time to enter an absorbing state $n+1$ from the set of n transient states $\{1, 2, \ldots, n\}$ of a Markov process. A phase type distribution is represented by a two tuple (α, T), where α is a substochastic vector of length n according to which the process selects the initial state from $\{1, 2, \ldots, n\}$ and $T = (t_{ij})$ is subgenerator matrix of order n such that $\begin{pmatrix} T & T^0 \\ 0 & 0 \end{pmatrix}$ is the infinitesimal generator matrix of the Markov process, given the column vector $T^0 = (\tau_1{}^o, \tau_2{}^o, \ldots, \tau_n{}^o)'$ satisfies the condition $T\mathbf{e} + T^0 = \mathbf{0}$. For more details about phase type distributions see Qi-Ming [11].

The symbol $\otimes$ denotes the Kronecker product of two matrices and $\oplus$ denotes Kronecker sum of two matrices. Given a $p \times q$ order matrix A and a $r \times s$ order matrix B, then $A \otimes B$ is a $pr \times qs$ order matrix, whose (i, j) th element will be $a_{ij}B$. Whereas if C and D are square matrices having order p and r respectively, then $C \oplus B = C \otimes I_r + I_p \otimes D$. Let I and $\mathbf{e}$ denotes, respectively, an identity matrix and a column vector of 1's of appropriate dimension.

This paper is structured as follows. In Sect. 2, we provide model descriptions. In Sect. 3, the mathematical model is developed, and its steady-state analytical solutions are derived. Various system performance measures are presented in Sect. 4. The impact of different system parameters on the performance measures is discussed numerically in Sect. 5. In Sect. 6, cost analysis of the models is performed.

2   Model Description

We consider two single server queuing models to which customers arrive according to a Markovian Arrival Process with parameter matrices D_0 and D_1 of dimension m. The service times follows a phase type distribution with representation (α, T) of order n. And the service rate is, $\mu = [\alpha(-T)^{-1}\mathbf{e}]^{-1}$. Both the queues are considered under N policy. The customers are served in the order of their arrival.

2.1   Model 1

In this model server requires a positive amount of time to start its service (that is, not instantaneous service). As soon as the total number of arrivals in the queue reaches the pre-determined threshold N $(1 \leq N < \infty)$, the server initiates the activation process. But it takes a random duration of time to get activated. When the sever is in activation mode there will be no service. Soon after completing the activation process, server begins its service. And the server is deactivated when all the customers present in the system are served. We assume that activation time of the server follows an exponential distribution with parameter θ.

Let $X_1(t)$ denotes the number of customers in the system at time t, $J_1(t)$ denotes the status of the server at time t,

$$J_1(t) = \begin{cases} 0 & \text{if the server is idle at time } t, \\ 1 & \text{if the server is in activation mode at time } t, \\ 2 & \text{if server is busy at time t.} \end{cases}$$

$S_1(t)$ denotes the phase of the service process when the server is busy, and $M_1(t)$ denotes the phase of the arrival process at time t.

Then $\{(X_1(t), J_1(t), S_1(t), M_1(t)) : t \geq 0\}$ is a four-dimensional Continuous Time Markov Chain with state space:

$$\Omega_1 = \bigcup_{i=0}^{\infty} l(i)$$

where $l(0) = \{(0, 0, *, 1), (0, 0, *, 2), \ldots, (0, 0, *, m)\}$
and for $1 \leq h \leq n$ and $1 \leq l \leq m$,
$l(i) = \{(i, 0, h, l) : 1 \leq i \leq N - 1\} \cup \{(i, 1, h, l) : i \geq N\} \cup \{(i, 2, h, l) : i \geq 1\}$
When $X_1(t) = 0$, $S_1(t)$ does not have any role and will not be traced. Until a customer arrives, this state component is considered as a frozen. In order to avoid confusion in coordinates of state space elements, we put '*' in the place of $S_1(t)$, when the number of customers in the system is zero. In this situation $J_1(t)$ always remains zero and the component, $M_1(t)$, of the arrival process only needs to be taken into account.

2.2 Model 2

In our previous model, the server requires activation time which is exponentially distributed with parameter θ. But the server may take long duration of time (even though finite) to get activated. This long waiting for service can cause many constraints to the system like time shortage, energy consumption, holding cost of customers etc. Taking into account the impact of these constraints on the system, it would be better to activate the server by giving an extra force. So we consider forced activation of the server as in [7], if the server does not get activated until the realization of a certain stage. In this model the server is forcefully activated, if the server does not get naturally activated up to the accumulation of a predetermined number $N + k$ ($k \geq 0$) of customers. As a result, the server will be in busy state when there are more than $N + k$ number of customers in the system.

Let $X_2(t)$ denotes the number of customers in the system at time t, $J_2(t)$ denotes the status of the server at time t, which is same as $J_1(t)$, $S_2(t)$ denotes the phase of the service process when the server is busy, and $M_2(t)$ denotes the phase of the arrival process at time t.

Then $\{(X_2(t), J_2(t), S_2(t), M_2(t)) : t \geq 0\}$ is a four-dimensional Continuous Time Markov Chain with state space:

$$\Omega_2 = \bigcup_{i=0}^{\infty} l(i)$$

where $l(0) = \{(0,0,*,1),(0,0,*,2),\ldots,(0,0,*,m)\}$
and for $1 \leq h \leq n$ and $1 \leq l \leq m$,
$l(i) = \{(i,0,h,l) : 1 \leq i \leq N-1\} \cup \{(i,1,h,l) : N \leq i \leq N+k\} \cup \{(i,2,h,l) : i \geq 1\}$

3   Steady-State Analysis

Using the lexicographical sequence for the states, the infinitesimal generator matrix Q of the QBD process in these models has the form,

$$Q = \begin{pmatrix} B_{00} & B_{01} & & & \\ B_{10} & Q_1 & Q_0 & & \\ & Q_2 & Q_1 & Q_0 & \\ & & \ddots & \ddots & \ddots \end{pmatrix}$$

where
for Model 1,

$$B_{00} = \begin{matrix} 0 \\ 1 \\ 2 \\ \vdots \\ N-1 \end{matrix} \begin{pmatrix} D_0 & Q_{0,1} & & & \\ Q_{1,0} & Q_{1,1} & Q_0 & & \\ & Q_2 & Q_{2,2} & Q_0 & \\ & & \ddots & \ddots & \ddots \\ & & & Q_2 & Q_{N-1,N-1} & Q_0 \end{pmatrix}$$

where the submatrices of B_{00} are as shown below.

$\star$ $Q_{0,1} = \begin{bmatrix} \alpha \otimes D_1 & \mathbf{0} \end{bmatrix}$ $Q_{1,0} = \begin{bmatrix} \mathbf{0} \\ T^0 \otimes I \end{bmatrix}$

$\star$ $Q_{i,i} = \begin{bmatrix} T \oplus D_0 + diag(\tau_1^o, \tau_2^o, \ldots, \tau_n^o) \otimes I & \mathbf{0} \\ \mathbf{0} & T \oplus D_0 \end{bmatrix}$ for $1 \leq i \leq N-1$

$\star$ $Q_0 = \begin{bmatrix} I \otimes D_1 & \mathbf{0} \\ \mathbf{0} & I \otimes D_1 \end{bmatrix}$ $Q_2 = \begin{bmatrix} \mathbf{0} & \mathbf{0} \\ \mathbf{0} & T^0\alpha \otimes I \end{bmatrix}$

$\star$ $Q_1 = \begin{bmatrix} T \oplus D_0 + diag(\tau_1^o, \tau_2^o, \ldots, \tau_n^o) \otimes I - \theta I & \theta I \\ \mathbf{0} & T \oplus D_0 \end{bmatrix}$

$B_{01} = \begin{bmatrix} \mathbf{0} \\ Q_2 \end{bmatrix}$ $B_{10} = \begin{bmatrix} \mathbf{0} & Q_0 \end{bmatrix}$

for Model 2,

$$B_{00} = \begin{array}{c} 0 \\ 1 \\ 2 \\ \vdots \\ N \\ \vdots \\ N+k-1 \\ N+k \end{array} \left(\begin{array}{cccccccc} D_0 & Q_{0,1} & & & & & & \\ Q_{1,0} & Q_{1,1} & Q_{1,2} & & & & & \\ & Q_{2,1} & Q_{2,2} & Q_{2,3} & & & & \\ & & \ddots & \ddots & & \ddots & & \\ & & & Q_{N,N-1} & Q_{N,N} & & Q_{N,N+1} & \\ & & & & \ddots & & \ddots & \\ & & & & & & \ddots & \\ & & & & Q_{N+k-1,N+k-2} & Q_{N+k-1,N+k-1} & Q_{N+k-1,N+k} & \\ & & & & & Q_{N+k,N+k-1} & Q_{N+k,N+k} & Q_{N+k,N+k} \end{array} \right)$$

where the submatrices of B_{00} are as shown below.

$\star\ Q_{0,1} = \begin{bmatrix} \alpha \otimes D_1 & 0 \end{bmatrix} \qquad Q_{1,0} = \begin{bmatrix} 0 \\ T^0 \otimes I \end{bmatrix}$

$\star\ Q_{i,i-1} = \begin{bmatrix} 0 & 0 \\ 0 & T^0 \alpha \otimes I \end{bmatrix} \quad$ for $2 \le i \le N+k$

$\star\ Q_{i,i} = \begin{bmatrix} T \oplus D_0 + diag(\tau_1^o, \tau_2^o, \ldots, \tau_n^o) \otimes I & 0 \\ 0 & T \oplus D_0 \end{bmatrix}$ for $1 \le i \le N-1$

$\star\ Q_{i,i} = \begin{bmatrix} T \oplus D_0 + diag(\tau_1^o, \tau_2^o, \ldots, \tau_n^o) \otimes I - \theta I & \theta I \\ 0 & T \oplus D_0 \end{bmatrix}$ for $N \le i \le N+k-1$

$\star\ Q_{i,i} = \begin{bmatrix} T \oplus D_0 + diag(\tau_1^o, \tau_2^o, \ldots, \tau_n^o) \otimes I + I \otimes diag\left(\sum_{i=1}^{m} d_{1i}{}^{(1)}, \ldots, \sum_{i=1}^{m} d_{mi}{}^{(1)} \right) - \theta I & \theta I \\ 0 & T \oplus D_0 \end{bmatrix}$

for $i = N+k$

$\star\ Q_{i,i+1} = \begin{bmatrix} I \otimes D_1 & 0 \\ 0 & I \otimes D_1 \end{bmatrix} \quad$ for $1 \le i \le N+k-1$

$Q_0 = I \otimes D_1 \qquad\qquad Q_1 = T \oplus D_0 \qquad\qquad Q_2 = T^0 \alpha \otimes I$

$B_{01} = \begin{bmatrix} 0 \\ Q_2 \end{bmatrix} \qquad\qquad B_{10} = \begin{bmatrix} 0 & Q_0 \end{bmatrix}$

The structure of Q indicates that these two models can be studied as a Level Independent Quasi-Birth-Death (LIQBD) process. We apply Matrix Geometric Method for analysing them.

3.1   Stability Condition

Let π denote the steady-state probability vector of the generator $Q_0 + Q_1 + Q_2$. That is, $\pi(Q_0 + Q_1 + Q_2) = 0$, $\pi e = 1$. The LIQBD description of the models indicates that the queuing system is stable (see, Neuts [12]) if and only if

$$\pi Q_0 e < \pi Q_2 e$$

3.2 Stationary Distribution

The stationary distribution of the Markov chain under consideration is obtained by solving the set of equations

$$\mathbf{x}Q = 0; \quad \mathbf{x}\mathbf{e} = 1 \tag{1}$$

where Q is the infinitesimal generator of the LIQBD describing the above process. Then $\mathbf{x}$ can be partitioned as,

$$\mathbf{x} = (\mathbf{x_0}, \mathbf{x_1}, \mathbf{x_2}, \ldots) \tag{2}$$

From $\mathbf{x}Q = 0$, we obtain the following equations;

$$\mathbf{x_0}B_{00} + \mathbf{x_1}B_{10} = 0$$
$$\mathbf{x_0}B_{01} + \mathbf{x_1}Q_1 + \mathbf{x_2}Q_2 = 0$$
$$\mathbf{x_{i-1}}Q_0 + \mathbf{x_i}Q_1 + \mathbf{x_{i+1}}Q_2 = 0, \ i \geq 2$$

By Matrix Geometric Method (Neuts [12]), the sub vectors $\mathbf{x_i}$'s has the form

$$\mathbf{x_i} = \mathbf{x_1}R^{i-1}, \ i \geq 2$$

We obtain the rate matrix R by solving the matrix quadratic equation,

$$R^2 Q_2 + R Q_1 + Q_0 = O \tag{3}$$

and $\mathbf{x_0}$ and $\mathbf{x_1}$ satisfies

$$(\mathbf{x_0}, \mathbf{x_1}) \begin{pmatrix} B_{00} & B_{01} \\ B_{10} & Q_1 + RQ_2 \end{pmatrix} = (0, 0)$$

The normalizing condition of (1) results in

$$\mathbf{x_0}\mathbf{e} + \mathbf{x_1}(I - R)^{-1}\mathbf{e} = 1.$$

After obtaining the rate matrix R, the vector $\mathbf{x}$ can be determined by exploiting the special structure of the coefficient matrices.

For Model 1, sub-vectors of $\mathbf{x_0}$ in (2) can be expressed as,

$$\mathbf{x_0} = (\mathbf{x_0^*}, \mathbf{x_1^*}, \ldots, \mathbf{x_{N-1}^*})$$

where the row vectors, $\mathbf{x_0^*}$ has dimension m, $\mathbf{x_1^*}, \ldots, \mathbf{x_{N-1}^*}$ are of dimension 2nm and the sub-vectors $\mathbf{x_1}, \mathbf{x_2}, \ldots$ of $\mathbf{x}$ are of dimension 2nm.

For use in the sequel, we subdivide

$$\mathbf{x_i^*} = (\mathbf{u_i^*}, \mathbf{v_i^*}), \ 1 \leq i \leq N - 1 \text{ and } \mathbf{x_i} = (\mathbf{w_i}, \mathbf{v_i}), \ i \geq 1$$

where $\mathbf{u_i^*}$, $\mathbf{v_i^*}$, $\mathbf{w_i}$ and $\mathbf{v_i}$ are of dimension nm.

For Model 2, sub-vectors of $\mathbf{x_0}$ in (2) can be expressed as,

$$\mathbf{x_0} = (\mathbf{x_0}^*, \mathbf{x_1}^*, \ldots, \mathbf{x_N}^*, \ldots, \mathbf{x_{N+k}}^*)$$

where the row vectors, $\mathbf{x_0}^*$ has dimension m, $\mathbf{x_1}^*, \mathbf{x_2}^*, \ldots, \mathbf{x_N}^*, \ldots, \mathbf{x_{N+k}}^*$ are of dimension 2nm and the sub-vectors $\mathbf{x_1}, \mathbf{x_2}, \ldots$ of $\mathbf{x}$ are of dimension nm.

For use in the sequel, we subdivide

$$\mathbf{x_i}^* = (\mathbf{u_i}^*, \mathbf{v_i}^*),\ 1 \leq i \leq N-1 \text{ and } \mathbf{x_i}^* = (\mathbf{w_i}^*, \mathbf{v_i}^*),\ N \leq i \leq N+k$$

where $\mathbf{u_i}^*$, $\mathbf{v_i}^*$ and $\mathbf{w_i}^*$ has dimension nm.

4   System Performance Measures of the Models

In this section, we evaluate important performance characteristics of the models discussed in the previous sections. The measures are given below along with their formula for computation.

– The probability that the server is idle, $p_{idle} = \mathbf{x_0}^* \mathbf{e} + \sum_{i=1}^{N-1} \mathbf{u_i}^* \mathbf{e}$

– The probability that the server is in activation mode,

  ⋆ for Model 2, $p_{act} = \sum_{i=1}^{\infty} \mathbf{w_i} \mathbf{e}$

  ⋆ for Model 3, $p_{act} = \sum_{i=N}^{N+k} \mathbf{w_i}^* \mathbf{e}$

– The probability that the server is busy,

  ⋆ for Model 2, $p_{busy} = \sum_{i=1}^{N-1} \mathbf{v_i}^* \mathbf{e} + \sum_{i=1}^{\infty} \mathbf{v_i} \mathbf{e}$

  ⋆ for Model 3, $p_{busy} = \sum_{i=1}^{N+k} \mathbf{v_i}^* \mathbf{e} + \sum_{i=1}^{\infty} \mathbf{x_i} \mathbf{e}$

– The expected number of customers in the system when server is idle,

$$E_{idle} = \sum_{i=1}^{N-1} i \mathbf{u_i}^* \mathbf{e}$$

– The expected number of customers in the system when server is in activation mode,

  ⋆ for Model 2, $E_{act} = N + \sum_{i=1}^{\infty} (N-1+i)\mathbf{w_i} \mathbf{e}$

  ⋆ for Model 3, $E_{act} = N + \sum_{i=N}^{N+k} i \mathbf{w_i}^* \mathbf{e}$

– The expected number of customers in the system when server is busy,

  ⋆ for Model 2, $E_{busy} = \sum_{i=1}^{N-1} i \mathbf{v_i}^* \mathbf{e} + \sum_{i=1}^{\infty} (N-1+i)\mathbf{v_i} \mathbf{e}$

  ⋆ for Model 3, $E_{busy} = \sum_{i=1}^{N+k} i \mathbf{v_i}^* \mathbf{e} + \sum_{i=1}^{\infty} (N+k+i)\mathbf{x_i} \mathbf{e}$

5   Numerical Examples

In this section we consider several representative examples to illustrate the models numerically.

For arrival process we consider the following three sets of matrices for D_0 and D_1.

⋆ MAP with positive correlation (MAP^p)

$$D_0 = \begin{pmatrix} -2 & 2 & 0 \\ 0 & -2 & 0 \\ 0 & 0 & -450.5 \end{pmatrix} \text{ and } D_1 = \begin{pmatrix} 0 & 0 & 0 \\ 1.98 & 0 & 0.02 \\ 4.505 & 0 & 445.995 \end{pmatrix}.$$

⋆ MAP with negative correlation (MAP^n)

$$D_0 = \begin{pmatrix} -2 & 2 & 0 \\ 0 & -2 & 0 \\ 0 & 0 & -450.5 \end{pmatrix} \text{ and } D_1 = \begin{pmatrix} 0 & 0 & 0 \\ 0.02 & 0 & 1.98 \\ 445.955 & 0 & 4.505 \end{pmatrix}.$$

⋆ MAP with zero correlation (Hyperexponential) (MAP^0)

$$D_0 = \begin{pmatrix} -10 & 0 \\ 0 & -1 \end{pmatrix} \text{ and } D_1 = \begin{pmatrix} 9 & 1 \\ 0.9 & 0.1 \end{pmatrix}.$$

The three MAP processes described above are normalized to have an arrival rate 1. The arrival process denoted as MAP^p has correlated arrivals with correlation between two consecutive inter-arrival times as 0.4889, the arrival process corresponding to the one denoted as MAP^n has a negative correlation with value -0.4889 and the arrival process denoted as MAP^0 has non correlated arrivals. For the service distribution we consider the following phase type distribution.

Let $\alpha = (0.3, \ 0.7)$, $T = \begin{pmatrix} -16 & 8 \\ 10 & -19 \end{pmatrix}$, $T^0 = \begin{pmatrix} 8 \\ 9 \end{pmatrix}$.

In this section, we examine how each one of the system performance measures behave under various scenarios. We examine the effect of variation of N and θ on the system measures: expected number of customers when server idle (E_{idle}), expected number of customers when server is under activation process (E_{act}) and expected number of customers when server busy (E_{busy}). Since other performance measures shares more or less same behaviour with E_{idle}, E_{act} and E_{busy}, we are not displaying it. For each model, we carried out many numerical experiments by varying different parameters while fixing others as mentioned in each table. We summarize the observations based on the Tables 1 and 2 as follows.

– Table 1: With the increase of N, the duration of idle mode of the server increases. It would result in an increase of E_{idle} for all models with positive, negative, and non correlated arrivals. The number of customers going to the server in activation mode will increase as N rises. As a result of this, E_{act} increases for both models with all types of arrivals. With the increase of N, the number of customers accumulated in the busy server will also increase. So the server remains busy until all the accumulated customers are served. Thus, E_{busy} is an increasing function of N for all models with positive, negative,

Table 1. Effect of N on $E_{idle}, E_{act}, E_{busy}$ fix $\theta = 5$ and $k = 2$

MAP^p

N	E_{idle}		E_{act}		E_{busy}	
	Model 1	Model 2	Model 1	Model 2	Model 1	Model 2
2	0.3802	0.3804	2.2356	2.1633	15.4211	15.2682
4	1.1494	1.1499	4.2340	4.1632	15.6505	15.5011
6	1.9100	1.9109	6.2332	6.1636	15.8798	15.7325
8	2.6626	2.6638	8.2329	8.1644	16.1082	15.9625
10	3.4089	3.4104	10.2329	10.1653	16.3356	16.1913
12	4.1499	4.1517	12.2331	12.1664	16.5621	16.4188

MAP^n

N	E_{idle}		E_{act}		E_{busy}	
	Model 1	Model 2	Model 1	Model 2	Model 1	Model 2
2	0.0092	0.0092	2.3333	2.3313	0.4602	0.4530
4	0.8003	0.8009	4.3263	4.3252	0.6888	0.6840
6	1.5902	1.5910	6.3216	6.3208	0.9213	0.9172
8	2.3733	2.3741	8.3184	8.3177	1.1546	1.1508
10	3.1517	3.1526	10.3162	10.3156	1.3884	1.3848
12	3.9274	3.9283	12.3145	12.3140	1.6225	1.6190

MAP^0

N	E_{idle}		E_{act}		E_{busy}	
	Model 1	Model 2	Model 1	Model 2	Model 1	Model 2
2	0.0676	0.0724	2.2570	2.2285	4.4245	3.9411
4	0.2821	0.2931	4.2776	4.2570	4.9458	4.5570
6	0.5539	0.5690	6.2957	6.2794	5.5393	5.1957
8	0.8593	0.8777	8.3104	8.2968	6.1578	5.8424
10	1.1860	1.2071	10.3220	10.3105	6.7851	6.4897
12	1.5266	1.5500	12.3313	12.3213	7.4146	7.1344

and zero correlated arrivals. We can observe that E_{act} and E_{busy} takes lesser values in Model 2 compared to Model 1 with all types of arrivals. Whereas E_{idle} has higher values in Model 2 when compared with Model 1. This is due to the impact of forced activation in Model 2. When comparing the measure E_{busy} among all the arrival processes considered, we can see that positive correlated arrivals yields a much higher value than others in both models.

– Table 2: An increase in activation rate θ reduces the mean duration of the activation time of the server, resulting in faster activation of the server. Once the threshold value N is reached and the activation of server happens, the customers are cleared out quickly. Hence there will be increase in E_{idle} for all models with positive, negative, and zero correlated arrivals. When θ increases, the measure E_{act} decreases for both models as expected with all the three arrival process considered. In Model 1, as θ increases activation of server happens at a faster rate, which results in decrease of E_{busy} with all the arrival processes. In Model 2, E_{busy} increases initially and then decreases for negative correlated arrivals. However, E_{busy} increases in Model 2 with positive and non correlated arrivals. Also, with the variation of θ it can be observed that E_{act} and E_{busy} takes lesser values in Model 2 compared to Model 1 with all types of arrivals. Whereas E_{idle} has higher values in Model 2 when compared with Model 1. This is due to the impact of forced activation in Model 2.

Table 2. Effect of θ on $E_{idle}, E_{act}, E_{busy}$ fix $N = 5$ and $k = 2$

MAP^p

θ	E_{idle}		E_{act}		E_{busy}	
	Model 1	Model 2	Model 1	Model 2	Model 1	Model 2
0.5	1.1201	1.1505	7.9262	6.4452	16.5425	14.4175
1	1.3179	1.3281	6.4080	5.7810	16.1239	15.3435
2	1.4444	1.4471	5.6608	5.4020	15.9109	15.5834
3	1.4915	1.4929	5.4190	5.2703	15.8341	15.6101
4	1.5159	1.5169	5.3018	5.2036	15.7923	15.6154
5	1.5308	1.5315	5.2335	5.1634	15.7652	15.6170

MAP^n

θ	E_{idle}		E_{act}		E_{busy}	
	Model 1	Model 2	Model 1	Model 2	Model 1	Model 2
0.5	0.9632	1.0259	8.0964	7.2988	1.5587	0.8082
1	1.2222	1.2549	6.5610	6.3626	1.1726	0.9045
2	1.4105	1.4220	5.7895	5.7474	1.0002	0.9242
3	1.4859	1.4911	5.5304	5.5145	0.9482	0.9159
4	1.5262	1.5290	5.4000	5.3922	0.9238	0.9070
5	1.5512	1.5529	5.3214	5.3170	0.9098	0.9000

MAP^0

θ	E_{idle}		E_{act}		E_{busy}	
	Model 1	Model 2	Model 1	Model 2	Model 1	Model 2
0.5	0.2015	0.2888	8.6335	7.0608	10.8331	3.8574
1	0.2799	0.3456	6.7072	6.2046	7.6936	4.4505
2	0.3490	0.3877	5.7912	5.6527	6.1519	4.7669
3	0.3812	0.4066	5.5041	5.4445	5.6430	4.8483
4	0.4001	0.4180	5.3668	5.3356	5.3895	4.8717
5	0.4127	0.4259	5.2871	5.2689	5.2377	4.8746

6   Cost Analysis of the Models

The cost function TEC_i (Total Expected Cost per unit time) for each Model i, $i = 1, 2$ is defined as follows,

$$TEC_1 = \mu c_s + p_{idle} E_{idle} c_{h_0} + p_{act} E_{act} c_{h_1} + p_{idle} c_I + \left(\frac{\mu - \lambda}{N + \frac{\lambda}{\theta}} \right) c_a$$

$$TEC_2 = \mu c_s + p_{idle} E_{idle} c_{h_0} + p_{act} E_{act} c_{h_1} + p_{idle} c_I + \left(\frac{\theta}{\theta + \frac{\lambda}{k}} \right) \left(\frac{\mu - \lambda}{N + \frac{\lambda}{\theta}} \right) c_f$$

where

* $\star$ c_s = cost of service per unit time,
* $\star$ c_I = cost of keeping the server idle per unit time,
* $\star$ c_a = cost of activation of the server,
* $\star$ c_f = cost of forced activation of the server,
* $\star$ c_{h_0} = cost of holding each customer in the system when the server is idle per unit time,
* $\star$ c_{h_1} = cost of holding each customer in the system when the server is in activation mode per unit time.

We would like to find the optimum value of N, N^* that minimizes the cost function TEC_i. We fix $k = 2$, $\theta = 5$, $c_s = 2$, $c_{h_0} = 1.9$, $c_{h_1} = 0.5$, $c_I = 0.5$, $c_a = 5$, $c_f = 5.2$ correspondingly for each model.

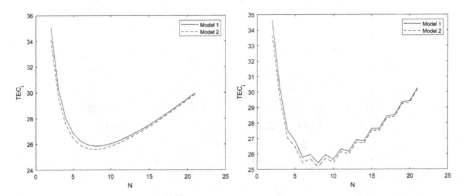

(a) MAP with positive correlated arrivals (b) MAP with negative correlated arrivals

Fig. 1. MAP with correlated arrivals

With the numerical parameters assumed here we have the following observations from Fig. 1 and Fig. 2.

– For positive correlated arrivals the optimal value of N, N^* is obtained as 8 for both models. And for the negative correlated arrivals, graph shows fluctuating

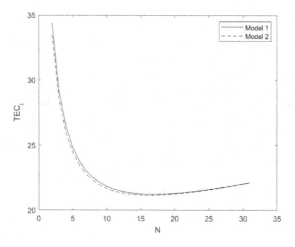

Fig. 2. MAP with zero correlated arrivals

behaviour for both models, which was surprising. However, in this case the minimum value taken by TEC_i for $i = 1, 2$ is at $N = 8$. The value of N^* is 17 and 16 respectively for Model 1 and Model 2, with zero correlated arrivals. When comparing the Total Expected Cost between different types of arrivals considered here, we can observe that it is maximum for positive correlated arrivals and minimum for zero correlated arrivals in both models.

– By comparing these two models, the impact of forced activation on the total expected cost is evident. For all values of N and with all types of arrivals, the total expected cost to the system is minimum in Model 2. This is due to the forced activation implemented on Model 2. Even though the cost of forced activation of the server is higher than the cost of natural activation of the server, the total cost of the system is reduced.

7   Conclusion

In this paper, we studied two single server queues under N policy with MAP arrivals and phase type services. We have considered different activation strategies of the server in these two models. All the models are exhaustively analysed using Matrix Geometric Method. The influence of various parameters on the system measures are also investigated through numerical examples including different types of MAP arrivals. A suitable cost function is constructed for each model. We have demonstrated through cost analysis that, in queuing systems where the server requires activation time before service, forced activation of the server would be helpful in lowering the total cost of the system. This notion can be applied to many real life queuing systems.

Acknowledgement. Greeshma Joseph's research is supported by the Kerala State Council for Science, Technology and Environment(KSCSTE/2092/2019-FSHP-MAIN).

References

1. Krishnamoorthy, A., Ushakumari, P.V.: k-out-of-n: G system with repair: the D-policy. Comput. Oper. Res. **28**, 973–981 (2001)
2. Chakravarthy, S.R., Krishnamoorthy, A., Ushakumari, P.V.: A k-Out-of-n reliability system with an unreliable server and phase type repairs and services: the (N, T) policy. J. Appl. Math. Stoch. Anal. **14**, 361–380 (2001)
3. Yadin, M., Naor, P.: Queueing system with a removable service station. Oper. Res. Q. **14**, 393–405 (1963)
4. Baker, K.R.: A note on operating policies for the queue M/M/1 with exponential start up. Inf. Syst. Oper. Res. **11**, 71–72 (1973)
5. Tian, N., Zhao, X., Wang, K.: The M/M/1 queue with single working vacation. Int. J. Inf. Manage. Sci. **19**(4), 621–634 (2008)
6. Sreenivasan, C., Chakravarthy, S.R., Krishnamoorthy, A.: MAP/PH/1 queue with working vacations, vacation interruptions and N policy. Appl. Math. Model. **37**, 3879–3893 (2013)
7. Joseph, G., Jacob, V.: A comparative study of queuing systems with variant of activation times and impatience under N policy. J. Indian Soc. Probab. Stat. (accepted) (2024)
8. Neuts, M.F.: A versatile Markovian point process. J. Appl. Probab. **16**(4), 764–779 (1979)
9. Chakravarthy, S.R.: Markovian arrival processes. In: Wiley Encyclopedia of Operations Research and Management Science (2011)
10. Neuts, M.F.: Probability Distributions of Phase Type. Liber Amicorum Prof. Emeritus H, Florin (1975)
11. He, Q.: Fundamentals of Matrix-Analytic Methods. Springer, New York (2014).https://doi.org/10.1007/978-1-4614-7330-5
12. Neuts, M.F.: Matrix-Geometric Solutions in Stochastic Models- An Algorithmic Approach, 2nd edn. Dover Publications Inc., New York (1994)
13. Dudin, Alexander N., Klimenok, Valentina I., Vishnevsky, Vladimir M.: The Theory of Queuing Systems with Correlated Flows. Springer, Cham (2020). https://doi.org/10.1007/978-3-030-32072-0
14. Jacob, V.: Analysis of customer induced interruption in a retrial queuing system with classical retrial policy. Int. J. Appl. Eng. Res. **15**(5), 445–451 (2020)
15. Neuts, M.F., Rao, B.M.: Numerical investigation of a multiserver retrial model. Queueing Syst. **7**, 169–190 (1990)
16. Lucantoni, D.M.: New results on the single server queue with a batch Markovian arrival process. Commun. Stat. Stoch. Model. **7**(1), 1–46 (1991)

On Asymptotic Insensitivity of Reliability Function of a 2-out-of-n Model Under Quick Recovery of Its Components

Vladimir Rykov[1,2,3] and Nika Ivanova[1,4(✉)]

[1] Peoples' Friendship University of Russia (RUDN University),
6 Miklukho-Maklaya Street, Moscow 117198, Russia
nm_ivanova@bk.ru
[2] Gubkin Russian State Oil and Gas University, 65 Leninsky Prospekt,
Moscow 119991, Russia
[3] Institute for Transmission Information Problems (Named after A.A.
Kharkevich) RAS, Bolshoy Karetny, 19, GSP-4, Moscow, Russia
[4] V.A. Trapeznikov Institute of Control Sciences of Russian Academy of Sciences,
65 Profsoyuznaya Street, Moscow 117997, Russia

Abstract. In some previous papers, closed-form representations for the reliability characteristics of k-out-of-n models with an exponential distribution of components lifetimes and arbitrary distributions of their repair times have been found. In those investigations, the markovization method was applied for calculation of the main reliability characteristics, including the reliability function in terms of Laplace transform. In the recent paper, the reliability function is calculated for a repairable 2-out-of-n model in the same assumptions about life and repair times distributions. The problem of its sensitivity in case of quick recovery of system's components and exponentially distributed repair time is discussed with the same example. It is proved that the reliability function in scale of its mean time to failure is an exponential one. An estimation of convergence rate is also obtained.

Keywords: k-out-of-n model · reliability · sensitivity analysis · quick recovery

1 Introduction

Many real technical systems can be described with a k-out-of-n model, which represents a system consists of n components and fails due to its k components' failure. The k-out-of-n model is an example of a redundant system, and it is of interest from both theoretical and practical points of view [1]. From the theoretical point of view, this provides extensive opportunities for creation and application of new mathematical methods and algorithms. There is a broad bibliography on the study of various types of k-out-of-n models now (see, for example, the book by Kuo and Zuo [2]). Among them, consecutive, weighted, binary and

© The Author(s), under exclusive license to Springer Nature Switzerland AG 2024
V. M. Vishnevskiy et al. (Eds.): DCCN 2023, LNCS 14123, pp. 185–196, 2024.
https://doi.org/10.1007/978-3-031-50482-2_15

other k-out-of-n models stand out. These models are considered under various assumptions about life and repair time distributions, the dependence of components, types of their restoration, and redistribution of the load from the failed components to survivors. See the review of these investigations in [3]. From the practical point of view, there is a huge number of real applications where such models can be used.

A high reliability of any system can be achieved through redundancy and "quick" recovery of its failed components. A number of papers have been devoted to the study of various redundant systems under quick recovery. In particular, B. Gnedenko [4,5] considered a double redundant system with Poisson input flow and an arbitrarily distributed repair time with the help of the theory of regenerative random processes. A. Solov'ev [6] investigated a cold redundant system with arbitrarily distributed life and repair time, using the theory of analytic functions. Later in [7,8] a similar approach to the analysis of complex systems under quick recovery is proposed. A number of papers on this topic are presented by Ya. Genis [9–11], where some estimates of system reliability indicators were obtained under broad assumptions regarding the maintenance and operation mode, failure criteria, types of redundancy, lifetime distribution of system's components and others.

As a continuation of these studies, the issue of the convergence rate of system lifetime distribution, including its convergence to the exponential one, is raised. In fact, the calculation of system lifetime distribution can be succeeded only under very strong assumptions about life and repair time distribution of system's components. As a part of the study of various redundant reliability systems, some works by I. Kovalenko and V. Kalashnikov [12,13] are dedicated to this problem. The direction of such research is still relevant today. Recent studies of some queuing and reliability systems have been aimed at estimating the rate of convergence of their time-dependent characteristics to limit functions [14–16].

This paper continues a series of investigations of k-out-of-n models and their reliability characteristics in case of exponentially distributed components lifetime and arbitrary distributed repair time [17–19]. The current study is devoted to the calculation of the reliability function of a 2-out-of-n model, for which some previous results obtained using the method of supplementary variables are presented. The closed-form representation of the reliability function is obtained by passing to the inverse Laplace transform (LT) in the case of exponentially distributed repair time, and the problem of its sensitivity is discussed. The analysis of convergence rate of the reliability function to the exponential one under quick recovery of its components is also performed.

The paper is organized as follows. In the next section, some notations, assumptions as well as problem setting will be done. Section 3 deals with the calculation of reliability function of considered k-out-of-n model with the help of markovization method. Further investigation is presented on an example of

2-out-of-n model, for which in Sect. 4 the closed-form representation of the relia-bility function in case of exponentially distributed repair time of system's compo-nents is given. Section 5 is devoted to study of the reliability function convergence to the exponential one. The paper ends with some conclusions.

2 Notations, Assumptions, and Problem Setting

Consider a k-out-of-n hot redundant model, which consists from n simultane-ously worked components and fails due to any k components' failure. Suppose that

- component's failure arises according to a Poisson flow with intensity α and mean time $a = \alpha^{-1}$, thus, the system failure intensity in its state i is $\lambda_i = (n-i)\alpha$, $(i = \overline{0, k-1})$;
- failed system's components are repaired by a single repair facility, repair times of components are independent identically distributed (i.i.d.) ran-dom variables (r.v.) B_i and their common cumulative distribution function (c.d.f.) $B(t) = \mathbb{P}\{T \le t\}$ is absolutely continuous with probability den-sity function (p.d.f.) $b(t) = B'(t)$, mean $b = \int\limits_0^\infty (1 - B(x))dx < \infty$ and LT $\tilde{b}(s) = \int\limits_0^\infty e^{-sx}b(x)dx$ with a real variable which is a downward convex func-tion [20];
- the system's states space is denoted by E, where each state means the number of failed components, $E = \{0, 1, ..., k\}$.

   To perform reliability analysis, we describe the system behavior as a random process $J = \{J(t),\ t \ge 0\}$ on the space set E:

$$J(t) = j\ (j \in E),\ \text{if the system is in state } j \text{ in time } t.$$

Suppose that in the initial time of system's working all its components are oper-ational, $J(0) = 0$. Denote by $T = \inf\{t :\ J(t) = k\}$ the system lifetime.

   In this paper we calculate time-dependent system state probabilities (t.d.s.s.p.'s)

$$\pi_j(t) = \mathbb{P}\{J(t) = j\},\ j \in E,$$

the system reliability function $R(t)$

$$R(t) = \mathbb{P}\{T > t\}$$

and mean system lifetime

$$\mathbb{E}[T] = \int_0^\infty R(t)dt.$$

The properties of asymptotic insensitivity of the reliability function in case of quick components' recovery are also investigated.

3   Calculation of Reliability Function of k-out-of-n Model

Denote r.v. W_i time to the first system failure and its c.d.f. by

$$W(t) = \mathbb{P}\{T \leq t\} = \mathbb{P}\{J(t) = k\}$$

with corresponding p.d.f. $w(t)$ and LT $\tilde{w}(s) = \int_0^\infty e^{-st} w(t) dt$. In this paper, we calculate the reliability function as the probability of absorbing state by the expression,

$$R(t) = \mathbb{P}\{T > t\} = 1 - \mathbb{P}\{T \leq t\} = 1 - \pi_k(t) = 1 - W(t), \tag{1}$$

that provides its closed-form representation.

To study considered k-out-of-n model and to obtain its reliability function, use the so-called markovization method based on introduction of supplementary variables [21]. Thus, consider a two-dimensional process

$$Z(t) = \{J(t), X(t)\}, \quad t \geq 0,$$

where

- $J(t)$ represents the number of failed components at time t,
- the supplementary variable $X(t)$ means elapsed repair time, that is the time spent by the repair facility for restoration of the component being repaired.

Thus, $\mathcal{E} = \{0, (i, x), k : i = 1, ..., k - 1, x \in \mathbb{R}_+\}$ is a discrete-continuous states space of the process $Z(t)$, where

- 0 means that no one component failed;
- (i, x) means that i $(i = 1, ..., k - 1)$ system's components have failed, and the elapsed time of repaired component is x. At that, the boundary states of the process are $(1, 0)$ when the process passes from the state 0 to 1 and from 2 to 1; $(i, 0)$ for the process transition from the state $i + 1$ to i $(i = 2, ..., k - 2)$;
- k means the absorbing state, that is the failure of k components.

Denote by $\beta(x) = \frac{b(x)}{1 - B(x)}$ a conditional repair density of components, given elapsed repair time is x. The transition graph of the process $Z(t)$ with absorption at the state k is presented in Fig. 1.

Due to the supplementary variable method, the process $Z(t)$ is a Markov one. Suppose that its micro-states' p.d.f.'s with respect to the supplementary variable in domain $0 \leq x \leq t < \infty$ exist and denote them by

$$\pi_j(t; x) dx = \mathbb{P}\{J(t) = j, \ x < X(t) \leq x + dx\} \ (j = \overline{1, k - 1}),$$

at that corresponding macro-states probabilities are

$$\pi_j(t) = \mathbb{P}\{J(t) = j\} = \int_0^t \pi_j(t; x) dx \ (j = \overline{1, k - 1}).$$

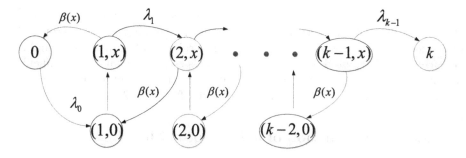

Fig. 1. Transition graph of the process $Z(t)$ with absorption

According to the presented transition graph, one can compose a system of Kolmogorov partial differential equations, which allows one to calculate t.d.s.s.p.'s of the process micro-states. For this, construct a system of finite-difference equations which are not given earlier in [17]. Comparing the corresponding probabilities of the process states at close times t and $t + \Delta$, the following equations hold,

$$\pi_0(t + \Delta) = \mathbb{P}\{J(t + \Delta) = 0\}$$
$$= (1 - \lambda_0\Delta)\pi_0(t) + \int_0^t \beta(x)dx\,\pi_1(t, x)\Delta + o(\Delta),$$
$$\pi_1(t + \Delta, x + \Delta)dx = \mathbb{P}\{J(t + \Delta) = 1,\ x + \Delta < X(t + \Delta) < x + \Delta + dx\}$$
$$= (1 - \lambda_1\Delta)(1 - \beta(x)\Delta)\pi_1(t, x) + o(\Delta),$$
$$\pi_i(t + \Delta, x + \Delta)dx = \mathbb{P}\{J(t + \Delta) = i,\ x + \Delta < X(t + \Delta) < x + \Delta + dx\}$$
$$= (1 - \lambda_i\Delta)(1 - \beta(x)\Delta)\pi_i(t, x) + \lambda_{i-1}\pi_{i-1}(t, x)\Delta + o(\Delta),$$
$$i = \overline{2, k - 2},$$
$$\pi_k(t + \Delta) = \mathbb{P}\{J(t + \Delta) = k\} = \int_0^t \lambda_{k-1}\Delta\,\pi_{k-1}(t, x)dx + o(\Delta),$$

$$(2)$$

with initial conditions which hold from the assumption that at the beginning of the system operation all its components are operational,

$$\pi_0(0) = 1, \quad \pi_i(0, x) = 0, \quad i = \overline{1, k - 1}, \quad \forall x \geq 0,$$

and boundary conditions,

$$\pi_1(t + \Delta, 0)\Delta = \lambda_0\pi_0(t)\Delta + \int_0^t \pi_2(t, x)\beta(x)dx\,\Delta + o(\Delta),$$
$$\pi_i(t + \Delta, 0)\Delta = \int_0^t \pi_{i+1}(t, x)\beta(x)dx\,\Delta + o(\Delta), \quad i = \overline{2, k - 2}.$$

$$(3)$$

For further calculation as $\Delta \longrightarrow 0$ in formulas (2)–(3) one can obtain the system of Kolmogorov forward partial differential equations for the process micro-states' probabilities in the scope $0 < x < t < \infty$, which is presented in [17]. The algorithm of its solution based on the method of characteristics was proposed to use for the calculation of t.d.s.s.p.'s. In general case of parameters k and n and arbitrary distributed repair times of system's components it's quite difficult to calculate the reliability function in its closed-form representation. So, to find the analytical solution of such a system, consider further a 2-out-of-n model.

4   Reliability Function of a 2-out-of-n Model

4.1   Calculation of Distribution of Time to the First System Failure in Terms of LT

The system of Kolmogorov forward partial differential equations for process $Z(t)$ with absorbing state $k = 2$ in the scope $0 < x < t < \infty$ for 2-out-of-n model has the following form

$$\frac{d}{dt}\pi_0(t) = -\lambda_0\pi_0(t) + \int_0^t \beta(x)\pi_1(t, x)dx,$$

$$\left(\frac{\partial}{\partial t} + \frac{\partial}{\partial x}\right)\pi_1(t; x) = -(\lambda_1 + \beta(x))\pi_1(t, x), \tag{4}$$

$$\frac{d}{dt}\pi_2(t) = \lambda_1 \int_0^t \pi_1(t, x)dx,$$

jointly with initial

$$\pi_0(0) = 1, \tag{5}$$

and boundary conditions

$$\pi_1(t, 0) = \lambda_0\pi_0(t). \tag{6}$$

According to the algorithm, proposed in [17], obtain LT of reliability function of a 2-out-of-n model. The solution of the second equation of the system (4) is

$$\pi_1(t, x) = h(t - x)e^{-\lambda_1 x}(1 - B(x)).$$

By applying the boundary condition (6), whence obtain

$$\pi_1(t, 0) \equiv h(t) = \lambda_0\pi_0(t) \quad \Longrightarrow \quad \tilde{h}(s) = \lambda_0\tilde{\pi}_0(s).$$

From the first equation of (4) using the initial condition (5) in terms of LT it follows,

$$s\tilde{\pi}_0(s) - 1 = -\lambda_0\tilde{\pi}_0(s) + \tilde{h}(s)\tilde{b}(s + \lambda_1),$$

and therefore, taking into account the expression for $\tilde{h}(s)$, it holds,

$$\tilde{\pi}_0(s) = (s + \lambda_0(1 - \tilde{b}(s + \lambda_1)))^{-1}.$$

From the last equation of the system (4), taking into account the expression for $\tilde{h}(s)$, calculate

$$s\pi_2(s) = \lambda_1 \frac{1 - \tilde{b}(s + \lambda_1)}{s + \lambda_1} \tilde{h}(s) = \frac{\lambda_0\lambda_1(1 - \tilde{b}(s + \lambda_1))}{(s + \lambda_1)(s + \lambda_0(1 - \tilde{b}(s + \lambda_1)))}.$$

Due to $\tilde{\pi}_2(s) = \tilde{W}(s) = \frac{1}{s}\tilde{w}(s)$ the last expression represents LT of p.d.f. of the time to the first system failure $\tilde{w}(s)$,

$$\tilde{w}(s) = \frac{\lambda_0\lambda_1(1 - \tilde{b}(s + \lambda_1))}{(s + \lambda_1)(s + \lambda_0(1 - \tilde{b}(s + \lambda_1)))}. \tag{7}$$

According to (1) obtain the reliability function in terms of LT,

$$\tilde{R}(s) = \frac{1}{s}\left[1 - \tilde{w}(s)\right] = \frac{s + \lambda_0(1 - \tilde{b}(s + \lambda_1)) + \lambda_1}{(s + \lambda_1)(s + \lambda_0(1 - \tilde{b}(s + \lambda_1)))}, \tag{8}$$

from where it is easy to obtain mean system lifetime,

$$\mathbb{E}[T] = \tilde{R}(0) = \frac{1}{\lambda_1} + \frac{1}{\lambda_0(1 - \tilde{b}(\lambda_1))}.$$

Thus, the explicit form of the mean system lifetime is obtained.

The calculation of the reliability function in final form is quite difficult task and can be calculated directly in case of fractional rational LT of components repair time distributions.

4.2   An Example: Exponential Distribution of Components' Repair Time

Let the component repair time has an exponential distribution with parameter β, that leads to LT $\tilde{b}(s) = \dfrac{\beta}{s + \beta}$. Then, the following corollary holds.

Corollary 1. *The reliability function $R(t)$ and the mean system lifetime $\mathbb{E}[T]$ of 2-out-of-n model with exponentially distributed life and repair time take the forms,*

$$R(t) = 1 - \frac{n(n-1)\alpha^2}{\hat{s}_1 - \hat{s}_2} \cdot \left[\frac{1 - e^{\hat{s}_2 t}}{\hat{s}_2} - \frac{1 - e^{\hat{s}_1 t}}{\hat{s}_1}\right], \quad \mathbb{E}[T] = \frac{\beta + \alpha(2n-1)}{n(n-1)\alpha^2},$$

where

$$\hat{s}_{1,2} = \frac{1}{2}\left[-\alpha(2n-1) - \beta \pm \sqrt{\alpha^2 + 2\alpha\beta(2n-1) + \beta^2}\right].$$

Proof. Apply LT $\tilde{b}(s) = \dfrac{\beta}{s + \beta}$ and the reverse substitution $\lambda_i = (n-i)\alpha$, $i = 0, 1$ into (7),

$$\tilde{w}(s) = \frac{n(n-1)\alpha^2}{s^2 + s(\beta + \alpha(2n-1)) + n(n-1)\alpha^2}. \tag{9}$$

Decompose this expression into simple fractions:

$$\tilde{w}(s) = \frac{n(n-1)\alpha^2}{\hat{s}_1 - \hat{s}_2} \cdot \left[\frac{1}{s - \hat{s}_1} - \frac{1}{s - \hat{s}_2}\right],$$

where $\hat{s}_1$, $\hat{s}_2$ – the roots of the characteristic equation in the scope $\hat{s}_2 < \hat{s}_1 < 0$:

$$\hat{s}_{1,2} = \frac{1}{2}\left[-\alpha(2n-1) - \beta \pm \sqrt{\alpha^2 + 2\alpha\beta(2n-1) + \beta^2}\right].$$

By the application of reverse LT, one can obtain,

$$w(t) = \frac{n(n-1)\alpha^2}{\hat{s}_1 - \hat{s}_2} \cdot \left[e^{\hat{s}_1 t} - e^{\hat{s}_2 t}\right].$$

Using the definition of the reliability function (1) it should be calculated by the following,

$$R(t) = 1 - \int_0^t w(u)du = 1 - \frac{n(n-1)\alpha^2}{\hat{s}_1 - \hat{s}_2} \cdot \left[\frac{1 - e^{\hat{s}_2 t}}{\hat{s}_2} - \frac{1 - e^{\hat{s}_1 t}}{\hat{s}_1}\right],$$

thus mean system lifetime $\mathbb{E}[T]$ has the form,

$$\mathbb{E}[T] = \int_0^\infty (1 - W(t))dt = \frac{\beta + \alpha(2n-1)}{n(n-1)\alpha^2}.$$

The same result can be obtained by the use of a simple birth and death process with further calculation of reliability function.   □

5   Investigation of the Convergence of the Reliability Function to the Exponential One

For a particular case of 2-out-of-n model with exponentially distributed repair times of its components, one can estimate and prove the convergence of the reliability function to the exponential one on the scale of its mean system lifetime. Denote $\rho = \frac{\beta}{\alpha}$ relative speed of component's recovery.

Theorem 1. *As $\rho = \frac{\beta}{\alpha} \to \infty$ there is a uniform convergence of the reliability function of 2-out-of-n model on the scale of its mean system lifetime: $\hat{R}(t) \to e^{-t}$, and the rate of convergence is of the order ε:*

$$|\hat{R}(t) - e^{-t}| < \varepsilon,$$

where $\varepsilon = \dfrac{n(n-1)}{(\rho + 2n - 1)^2}$.

Proof. Let us write out LT $\tilde{w}(s)$ (9) on the scale of its mean lifetime:

$$\tilde{w}(s/\mathbb{E}[T]) = \frac{n(n-1)\alpha^2}{s^2\left(\dfrac{n(n-1)\alpha^2}{\beta+\alpha(2n-1)}\right)^2 + n(n-1)\alpha^2\,s + n(n-1)\alpha^2}$$

$$= \frac{1}{s^2\dfrac{n(n-1)\alpha^2}{(\beta+\alpha(2n-1))^2} + s + 1} = \frac{1}{\varepsilon s^2 + s + 1}, \tag{10}$$

where $\varepsilon = \frac{n(n-1)\alpha^2}{(\beta+\alpha(2n-1))^2} = \frac{n(n-1)}{(\rho+2n-1)^2} \to 0$ for all $n > 1$ as $\rho \to \infty$. Expand the expression (10) into simple fractions at $\varepsilon \to 0$, then

$$\tilde{w}(s/\mathbb{E}[T]) = \frac{1}{\sqrt{1-4\varepsilon}}\left(\frac{1}{s - s_2'} - \frac{1}{s - s_1'}\right), \tag{11}$$

where s_1', $s_2' = \frac{-1 \mp \sqrt{1-4\varepsilon}}{2\varepsilon}$ – the roots of the characteristic equation. By expanding $\sqrt{1-4\varepsilon}$ into a Taylor series, $\sqrt{1-4\varepsilon} \approx 1 - 2\varepsilon$, the roots of Eq. (11) take the values:

$$s_1' = \frac{-1 - (1-2\varepsilon)}{2\varepsilon} + o(\varepsilon^2) \xrightarrow{\varepsilon\to 0} -\infty,$$

$$s_2' = \frac{-1 + (1-2\varepsilon)}{2\varepsilon} + o(\varepsilon^2) \xrightarrow{\varepsilon\to 0} -1.$$

Then passing to the inverse LT in (11) we get

$$\hat{w}(t) = \frac{1}{\sqrt{1-4\varepsilon}}\left(e^{s_2't} - e^{s_1't}\right) \xrightarrow{\varepsilon\to 0} e^{-t}. \tag{12}$$

Using formula (12) by definition $\hat{R}(t) = 1 - \hat{W}(t \cdot \mathbb{E}[T])$ calculate the reliability function in the new timescale:

$$\hat{R}(t) = 1 - \hat{W}(t \cdot \mathbb{E}[T]) = 1 - \int_0^t \hat{w}(u)\,du$$

$$= \frac{s_1'(1 - e^{s_2't} + s_2'\sqrt{1-4\varepsilon}) - s_2'(1 - e^{s_1't})}{s_1's_2'\sqrt{1-4\varepsilon}}.$$

Putting the obtained roots s_1', s_2', the system reliability function on the scale of its mean lifetime takes the form:

$$\hat{R}(t) = \frac{e^{-\frac{(1+\sqrt{1-4\varepsilon})t}{2\varepsilon}}}{2\sqrt{1-4\varepsilon}}\left[\sqrt{1-4\varepsilon} + (1+\sqrt{1-4\varepsilon})e^{\frac{\sqrt{1-4\varepsilon}t}{\varepsilon}} - 1\right]. \tag{13}$$

As $\varepsilon \to 0$ it gives

$$\hat{R}(t) = \frac{e^{-t}}{1-2\varepsilon} - \frac{\varepsilon e^{-t}}{1-2\varepsilon} - \frac{\varepsilon e^{-\frac{1-\varepsilon-\varepsilon^2}{\varepsilon}t}}{1-2\varepsilon} + o(\varepsilon^2) \xrightarrow{\varepsilon\to 0} e^{-t}. \tag{14}$$

Thus, formula (14) shows that the limit expression for the reliability function on the scale of its mean lifetime is e^{-t} and provides a uniform estimation for the rate of convergence to this limit expression.

To obtain a uniform absolute estimation of the convergence rate, consider:

$$
\left| \hat{R}(t) - e^{-t} \right| = \left| \frac{e^{-\frac{(1+\sqrt{1-4\varepsilon})t}{2\varepsilon}}}{2\sqrt{1-4\varepsilon}} \left(\sqrt{1-4\varepsilon} + (1+\sqrt{1-4\varepsilon})e^{\frac{\sqrt{1-4\varepsilon}t}{\varepsilon}} - 1 \right) - e^{-t} \right|
$$

$$
= \left| \frac{e^{-t}}{1-2\varepsilon} - \frac{\varepsilon e^{-t}}{1-2\varepsilon} - \frac{\varepsilon e^{-t+t(2-1/\varepsilon)}}{1-2\varepsilon} - e^{-t} + o(\varepsilon^2) \right|
$$

$$
\leq \left| e^{-t} \left(\frac{1}{1-2\varepsilon} - \frac{\varepsilon}{1-2\varepsilon} - 1 \right) \right| = \left| e^{-t} \frac{\varepsilon}{1-2\varepsilon} \right|.
$$

Since $e^{-t} \leq 1$, we finally have:

$$
\left| \hat{R}(t) - e^{-t} \right| \leq \frac{\varepsilon}{1-2\varepsilon} < \varepsilon.
$$

$\square$

As a graphical representation of the results of Theorem 1, consider uniform convergence of $\hat{R}(t) - e^{-t}$ on the scale of the mean lifetime of a 2-out-of-6 model to the limit function e^{-t} for different values of the relative recovery rate ρ. For this, define $\rho = 1, 5, 10, 100$ (see Fig. 2).

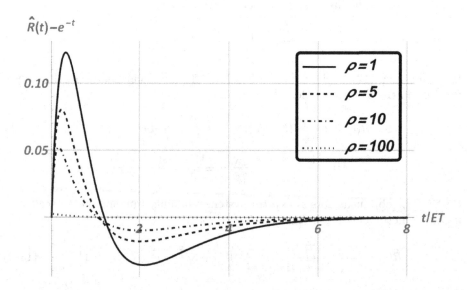

Fig. 2. Uniform convergence of the reliability function $\hat{R}(t)$ of 2-out-of-6 model on the scale of its mean system lifetime to e^{-t} as ρ grows

It can be seen from the figure that the deviation of the curves does not exceed ε and tends to zero with increasing ρ over all values of $\frac{t}{ET}$. ε for each case are presented in Table 1.

Table 1. ε with increasing ρ

	$\rho = 1$	$\rho = 5$	$\rho = 10$	$\rho = 100$
ε	$0,208333$	$0,117188$	$0,068027$	$0,002435$

6   Conclusion

In this paper, the reliability function of 2-out-of-n model with arbitrary distribution of its components repair times is considered. The distribution of time to the first system failure and the reliability function is obtained in terms of LT. The mean lifetime of the system is calculated in explicit form. The exact formulas for the reliability function and mean system lifetime are obtained in the case of the exponential distribution of the repair time. For this model, the convergence of the reliability function on the scale of its mean lifetime to the exponential one is studied, and an estimate of convergence rate is obtained.

Further research will be devoted to the issue of studying the reliability function with an arbitrary distribution of repair time in explicit form. It is proposed to study the conversion formula for some repair time distribution functions, as well as to analyze the sensitivity of reliability characteristics to the shape of repair time distribution of system components.

Acknowledgments. This publication has been supported by the RUDN University Strategic Academic Leadership Program (V. Rykov, supervision and problem setting, N. Ivanova, review and analytic results). The publication has been partially funded by RSF project No. 22-49-02023 (N. Ivanova, formal analysis, validation).

References

1. Trivedi, K.: Probability and Statistics with Reliability, Queuing and Computer Science Applications, 2nd edn. Wiley, New York (2016)
2. Kuo, W., Zuo, M.: Optimal Reliability Modeling: Principles and Applications. Wiley, Hoboken (2003)
3. Vishnevsky, V., Selvamuthu, D., Rykov, V., Kozyrev, D., Ivanova, N., Krishnamoorthy, A.: Reliability Assessment of Tethered High-Altitude Unmanned Telecommunication Platforms - k-out-of-n Reliability Models and Applications. Springer, Singapore (2023)
4. Gnedenko, B.: On cold double redundant system. Izv. AN SSSR. Texn. Cybern. **4**, 3–12 (1964)
5. Gnedenko, B.: On duplication with renewal. Izv. AN SSSR. Texn. Cybern. **5**, 111–118 (1964)

6. Solov'ev, A.: Asymptotic distribution of the lifetime of a duplicated element. Izv. AN SSSR. Texn. Cybern. **5**, 119–121 (1964)

7. Korolyuk, V., Turbin, A.: Mathematical Foundations of Phase Enlargement Complex Systems. Naukova Dumka, Kiev (1978)

8. Korolyuk, V., Turbin, A.: Phase Enlargement of Complex Systems. Vitha Shkola, Kiev (1978)

9. Genis, Y.: The failure rate of a renewable system with arbitrary distributions of the duration of failure-free operation of its elements. Soviet J. Comput. Syst. Sci. **23**(5), 126–131 (1986)

10. Genis, Y.: Reliability and risk assessment of systems of protection and blocking with fast restoration. Reliab. Theory Appl. **3**(1(8)), 41–57 (2008)

11. Genis, Y.: On reliability of systems with periodic maintenance under rare failures of its elements. Autom. Remote Control **71**(7), 1337–1345 (2010)

12. Kovalenko, I.: On the asymptotic consolidation of random processes. Kibernetika **6**, 87–95 (1980)

13. Kalashnikov, V.: Geometric Sums: Bounds for Rare Events with Applications: Risk Analysis, Reliability, Queueing. Springer, Dordrecht (1997). https://doi.org/10.1007/978-94-017-1693-2

14. Kozyrev, D.V.: Analysis of asymptotic behavior of reliability properties of redundant systems under the fast recovery. RUDN J. Math. Inf. Sci. Phys. **3**, 49–57 (2011)

15. Zverkina, G.A.: On some extended Erlang-Sevastyanov queueing system and its convergence rate. J. Math. Sci. **254**, 485–503 (2021). https://doi.org/10.1007/s10958-021-05320-7

16. Zverkina, G.A.: On the exponential convergence rate of the distribution for some nonregenerative reliability system. J. Math. Sci. **262**, 493–503 (2022). https://doi.org/10.1007/s10958-022-05830-y

17. Rykov, V., Kozyrev, D., Filimonov, A., Ivanova, N.: On reliability function of a k-out-of-n system with general repair time distribution. Probab. Eng. Inf. Sci. **35**, 885–902 (2021). https://doi.org/10.1017/S0269964820000285

18. Rykov, V., Ivanova, N.: Reliability and sensitivity analysis of a repairable k-out-of-n:f system with general life- and repair times distributions. In: Baraldi, P., Di Maio, F., Zio, E. (eds.) Proceedings of the 30th European Safety and Reliability Conference and the 15th Probabilistic Safety Assessment and Management Conference (2020). https://doi.org/10.3850/978-981-14-8593-05750-cd

19. Rykov, V., Ivanova, N., Kozyrev, D., Milovanova, T.: On reliability function of a k-out-of-n system with decreasing residual lifetime of surviving components after their failures. Mathematics **10**, 4243 (2022). https://doi.org/10.3390/math10224243

20. Feller, W.: An Introduction to Probability Theory and Its Applications. Wiley, New York (1957)

21. Cox, D.: The analysis of non-Markovian stochastic processes by the inclusion of supplementary variables. Math. Proc. Cambridge Philos. Soc. **51**, 433–441 (1955). https://doi.org/10.1017/S0305004100030437

On the Variance Reduction Methods for Estimating the Reliability of the Multi-phase Gaussian Degradation System

Oleg Lukashenko$^{(\boxtimes)}$

Institute of Applied Mathematical Research of the Karelian Research Centre of RAS, Petrozavodsk, Russia
lukashenko@krc.karelia.ru

Abstract. We consider the Gaussian multi-phase degradation model. The primary target quantity is reliability defined as the tail distribution of the system lifetime. When the required performance measure has no closed-form expression one has to rely on simulation methods. Since the reliability is usually extremely small the problem of the rare-event simulation arises. To tackle this problem a few variance reduction methods have been applied. Some numerical experiments are performed to compare the relative error of the proposed estimators.

Keywords: Reliability · Degradation process · Gaussian process · Variance reduction methods

1 Introduction

The development and evaluation of the models describing the degradation process is an actual research area in reliability analysis since such models give an opportunity to predict fault tolerance. Thus, intensive attention to the aging and degradation models for technical and biological systems has been attracted.

The deterioration system usually operates in a random environment, thus it seems quite natural to describe the degradation in terms of some stochastic process. In this regard, degradation models based on the Gaussian process seem to be quite natural due to the large number of small effects having an influence on the degradation dynamic. Such models based on the Wiener process being the most well-recognized Gaussian process have been extensively studied in the literature and were previously considered in several works, see for example [1–3,10,14,16–19] and references therein. Due to many reasons, some practical systems cannot obey the Markov property. Hence, general Gaussian processes have been considered in [16,20,21] in order to capture more complicated dependence structures including long-range dependence (LRD). The LRD property has been revealed in the degradation processes of turbofan engines, blast furnace walls,

The study was supported by the Russian Science Foundation, project 21-71-10135.

and Li-Ion batteries [21]. Degradation models based on the Gaussian processes have been successfully applied for different practical issues, such as modeling the fatigue crack growth [9,16], modeling of wearing of hard disk drives [15], gyros drift modeling [13], degradation of the light emitting diode (LED) lamps [16] and some others.

The standard models assume the fixed mean degradation rate. But such an assumption can be not suitable for practical needs. The multi-phase Wiener degradation system with a deterministic sequence of change points has been proposed in [3]. The more general case of general Gaussian with stationary and possibly dependent increments was considered in [8] where the required performance measures were estimated via the Conditional Monte Carlo method based on the properties of the so-called Gaussian Bridge process (Bridge Monte Carlo method). In both papers [3,8] it was assumed that change points are deterministic. The more realistic assumption deals with the case of random change point points depending on the trajectory of the corresponding degradation process. The main contribution of this paper can be summarized as follows: 1) a comparative analysis of a few variance reduction methods in terms of the relative error is provided 2) the slight generalization of the multi-phase degradation model to the case when change points form a sequence of the successive hitting times of the underlying degradation process.

The rest of the paper is organized as follows. Section 2 provides a description of the proposed multi-phase degradation model. Section 3 contains a general description of the reliability estimation via the Monte Carlo simulation as a rare-event estimation problem. Section 4 is devoted to the variance reduction techniques when estimating performance measures of the considered reliability model via Monte Carlo simulation. A few numerical examples presented in Sect. 5 demonstrate the quality of the proposed estimators. Finally, a few concluding remarks are given in Sect. 6.

2   Degradation Model

Following the previous research [8], let's consider the multi-phase degradation model with a sequence of the change points $\tau_1 < \cdots < \tau_n$. Each degradation phase (τ_{i-1}, τ_i), $i = 1, ..., n$ is characterized by the mean degradation rate m_i.

The degradation dynamic of the considered reliability system is defined by the stochastic process $\{A(t), t \in \mathcal{T}\}$ satisfying the following model

$$A(t) = \Lambda(t) + X(t), \tag{1}$$

where random fluctuations are described by the centered Gaussian process $\{X(t), t \in \mathcal{T}\}$ with a covariance function

$$\Gamma(t, s) := \mathbb{E}\left[X(t)X(s)\right].$$

The drift $\Lambda(t)$ is a piece-wise linear function, namely

$$\Lambda(t) = \sum_{i=1}^{n} m_i \cdot t \cdot I(\tau_{i-1} < t < \tau_i),$$

where I denotes the indicator function.

Let's suppose that a failure of the system will occur if the degradation process reaches a certain boundary degradation level, which, in general, is unknown. Then, for a given boundary threshold D, the lifetime of the considered system is defined as follows [3,6]:

$$T := \min\{t : A(t) \geq D\}. \tag{2}$$

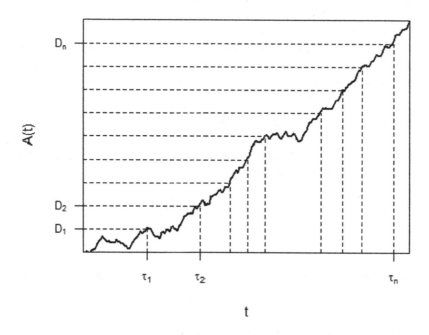

Fig. 1. The typical trajectory of the degradation process.

In general, we are interested in estimating the reliability of the system being defined as the tail distribution of the lifetime T:

$$R(u) := \mathbb{P}(T > u). \tag{3}$$

We consider the following cases:

1. $\{\tau_i\}$ is deterministic sequence;
2. $\{\tau_i\}$ is a sequence of random variables defined as successive hitting times

$$\tau_i = \min\{t : A(t) \geq D_i\},$$

where $D_1 < D_2 < ...$ is increasing sequence of the thresholds. Note that in this case the drift Λ actually depends on the path of the process X, i.e.

$$\Lambda(t) = \Lambda(t, X(t), \ t \in \mathcal{T}).$$

Thus, we call the sequence of change points to be *path-dependent* [12]. The typical sample path of the degradation process with the path-dependent thresholds is shown in Fig. 1.

3   Monte Carlo Estimation

This research is aimed at estimating the system reliability of the degradation system with changing mean rates via Monte Carlo (MC) simulation. In this section, we describe a general problem arising in the estimation of a small probability $R(u)$ (which is true when $u \to \infty$).

Denote by Z_u an estimator of $R(u)$, that means $\mathbb{E}Z_u = R(u)$. To estimate $R(u)$ by MC simulation, one has to generate the independent and identically distributed (i.i.d) copies of the random variable Z_u and calculate the sample mean

$$\widehat{R}_u := \frac{1}{N} \sum_{n=1}^{N} Z_u^{(n)}. \tag{4}$$

The typical measure of the goodness of the estimator is expressed by the *relative error* (RE) defined as follows:

$$\mathrm{RE}\left[\widehat{R}_u\right] := \frac{\sqrt{\mathrm{Var}\left[\widehat{R}_u\right]}}{\mathbb{E}\left[\widehat{R}_u\right]}. \tag{5}$$

The standard MC approach is based on the indicator of the target event, i.e.

$$Z_u^{\mathrm{MC}} = I(T \geq u),$$

which corresponds to the naive Monte Carlo. It is straightforward to show that the RE of the standard MC estimator diverges for small values of the target probability and the sample size required to get a suitable RE is inversely proportional to $R(u)$.

4   Variance Reduction Techniques

There are some rare event simulation techniques [7,11] which aim at modifying the estimator (4) in order to reduce its variance, hence requiring less sample size for the desired accuracy. Some of them are discussed below.

4.1   Conditional Monte Carlo

In this subsection, we assume that the sequence of change points $\{\tau_i\}$ is deterministic. Hence in this case the drift $\Lambda(t)$ being a piece-wise linear deterministic function does not depend on the process $\{X(t), t \in \mathcal{T}\}$.

Following [4,5] let consider the so-called bridge process

$$Y(t) = X(t) - \psi(t)X(s), \tag{6}$$

where s is a prefixed time instant, ψ is expressed in terms of the covariance function of the process X:

$$\psi(t) := \frac{\Gamma(t,s)}{\Gamma(s,s)}.$$

The BMC (Bridge Monte Carlo) estimator is defined as follows (see [8] for the derivation):

$$Z_u^{BMC} = \Psi\left(\frac{\overline{Y}}{\sqrt{\Gamma(s,s)}}\right), \tag{7}$$

where Ψ denotes the distribution function of a standard normal variable and

$$\overline{Y} := \inf_{t\in[0,u]} \frac{D - Y(t) - \Lambda(t)}{\psi(t)}. \tag{8}$$

Note that

$$\Psi\left(\frac{\overline{Y}}{\sqrt{\Gamma(s,s)}}\right) = \mathbb{E}\left[I\left(X(s) \leq \overline{Y}\,\middle|\, \overline{Y}\right)\right],$$

thus, the BMC method is a particular case of the conditional Monte Carlo method which always provides variance reduction [7,11].

Unfortunately, the BMC approach cannot be applied when the change points τ_i depend on the process $\{X(t),\, t \in \mathcal{T}\}$ since the drift term $\Lambda(t)$ will also depend on $\{X(t),\, t \in \mathcal{T}\}$.

4.2 Importance Sampling

Another popular method for variance reduction is importance sampling (IS) aimed at changing the probability measure so that the target rare event becomes more likely to occur.

Let us restrict ourselves to the finite-dimensional case (enough for the simulation needs) when $\mathcal{T} = \{t_1, ..., t_L\}$, where $t_L = u$. Denote the random vectors $X = (X(t_1), ..., X(t_L))$, $\Lambda = (\Lambda(t_1), ..., \Lambda(t_L))$. Let $f(x)$ be a probability density function (pdf) of X and

$$h_u(x) = I(\Lambda(t) + x(t) \leq D,\, t \in \mathcal{T}), \quad x \in \mathbb{R}^L.$$

Having some proposal pdf $g(x)$, the target probability is

$$R(u) = \int h_u(x)\frac{f(x)}{g(x)}g(x)dx = \mathbb{E}_g\left[h_u(X)\frac{f(X)}{g(X)}\right], \tag{9}$$

Thus,

$$Z_u^{IS} = h_u(X)\frac{f(X)}{g(X)}, \quad X \sim g, \tag{10}$$

is the unbiased estimator of $R(u)$.

Optimal Proposal Distribution. It is quite straightforward to show (see for example [7]) that the optimal density g_* which provides the zero variance of the estimator has the following form:

$$g_*(x) = \frac{h_u(x)f(x)}{R(u)}, \tag{11}$$

which corresponds to the truncated multivariate normal distribution.

However, it is not implementable in practice because it requires knowledge of the target quantity $R(u)$. Nevertheless, given above expression provides an insight into the general form of a good proposal distribution. Thus, one can try to use the pdf of the multivariate normal distribution with the same mode as (11) after solving the corresponding optimization problem:

$$h_u(x)f(x) \to \max_x. \tag{12}$$

Note that

$$f(x) = \frac{1}{(2\pi)^{L/2}|\Sigma|^{1/2}} \exp\left(-\frac{1}{2}x^T \Sigma^{-1} x\right), \quad x \in \mathbb{R}^L,$$

where the covariance matrix

$$\Sigma = \begin{pmatrix} \Gamma(t_1,t_1) & \Gamma(t_1,t_2) & \dots & \Gamma(t_1,t_L) \\ \Gamma(t_2,t_1) & \Gamma(t_2,t_2) & \dots & \Gamma(t_2,t_L) \\ \vdots & \vdots & \ddots & \vdots \\ \Gamma(t_L,t_1) & \Gamma(t_L,t_2) & \dots & \Gamma(t_L,t_L) \end{pmatrix}.$$

When change points are deterministic one can obtain that the optimization problem (12) reduces to the following quadratic optimization problem

$$\begin{aligned} &\text{minimize} \quad x^T \Sigma^{-1} x, \quad x \in \mathbb{R}^L, \\ &\text{subject to} \quad x_i + \Lambda(t_i) \le D, \ i = 1, \dots, L. \end{aligned} \tag{13}$$

Thus, described above proposal of the IS estimator corresponds to twisting the Gaussian process X with the deterministic drift determined as a solution of the problem (13).

In the general case of path-dependent change points the constrained optimization problem (12) can be complicated. Therefore, additional computational efforts are required in order to approximate the optimal proposal distribution.

4.3   Control Variables

Let Z_u be any estimator of the target quantity. A further improvement can be achieved by using a *control variable* Y [11],

$$Z_u^{\mathrm{CV}} = Z_u + c(Y - \mathbb{E}[Y]), \tag{14}$$

where constant

$$c = -\frac{\mathbb{Cov}(Z_u, Y)}{\mathbb{Var}[Y]},$$

can be estimated using the i.i.d. copies of the random variables Z_u and Y. Note that the constant c provides the *minimum variance of Z^{CV}*. The control variable Y should be chosen correlated with the random variable Z_u in order to provide substantial variance reduction. In our case, one can set

$$Y = \sum_{i=1}^{L} A(t_i).$$

5   Simulation Results

In this section, we provide a simulation analysis of the accuracy of the proposed in the previous section estimators for the two-phase degradation process.

All experiments were conducted for the case when the process X is the fractional Brownian motion (FBM) being the most remarkable example of the long-range dependent process. The covariance function of the FBM has the following expression

$$\Gamma(t, s) := \frac{1}{2}\left(t^{2H} + s^{2H} - |t - s|^{2H}\right).$$

Let's describe the simulation procedure. It is enough to simulate the FBM over the interval $[0, u]$ for the considered problem. To estimate the desired measures one has to generate the sample paths of the FBM, i.e. the realizations of random vectors:

$$(X(t_1), ...X(t_L)),$$

where $t_1, ..., t_L$ is a uniform partition of the interval $[0, u]$.

In all experiments described below $N = 10000$ trajectories of the FBM with Hurst parameter $H = 0.7$ were generated.

5.1   Deterministic Change Points

In this subsection, we consider the two-phase degradation process with the deterministic change point $\tau_1 = u/2$.

We study the quality of the following estimators:

– The standard MC estimator $\widehat{R}_u^{MC}$;
– The BMC estimator $\widehat{R}_u^{BMC}$
– The IS estimator $\widehat{R}_u^{IS}$
– Combination of the IS estimator with the control variable $\widehat{R}_u^{CV}$

both calculated as the sample mean of corresponding random variables Z^{MC}, Z^{BMC}, Z^{IS}, Z^{CV}. Then, the relative errors of the estimators are calculated as the sample coefficients of variation (see formula (5)).

Table 1. Comparative study of the estimators in case of deterministic change point.

u	RE($\widehat{R}_u^{MC}$)	RE($\widehat{R}_u^{BMC}$)	RE($\widehat{R}_u^{IS}$)	RE($\widehat{R}_u^{CV}$)
40	2.49e−01	2.31e−03	1.94e−02	1.91e−02
50	4.99e−01	2.91e−03	2.13e−02	2.11e−02
60	−	3.49e−03	2.32e−02	2.30e−02
70	−	3.98e−03	2.39e−02	2.35e−02
80	−	4.32e−03	2.55e−02	2.54e−02
90	−	4.77e−03	2.68e−02	2.66e−02
100	−	5.08e−03	2.80e−02	2.77e−02
110	−	5.28e−03	2.89e−02	2.88e−02
120	−	5.67e−03	2.97e−02	2.95e−02
130	−	5.75e−03	3.04e−02	3.03e−02
140	−	6.03e−03	3.14e−02	3.11e−02
150	−	6.23e−03	3.17e−02	3.13e−02

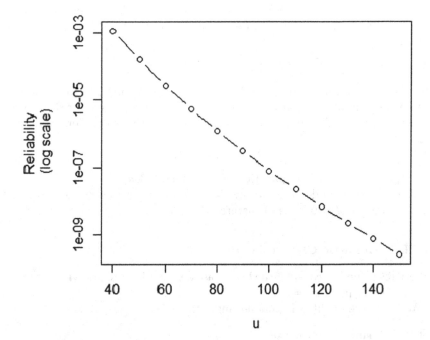

Fig. 2. Reliability of the two-phase FBM degradation model with the deterministic change point.

The proposal distribution of the IS estimator leads to the following generation procedure: one has to generate sample paths of the FBM with a deterministic drift determined as a solution of the optimization problem (13).

The following values of the parameters were used in the simulation: the mean degradation rates are $m_1 = 1, m_2 = 1.5$ and the failure threshold is $D = 20$. The conditioning point for the BMC estimator $s = u$. To verify the effectiveness of the proposed estimators, we considered the dependence of the relative error on the rarity parameter u. The numerical results are presented in Table 1.

These results demonstrate that the best performance exhibits the BMC estimator. The obtained reliability function is shown in Fig. 2. However, as was already mentioned, this method cannot be applied in case when change points depend on the process X (which is the case when change points are successive hitting times).

5.2   Path-Dependent Change Points

Table 2. Comparative study of the estimators in case of random change point.

u	RE($\widehat{R}_u^{MC}$)	RE($\widehat{R}_u^{IS}$)	RE($\widehat{R}_u^{CV}$)
40	0.0378	0.0141	0.0132
50	0.0623	0.0157	0.0151
60	0.0975	0.0172	0.0172
70	0.1539	0.0188	0.0183
80	0.2129	0.0197	0.0192
90	0.2580	0.0206	0.0200
100	0.4471	0.0218	0.0213
110	0.5772	0.0228	0.0226
120	–	0.0231	0.0226
130	0.7070	0.0240	0.0236
140	–	0.0251	0.0246
150	–	0.0258	0.0252

In this subsection, we consider the two-phase degradation process with the random path-dependent change point:

$$\tau_1 = \min\{t : A(t) \geq D_1\},$$

where $D_1 < D$ is a given intermediate threshold.

We study the quality of the following estimators:

- The standard MC estimator $\widehat{R}_u^{MC}$;
- The IS estimator $\widehat{R}_u^{IS}$
- Combination of the IS estimator with the control variable $\widehat{R}_u^{CV}$

Since calculating the mode of the optimal proposal of the IS estimator is difficult we try the following heuristic approach. First, the preliminary simulation procedure is conducted in order to estimate the mean value of the change point $\hat{\tau}_1 = \mathbb{E}\tau_1$. Then, we use the same proposal as for the case of the given deterministic change point $\hat{\tau}_1$ that was described in the previous subsection.

The rest values of the parameters are the following: the mean degradation rates are $m_1 = 1.5, m_2 = 1$; the failure threshold is $D = 20$; the intermediate threshold $D_1 = 10$. The comparative study is presented in Table 2. The reliability function obtained via the control variable method is shown in Fig. 3.

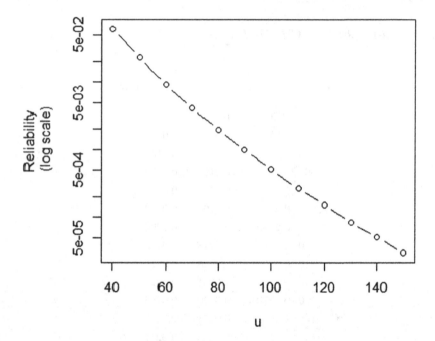

Fig. 3. Reliability of the two-phase FBM degradation model with the random change point.

6   Conclusion

In this paper, we have considered the multi-phase Gaussian degradation model with a sequence of change points. Each degradation phase is characterized by the mean rate which in general can depend on the degradation history. We provide a comparative study of some variance reduction methods in terms of relative error for the particular case of the deterministic change points. The obtained numerical results indicate that the best performance has the BMC estimator which is a particular case of the more general conditional Monte Carlo method. In case when the change points are random with a distribution depending on the trajectory of the underlying degradation process, the IS method has been applied. The development of IS estimators needs further investigation. Thus, one can try to approximate the optimal proposal distribution more precisely using the variational approximation technique and the so-called cross entropy method.

References

1. Chetvertakova, E.S., Chimitova, E.V.: The Wiener degradation model in reliability analysis. In: 2016 11th International Forum on Strategic Technology (IFOST), pp. 488–490 (2016)
2. Doksum, K.A., Höyland, A.: Models for variable-stress accelerated life testing experiments based on Wiener processes and the inverse Gaussian distribution. Technometrics **34**(1), 74–82 (1992)
3. Gao, H., Cui, L., Kong, D.: Reliability analysis for a Wiener degradation process model under changing failure thresholds. Reliab. Eng. Syst. Saf. **171**, 1–8 (2018). https://doi.org/10.1016/j.ress.2017.11.006
4. Giordano, S., Gubinelli, M., Pagano, M.: Bridge Monte-Carlo: a novel approach to rare events of Gaussian processes. In: Proceedings of the 5th St. Petersburg Workshop on Simulation, pp. 281–286, St. Petersburg, Russia (2005)
5. Giordano, S., Gubinelli, M., Pagano, M.: Rare events of Gaussian processes: a performance comparison between Bridge Monte-Carlo and Importance Sampling. In: Next Generation Teletraffic and Wired/Wireless Advanced Networking, pp. 269–280, St. Petersburg, Russia (2007)
6. Kahle, W., Lehmann, A.: The Wiener Process as a Degradation Model: Modeling and Parameter Estimation, pp. 127–146. Birkhäuser Boston, Boston, MA (2010). https://doi.org/10.1007/978-0-8176-4924-1_9
7. Kroese, D.P., Taimre, T., Botev, Z.I.: Handbook of Monte Carlo Methods. Wiley, New York (2011)
8. Lukashenko, O.: On the reliability estimation of the Gaussian multi-phase degradation system. In: Vishnevskiy, V.M., Samouylov, K.E., Kozyrev, D.V. (eds.) Distributed Computer and Communication Networks: Control, Computation, Communications, pp. 410–421. Springer, Cham (2022). https://doi.org/10.1007/978-3-031-23207-7_32
9. Park, C., Padgett, W.: New cumulative damage models for failure using stochastic processes as initial damage. IEEE Trans. Reliab. **54**(3), 530–540 (2005). https://doi.org/10.1109/TR.2005.853278
10. Prakash, G., Kaushik, A.: A change-point-based Wiener process degradation model for remaining useful life estimation. Saf. Reliab. **39**(3–4), 253–279 (2020). https://doi.org/10.1080/09617353.2020.1801165
11. Ross, S.M.: Simulation. Elsevier, New York (2006)
12. Si, X.S., Wang, W., Chen, M.Y., Hu, C.H., Zhou, D.H.: A degradation path-dependent approach for remaining useful life estimation with an exact and closed-form solution. Eur. J. Oper. Res. **226**(1), 53–66 (2013). https://doi.org/10.1016/j.ejor.2012.10.030, https://www.sciencedirect.com/science/article/pii/S0377221712007953
13. Si, X.S., Wang, W., Hu, C.H., Chen, M.Y., Zhou, D.H.: A Wiener-process-based degradation model with a recursive filter algorithm for remaining useful life estimation. Mech. Syst. Signal Process. **35**(1), 219–237 (2013). https://doi.org/10.1016/j.ymssp.2012.08.016

14. Wang, X., Jiang, P., Guo, B., Cheng, Z.: Real-time reliability evaluation for an individual product based on change-point gamma and Wiener process. Qual. Reliab. Eng. Int. **30**, 513–525 (2014)
15. Wang, Y., Ye, Z.S., Tsui, K.L.: Stochastic evaluation of magnetic head wears in hard disk drives. IEEE Trans. Magn. **50**(5), 1–7 (2014). https://doi.org/10.1109/TMAG.2013.2293636
16. Wang, Z., et al.: A generalized degradation model based on Gaussian process. Microelectron. Reliab. **85**, 207–214 (2018). https://doi.org/10.1016/j.microrel.2018.05.001
17. Wei-an, Y., Bao-wei, S., Gui-lin, D., Yi-min, S.: Real-time reliability evaluation of two-phase Wiener degradation process. Commun. Stat. - Theory Methods **46**, 176–188 (2016). https://doi.org/10.1080/03610926.2014.988262
18. Whitmore, G.: Estimating degradation by a Wiener diffusion process subject to measurement error. Lifetime Data Anal. **1**(3), 307–319 (1995). https://doi.org/10.1007/BF00985762
19. Whitmore, G., Schenkelberg, F.: Modelling accelerated degradation data using Wiener diffusion with a time scale transformation. Lifetime Data Anal. **3**, 27–45 (1997). https://doi.org/10.1023/A:1009664101413
20. Zhang, H., Chen, M., Shang, J., Yang, C., Sun, Y.: Stochastic process-based degradation modeling and RUL prediction: from Brownian motion to fractional Brownian motion. Sci. China Inf. Sci. **64**(7), 171201 (2021). https://doi.org/10.1007/s11432-020-3134-8
21. Zhang, H., Zhou, D., Chen, M., Xi, X.: Predicting remaining useful life based on a generalized degradation with fractional Brownian motion. Mech. Syst. Signal Process. **115**, 736–752 (2019). https://doi.org/10.1016/j.ymssp.2018.06.029

Multiphase Queuing System of Blocking Queues and a Single Common Orbit Retrial Queue with Limited Buffer

Vladimir Vishnevsky[1] , Olga Semenova[1] , Minh Cong Dang[2,3](✉) ,
and Van Hieu Nguyen[2]

[1] Institute of Control Sciences of Russian Academy of Sciences,
Profsoyuznaya Str. 65, Moscow 117997, Russia
[2] Moscow Institute of Physics and Technology, Institutskiy per. 9, Dolgoprudny,
Moscow Region 141701, Russia
{hieu.nguyen, dang.mk}@phystech.edu
[3] University of Engineering and Technology, Vietnam National University Hanoi,
Xuan Thuy Str. 144 E3, Hanoi, Vietnam

Abstract. We consider a multiphase queuing system of type M/M/1/k
→ •/M/1/k → · · · → •/M/1/k with an arbitrary number of nodes and
a common retrial queue of type •/M/1/k for other queues in the system.
This kind of queuing system can serve as a model for wireless networks
of asymmetric nodes, where a node with weak processing power can act
as a temporary waiting place for blocked packets. An analytical model
of the system is investigated and various performance measures of the
system are derived. Comparisons of numerical results from the analytical
model with simulation results are also performed. We then utilize the
simulation model to explore the impact of the orbit node on system
performance. Simulation results point out that in the situation of network
saturation, the existence of the orbit queue brings little improvement to
the performance of the system. On the other hand, in the case of near-
saturation network and the case of low-utilization network with queues
of small buffer size, the orbit node can improve the packet loss rate
significantly, with the trade-off of increasing average response time.

Keywords: Multiphase queuing system · Retrial queue · Tandem
queue · Queuing network

1 Introduction

In the field of networking and telecommunications, queuing models are widely
used for evaluating network performance. With the increasing prevalence of wireless technology in recent decades, wireless networks play an important role in

The reported study was funded by the Russian Science Foundation within scientific
project No. 22-49-02023 "Development and study of methods for reliability enhancement of tethered high-altitude unmanned telecommunication platforms".

our telecommunication systems, and performance evaluation of wireless networks becomes an extensively researched subject. A particular application of wireless networks is in road or bridge monitoring systems [11,13]. Due to the linear structure of monitored objects, wireless networks of monitoring nodes with linear topology could be employed suitably. In those kinds of networks, packets are transmitted through a sequence of wireless nodes before reaching the base station. Correspondingly, in queuing theory, multiphase (or tandem) queues are systems where jobs are serviced by a sequence of queues. Because of that similarity, multiphase queueing models can be used for assessing the performance of wireless networks with linear topology [4]. Various multiphase queueing models for evaluating wireless network performance have been thoroughly researched in [9,11–13]. In [13], a multiphase queuing system $MAP/M/1/k \to \bullet/M/1/k \to \cdots \to \bullet/M/1/k$ with cross-traffic is considered, while in [9,11,12], the authors investigated multiphase queueing system of type $MAP/PH/1/k \to \bullet/PH/1/k \to \cdots \to \bullet/PH/1/k$ with cross-traffic. For those considered models, each queue is equipped with limited buffer size and thus packets blocked from entering the queue are lost. Therefore, in this work, we consider the option of a multiphase queueing system with retrial queue.

Retrial queuing models are another extension of standard queuing models. In classical queuing systems, if a job is blocked from entering the waiting queue, it will leave and never return to the system. In retrial models, however, a job blocked from entering the system will wait at an area called orbit and retry entering the system again after some period of time [1]. Such a situation happens frequently in reality, for example, on the Internet due to the re-transmission mechanism of TCP protocol. Existing literature on retrial queuing models is extensive [1,6], however, much attention is paid to single station systems while researches on multiple stations systems [2,3,5,7,8,10] are limited in comparison. In [2], the author considered a system of n blocking queues in tandem with a common orbit queue of infinite buffer size. Approximation methods based on mean value analysis and on fixed point approach are used to obtain performance characteristics. Exact analytical results were obtained for a system consisting of one single M/M/1/1 primary queue and associated M/M/1/∞ retrial queue [3].

In this work, we investigated a multiphase queueing system of type M/M/1/k $\to \bullet/M/1/k \to \cdots \to \bullet/M/1/k$ with an arbitrary number of nodes (phases) and one common orbit queue. Packets blocked at any phase are transmitted to the orbit queue, waiting for their turns to be transmitted back to the first node. In [2,3], the orbit queue has infinite buffer size and therefore blocked packets can wait indefinitely at the orbit queue without being lost. In our work, however, the orbit queue has limited buffer capacity and therefore, the probability of lost packets still exists. This was motivated by the fact that in mobile and wireless networks, nodes are usually limited in power and processing capacity, and we can relegate some of them as orbit nodes. Moreover, with every node in the network having limited buffer size, analytical results can be obtained to verify the correctness of simulation results.

2   System Description

We consider a network of n nodes, of which $n-1$ nodes are connected in linear topology and the remaining node serves as orbit, i.e. packets blocked from entering other nodes will be redirected to the orbit ($n \geq 2$). We denote the orbit node as node 0 and the other nodes as nodes 1 to $n-1$, according to their connecting order, e.g. output flow of node 1 will enter node 2. Outgoing packets from node $n-1$ (the last node) will leave the system and outgoing packets from the orbit node will return to node 1 (the first node). Each network node i ($i = \overline{0, n-1}$) is equipped with limited buffer capacity k_i and a single server with exponential distributed service time of parameter μ_i. New packets arrive at the system at node 1 and their arrival process is a Poisson flow of parameter λ. A schematic description of the system is given in Fig. 1.

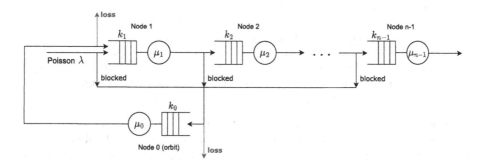

Fig. 1. System description

Blocked packets at each node are handled according to following protocol:

- For a new packet arriving at the system at node 1, if it is blocked, then it will be transmitted to node 0.
- If a packet is blocked at node i, $1 < i \leq n-1$, then it will also be transmitted to node 0.
- Blocked packets at node 0 are dropped and exit from the system permanently.
- For packets return to node 1 from node 0, if they are blocked, they will also be dropped and leave the system permanently.

Thus, a packet can potentially be dropped in two cases, first when it is blocked from entering node 0 and second when it is blocked after returning from node 0 to node 1. These two cases correspond to two loss flows depicted in Fig. 1.

3   Global Balance Equations of the System and Performance Measures

3.1   Global Balance Equations of the System

Let $X_i(t)$ denote the number of packets at node i at time $t \geq 0$, $0 \leq X_i(t) \leq k_i + 1$, $i = \overline{0, n-1}$. Let $X(t) = (X_0(t), X_1(t), ..., X_{n-1}(t))$, the system process $X(t), t \geq 0$ is a continuous-time Markov process with state space $S = \{(x_0, x_1, ..., x_{n-1})\}$, where $x_i \in \{0, 1, ..., k_i + 1\}$.

We then proceed to determine the infinitesimal generator matrix Q of the process $X(t), t \geq 0$. Order the state space S lexicographically, i.e. by the following bijective map: $r(x_0, x_1, ..., x_{n-1}) = x_0 \prod_{j=1}^{n-1} k_j + x_1 \prod_{j=2}^{n-1} k_j + ... + x_{n-2} k_{n-1} + x_{n-1} + 1$

For each state $s = (x_0, x_1, ..., x_{n-1}) \in S$, the entries $Q_{r(s),v}$ of matrix are determined as follows:

1. If $x_1 \leq k_1$, $Q_{r(s),r(d_1)} = \lambda$ where $d_1 = (x_0, x_1 + 1, ..., x_{n-1})$.
2. If $x_1 = k_1 + 1$ and $x_0 \leq k_0$, $Q_{r(s),r(d_2)} = \lambda$ where $d_2 = (x_0 + 1, x_1, ..., x_{n-1})$.
3. For each $i = \overline{0, n-2}$, if $x_i > 0$ and $x_{i+1} \leq k_{i+1}$, $Q_{r(s),r(d_3)} = \mu_i$ where $d_3 = (x_0, ..., x_i - 1, x_{i+1} + 1, ..., x_{n-1})$.
4. For each $i = \overline{0, n-2}$, $x_i > 0$ and $x_{i+1} = k_{i+1} + 1$ and $x_0 \leq k_0$, $Q_{r(s),r(d_4)} = \mu_i$ where $d_4 = (x_0 + 1, ..., x_i - 1, x_{i+1}, ..., x_{n-1})$.
5. For each $i = \overline{0, n-2}$, if $x_i > 0$ and $x_{i+1} = k_{i+1} + 1$ and $x_0 = k_0 + 1$, $Q_{r(s),r(d_5)} = \mu_i$ where $d_5 = (x_0, ..., x_i - 1, ..., x_{n-1})$.
6. If $x_{n-1} > 0$, $Q_{r(s),r(d_6)} = \lambda$ where $d_6 = (x_0, x_1, ..., x_{n-1} - 1)$.
7. Aside from $v = r(s)$, $Q_{r(s),v}$ are set to 0 if they do not belong to any of the above conditions. The diagonal entry $Q_{r(s),r(s)} = -\sum_{v \in \{1,...,|S|\} \setminus \{r(s)\}} Q_{r(s),v}$.

The continuous-time Markov process $X(t), t \geq 0$ is irreducible with finite state space, therefore it is positive recurrent and has a unique final stationary probability distribution $\pi = (\pi_1, \pi_2, ..., \pi_{|S|})$, with π_u is the final probability of state $s = r^{-1}(u)$, $u = 1, 2, ..., |S|$.

The row vector π can be obtained by solving the system of balance equations:

$$Q^T \pi^T = 0 \tag{1}$$

With the last equation (row) replaced by normalization condition:

$$\sum_{u=1}^{|S|} \pi_u = 1. \tag{2}$$

From the possible outgoing transitions of every state, we can guarantee that there are no more than $n + 1$ of non-zero elements for every row of matrix Q, which leads to the number of non-zero elements (NNZ) of matrix Q is bounded by $|S|(n + 1)$. However, the byproduct matrices generated by the matrix factorization method during solving require significantly larger memory than Q. Therefore, the possible matrix Q we can solve realistically by the matrix factorization method on one computer is smaller than the matrix we can store in RAM. Iterative methods can be used to solve the linear system of (1) and (2) for larger Q, however, convergence to solution is not guaranteed. In Table 1, the solving time and memory usage for various sizes of S are given. As we can see from Table 1, the memory usage and solving time for large state space ($n = 3, k = 100$) are very high, therefore this method only is applicable for reasonable state space size.

Table 1. Solving Time and RAM Usage for various parameters on a computer of 32 GB RAM and i7-11700 processor, using PARDISO direct solver of Intel Math Library

n	k	$|S|$	Solving Time	RAM Usage
3	20	10648	0,5 s	106 MB
3	30	32768	1,6 s	212 MB
3	50	140608	8,7 s	1016 MB
3	60	238328	20,5 s	2008 MB
3	70	373248	40,5 s	3679 MB
3	80	551368	121 s	6850 MB
3	100	1061208	501 s	16075 MB

3.2   Performance Measures

From the final stationary probability π, we derive the following performance measures: average number of packets at each node, average output flow of each node, average staying time of packets at each node, average packet loss rate of system.

With each state $s = (x_0, x_1, ..., x_{n-1})$, let us denote $\pi(s) = \pi_{r(s)}$. Average number $L(i)$ of packets at each node i is

$$L(i) = \sum_{s \in S} \pi(s) x_i.$$

The average number L of packets in whole system is

$$L = \sum_{i=0}^{n-1} L(i).$$

Let us denote $S_{working}(i) = \{s \in S | x_i > 0\}$. Then the output rate $\lambda_{out}(i)$ from each node is

$$\lambda_{out}(i) = \sum_{s \in S_{working}(i)} \pi(s)\mu_i.$$

By applying Little's Law for throughput, we obtain the average staying time $W(i)$ at each node:

$$W(i) = \frac{L(i)}{\lambda_{out}(i)}.$$

The throughput of last node is the effective throughput of the system. Therefore, the average packet loss rate of whole system is

$$P_{loss} = 1 - \frac{\lambda_{out}(n-1)}{\lambda}.$$

For average response time of system with non-trivial cases of $n \geq 3$, we only provide an approximation by considering all output flows from each node and input flows to each node as Poisson flows and therefore, every arrival packet to a node observes the system as if it is an outside random observer. All the following calculations are based on that assumption.

Let us denote $\lambda_{out}(i, j)$ be the output rate of node i to j, for example, $\lambda_{out}(i, 0)$ is output rate from node i to orbit node. Then the probability of a packet from node i successfully entering node $i + 1$, $0 < i < n - 1$, is $\frac{\lambda_{out}(i, i+1)}{\lambda_{out}(i)} = \frac{\lambda_{out}(i+1)}{\lambda_{out}(i)}$.

Let $P_{success}(i)$ be the probability a packet starting from node 1 and successfully entering node i, $P_{divert}(i)$ be the probability a packet starting from node 1 but being blocked before reaching node i and entering orbit node instead. Then for every $i > 1$:

$$P_{success}(i) = \prod_{j=1}^{i-1} \frac{\lambda_{out}(j+1)}{\lambda_{out}(j)},$$

$$P_{divert}(i) = \sum_{j=1}^{i-1} P_{success}(j) \frac{\lambda_{out}(j, 0)}{\lambda_{out}(j)}$$

where

$$P_{success}(1) = 1.$$

and for every $0 < j < n - 1$:

$$\lambda_{out}(j, 0) = \sum_{s \in \{s \in S | x_j > 0, x_{j+1} > k_{j+1}, x_0 \leq k_0\}} \pi(s) \mu_j.$$

Correspondingly with $P_{divert}(i)$, let $W_{divert}(i)$ be the average time for a packet from arriving at node 1, being blocked before reaching node i, to entering orbit node. Then for every $i > 1$:

$$W_{divert}(i) = \frac{1}{P_{divert}(i)} \sum_{j=1}^{i-1} P_{success}(j) \frac{\lambda_{out}(j, 0)}{\lambda_{out}(j)} \sum_{u=1}^{j} W(u).$$

Let P_{return} be the probability of a packet starting from node 1 and returning to node 1 after visiting the orbit node exactly one time. Let W_{return} be the corresponding time for returning. Then

$$P_{return} = P_{divert}(n - 1) \frac{\lambda_{out}(0, 1)}{\lambda_{out}(0)},$$

$$W_{return} = W_{divert}(n - 1) + W(0)$$

where

$$\lambda_{out}(0,1) = \sum_{s \in \{s \in S \mid x_0 > 0, x_1 \leq k_1\}} \pi(s)\mu_0.$$

In Fig. 1, notice that the Poisson arrival flow to the system divides into three flows: the first flow enters at node 1, the second flow enters at orbit node, and the third flow is lost because of being blocked from entering node 1 and orbit node. Let λ_1 be the average packet rate of the first flow, and λ_0 be the average packet rate of the second flow respectively. Then:

$$\lambda_1 = \sum_{s \in \{s \in S \mid x_1 \leq k_1\}} \pi(s)\lambda,$$

$$\lambda_0 = \sum_{s \in \{s \in S \mid x_1 \leq k_1\}} \pi(s)\lambda.$$

We are now ready to derive the formula for the average response time of the system. Let denote $W_{success} = \sum_{i=1}^{n-1} W(i)$ and $P_{success} = P_{success}(n-1)$. Then the average response time W of the system is:

$$W = \frac{1}{1 - P_{loss}} \left(\frac{\lambda_1}{\lambda} W_1 + \frac{\lambda_0}{\lambda} W_0 \right) \tag{3}$$

where W_1 and W_0 are given as follows:

$$W_1 = \sum_{m=0}^{\infty} (P_{return} + P_{success})^m (mW_{return} + W_{success}), \tag{4}$$

$$W_0 = \sum_{m=0}^{\infty} (P_{return} + P_{success})^m (W(0) + mW_{return} + W_{success}) \tag{5}$$

W_1 and W_0 can be considered as average response time for packets starting at node 1 and node 0 respectively. For our considered system, we always have $P_{loss} > 0$ and thus $P_{return} + P_{success} < 1$, W_1, W_0 are convergent series, and W is finite value. Nonetheless, we remind that the formulas (3), (4), (5) are only approximations, due to the fact that packet flows in our system are correlated flows instead Poisson processes.

4   Simulation Model and Results

A packet simulation program was written in C++ for the aforementioned multiphase system, by which various performance measures are collected. In order to verify the correctness of both simulation model and analytical model, we generated four sets of synthetic data of 100 samples for the following cases: $n = 3, k = 20$ (i.e. 3 nodes and buffer size of each node is 20); $n = 3, k = 40$; $n = 4, k = 10$ and $n = 4, k = 20$. For each point in datasets, the parameters λ and μ_i are randomly generated according to these rules:

- Input flow parameter λ is greater than 0.0 and smaller than 100.0.
- For each parameter μ_i, a value r_i is chosen in the interval $(-4, 4)$. Then $\mu_i = e^{r_i}\lambda$. This ensures a wide range of possible values for the generated dataset.

In Table 2 comparison results between the simulation model and analytical model on four generated datasets are provided. As expected, differences between the two models are insignificant, aside from average response time W, due to the approximation nature of the formula (3). In Table 3, some concrete results for the dataset $n = 3, k = 40$ are provided. As we can infer from Table 3, the approximation formula (3) is mostly inaccurate in high-utilization cases.

Table 2. Comparison of results from simulation and analytical model

n	k	$\overline{\Delta L}, \%$	$\sigma_\Delta, \%$	$\overline{\Delta P_{loss}}$	σ_Δ	$\overline{\Delta W}, \%$	$\sigma_\Delta, \%$
3	20	0.1169	0.2065	0.0003	0.0004	11.91	14.75
3	40	0.1491	0.2767	0.0002	0.0003	10.34	14.40
4	10	0.0645	0.1142	0.0002	0.0003	9.19	13.32
4	20	0.0595	0.1172	0.0002	0.0004	8.65	12.05

In order to examine the effect of orbit queue on P_{loss} of the system, we performed the simulation for system parameters of $n = 5$, $\lambda = 30$, $\mu_i = 30, i = \overline{0,4}$, i.e. the case of saturation at node 1. All nodes have the same buffer size of k. Figure 2 depicted the relation between loss rate and buffer size in two cases of orbit node functions normally and orbit node is turned off.

We can see that the improvement from having the orbit queue in the case of saturation at node 1 is insignificant. The reason is that, when node 1 is saturated, most of the output packets from the orbit node are lost because of being blocked from entering node 1. Therefore we should examine the case of near-saturation network instead.

With the simulation parameters of $n = 5$, $\lambda = 30$, $\mu_i = 33, i = \overline{0,4}$ (near-saturation network), we obtained the result in Fig. 3. As expected, the improvement from orbit node in this case is much better than the previous case. We then proceeded to examine the effect of buffer size k_0 and service rate μ_0 on system performance separately. By fixing $k_i = 15, i = \overline{0,4}$ we obtained the results on Fig. 4. On the other hand, by fixing $k_0 = 40, k_i = 15, i = \overline{0,4}$ we obtained Fig. 5.

From Fig. 4, we can see that when the service rate μ_0 of orbit node is high ($\mu_0 = 33$), increasing the buffer size k_0 of orbit node does not bring much improvement in loss rate (from 0.096 at $k_0 = 0$ to oscillating around 0.088 for $k_0 = \overline{5,40}$). On the contrary, from Fig. 5, we can see that reducing service rate μ_0 can significantly improve the loss rate of system (the best value at $\mu_0 = 5$) with the cost of increasing average response time W. Therefore, the parameters of orbit node should be chosen carefully to achieve the desired trade-off.

Table 3. Simulation results in the dataset of $n = 3$ nodes with buffer size $k = 40$

λ	μ_0	μ_1	μ_2	Simulation				Analytical			
				P_{loss}	L	W	$L(0)$	P_{loss}	L	W	$L(0)$
39.2	26.8	57.1	187.0	0	2.44	0.062	0	3.00e−8	2.45	0.062	2.57e−7
8.37	0.15	11.79	133.1	0	2.51	0.300	0	3.20e−9	2.52	0.301	4.54e−5
64.9	53.7	89.0	85.6	0	5.81	0.090	4.64e−5	4.06e−7	5.85	0.090	1.56e−5
76.8	456.7	168.4	94.7	0	5.14	0.067	2.16e−5	1.01e−8	5.14	0.067	4.55e−5
19.8	2.51	119.3	20.7	1.16e−3	19.8	1.00	3.00	1.04e−3	19.5	0.99	2.79
52.3	25.5	1498	52.8	7.99e−3	35.3	0.680	8.47	7.99e−3	35.7	0.687	8.61
70.9	3130	72.2	201.2	1.63e−2	18.3	0.263	3.89e−4	1.64e−2	18.5	0.266	3.27e−4
14.0	104.7	9.73	21.7	0.306	39.7	4.08	0.051	0.304	39.7	4.07	0.050
13.0	29.9	8.01	256.2	0.382	40.0	4.96	0.309	0.381	40.0	4.97	0.310
3.42	12.0	31.8	1.43	0.578	81.6	45.8	39.7	0.579	81.6	40.3	39.7
77.5	10.9	348.0	13.4	0.827	82.0	4.45	40.9	0.827	82.0	3.57	40.9

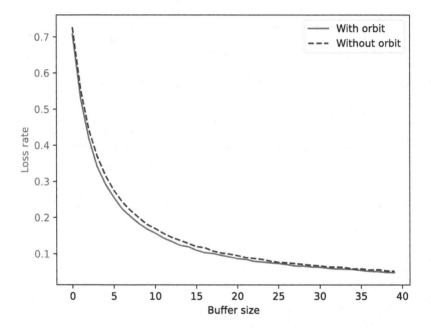

Fig. 2. Relation between loss rate and buffer size in the case of saturation at node 1

Finally, we consider the case of low-utilization network by setting simulation parameters to $n = 5$, $\lambda = 15$, $\mu_i = 30, i = \overline{0,4}$ and obtained the results in Fig. 6. Unsurprisingly, the orbit node has minimal effect on the system when the buffer size is enough ($k > 5$). However, in the condition of very small buffer sizes ($k < 5$), the improvement from having the orbit node is significant.

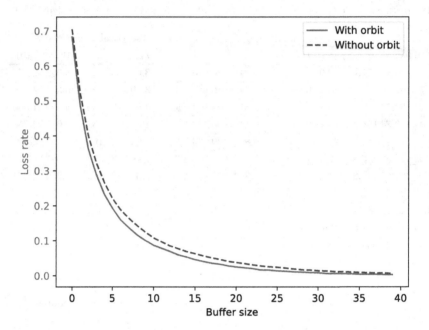

Fig. 3. Relation between loss rate and buffer size in near-saturation network

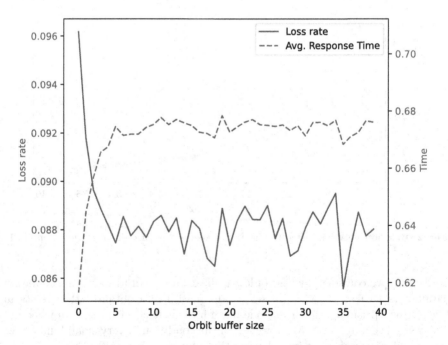

Fig. 4. Effect of orbit buffer size in near-saturation network

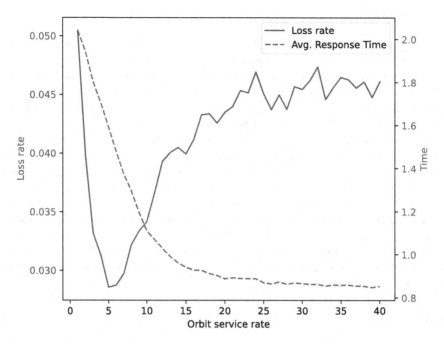

Fig. 5. Effect of orbit service rate in near-saturation network

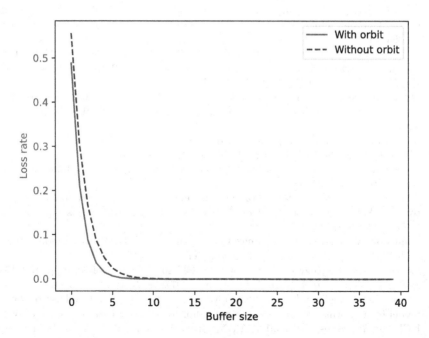

Fig. 6. Relation between loss rate and buffer size in low-utilization network

5   Conclusion

In this paper, we analyzed a multiphase system of $M/M/1/k \to \bullet/M/1/k \to \cdots \to \bullet/M/1/k$ with one additional $\bullet/M/1/k$ node acting as a common orbit retrial queue. We provided the infinitesimal generator matrix and corresponding balance equations of the system. A packet simulation was constructed for the system and was verified by comparison with analytical results obtained from solving balance equations of the system.

From simulation results, we noticed that the existence of an orbit node brings little improvement in the case of network saturation. On the other hand, in the case of near-saturation network and the case of low-utilization network with small buffers, the orbit node can improve the loss rate significantly, with the trade-off of increasing average response time.

A downside of our system is that the benefit of having an orbit node is greatest when the arrival flow is bursty, but the Poisson input flow in our system cannot model that aspect well. In future works, a system with MMPP or MAP input flow should be considered in order to evaluate better the effect of the orbit node for the cases of bursty traffics.

References

1. Artalejo, J.R., Gómez-Corral, A.: Retrial Queueing Systems. Springer, Heidelberg (2008). https://doi.org/10.1007/978-3-540-78725-9
2. Avrachenkov, K., Yechiali, U.: On tandem blocking queues with a common retrial queue. Comput. Oper. Res. **37**(7), 1174–1180 (2010). https://doi.org/10.1016/j.cor.2009.10.004
3. Avrachenkov, K., Yechiali, U.: Retrial networks with finite buffers and their application to internet data traffic. Probab. Eng. Inf. Sci. **22**(4), 519–536 (2008). https://doi.org/10.1017/s0269964808000314
4. Dudin, A., Klimenok, V., Vishnevsky, V.: The Theory of Queuing Systems with Correlated Flows. Springer, Heidelberg (2020). https://doi.org/10.1007/978-3-030-32072-0
5. Falin, G.I.: On a tandem queue with retrials and losses. Oper. Res. Int. J. **13**(3), 415–427 (2012). https://doi.org/10.1007/s12351-012-0126-x
6. Falin, G.: A survey of retrial queues. Queueing Syst. **7**(2), 127–167 (1990). https://doi.org/10.1007/bf01158472
7. Klimenok, V.I., Savko, R.C.: Tandem system with retrials and impatient customers. Autom. Remote. Control. **76**(8), 1387–1399 (2015). https://doi.org/10.1134/s0005117915080056
8. Klimenok, V., Dudina, O., Vishnevsky, V., Samouylov, K.: Retrial tandem queue with *BMAP*-input and Semi-Markovian service process. In: Vishnevskiy, V.M., Samouylov, K.E., Kozyrev, D.V. (eds.) DCCN 2017. CCIS, vol. 700, pp. 159–173. Springer, Cham (2017). https://doi.org/10.1007/978-3-319-66836-9_14
9. Larionov, A., Vishnevsky, V., Semenova, O., Dudin, A.: A multiphase queueing model for performance analysis of a multi-hop IEEE 802.11 wireless network with DCF channel access. In: Dudin, A., Nazarov, A., Moiseev, A. (eds.) Information Technologies and Mathematical Modelling. Queueing Theory and Applications, pp. 162–176. Springer, Cham (2019). https://doi.org/10.1007/978-3-030-33388-1_14

10. Phung-Duc, T.: An explicit solution for a tandem queue with retrials and losses. Oper. Res. Int. J. **12**(2), 189–207 (2011). https://doi.org/10.1007/s12351-011-0113-7
11. Vishnevsky, V., Dudin, A., Kozyrev, D., Larionov, A.: Methods of performance evaluation of broadband wireless networks along the long transport routes. In: Vishnevsky, V., Kozyrev, D. (eds.) Distributed Computer and Communication Networks, pp. 72–85. Springer, Cham (2016). https://doi.org/10.1007/978-3-319-30843-2_8
12. Vishnevsky, V., Larionov, A., Semenova, O., Ivanov, R.: State reduction in analysis of a tandem queueing system with correlated arrivals. In: Dudin, A., Nazarov, A., Kirpichnikov, A. (eds.) Information Technologies and Mathematical Modelling. Queueing Theory and Applications, pp. 215–230. Springer, Cham (2017). https://doi.org/10.1007/978-3-319-68069-9_18
13. Vishnevsky, V.M., Larionov, A.A., Semenova, O.V.: Performance evaluation of the high-speed wireless tandem network using centimeter and millimeter-wave channels. Probl. Upr. (4), 50–56 (2013). https://www.mathnet.ru/eng/pu800

Analysis of Procedures for Joint Servicing of Multiservice Traffic in Access Nodes

Mikhail S. Stepanov[1], Sergey N. Stepanov[1,2(✉)],
Margarita G. Kanischeva[1], and Fedor S. Kroshin[1]

[1] Department of Communication Networks and Commutation Systems, Moscow Technical University of Communication and Informatics, 8A, Aviamotornaya Street, Moscow 111024, Russia
{m.s.stepanov,m.g.kanishcheva,f.s.kroshin}@mtuci.ru
[2] Kotel'nikov Institute of Radio Engineering and Electronics of RAS, Mokhovaya 11-7, Moscow 125009, Russia
s.n.stepanov@mtuci.ru

Abstract. A generalized mathematical model of joint servicing of real-time traffic and data traffic in a multiservice access node has been constructed and investigated. The model takes into account the presence of a priority for real-time traffic, as well as the group nature of the arrival, elastic properties, the possibility of waiting and aging of the transmitted information for data traffic. The coming of requests for the transmission of real-time traffic and groups of files obeys the Poisson laws, and the service times have an exponential distribution. The definitions of the indicators of the quality of the joint service of incoming requests are formulated and a method for their evaluation based on the solution of a system of equilibrium equations is considered. The dependence of the performance measures on the studied features of the formation and servicing of traffic in multiservice access nodes is numerically studied. The use of the model for solving the problems of estimating the value of transmission resource required to serve the offered traffic for given QoS indicators and estimating the volume of the offered traffic offloaded in a situation of congestion to other access nodes in order to achieve the specified QoS indicators is considered.

Keywords: Multiservice traffic · Real-time traffic · Elastic traffic · Batch arrival of files · Aging of transmitted information · Properties of traffic servicing · Transmission resource planning

1   Introduction

Achieving the specified indicators of the quality of service of incoming requests for the provision of various types of services is the main task of modeling networks and communication systems. It is especially important to find its solution for access nodes that perform the function of concentrating multiservice subscriber traffic. Depending on the formulation, the problem to be solved can be classified into the following two categories [1–5, 7].

V. M. Vishnevskiy et al. (Eds.): DCCN 2023, LNCS 14123, pp. 222–235, 2024.
https://doi.org/10.1007/978-3-031-50482-2_18

- Problem of strategic planning. Development of software and analytical tools for estimating the volume of a resource that provides the required quality of service for given flows of information load.
- Problem of operational planning. Estimation of the maximum allowable amount of traffic load that can be served on a given resource with the required quality. This task is also called traffic offloading.

Both tasks are solved by constructing a mathematical model that takes into account the main features of the formation of incoming requests and the distribution of the access node transmission resource during their service. The list of features includes: the multiservice nature of incoming requests, the priority of requests for the transmission of real-time traffic, as well as the group nature of arrival, elastic properties, the possibility of waiting and aging of the transmitted information for data traffic. We will assume that data traffic consists of files containing the results of observations and other information messages similar to them in terms of properties. Depending on the problem statement, all of the above features may be present in the analyzed model or considered in a smaller aggregate. Some of the listed characteristics of the multiservice access node were taken into account in publications [7–14]. The novelty of this work consist in construction of a generalized mathematical model of an access node having mentioned features and the analysis of the possibilities of its use to provide the required QoS indicators in the joint service of traffic of modern communication applications.

The urgency of the formulated problems attracted the attention of specialists from various background that varies from the engineers involved in the design and maintenance of communication systems, to mathematicians. Giving a general description of the published work, it should be noted that most of the research represents engineering development, based on the results of simulation modeling and field measurements [1–5,7,8].

The theoretical analysis of the joint service of priority real-time traffic and elastic data traffic is usually carried out in the framework of Markov models [8–16,18,19]. Using this class of models, it is possible to determine the characteristics of servicing of incoming requests through the values of the stationary probabilities of the model, which are determined after compiling and solving the system of equilibrium equations [15–18,20]. The paper proposes to use the Gauss-Seidel iterative method for these purposes [11,13,14,18,19]. It makes it possible to solve systems of equilibrium equations with up to several million unknowns [18]. This is enough to explore the properties of joint servicing of multiservice traffic in access nodes and solve resource planning problems for values of input parameter close to real values. This paper is devoted to the solution of these problems.

A short outline of the paper is as follows. Section 2 describes the model. In the next two sections, we formulate the definitions of quality indicators for the joint service of incoming requests and consider a method for their evaluation based on the solution of a system of equilibrium equations. The use of the model for solving the problems of estimating the value of transmission resource required to serve the offered traffic for given QoS indicators and estimating the volume of the offered traffic offloaded in a situation of congestion to other access nodes in order to achieve the specified QoS indicators is considered in Sect. 5. In the last section, conclusions are formulated based on the results of the study.

2   Model Description

Let us denote by C the throughput of the multiservice access node provided by the used communication standard in bits per second. For the convenience of modeling, we introduce the concept of a virtual channel, which will be used to numerically characterize the resource for transmitting information provided to users. Let us denote by c the information transfer rate of one channel in bits per second. The choice of the value of c depends on the problem statement, for example, it may be the minimum requirement for the transfer rate required to service the ordered services.

The access node processes a Poisson flow of requests for the provision of real-time services of intensity λ_r, divided into n service categories. With probability $p_{r,k}$, the request belongs to the k-th category, requires c_k bits per second, and occupies the resource during random time that has an exponential distribution with the parameter $\alpha_{r,k}$, $k = 1, \ldots, n$. To build the model, the C and c_k are converted to the virtual channel format. The v total number of virtual channels (v.c.) is calculated from expression $v = \left\lfloor \frac{C}{c} \right\rfloor$ v.c. The requests of the kth flow require $b_k = \left\lceil \frac{c_k}{c} \right\rceil$ v.c. for their service, and the moments of their arrival form a Poisson flow of intensity $\lambda_{r,k} = \lambda_r p_{r,k}$, $k = 1, \ldots, n$.

Together with the real-time traffic, the access node processes the Poisson group flow of requests for the transmission of elastic data traffic in the form of files. Denote by λ_d the intensity of this flow. The appearance of each request means that with probability f_j it is necessary to transfer a group of j files, $j = 1, 2, \ldots, m_g$ and $\sum_{j=1}^{m_g} f_j = 1$. Denote by $\bar{b}$ the average number of files in one group $\bar{b} = \sum_{j=1}^{m_g} f_j j$. Files that are not accepted in service when they arrive may be waiting for a resource to be released. The number of waiting places will be denoted by w. The waiting time is limited by a random variable that has an exponential distribution with the parameter σ. If after this time the file has not been transferred, it is assumed that the transmitted information is outdated and the file is considered lost without renewal. The procedure for generating incoming requests can be considered in Fig. 4, if we put $p_r = p_d = 0$.

Let's assume that the size of each file has an exponential distribution with the mean value F expressed in bits, and the minimum amount of resource used to transfer one file is one channel. It is clear that the time until the end of the transmission of one file by one channel has an exponential distribution with the parameter α_d. Let's denote by ℓ the number of channels used to service real-time traffic, and by i_d we'll denote the total number of files being served and waiting. Files are served using $(v - \ell)$ channels. Let $i_d \leq v - \ell$ and $s = \left\lfloor \frac{v - \ell}{i_d} \right\rfloor$.

The procedure for sharing a resource between files under transmission follows the provisions of the Processor Sharing discipline. As a result, free channels are divided between i_d files according to the following rule. To serve each of the $(v - \ell - si_d)$ files, $(s+1)$ channels are used, and to serve each of the $((s+1)i_d - (v - \ell))$ files, s channels. Since all $(v - \ell)$ channels are busy, it is easy to show that the time until the end of the transmission of one of the i_d files being served has an exponential distribution with the parameter $(v - \ell)\alpha_d$. It is also clear that when the inequality $i_d > v - \ell$ is fulfilled, in the considered case $(v - \ell)$ files are transmitted using the capabilities of one channel, and $(i_d - v + \ell)$ files will be waiting for service to start.

An incoming request for real-time traffic transmission has priority in occupancy of the resource, reducing, if necessary, the number of channels used for file transmission to the value of one channel. Let us denote by $i_{r,k}(t)$, $k = 1, \ldots, n$ the number of requests of the k-th flow for the transmission of traffic of real-time services that are in service at time t, and denote by $i_d(t)$ the number of files being served and waiting at time t. The dynamics of changes in the number of requests located in the access node at different stages of service is described by the Markov process $r(t) = (i_{r,1}(t), \ldots, i_{r,n}(t), i_d(t))$ defined on the finite state space S, which includes the states $(i_{r,1}, \ldots, i_{r,n}, i_d)$, with components

$$i_{r,1} = 0, 1, \ldots, \left\lfloor \frac{v}{b_1} \right\rfloor; \quad i_{r,2} = 0, 1, \ldots, \left\lfloor \frac{v - i_{r,1}b_1}{b_2} \right\rfloor;$$

$$\cdots\cdots\cdots\cdots\cdots\cdots\cdots\cdots\cdots\cdots\cdots\cdots \tag{1}$$

$$i_{r,n} = 0, 1, \ldots, \left\lfloor \frac{v - i_{r,1}b_1 - \ldots - i_{r,n-1}b_{n-1}}{b_n} \right\rfloor;$$

$$i_d = 0, 1, \ldots, v + w - i_{r,1}b_1 - \ldots - i_{r,n}b_n.$$

3   Performance Measures

The quality of service for the requests of the k-th flow for the transmission of real-time traffic is determined by the portion of lost requests $\pi_{r,k}$ and the average number of busy virtual channels $m_{r,k}$. The value of the last characteristic makes it possible to calculate the average number of requests of the kth flow being serviced $y_{r,k} = m_{r,k}/b_k$ and the average amount of bandwidth occupied by them $z_{r,k} = m_{r,k}c$.

The quality of service for the requests of elastic traffic is determined by the portion of the time spent by the allocated resource in the state of unavailability for receiving of the incoming file $\pi_{d,t}$; by the proportion of files lost due to the presence of the allocated resource in a state of unavailability for receiving an incoming file for transfer or waiting $\pi_{d,c}$; by the portion of files lost for all reasons analyzed in the model π_d; by the average number of virtual channels occupied by elastic traffic m_d; by the average bandwidth of the access node used by elastic traffic $z_d = m_d c$; by the average time for file to be in access node before leaving h_d; by the average number of virtual channels b used to transmit one file; by average bit rate used for file transmission $c_d = bc$; by the average number of files waiting L_q; by the average number of files being on servicing L_s; by the average number of files being on waiting and servicing $L = L_q + L_s$.

The introduced performance measures can be calculated if stationary probabilities $p(i_{r,1}, \ldots, i_{r,n}, i_d)$ of states $(i_{r,1}, \ldots, i_{r,n}, i_d) \in S$ are known. For the state

$(i_{r,1}, \ldots, i_{r,n}, i_d)$ let ℓ denote the number of virtual channels used to service real-time traffic $\ell = i_{r,1}b_1 + \cdots + i_{r,n}b_n$ and assume that the maximum size of a group of incoming files is determined from the expression $m_g = v + w$. The introduced performance measures of servicing real time traffic are looking as follows:

$$
\begin{aligned}
\pi_{r,k} &= \sum_{\left\{(i_{r,1},\ldots,i_{r,n},i_d)\in S \mid \ell+i_d+b_k>v\right\}} p(i_{r,1}, \ldots, i_{r,n}, i_d); \\
m_{r,k} &= \sum_{(i_{r,1},\ldots,i_{r,n},i_d)\in S} p(i_{r,1}, \ldots, i_{r,n}, i_d)i_{r,k}b_k; \\
& k = 1, \ldots, n.
\end{aligned}
\tag{2}
$$

Let us define the performance characteristics of elastic data transfer. The value of $\pi_{d,t}$ is summed up from the values of the probabilities of states, for which the coming file is either going to wait or lost. The expression for $\pi_{d,t}$ is as follows

$$
\pi_{d,t} = \sum_{\left\{(i_{r,1},\ldots,i_{r,n},i_d)\in S \mid \ell+i_d\geq v\right\}} p(i_{r,1}, \ldots, i_{r,n}, i_d).
$$

Now let us find the expression for $\pi_{d,c}$. Let us denote as Λ_b the intensity of the flow of files that were lost due to the stay of the allocated resource in the state when $\ell + i_d = v + w$. Let us find the average number of lost files for an arbitrary state $(i_{r,1}, \ldots, i_{r,n}, i_d - i)$ where $(i_{r,1}, \ldots, i_{r,n}, i_d) \in S$, $\ell + i_d = v + w$ and $i = 0, 1, \ldots, i_d$. The arrival of a group containing $i + 1$, $i + 2, \ldots, v + w$ files results in a loss of $1, 2, \ldots, v + w - i$ files, respectively. Average number of files lost in state $(i_{r,1}, \ldots, i_{r,n}, i_d - i)$, determined from expression

$$
f_{i+1} \cdot 1 + f_{i+2} \cdot 2 + \ldots + f_{v+w} \cdot (v + w - i) = \sum_{j=i+1}^{v+w} f_j(j - i).
$$

After averaging over all $i = 0, 1, \ldots i_d$, we get an expression for Λ_b

$$
\Lambda_b = \lambda_d \sum_{i=0}^{i_d} \sum_{\left\{(i_{r,1},\ldots,i_{r,n},i_d)\in S \mid \ell+i_d=v+w\right\}} p(i_{r,1}, \ldots, i_{r,n}, i_d - i) \sum_{j=i+1}^{v+w} f_j (j - i).
$$

The total intensity of files coming on the transmission, determined from relation $\lambda_d \bar{b}$. Final expression for $\pi_{d,c}$ is looking as follows

$$
\pi_{d,c} = \tfrac{1}{\bar{b}}
$$

$$
\times \left(\sum_{i=0}^{i_d} \sum_{\left\{(i_{r,1},\ldots,i_{r,n},i_d)\in S \mid \ell+i_d=v+w\right\}} p(i_{r,1}, \ldots, i_{r,n}, i_d - i) \sum_{j=i+1}^{v+w} f_j (j - i) \right).
\tag{3}
$$

To estimate the total portion of file losses π_d it is necessary to divide the intensity of files lost for all reasons analyzed in the model to the intensity of files coming. We obtain the following result

$$\pi_d = \frac{1}{\lambda_d b}$$

$$\times \left(\Lambda_b + \sum_{\left\{ (i_{r,1},\ldots,i_{r,n},i_d) \in S \mid \ell+i_d>v \right\}} p(i_{r,1},\ldots,i_{r,n},i_d)(\ell+i_d-v)\sigma \right). \tag{4}$$

Other performance measures of files servicing are define as follows

$$m_d = \sum_{\left\{ (i_{r,1},\ldots,i_{r,n},i_d) \in S \mid i_d>0 \right\}} p(i_{r,1},\ldots,i_{r,n},i_d)(v-\ell);$$

$$L_q = \sum_{\left\{ (i_{r,1},\ldots,i_{r,n},i_d) \in S \mid \ell+i_d>v \right\}} p(i_{r,1},\ldots,i_{r,n},i_d)(\ell+i_d-v);$$

$$L_s = \sum_{\left\{ (i_{r,1},\ldots,i_{r,n},i_d) \in S \mid \ell+i_d\leq v \right\}} p(i_{r,1},\ldots,i_{r,n},i_d)i_d \tag{5}$$

$$+ \sum_{\left\{ (i_{r,1},\ldots,i_{r,n},i_d) \in S \mid \ell+i_d>v \right\}} p(i_{r,1},\ldots,i_{r,n},i_d)(v-\ell);$$

$$h_d = \frac{L}{\lambda_d b(1-\pi_d)}, \quad \bar{b} = \frac{m_d}{L_s}.$$

The introduced characteristics of servicing incoming requests are related by relations, which are the laws of conservation of the intensities of flows of requests arriving and served by the access node. The number of these relations is equal to the number of flows $n+1$. They are looking as follows:

$$\lambda_{r,k} b_k = \lambda_{r,k} \pi_{r,k} b_k + m_{r,k} \alpha_{r,k}; \quad k=1,\ldots,n;$$
$$\lambda_d \bar{b} = \lambda_d \pi_{d,c} \bar{b} + m_d \alpha_d + L_q \sigma. \tag{6}$$

To prove the relations (6), it suffices to multiply the system of equilibrium equations (7) successively by $i_{r,k}$, $k=1,\ldots,n$ and i_d and then add obtained expressions for all $(i_{r,1},\ldots,i_{r,n},i_d) \in S$. Using the obtained relations, it is possible to construct alternative expressions for some characteristics, which simplify their calculation or measurement. In particular, if $w=0$ for h_d and $\bar{b}$ such expressions has the form

$$h_d = \frac{L_s}{m_d \alpha_d}, \quad \bar{b} = \frac{1}{h_d \alpha_d}.$$

Another important field of application of conservation laws is the verification of the correctness of the solution of a system of state equations.

4 System of State Equations

In order to evaluate the introduced performance measures according to the definition introduced above, it is necessary to compose and solve a system of state equations. Let us write down system of state equations in the form of a single equality, which with help of indicator function is convenient to use for realization of the iterative Gauss-Seidel or Jacobi method. By equating the intensities of transition of the process $r(t)$ in and out of arbitrary state of the model $(i_{r,1}, \ldots, i_{r,n}, i_d)$ we obtain the relation

$$
P(i_{r,1}, \ldots, i_{r,n}, i_d)\left\{ \sum_{k=1}^{n}\left(\lambda_{r,k} I(\ell + i_d + b_k \leq v) + i_k \alpha_{r,k} I(i_{r,k} > 0) \right) \right.
$$
$$
\left. + \lambda_d I(\ell + i_d + 1 \leq v + w) + (v - \ell)\alpha_d I(i_d > 0) + (\ell + i_d - v)\sigma I(\ell + i_d > v) \right\}
$$
$$
= \sum_{k=1}^{n} P(i_{r,1}, \ldots, i_{r,k} - 1, \ldots, i_{r,n}, i_d)\lambda_{r,k} I(i_{r,k} > 0, \ell + i_d \leq v)
$$
$$
+ \sum_{i=1}^{i_d} P(i_{r,1}, \ldots, i_{r,n}, i_d - i)\lambda_d \left(f_i + I(\ell + i_d = v + w) \sum_{j=i+1}^{v+w} f_j \right)
$$
$$
+ \sum_{k=1}^{n} P(i_{r,1}, \ldots, i_{r,k} + 1, \ldots, i_{r,n}, i_d)(i_{r,k} + 1)\alpha_{r,k}
$$
$$
\times \left(I(\ell + b_k + i_d \leq v) + I(\ell + b_k + i_d > v, \ell + b_k \leq v, \ell + b_k + i_d \leq v + w) \right)
$$
$$
+ P(i_{r,1}, \ldots, i_{r,n}, i_d + 1) \times \left((v - \ell)\alpha_d I(\ell + i_d + 1 \leq v + w) \right.
$$
$$
\left. + (\ell + i_d + 1 - v)\sigma I(\ell + i_d + 1 \leq v + w, \ell + i_d + 1 > v) \right);
$$
$$
\sum_{(i_{r,1}, \ldots, i_{r,n}, i_d) \in S} p(i_{r,1}, \ldots, i_{r,n}, i_d) = 1.
$$

(7)

In the above expression, the indicator function $I(\cdot)$ is determined from the relation

$$
I(\cdot) = \begin{cases} 1, \text{ if the condition formulated in parentheses is fulfilled;} \\ 0, \text{ in opposite case.} \end{cases}
$$

The matrix of the system of state equations does not have any properties that would allow the use of recursive or matrix methods for its solution. In this case, the solution of the (7) can be obtained by the iterative Gauss-Seidel method for the number of unknowns in the (7) not exceeding several millions. Details of the use of this algorithm in solving the (7) can be found in [19].

5 Numerical Assessment

The constructed model of the access node and a convenient algorithm for calculating the characteristics of coming requests servicing make it possible to investigate the dependence of the estimate of the minimum required throughput of the access node and the maximum allowable load on the features of the formation of request flows and resource distribution taken into account in the model. Let us consider the model for the following fixed values of input parameters: $C = 80\,\text{Mbps}$; $n = 2$; $c_1 = 2\,\text{Mbps}$; $c_2 = 5\,\text{Mbps}$. Based on the assumptions made, we get the structural parameters of the model: $c = 1$ Mbps; $v = 80$ v.c.; $b_1 = 2$ v.c.; $b_2 = 5$ v.c. Let's assume that $F = 80$ Mbit. The average file transfer time by one channel is $80\,\text{s}$. When performing calculations, this time will be taken as a unit. From this assumption $\alpha_d = 1$. For real-time services, the service time parameters are selected from the $\alpha_1 = 0{,}5$ and $\alpha_2 = 0{,}5$. Let us denote by ρ the offered load per on virtual channel. The value of ρ is defined as follows

$$\rho = \frac{1}{v}\left(\frac{\lambda_{r,1}}{\alpha_{r,1}}b_1 + \frac{\lambda_{r,2}}{\alpha_{r,2}}b_2 + \frac{\lambda_d}{\alpha_d}\bar{b}\right). \tag{8}$$

When carrying out further calculations, we will assume that all three traffic flows considered in the model create the same offered load. From this statement we obtain relations for determining the intensities of input flows

$$\lambda_{r,1} = \frac{v\rho\alpha_{r,1}}{3b_1}; \quad \lambda_{r,2} = \frac{v\rho\alpha_{r,2}}{3b_2}; \quad \lambda_d = \frac{v\rho\alpha_d}{3\bar{b}}. \tag{9}$$

Let us start from the use of the constructed model to solve the problem of estimating the value of transmission resource required to serve the offered traffic for given QoS indicators. Assume that resource sufficiency is determined from the condition $\max(\pi_{r,1}, \pi_{r,2}, \pi_d) < \pi$, where π—prescribed value of losses. Let us show that ignoring the group nature of file arrivals leads to a strong underestimation of the required node throughput.

Assume that in the first model of data traffic generation there is only one file in the group of incoming files. It means that $f_1 = 1$. Let us take $\rho = 1$, $\pi = 0{,}05$ and $w = 0$. For chosen values of parameters offered load Λ_d of elastic traffic is calculated from expression $\Lambda_d = \frac{\lambda_d \bar{b}}{\mu_d} = 26{,}27$ v.c. Denote by d the variance of the number of files in the incoming group. For chosen model of group arrival

$$\bar{b} = \sum_{j=1}^{v+w} f_j j = 1; \qquad d = \sum_{j=1}^{v+w} f_j(j - \bar{b})^2 = 0. \tag{10}$$

The results of estimating the number of virtual channels required to service the proposed traffic in accordance with the selected values of the input parameters are shown in Fig. 1 (curve 1). It takes 89 v.c. to satisfy the inequality $\max(\pi_{r,1}, \pi_{r,2}, \pi_d) < 0{,}05$. Note that the data traffic generation model used corresponds to the classical Poisson model.

Let's consider two additional models of formation of grouped elastic data. The second model is $f_i = 1/30$, $i = 1, 2, \ldots, 30$ ($\bar{b} = 15{,}5$; $d = 74{,}92$) and the third model is $f_1 = 0{,}5$ and $f_{30} = 0{,}5$ ($\bar{b} = 15{,}5$; $d = 210{,}25$). The remaining parameters of the model are the same as those used for the calculation of the data used in drawing of curve 1 in Fig. 1, except for λ_d, the value of which is chosen so as to keep the potential resource load by elastic data at the level of 26,27 v.c.

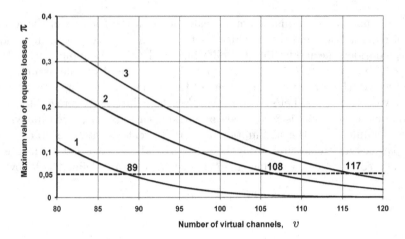

Fig. 1. The results of finding the minimum value of v to satisfy the inequality $\max(\pi_{r,1}, \pi_{r,2}, \pi_d) < 0{,}05$ for different models of data traffic generation.

The results of estimating the number of virtual channels required to service the proposed traffic in accordance with the selected values of the input parameters are shown in Fig. 1 (curve 2 and 3). For the second model it takes 108 v.c. and for the third model—117 v.c. The above data show that for the same values of offered traffic parameters an increase in the impulsive nature of file arrivals that is measured by variance of the number of files in the incoming group significantly increases the need for node bandwidth, which provides a given level of losses.

The negative effects of the bursty nature of file arrivals can be mitigated by allowing blocked files to wait for service in the buffer. The results of solving the problem of estimating the required throughput of a node are shown in Fig. 2. The fixed model parameters are the same as those used in Fig. 1 except parameters of waiting. Curve 1 shows the solution of the problem for the Poisson model of file arrival without the possibility of waiting. Curves 2 and 3 show the solution of the problem for the model with batch arrival of files $f_1 = 0{,}5$ and $f_{30} = 0{,}5$ and $w = 10$ (curve 2) and $w = 30$ (curve 3). The value of $\sigma = 0{,}1$.

The data presented in Fig. 2 shows that the use of waiting does indeed reduce the need for the resource (compare curve 3 Fig. 1 and curve 2 Fig. 2). However, with a further increase in the number of waiting places, a decrease in demand of transmission resource is no longer observed. The reasons why this happens can be seen from the results of calculation presented in Fig. 3, which shows the dependence of the losses of requests $\pi_{r,1}$, $\pi_{r,2}$, π_d curves 1,2,3 respectively on increase in w. The fixed input parameters are the same as those used in the calculation of the data presented in Fig. 1. The arrival of files is given by the probabilities $f_1 = 0{,}5$ and $f_{30} = 0{,}5$.

From the presented data, it can be seen that an increase in w reduces the loss of files and thereby increases the loss of requests for the transmission of real-time traffic. Thus, the use of waiting to reduce the need for information

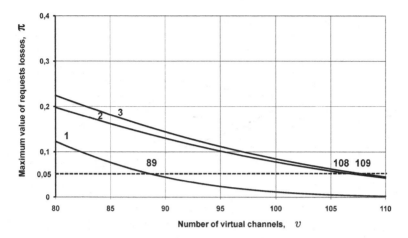

Fig. 2. The results of finding the minimum value of v to satisfy the inequality $\max(\pi_{r,1}, \pi_{r,2}, \pi_d) < 0{,}05$ for differen models of data traffic generation and possibility of waiting.

transfer resource should be done with some caution. The presence of a positive effect depends on the ratio between the volumes of real-time traffic and elastic data. Possible scenarios for the use of waiting must be evaluated using the model constructed and investigated in this work, or similar to it.

Concluding the discussion of the peculiarities of solving the problem of estimating the value of transmission resource required to serve the offered traffic for given QoS indicators, we note the following. The algorithm for selecting the required resource value depends on the structural parameters of the model. If the number of unknowns in the system of state equations does not exceed several millions, then the results of solving the system of state equations by the Gauss-Seidel algorithm are used to estimate the QoS indicators. If the number of unknowns in the system of state equations exceeds this value, then the decomposition method proposed in [19] is used to evaluate the characteristics of the model. It is based on a joint numerical analysis of two particular cases of the model under study: when the access node resource is occupied only by real-time service traffic or only by elastic data traffic. In each of the listed cases, efficient recursive algorithms are used to calculate the characteristics. It can be shown that the results of calculating the characteristics will be asymptotically exact in the region of low losses. This corresponds to the range of load parameters, where the problem of estimating the required throughput of the access node is solved.

Let's turn to the task of offloading traffic. Usually it is solved in a situation of local overloads of the access node [21]. Let's assume that the available traffic leads to exceeding the level of loss of requests. Excess traffic is directed to other access nodes. For wireless networks, these may be communication systems operating in unlicensed frequency bands. Let us denote by p_r the share of requests for servicing real-time services, by p_d we will denote the share of requests for the

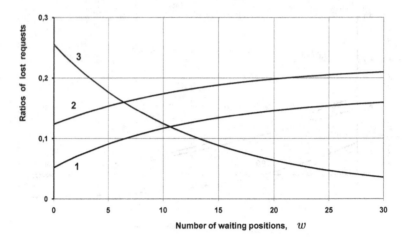

Fig. 3. The ratios of losses as function of number of waiting positions.

transfer of elastic data that will be redirected for servicing to other access nodes so that the amount of losses in servicing the remaining part of the traffic on a given resource does not exceed normative values. The procedure for offloading traffic is shown in Fig. 4.

Let us give a numerical example illustrating the algorithm for estimating p_r and p_d. Let us consider the model of an overloaded access node with the

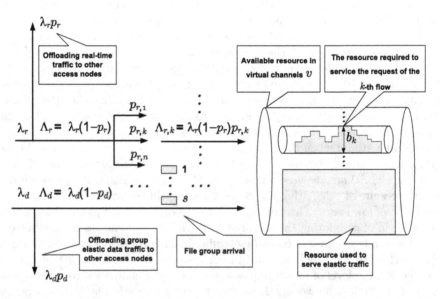

Fig. 4. The procedure for offloading traffic in a multiservice access node when servicing real-time service traffic and elastic data

following parameters: $v = 60$ v.c., $w = 10$, $n = 2$; $\lambda_r = 18$ req./s; $p_{r,1} = 5/6$; $p_{r,2} = 1/6$; $b_1 = 1$ v.c.; $b_2 = 5$ v.c.; $\alpha_1 = \alpha_2 = \alpha_d = 1$; $f_i = 1/30$, $i = 1, \ldots, 30$, $\sigma = 0,1$. Let's assume for simplicity that the upload is only for real-time traffic, i.e. $p_d = 0$. After the implementation of the offloading procedure, the access node serves the flows of requests with the following rates: $\Lambda_{r,1} = \lambda_r(1 - p_r)p_{r,1}$; $\Lambda_{r,2} = \lambda_r(1 - p_r)p_{r,2}$; $\Lambda_d = \lambda_d$. Changing p_r, we find the value $p_r = 0,68$, at which the required level of requests losses $\max(\pi_{r,1}, \pi_{r,2}, \pi_d) < \pi$ is reached. In this case $\pi = 0,05$. The results of estimating the share of offloaded traffic of real-time services are shown in Fig. 5.

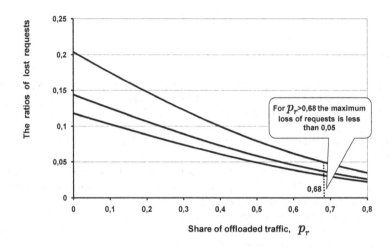

Fig. 5. Results of evaluating the share of offloaded traffic of real-time services

6 Conclusion

A generalized mathematical model of joint servicing of real-time traffic and elastic data traffic in a multiservice access node has been constructed and investigated. The model takes into account the presence of a priority for real-time traffic, as well as the group nature of the arrival, elastic properties, the possibility of waiting and aging of the transmitted information for data traffic. The arrivals of requests for the transmission of real-time traffic and groups of files obeys the Poisson laws, and the service times have an exponential distribution. The change of model states is described by a multidimensional Markov process with a finite state space. The definitions of the performance measures of the joint service of incoming requests are formulated and a method for their evaluation based on the solution of a system of equilibrium equations is considered. The constructed model and the results of its analysis can be used to solve the problems of estimating the value of the transmission resource required to serve offered traffic for given QoS indicators and estimating the volume of the offered traffic offloaded in a situation of congestion to other access nodes in order to achieve the specified QoS indicators. Numerical examples illustrating this point are given. Additional results obtained for this model include approximate evaluation methods based on the implementation of model decomposition principles and the use of a system of simplified equilibrium equations are proposed [19]. It has been established that the obtained estimates of the characteristics are asymptotically exact in the region of small losses [19,20]. The results obtained were also used to solve the problems of estimating the necessary information transmission resource of access nodes in satellite communication networks [22], cloud computing systems [23], and in reference and information services operating under overload conditions [24].

References

1. Study on scenarios and requirements for next generation access technologies. 3GPP Technical report (TR) 138.913 version 15.0.0 Release 15 (2018)
2. System architecture for the 5G System. 3GPP Technical Specification (TS) 123.501 version 15.9.0. Release 15 (2020)
3. Network Slice Selection Services. 3GPP Technical Specification (TS) 129.531 version 15.5.0. Release 15 (2019)
4. Ericsson. Network slicing: a go-to-market guide to capture the high revenue potential (2016). https://www.ericsson.com/assets/local/digital-services/network-slicing/network-slicing-value-potential.pdf
5. Nokia. Dynamic end-to-end network slicing for 5G. White Paper (2017)
6. Marquez, C., et al.: Resource sharing efficiency in network slicing. IEEE Trans. Netw. Serv. Manage. **16**(3), 909–923 (2019)
7. Iversen, V.B.: Teletraffic Engineering and Network Planning. Technical University of Denmark (2010)
8. Ross, K.W.: Multiservice Loss Models for Broadband Telecommunications Networks. Springer, Cham (1995). https://doi.org/10.1007/978-1-4471-2126-8
9. Begishev, V., et al.: Resource allocation and sharing for heterogeneous data collection over conventional 3GPP LTE and emerging NB-IoT technologies. Comput. Commun. **120**(2), 93–101 (2018)
10. Gudkova, I.A., Samouylov, K.E.: Modelling a radio admission control scheme for video telephony service in wireless networks. In: Andreev, S., Balandin, S., Koucheryavy, Y. (eds.) NEW2AN/ruSMART -2012. LNCS, vol. 7469, pp. 208–215. Springer, Heidelberg (2012). https://doi.org/10.1007/978-3-642-32686-8_19
11. Stepanov, S.N., Stepanov, M.S.: Efficient algorithm for evaluating the required volume of resource in wireless communication systems under joint servicing of heterogeneous traffic for the Internet of Things. Autom. Remote. Control. **80**(8), 1970–1985 (2019)
12. Bonald, T., Virtamo, J.: A recursive formula for multirate systems with elastic traffic. IEEE Commun. Lett. **9**(8), 753–755 (2005)
13. Stepanov, S.N., Stepanov, M.S., Andrabi, U., Ndayikunda, J.: The analysis of resource sharing for heterogenous traffic streams over 3GPP LTE with NB-IoT functionality. In: Vishnevskiy, V.M., Samouylov, K.E., Kozyrev, D.V. (eds.) DCCN 2020. LNCS, vol. 12563, pp. 422–435. Springer, Cham (2020). https://doi.org/10.1007/978-3-030-66471-8_32

14. Stepanov, M.S., Stepanov, S.N., Andrabi, U.M., Petrov, D.S., Ndayikunda, J.: The increasing of resource sharing efficiency in network slicing implementation. In: Vishnevskiy, V.M., Samouylov, K.E., Kozyrev, D.V. (eds.) Distributed Computer and Communication Networks. DCCN 2021. CCIS, LNCS, vol. 1552, pp. 18–35. Springer, Cham (2022). https://doi.org/10.1007/978-3-030-97110-6_2

15. Fortet, R., Grandjean, Ch.: Congestion in a loss system when some calls want several devices simultaneously. Electr. Commun. **39**(4), 513–526 (1964)

16. Basharin, G.P., Gaidamaka, Y.V., Samouylov, K.E.: Mathematical theory of teletraffic and its application to the analysis of multiservice communication of next generation networks. Autom. Remote. Control. **47**(2), 62–69 (2013)

17. Stasiak, M., Sobieraj, M., Zwierzykowski, P.: Modeling of multi-service switching networks with multicast connections. IEEE Access **10**, 5359–5377 (2022)

18. Stepanov, S.N.: Teletraffic Theory: Concepts, Models, Applications. Goriachay Linia-Telecom, Moscow (2015). (in Russian)

19. Stepanov, S.N., Stepanov, M.S.: Methods for estimating the required volume of resource for multiservice access nodes. Autom. Remote Control **81**(12), 2244–2261 (2020)

20. Stepanov, S.N., Stepanov, M.S.: Planning transmission resource at joint servicing of the multiservice real time and elastic data traffics. Autom. Remote. Control. **78**(11), 2004–2015 (2017)

21. Chen, J., Chang, Z., Guo, X., Li, R., Han Z., Hamalainen, T.: Resource allocation and computation offloading for multi-access edge computing with Fronthaul and Backhaul constraints. IEEE Trans. Veh. Technol. **70**(8), 8037–8049 (2021)

22. Stepanov, S.N., Andrabi, U.M., Stepanov, M.S., Ndayikunda, J.: Reservation based joint servicing of real time and batched traffic in inter satellite link. In: Proceedings of 2020 Systems of Signals Generating and Processing in the Field of on Board Communications, Moscow, Russia, pp. 1–5 (2020). https://doi.org/10.1109/IEEECONF48371.2020.9078542

23. Volkov, A.O., Korobkina, A.V., Stepanov, S.N.: Development of model and algorithms for servicing real-time and data traffic in a cloud computing system. In: Proceedings of 2022 Systems of Signals Generating and Processing in the Field of on Board Communications, pp. 1–6 (2022). https://doi.org/10.1109/IEEECONF53456.2022.9744289

24. Stepanov, S., Stepanov, M.: Estimation of the performance measures of a group of servers taking into account blocking and call repetition before and after server occupation. Mathematics **9**(21), 2811 (2021). https://doi.org/10.3390/math9212811

Recovery of Real-Time Clusters with the Division of Computing Resources into the Execution of Functional Queries and the Restoration of Data Generated Since the Last Backup

V. A. Bogatyrev[1,3] , S. V. Bogatyrev[2,3] , and A. V. Bogatyrev[2(✉)]

[1] Department Information Systems Security, Saint-Petersburg State University
of Aerospace Instrumentation, Saint Petersburg, Russia
[2] Yadro Cloud Storage Development Center, Saint Petersburg, Russia
gangleon@gmail.com
[3] ITMO University, Saint Petersburg, Russia
vabogatyrev@itmo.ru

Abstract. The possibilities of increasing the readiness of fault-tolerant cluster systems for the timely execution of functional requests are investigated. Duplicated systems containing two computers and two two-input memory nodes are considered as cluster nodes, which ensures direct accessibility of any memory node for two computers. For accelerated recovery of duplicated cluster nodes, the results of the last backup are preliminarily entered into the memory node intended to replace the failed one. As a result of this decision, after replacing a failed memory node, its information recovery requires only the entry of up-to-date information generated after the last backup. Up-to-date data can be replicated from a healthy storage node of the same cluster node that contains data generated since the last backup.

The purpose of this article is to investigate the possibilities of increasing the readiness of a cluster system for the timely execution of functional requests based on the rationale for the distribution of computing resources of duplicate cluster nodes for restoring actual data and for executing functional requests. Variants of division of computing resources into information recovery of actual data and execution of functional queries are considered.

Markov models of duplicated cluster nodes are proposed, on the basis of which the dependences of the system readiness for timely execution of requests on resource allocation options that have retained the operability of computing nodes to perform functional tasks and restore information in memory, including the stages of entering the results of the last backup and replicating relevant information from memory of healthy nodes.

The risks of using information based on the results of the last backup in the process of servicing functional requests are assessed. The effectiveness of the proposed solutions for restoring the system after failures is evaluated by the intensity of profit from servicing information requests, taking into account the risks of using outdated information.

V. M. Vishnevskiy et al. (Eds.): DCCN 2023, LNCS 14123, pp. 236–250, 2024.
https://doi.org/10.1007/978-3-031-50482-2_19

Keywords: reliability · replication · Markov model · cluster · fault tolerance · real time · backup

1   Introduction

Recently, research has been actively conducted for infocommunication systems and networks aimed at ensuring their high fault tolerance and availability, with low computational delays while supporting the continuity of the computational process [1–4]. In this regard, it is important to analyze the capabilities of real-time cluster computer systems to ensure fault tolerance and a high probability of readiness at an arbitrary moment to service the flow of required functions. When researching such systems, it is important to take into account restrictions on service delays and the inadmissibility of interruptions in computing processes with the loss of unique data accumulated during its operation [5–7].

The study of fault tolerance, reliability and timeliness of servicing functional requests is associated with the analysis of transmission processes through the network [8,9] and their maintenance and storage directly in the cluster [10–16].

To ensure the required fault tolerance and readiness of the cluster at the level of organization of its nodes, it is advisable to build them in the form of duplicated computing systems, equipped with two calculators and two dual-input memory nodes. Such nodes provide direct accessibility of any memory node for two calculators, which increases the flexibility of the structure and the fault tolerance of the system during the accumulation of failures and its degradation [17,18]. The choice of the best configurations, organization of computational processes and maintenance disciplines of duplicated computer systems requires substantiation in their simulation and/or analytical modeling.

Markov models for the reliability of duplicated cluster nodes while maintaining functioning based on the migration of computers are proposed in [17,18]. For computer systems containing memory nodes, a feature is the need for a phased restoration of memory: first physical, and then informational [17,18].

To restore information, replicas of data generated during the execution of functional queries and stored in the working memory nodes of the cluster can be used. Migrating replicas within a redundant cluster node requires the use of compute node resources. Since the computing resources of the nodes are also required to service the flow of functional requests, the problem of efficient distribution of computing resources arises. The distribution of computing resources between the processes of information recovery and traffic maintenance should be aimed at reaching a compromise to reduce the delays of the computing process and increase the availability of the system.

Studies of the distribution of resources of serviceable computing nodes for restoring information in memory and for solving functional problems were carried out in [18]. The studies [18] are aimed at increasing the probability of timely servicing of the traffic of functional tasks. At present, the possibility of recovering memory by first entering the results of backup into a node designed to replace failed memory is not studied. After the backup, the full use of the restored memory requires only loading the actual data accumulated after the backup. In informational memory recovery, it is of interest to study the possibility of combining the actualization process with the execution of an incoming stream of functional tasks.

Replication of actual data can be carried out from a healthy memory node of the same cluster node. When combining several duplicate nodes in a cluster with replication of generated data in them, after replacing two memory nodes of a cluster node, it is possible to load them with up-to-date information from other cluster nodes. The restored memory node with the backup results entered into it (before they are updated) can be used to service functional requests, which increases the likelihood of their timely service, but at the same time can lead to unreliable results due to access to irrelevant (outdated) data.

Thus, the goal of the work is to increase the readiness of the cluster system for the timely execution of functional requests based on the rationale for the distribution of computing resources of duplicated cluster nodes for restoring actual data and for executing functional requests. Restoration of actual data is carried out after the physical restoration of failed memory nodes with preliminary loading of the last backup data into them.

2   Organization of a Cluster Node with the Restoration of Actual Data Generated After Backup

Let us consider the organization of a cluster node composed by two calculators (B) and two two-input memory nodes (M). Combining nodes is implemented through a switch (S). The structure of the considered cluster is shown in Fig. 1 [18].

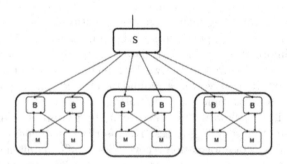

Fig. 1. Cluster of duplicated computer systems with two-input memory nodes.

Two calculators can work in the load sharing mode and in the duplicated calculation mode. In the first case, the load of calculators is reduced, which leads to a decrease in the average delays in the execution of functional requests, and an increase in the probability of timely execution of requests. In the second case, when comparing the results on two calculators, the reliability of calculations increases, accompanied by a decrease in the probability of their execution in the required time. If, in case of duplicated calculations, the first result obtained in time is issued without waiting for the second result for comparison, then the probability of timely execution of queries at low loads can increase. However, with such an organization of duplicated calculations, there is a load limit, above which the probability of executing queries in the required time decreases.

In duplicated cluster nodes in two memory nodes, duplication (replication) of information accumulated during the operation of the system can be carried out. With a two-input memory, direct access to the redundant data is possible from any calculator, which simplifies data recovery after replacing one of the memory nodes. To reduce the time of information recovery after the replacement of a failed memory node in the system, backups are periodically performed. To replace a failed memory node, a node is preliminarily prepared, into which the results of the last backup are applied. After connecting a backup memory block based on replication from a healthy node, the information accumulated since the last backup is updated. Data replication requires the use of resources of healthy computers of the duplicated cluster node. If information is lost in two memory nodes, then a more complex and lengthy procedure for recovering information is provided.

Let's consider options for dividing the resources of calculators into information recovery of actual data and execution of functional queries.

With the performance of two calculators and the presence of one memory node with actual and one with outdated data, we select the following options for distributing computing resources of a cluster node:

A1 - one calculator is always allocated for informational recovery of actual data, and the second one for traffic maintenance, in which access to the memory node with outdated data is prohibited.

A2 - both calculators are involved in servicing traffic when using one memory node with up-to-date information. Information recovery is implemented in the absence of functional requests.

A3 - one calculator is always allocated for informational recovery of actual data, the share of the resources of the second calculator allocated for informational recovery a_2, and for the execution of functional requests $(1-a_2)$. Accessing memory with irrelevant data is prohibited.

A4 - one calculator is always allocated for servicing functional requests, the share of the resources of the second calculator allocated for informational restoration of actual data a_2, and for the execution of functional requests $(1-a_2)$. When executing functional queries, only access to memory with actual data is allowed;

If one of the computers fails, its priority recovery is possible, using the resources of a workable computer for information recovery and solving functional problems.

If there is one memory node with up-to-date and one with out-of-date data, we will single out the following options for distributing computing resources:

B1 - a workable calculator is always allocated exclusively for informational recovery of actual data. Function requests are not serviced.

B21 - a workable computer is involved in traffic maintenance. The use of memory with not up-to-date information is prohibited. Informational recovery of actual data in the memory containing the results of the last backup is not implemented

B22 - a workable calculator is involved in traffic maintenance. The use of memory with not up-to-date information is prohibited. Informational recovery of actual data in memory containing the results of the last backup is implemented in the absence of functional requests.

B3 - The share of the resources of a workable computer allocated for the restoration of actual data a1, and for the execution of functional requests $(1-a_1)$. When executing functional queries, memory accesses are allowed only with actual data. Options B21 and B22 are a special case of option B3.

Let us now consider the cases when there is not a single memory node with up-to-date data in the cluster node. For such states, the service of functional requests can be blocked, which eliminates erroneous service when accessing outdated data.

If the system identifies requests to require the use of data generated after or before backup, then blocking the service is carried out only for requests that require up-to-date data that has not yet been restored. For requests to data generated before backup, the above options for dividing computing resources into information recovery and servicing functional requests are possible.

If the system does not implement the identification of requests to require the use of data generated after or before backup, then servicing the flow of functional requests is associated with the risk of accessing up-to-date data that has not yet been restored.

To justify the choice of the options considered, it is required to build a model that allows one to find the probabilities of possible states of the system and for each state to determine the probabilities of timely execution of requests for each variant of organizing duplicated cluster nodes.

3 Duplicated Cluster Node Model

A feature of the developed duplicated model is the preliminary entry into the memory nodes intended to replace failed memory nodes of the results of the last backup and various options for sharing the resources of operable computers for servicing functional requests and for restoring up-to-date information generated after backup.

For simplicity and clarity of calculations to justify the choice of options for organizing duplicated cluster nodes, we will focus on building a Markov model as the simplest model that allows us to take into account the features of organizing and servicing duplicated cluster nodes. The developed Markov models should take into account the above cases of combining the use of computing resources for servicing functional requests and restoring actual data obtained after the last backup.

The model should take into account the risks associated with the use of irrelevant data generated after backup in the computing process. The state and transition diagrams for the proposed Markov model of the system under study are shown in Figs. 2, 3, and 4 for various options for information recovery with the division of computing resources into solving functional problems and data recovery after replacing failed memory nodes. The state of the cluster node is set by the matrix $S_{2\times2}$, the first row of which displays the performance of the calculators, and the second of the memory nodes.

The performance of the memory nodes and the calculator is designated as "1", and the failure as "0". The s denotes the state of the memory after the replacement of the failed memory node with the node with the data of the last backup. The diagrams in Figs. 2, 3, and 4 show the probabilities P0, P1,..., P16 of the cluster node being in the states 0, 1,...,16.

Due to the symmetry of the node structure, as a result of the use of two-input memory, the mapping states of the matrix $S_{2\times2}$ are invariant to the permutation of row elements. So, for example, for the state "4" the following matrices are equivalent:

$$\begin{bmatrix} 0 & 1 \\ 0 & 1 \end{bmatrix}, \begin{bmatrix} 1 & 0 \\ 1 & 0 \end{bmatrix}, \begin{bmatrix} 0 & 1 \\ 1 & 0 \end{bmatrix}, \begin{bmatrix} 1 & 0 \\ 0 & 1 \end{bmatrix}.$$

For the state "7", the cases displayed by the matrices are equivalent:

$$\begin{bmatrix} 0 & 0 \\ s & 0 \end{bmatrix}, \begin{bmatrix} 1 & 0 \\ 0 & s \end{bmatrix}, \begin{bmatrix} 0 & 1 \\ s & 0 \end{bmatrix}, \begin{bmatrix} 0 & 1 \\ 0 & s \end{bmatrix}.$$

For states in which there is at least one operable calculator and current data is stored in one memory node, the last backup data is entered in the other, the restoration of actual data in it is possible by replicating the corresponding data of the memory node that has retained its operability. In the absence of up-to-date data in two memory nodes of a cluster node, restoring the data generated after backup requires more complex (longer) procedures.

The diagrams in Figs. 2, 3, and 4 show the failure rates of the computing node and memory Λ_1, Λ_2, the intensity of recovery of the computer - μ_1 and memory with the preliminary entry of the results of the last backup μ_2. In the case of using the resources of one computer to restore the actual information, the intensity of this restoration based on replication from a working memory node is equal to μ_3. In the case of using the calculator also for servicing functional requests, the intensity of the transition of their state "3" to the initial state "0" is equal to $a_2\mu_3$, while $a_3 \leq 1$.

Of the states with the absence of actual data in two memory nodes (state 8), with the operability of two calculators, there are two options for restoring data generated after backup. Figures 2 and 3 shows a variant of restoring actual data first in one memory node (transition to state 3) with their further replication to another memory node with a transition from state 3 to state 0. In Fig. 4 shows the variant with the restoration of actual data with their entry directly into two memory nodes, which corresponds to a direct transition from state 8 to the initial state 0.

With the loss of relevant information in two memory bonds, the transition from state 8 to state 3 according to Figs. 1 and 2 is realized with an intensity of μ_4. The transition from state 8 to state 0 according to Fig. 3 is carried out with intensity $\mu_5 = a_3\mu_4$, and $a_3 \leq 1$. It should be noted that the intensity of the transition from state 3 to the initial state 0 depends on the options A1, A2,..., A4 of the combined use of computing resources for servicing traffic and restoring actual data during replication from healthy memory nodes.

Figure 2 shows the simplest version of the state diagram and transitions of the Markov model, when in state "3" the variant A1 of the distribution of computing resources of the cluster node is implemented, and in state "5" only the computer is restored according to variant B1.

Figure 3 shows the restoration of a duplicated cluster node, when in state "3" one of the options A1, A2, A3, A4 is implemented with the distribution of computing resources of the cluster node, and in state "5" one of the options B1, B21 is restored, B22, B3.

The set of states of the cluster node according to the possibilities of servicing the flow of functional tasks will be divided into states in which two-channel or single-channel servicing of the request flow is possible, states with a risk associated with outdated data in memory, and also states in which the servicing of functional requests is impossible.

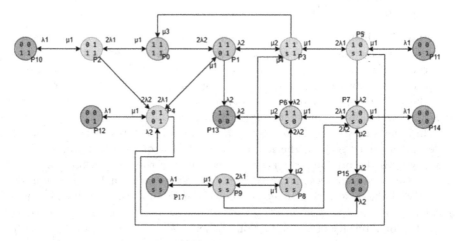

Fig. 2. Recovery of a duplicated cluster node with a combination of options A1 and B1.

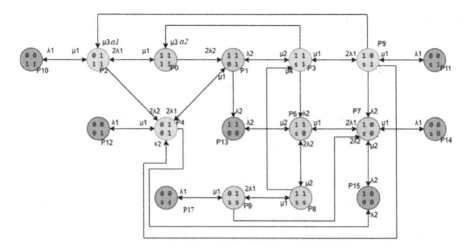

Fig. 3. Recovery of a duplicated cluster node, with various options for distributing computing resources.

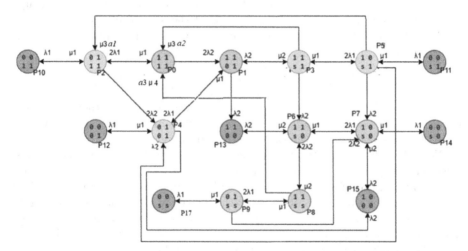

Fig. 4. Restoring actual data lost in two memory nodes of a duplicated cluster node with a direct transition from state 8 to the initial state 0, taking into account various options for distributing computing resources for information recovery and executing functional queries.

According to the diagrams presented in Figs. 1, 2, 3, and 4, according to known rules, a system of algebraic equations is compiled [19], solving which, by known methods, the probabilities of all states of the duplicated cluster node are determined. The stationary availability factor of a duplicated node is determined by summing up the probabilities of its operable states, under which the execution of functional requests is realized [18].

$$K = \sum_{i \in W} P_i.$$

where W is the set of operable states of the cluster node, in which its computing resources are allocated to service the flow of functional requests.

So for the options for organizing cluster nodes, affixed by the models in Figs. 2 and 3, if states with irrelevant data in memory are not used in the computational process, the availability factor will be equal to

$$K = P_0 + P_1 + P_2 + P_3 + P_4 + P_5.$$

If the cluster node is represented by the model in Fig. 3, while in state "5" the only working computer is completely allocated for information recovery without servicing functional requests

$$K = P_0 + P_1 + P_2 + P_3 + P_4.$$

If the system identifies requests with access to data generated before backup, and at the same time in the states 6, 7, 8, 9 the service of such requests is implemented, then to organize the system according to Fig. 3 availability factor is defined as

$$K_3 = P_0 + P_1 + P_2 + P_3 + P_4 + P_5 + P_6 + P_7 + P_8 + P_9.$$

Note that if requests are not identified by necessity during their execution using actual data, then the use of states 6, 7, 8, 9 in the process of servicing requests is associated with the risks of erroneous service.

Note that the availability factor does not take into account different delays when servicing a request by one or two calculators or when dividing the resources of calculators into information recovery and servicing functional requests. This difference can be taken into account by the efficiency retention coefficient [19], but it does not allow estimating the probabilities of timely servicing of requests, which is important for real-time systems. This estimate can be obtained on the basis of the coefficient of readiness for the timely fulfillment of functional requests K_c proposed in [18].

However, the coefficient K_c proposed in [18] does not take into account the risks associated with serving functional requests that require data generated since the last backup. To overcome this shortcoming, it is proposed to evaluate the efficiency of the system by the intensity of profit received from servicing functional requests. A comprehensive performance indicator based on the intensity of profit, having a physical meaning, allows you to resolve the contradictions on the use of the above indicators.

The intensity of profit from servicing functional requests for the system according to Fig. 2 when servicing functional requests in states 6, 7, 8, 9 without identifying the requirement to use actual data is estimated as

$$S = \Lambda\{[R_2B_2 + R_1B_1 + R_3B_2 + aR_4B1]c_1 +$$
$$+[R_1(1 - B_2) + R_2(1 - B_1) + aR3(1 - B_2) + aR_4(1 - B_1)]c_2 + \quad (1)$$
$$+R_0c_3 + (1 - a)R_5c_4\}$$

where R_1 and R_2 are the probability of states in which one or two calculators are involved in servicing functional requests, while there are no risks of accessing outdated data

$$R_2 = P_0 + P_1.$$

If in state "5" the resources of the calculator are involved exclusively in the restoration of actual data and are not used to solve functional problems, then

$$R_1 = P_2 + P_3 + P_4,$$

R_3 and R_4 are the probability of states in which one or two calculators are involved in servicing functional requests, while there are risks of accessing outdated data

$$R_4 = P_7 + P_9,$$
$$R_3 = P_6 + P_8$$

R_0 is the probability of an inoperable state of a duplicated cluster node or a state in which the service of functional requests is not carried out.

So if able

$$R_0 = P_{10} + P_{11} + P_{12} + P_{13} + P_{14} + P_{15} + P_{16} + P_5,$$

$$R_5 = R_3 + R_4.$$

Λ is the intensity of the flow of functional tasks,

c_1 - profit from timely servicing of requests,

c_2 - penalty for not timely servicing requests,

c_3 - penalty for refusing to service requests,

c_4 - penalty for servicing using outdated data,

B_1 and B_2 are the probabilities of timely execution of requests for a time less than the maximum allowable t_0, respectively, with two and one operable calculator,

$$B_1 = 1 - \Lambda v \exp(t_0(\Lambda - v^{-1})),$$

$$B_2 = 1 - \frac{\Lambda}{2} v \exp(t_0(\frac{\Lambda}{2} - v^{-1})),$$

where v is the average execution time of a functional request.

When not serviced in states 6, 7, 8, 9 of functional requests, there are no risks of erroneous calculations when accessing irrelevant data. In this case, the intensity of profit is estimated as

$$S = \Lambda\{[R_2B_2 + R_1B_1]c_1 + [R_1(1 - B_2) + R_2(1 - B_1)]c_2 + R_0c_3\} \quad (2)$$

At the same time, if in state "5" the resources of the calculator are involved exclusively in the restoration of actual data and are not used to solve functional problems

$$R_0 = P_{10} + P_{11} + P_{12} + P_{13} + P_{14} + P_{15} + P_{16} + P_5 + P_6 + P_7 + P_8 + P_9.$$

If the system identifies requests to require the use of data generated after or before backup, then blocking the service is carried out only for requests that require up-to-date data that has not yet been restored.

$$S = \Lambda\{[R_2B_2 + R_1B_1 + R_3B_2 + aR_4B_1]c_1 +$$
$$+[R_1(1 - B_2) + R_2(1 - B_1) + aR_3(1 - B_2) + aR_4(1 - B_1)]c_2 \qquad (3)$$
$$+(R_0 + [P_6 + P_7 + P_8 + P_9](1 - a))c_3\}$$

At the same time, if in state "5" the resources of the calculator are involved exclusively in the restoration of actual data and are not used to solve functional problems

$$R_0 = P_{10} + P_{11} + P_{12} + P_{13} + P_{14} + P_{15} + P_{16} + P_5 + P_6 + P_7 + P_8 + P_9.$$

4   Calculation Example

Let us estimate the intensity of profit from servicing functional requests for the considered options for organizing duplicated cluster nodes. The results of calculating the profit from servicing functional requests are shown in Fig. 5 for a system whose state and transition diagrams correspond to Fig. 2. For the analyzed case, in the states 6, 7, 8, 9, the service of functional requests is implemented without identifying the requirement to use the data generated after the backup.

Curves 1–3 correspond to the intensity of income from servicing functional requests, the maximum allowable waiting time of which should not exceed $t_0 = v, 2v, 4v$. The calculation is made according to the formula (1) at $v = 0.1$ s, $\Lambda_1 = \Lambda_2 = 10^{-4}$ 1/h. $\mu_1 = 1$ 1/h, $\mu_2 = 1$ 1/h, $\mu_3 = 0.2$ 1/h, $\mu_4 = 0.1$ 1/h, $c_1 = 30$ c.u., $c_2 = -10$ c.u., $c_3 = -20$ c.u., $c_4 = -50$ c.u.

The calculation was carried out in the Mathcad 15 computer mathematics system.

Analysis of the results shows the existence of a boundary value for the intensity of the flow of functional requests, when exceeded, it becomes appropriate to block the servicing of functional requests.

Let us compare the considered case, in which there are possible risks of accessing irrelevant data when executing functional queries with the case when these risks are excluded when functional queries are not served in states 6, 7, 8, 9. In this case, the intensity of profit is estimated by Formula (2). The results of calculating the difference in the intensity of profit for the two compared cases are shown in Figs. 6, 7, and 8. Curves 1–3 in Figs. 6, 7, and 8 show the required differences in obtaining profits at $t_0 = v, 2v, 4v$ for the compared cases. For Figs. 6, 7, and 8 the calculations were performed when accessing functional queries to data generated after backup, with probabilities $a = 0.99, 0.9, 0.7$.

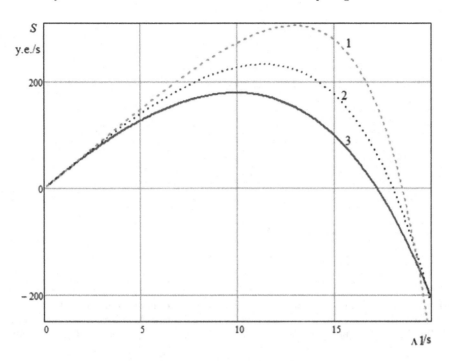

Fig. 5. Profit from servicing functional requests when servicing functional requests without identifying data access before it is updated.

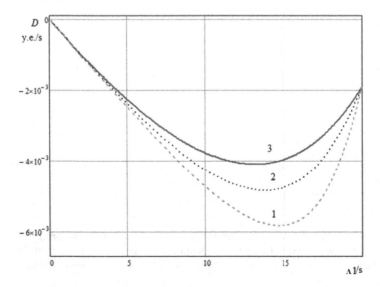

Fig. 6. The results of calculating the difference in the intensity of profit for the two compared cases with and without the exclusion of the risks of accessing outdated data. Calling functional queries to data generated after backup, with probabilities $a = 0.99$.

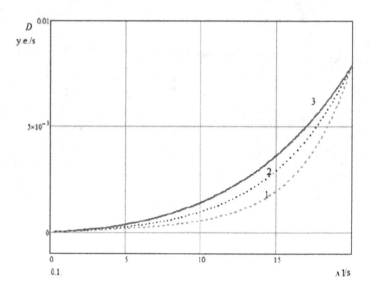

Fig. 7. The results of calculating the difference in the intensity of profit for the two compared cases with and without the exclusion of the risks of accessing outdated data. Calling functional queries to data generated after backup, with probabilities $a = 0.9$.

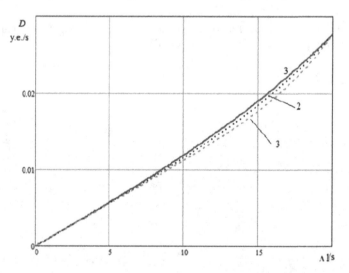

Fig. 8. The results of calculating the difference in the intensity of profit for the two compared cases with and without the exclusion of the risks of accessing outdated data. Calling functional queries to data generated after backup, with probabilities $a = 0.7$.

5   Conclusion

For computing systems of a cluster architecture built on the basis of duplicated nodes containing two computers and two nodes of dual-input memory, the possibilities of increasing the readiness of the cluster for the timely execution of requests that are critical to service delays are analyzed, taking into account the multi-stage information recovery of memory after failures.

Markov models of duplicated cluster nodes are proposed, on the basis of which the dependences of the system readiness for timely execution of requests on resource allocation options that have retained the operability of computing nodes to perform functional tasks and restore information in memory, including the stages of entering the results of the last backup and replicating relevant information from memory of healthy nodes.

The possibilities of increasing the readiness of the cluster system for the timely execution of functional requests based on the preliminary entry into the memory node intended to replace the failed one, the results of the last backup, are studied.

The effectiveness of the proposed solutions for restoring the system after failures is estimated by the intensity of profit from servicing information requests, taking into account the risks of accessing irrelevant information by information.

References

1. Kim, S., Choi, Y.: Constraint-aware VM placement in heterogeneous computing clusters. Cluster Comput. **23**, 71–85 (2020). https://doi.org/10.1007/s10586-019-02966-6
2. Krasnobaev, V., Kuznetsov, A., Kiian, A., Kuznetsova K.: Fault tolerance computer system structures functioning in residue classes. In: Proceedings of the 11th IEEE International Conference on Intelligent Data Acquisition and Advanced Computing Systems: Technology and Applications, IDAACS 2021, vol. 11, pp. 471–474 (2021)
3. Machida, F., Kawato, M., Maeno, Y.: Redundant virtual machine placement for fault-tolerant consolidated server clusters. In: IEEE Network Operations and Management Symposium, pp. 32–39. IEEE Press, Osaka (2010). https://doi.org/10.1109/NOMS.2010.5488431.020.71-85
4. Tatarnikova, T.M., Poymanova, E.D.: Differentiated capacity extension method for system of data storage with multilevel structure. Scient. Tech. J. Inform. Technol. Mech. Optics **20**(1), 66–73 (2020). https://doi.org/10.17586/2226-1494-2020-20-1-66-73
5. Stepanov, N., Turlikov, A., Begishev, V., Koucheryavy, Y., Moltchanov, D.: Accuracy assessment of user micromobility models for thz cellular systems: mmNets 202. In: Proceedings of the 5th ACM Workshop on Millimeter-Wave and Terahertz Networks and Sensing Systems, Part of ACM MobiCom 2021, vol. 5, pp. 37–42 (2021)
6. Shubinsky, I.B., Rozenberg, I.N., Papic, L.: Adaptive fault tolerance in real-time information systems Reliability. Theory Appli. **12**(1)(44), 18–25 (2017)

7. Tatarnikova, T.M., Sikarev, I.A., Bogdanov, P.Y., et al.: LBotnet attack detection approach in IoT networks. Aut. Control Comp. Sci. **56**, 838–846 (2022). https://doi.org/10.3103/S0146411622080259

8. Bogatyrev, V.A.: Increasing the fault tolerance of a multi-trunk channel by means of inter-trunk packet forwarding. Autom. Control. Comput. Sci. **33**(2), 70–76 (1999)

9. Bogatyrev, V.A.: An interval signal method of dynamic interrupt handling with load balancing. Autom. Control. Comput. Sci. **34**(6), 51–57 (2000)

10. Wang, X., Carey, M.J., Tsotras, V.J.: Subscribing to big data at scale. Distrib. Parallel Databases **40**(2), 475–520 (2022)

11. Bartocci, E., Manjunath, N., Mariani, L., Mateis, C., Ničković, D.: Cpsdebug: automatic failure explanation in cps models. Inter. J. Softw. Tools Technol. Trans. (STTT) **23**(5), 783–796 (2021)

12. Attkan, A., Ranga, V.: Cyber-physical security for IOT networks: a comprehensive review on traditional, blockchain and artificial intelligence based key-security. Complex Intell. Syst. (2022)

13. Belcastro, L., Cantini, R., Marozzo, F., Orsino, A., Talia, D., Trunfio, P.: Programming big data analysis: principles and solutions. J. Big Data **9**(1), 1–50 (2022)

14. Kshirsagar, P.S., Pujar, A.M.: Resource allocation strategy with lease policy and dynamic load balancing. Inter. J. Mod. Educ. Comput. Sci. **9**(2), 27–33 (2017)

15. Ed-daoudy, A., Maalmi, K.H.: A new internet of things architecture for real-time prediction of various diseases using machine learning on big data environment. J. Big Data **6**(1), 1–25 (2019)

16. Saxena D., Singh A.K.: Ofp-tm: an online vm failure prediction and tolerance model towards high availability of cloud computing environments. J. Supercomput. 2022, **78**(6), 8003–8024

17. Bogatyrev, V.A., Bogatyrev, S.V., Bogatyrev, A.V.: Reliability and timeliness of servicing requests in infocommunication systems, taking into account the physical and information recovery of redundant storage devices. In: 2022 International Conference on Information, Control, and Communication Technologies (ICCT), pp. 1–4 (2022)

18. Bogatyrev, V.A., Bogatyrev, S.V., Bogatyrev, A.V.: Assessment of the readiness of a computer system for timely servicing of requests when combined with information recovery of memory after failures. Sci. Tech. J. Inform. Technol. Mech. Optics **23**(3). https://doi.org/10.17586/2226-1494-2023-23-3-

19. Polovko, A.M., Gurov S.V.: New Theories of Reliability, pp. 102–109 (2006)

Numerical Study of Queuing-Inventory Systems with Catastrophes Under Base Stock Policy

Agassi Melikov[1]([⊠]), Laman Poladova[2], and Edayapurath Sandhya[3]

[1] Baku Engineering University, Khirdalan City, Hasan Aliyev Street 120, AZ0101 Absheron, Azerbaijan
agassi.melikov@gmail.com
[2] Institute of Control Systems, 1141 Baku, Azerbaijan
[3] Jyothi Engineering College, APJ Abdul Kalam Technological University, Thrissur, Kerala, India

Abstract. We consider a single-server queuing-inventory system (QIS) with catastrophes under the base stock policy. Consumer customers (c-customers) that arrived to buy inventory, can be a form of queue in an infinite buffer. All items in the warehouse are destroyed if a catastrophe is occurring, but in such cases, the c-customers in the system (on the server or in the buffer) are still waiting to be restocked. Upon the arrival of the negative customer (n-customer), one c-customer is pushed out, if any. A hybrid sale rule is used: if upon arrival of the c-customer, the inventory level is zero, then according to the Bernoulli scheme, this customer is either lost (lost sale rule) or is joining the queue (backorder rule). The mathematical model of the investigated QIS is constructed as a two-dimensional Markov chain (2D MC). The ergodicity condition is established, and the matrix-analytic method (MAM) is used to calculate the steady-state probabilities of the constructed 2D MC. Formulas for performance measures are found and the results of numerical experiments are illustrated.

Keywords: queuing-inventory system · base stock policy · catastrophes · negative customers · matrix-analytic method

MSC: 60J28 · 60K25 · 90B05 · 90B22

1 Introduction

In a classical inventory management system (IMS) as demanded by consumer customers (c-customers) items are immediately delivered from the stock (if available). However, in the case of IMS with a service station, a demanded item is delivered to the c-customer after some positive service time. In such systems, the items are delivered not at the time of demand but after a random time of service causing the formation of queues. For the first time the IMSs with positive service times were called queuing-inventory systems (QIS) in the papers Schwarz and Daduna [28] and Schwarz et al. [29]. Note that the first papers devoted to models of QISs were published by Sigman and Simchi-Levi [30] and Melikov and Molchanov [24] independently of each other. After that paper's models of QISs have been the subjects of extensive research for the last three decades and substantial literature on the topic exists. For the current state of the art from both theoretical and practical points of view, one can refer to the detailed reviews by Bijvank and Vis [4], Krishnamoorthy, Manikanadan, and Laxmi [17], and Krishnamoorthy, Shajin, and Narayanan [18]. The main performance measures of QISs are related to both the inventory level and queue length distributions; these two kinds of measures impact each other. They depend significantly on the replenishment policy (RP) adopted. In the available literature, the following RPs are considered, see Krishnamoorthy, Shajin, and Narayanan [18].

(1) (s, S)-policy: in this policy, the replenishment size is that much to bring the level back to S at the replenishment epoch, where s is the recorded level and S is the maximum capacity of the warehouse; sometimes this policy is called "Up to S" policy.

(2) (s, Q)-policy: in this policy, the replenishment size is fixed and is equal to Q=S-s; in this policy to avoid repeated replenishment, it is assumed that $s < (S/2)$.

(3) Randomized policy: in this policy, the probability that replenishment size is n equal to p_n, such that $\sum_{n=1}^{S} p_n = 1$, where $p_S > 0$.

(4) Base stock policy: in this policy, a replenishment is called every time an item sells out; sometimes this RP is called either (S-1, S)-policy or one-to-one ordering policy. This policy is advised for bulky, expensive items with low demand and slow lead times.

The class of QIS, investigated in this paper, consists solely of Markovian systems with base stock RP. Based on this, below we briefly consider the related works in which models of QISs with such a replenishment policy are studied. Models of perishable QISs with Poisson flow and exponential lifetimes under lost sales are constructed by Kalpakam and Sapna [11,12]. For the case modified (S-1, S)-policy, Kalpakam, and Shanthi [13,14] studied the same system in which the distribution function of processing times is arbitrary. A modified (S-1, S)-policy means that orders for items are placed only at demand epochs which results in variable ordering quantity. Further, the modified (S-1, S)-policy for the case constant lead time was analyzed by Alizadeh et al. [2].

Gomathi, Jeganathan, and Anbazhagan [9] consider a single-server Markov QIS model with a finite capacity waiting room. An extension of this model for the case of the vacation of servers and impatient customers was considered by Kathiresan and Anbazhagan [15]. The authors obtained the joint probability distribution of the number of customers in the waiting hall, the inventory level, and the server status for the steady state case, and the optimization problem is solved.

Kathiresan and Anbazhagan [16] consider a single-server finite Markovian model with multiple vacations and negative customers. When the inventory becomes empty, the server takes a vacation and the vacation duration is exponentially distributed. The stationary distribution of the number of customers in the waiting hall, the inventory level, and the server status are calculated.

Anbazhagan, Wang, and Gomathi [3] consider a Markovian model with retrial demands. It is assumed that customers that arise during stock-out periods enter an orbit of infinite size. The joint probability distribution of the inventory level and the number of customers in the orbit are obtained. Various system performance measures in the steady state are derived. A similar model but with stock-dependent arrivals was considered in Abdul Reiyas and Jeganathan [1]. In Geetha Rani and Elango [8] it is assumed that the customers in the queue may become impatient.

Recently Otten and Daduna [27] consider a model of QIS with two unbounded queues for customers of different priority classes under lost sales. The authors derive sufficient conditions for the stability of the system. For special cases, they show that the condition is necessary as well. In addition, for the special case of zero service time, explicit relations to determine the stationary distribution are developed. For further literature, we refer to Otten [26].

Note that in mentioned papers using the MAP flows and PH distributions for both services and lead times are made possible through the well-established matrix-analytic methods (MAM) introduced by Neuts [25]. The development of MAM with applications in various stochastic models of QIS and reliability theory can be found in the books by Chakravarthy [5,6], Dudin, Klimenok and Vishnevsky [7], He [10], and Latouche and Ramaswami [19].

One of the main features of the QIS model studied here is that in addition to consumer customers, negative customers (n-customers) also enter the system. Note that n-customers do not require the inventory, but they force one c-customer out of the system, i.e. n-customers will not affect the system warehouse. Such situations arise in real QISs in which it is possible to campaign among the c-customers of this system so that they leave this system to buy stocks from another system. Even though such situations often occur in real life, such models have been little studied, see Kathiresan and Anbazhagan [16] and Melikov et al. [23].

Another feature of the investigated here QIS model is the possibility of catastrophes in the warehouse block (facility) of the system. It means that all items in the inventory are destroyed instantly at the same time when a catastrophe has occurred. Note that similar models were recently studied in the works of Melikov, Mirzayev, and Nair [20,21] and Melikov, Mirzayev, and Sztrik [22], where as a result of a destructive request, only a stock of a unit size is destroyed.

In addition, in our paper a hybrid sale policy is used, i.e. c-customers who arrive during zero stock level either are lost (lost sale scheme) or join the queue (backlog scheme) by Bernoulli trials.

To our best knowledge, only in a recent paper by Melikov et al. [23] models of QISs with catastrophes and negative customers under a hybrid sale policy are considered. They considered two well-known RPs: (s, S) and randomized replenishment policies. This paper is a continuation of the research that started in these works. Here we consider a similar model of QIS under the base stock policy.

2 Describing the Model

The main assumptions of the QIS model studied here are as follows:

- The maximum warehouse capacities equal to S, S $<\infty$.
- Homogeneous and positive c-customers arrive at the facility with one server according to the Poisson process at a rate of λ^+, and each c-customer needs a stock of unit size. The waiting room for c-customers has an infinite size.
- Consumer customers from the queue are selected for servicing according to their arrivals and their service times are assumed to be exponential with parameter μ.
- Along with c-customers, the system receives n-customers with rate λ^-. The influence of n-customers is as follows: (1) If at the moment of arrival of the n-customer, there is a queue of c-customers, then one c-customer is pushed out from the queue; (2) A n-customer can force out from the system even a c-customer, which is in the server if a queue is empty. In such cases the inventory level does not change, i.e. it is assumed that stocks are released after the completion of servicing a c-customer; (3) If there are no c-customers in the system, then the received n-customer does not affect the operation of the system.
- The hybrid sales scheme is used, i.e. if there are no stocks in the system upon arrival of c-customer, then, in accordance to the Bernoulli trials, it either with probability (w.p.) φ_1 joins the queue of infinite length(backorder sale scheme), or w.p. φ_2 leaves the system unserved(lost sale scheme), where $\varphi_1 + \varphi_2 = 1$. If the stock level is positive, then the arriving c-customer is queued w.p. 1.
- Catastrophes follow the Poisson flow with the rate κ, and at the moment of arrival of such an event, all the items in the stock are instantly destroyed. As a result of the catastrophes, even the item, which is at the status of release to the c-customer, is destroyed, and the c-customer whose service was interrupted is returned to the queue; in other words, the catastrophe only destroys the stocks of the system and does not force c-customers out of the system. If the inventory level is zero, then the disaster does not affect the operation of the system.
- Base replenishment policy is used. It means that a replenishment is called every time an item sells out and lead times are assumed to be exponential with parameter ν, $\nu < \infty$.

The task is to find the joint distribution of the number of c-customers in the system and the inventory level of the system, as well as to calculate the main performance measures of the system.

3   Stationary Distribution

Let X_t be the number of c-customers at the time and Y_t be the inventory level at the time. Then, the process $\{Z_t, t \geq 0\} = \{(X_t, Y_t), t \geq 0\}$ forms a continuous time Markov chain (CTMC) with state space E $= \{0, 1, \dots\} \times \{0, 1, \dots, S\}$.

Proposition 1. The generator G of the process $\{Z_t, t \geq 0\}$ has following form:

$$G = \begin{pmatrix} B & A_0 & O & \dots & O & \dots \\ A_2 & A_1 & A_0 & O & O & \dots \\ O & A_2 & A_1 & A_0 & O & \dots \\ O & O & A_2 & A_1 & A_0 & \dots \\ \vdots & \vdots & \ddots & \ddots & \ddots & \vdots \end{pmatrix} \tag{1}$$

where O denotes zero square matrices with dimension S+1, and all other block matrices are square matrices of the same dimension. Entities of the block matrices

$B = \| b_{ij} \|$ and $A_k = \| a_{ij}^{(k)} \|$, i,j $= 0, 1, \dots, S$ are given by

$$b_{ij} = \begin{cases} (S-i)\nu & \text{if } 0 \leq i \leq S-1, \quad j = i+1 \\ \kappa & \text{if } 0 < i \leq S, \quad j = 0 \\ -(S\nu + \lambda^+ \varphi_1) & \text{if } i = j = 0 \\ -\left((S-i)\nu + k + \lambda^+\right) & \text{if } 0 < i \leq S, \quad i = j, \\ 0 & \text{in other cases;} \end{cases} \tag{2}$$

$$a_{ij}^{(0)} = \begin{cases} \lambda^+ \varphi_1 & \text{if } i = j = 0 \\ \lambda^+ & \text{if } i \neq 0, \quad i = j \\ 0 & \text{in other cases;} \end{cases} \tag{3}$$

$$a_{ij}^{(1)} = \begin{cases} (S-i)\nu & \text{if } 0 \leq i \leq S-1, \quad j = i+1 \\ \kappa & \text{if } i > 0, \quad j = 0 \\ -(\lambda^- + S\nu + \lambda^+ \varphi_1) & \text{if } i = j = 0 \\ -\left((S-i)\nu + \kappa + \mu + \lambda^+ + \lambda^-\right) & \text{if } 0 < i \leq S, \quad i = j, \\ 0 & \text{in other cases;} \end{cases} \tag{4}$$

$$a_{ij}^{(2)} = \begin{cases} \lambda^- & \text{if } i = j \\ \mu & \text{if } > 0, \quad j = i-1 \\ 0 & \text{in other cases.} \end{cases} \tag{5}$$

Proof. The transition rate $(n_1, m_1) \rightarrow (n_2, m_2)$ is denoted as $q((n_1, m_1);$ $(n_2, m_2))$. By taking into account the assumptions made in Sect. 2, we conclude that the indicated parameters are calculated as follows:

(a) Transitions due to the arrival of c-customers:

$(n_1, m_1) \rightarrow (n_1 + 1, 0)$: the rate is $\lambda^+ \varphi_1$, for $m_1 = 0$;
$(n_1, m_1) \rightarrow (n_1 + 1, m_1)$: the rate is λ^+, for $0 < m_1 \leq S$.

(b) Transitions due to the arrival of n-customers:

$(n_1, m_1) \rightarrow (n_1 - 1, m_1)$: the rate is λ^-, for $n_1 > 0$.

(c) Transitions due to service completion of c-customers $(n_1, m_1) \rightarrow (n_1 - 1, m_1 - 1)$: the rate is μ, for $n_1 > 0, m_1 > 0$.

(d) Transitions due to catastrophes: $(n_1, m_1) \rightarrow (n_1, 0)$: the rate is κ, for $m_1 > 0$.

(e) Transitions due to replenishment: $(n_1, m_1) \rightarrow (n_1, m_1 + 1)$: the rate is $(S - m_1)\nu$, for $0 \leq m_1 < S$

All other transition pairs have a rate of zero.

So, we have the following relations:

$$q((n_1, m_1); (n_1 + 1, 0)) = \lambda^+ \varphi_1 \cdot \chi(m_1 = 0); \tag{6}$$

$$q((n_1, m_1); (n_1 + 1, m_1)) = \lambda^+ \cdot \chi(m_1 > 0); \tag{7}$$

$$q((n_1, m_1); (n_1 - 1, m_1)) = \lambda^- \cdot \chi(n_1 > 0); \tag{8}$$

$$q((n_1, m_1); (n_1 - 1, m_1 - 1)) = \mu \cdot \chi(n_1 > 0) \cdot \chi(m_1 > 0); \tag{9}$$

$$q((n_1, m_1); (n_1, 0)) = \kappa \cdot \chi(m_1 > 0); \tag{10}$$

$$q((n_1, m_1); (n_1, m_1 + 1)) = (S - m_1)\nu \cdot \chi(0 \leq m_1 < S). \tag{11}$$

Hereinafter, χ (A) is the indicator function of the event A, which is 1 if A is true and 0 otherwise. By considering a lexicographic order of the system's states, $(0,0),(0,1),\ldots,(0,S);(1,0),(1,1),\ldots,(1,S); \ldots; (i, 0), (i, 1), \ldots (i, S); \ldots$ from relations (6)-(11) we conclude that the generator G of the process $Z_t, t \geq 0$ might be represent via relations (1)-(5).

Proposition 2. The process $\{Z_t, t \geq 0\}$ is ergodic if and only if the following condition is fulfilled:

$$\lambda^+(1 - \varphi_2 \pi(0)) < \lambda^- + \mu(1 - \pi(0)), \tag{12}$$

where

$$\pi(m) = b_m \left(\sum_{i=0}^{S} b_i \right)^{-1} \tag{13}$$

and parameters b_m, m = 0, 1, ..., S, are calculated via the following reverse recursive formulas:

$$b_m = \begin{cases} \frac{1}{(S-m)\nu}\left(a_{S-m-1}b_{m+1} - \mu b_{m+2}\right) \text{ if } 0 \le m < S - 2 \\ \frac{1}{2\nu}\left(a_{S-m-1}b_{m+1} - \mu\right) \text{ if } m = S - 2 \\ \frac{a_0}{\nu} \text{ if } m = S - 1, \\ 1 \text{ if } m = S; \end{cases} \tag{14}$$

$$a_n = \mu + \kappa + n\nu, n = 1, 2, \ldots, S - 1;$$

Proof. By Neuts (1981), pp. 81-83, the process $\{Z_t, t \ge 0\}$ is ergodic if and only if

$$\pi A_0 e < \pi A_2 e, \tag{15}$$

where $\pi = \left(\pi(0), \pi(1), \ldots, \pi(S)\right)$, is the stationary probability vector that correspond to generator $A = A_0 + A_1 + A_2$ and e is column vector of dimension S+1 that contains only 1 s.

From relations (3)–(5) conclude that the nonzero entities of the matrix A are determined as follows:

$$a_{ij} = \begin{cases} -S\nu \text{ if } i = j = 0, \\ (S - i)\nu \text{ if } 0 < i \le S - 1, \quad j = i + 1, \\ \mu + \kappa \text{ if } i = 1, j = 0, \\ \kappa \text{ if } i > 1, \quad j = 0, \\ -\mu \text{ if } i > 0, \quad j = 1, \\ \mu \text{ if } i \ge 2, \quad j = i - 1. \end{cases} \tag{16}$$

In other words, we have balance equations for stationary probability vector π :

$$\pi A = \mathbf{0}, \pi e = 1, \tag{17}$$

where $\mathbf{0}$ is the null row vector of dimension S+1.

Using the reverse recursive procedure, we obtain the balance Eq. (17) with a solution (13).

The stationary probability vector of the generator A given by (13), allows us to derive the ergodicity (stability) condition of the process $\{Z_t, t \ge 0\}$. So, by using relations (13), from the matrices A_0 and A_2, we have

$$\pi A_0 e = \lambda^+ \varphi_1 \pi(0) + \lambda^+ \sum_{m=1}^{S} \pi(0) = \lambda^+ \varphi_1 \pi(0) + \lambda^+(1 - \pi(0)) = \lambda^+(1 - \varphi_2 \pi(0)),$$

and

$$\pi A_2 e = \lambda^- \sum_{m=0}^{S} \pi(m) + \mu \sum_{m=1}^{S} \pi(m) = \lambda^- + \mu(1 - \pi(0)).$$

Thus, relation (15) is equivalent to the inequality (12).

Note 1. As in Melikov et al. [23], here the established ergodicity condition (12) has the following probabilistic meaning: the rate of c-customers entering the system must be less than the total rate of n-customers and the rate of served c-customers. We also conclude from (12) that, in the general case, the stability condition for this model depends on the size of the system warehouse, the rates of arrivals, catastrophes, service, and replenishment. However, in some special cases, the stability condition does not depend on all parameters. For instance, since the left side (LS) of (12) is less than λ^+ and the right side (RS) of (12) is more than λ^- so we conclude the system is stable if $\lambda^+ < \lambda^-$. In other words, the last condition is a rougher one, i.e. if $\lambda^+ < \lambda^-$ then the condition (12) does not need to be checked.

   Note 2. Consider the following special cases.

(i) If $\varphi_2 = 1$ (i.e. when purely lost sale scheme is used) and $\lambda^- = 0$ (i.e. when there are not n-customers) from (12) we find the ergodicity condition for the single-server Markovian queuing system, i.e., $\lambda^+ < \mu$, i.e. in this case, the ergodicity condition of the system does not depend on the storage size of the system, the rate of catastrophes, and the replenishment rate.

(ii) If $\varphi_2 = 1$ and $\lambda^- > 0$ then the ergodicity condition is depending on all indicated parameters of the system, see formulas (14).

(iii) If $\varphi_2 = 0$ (pure backorder scheme is used) then the ergodicity condition is depending on all indicated parameters of the system even for case $\lambda^- = 0$, see formulas (14).

Steady-state probabilities that correspond to the generator matrix G we denote by $p = (p_0, p_1, p_2, \ldots)$, where $p_n = (p(n,0), p(n,1), \ldots, p(n,S)), n = 0, 1, \ldots$. Under the ergodicity condition (12) desired steady-state probabilities are determined from the following equations:

$$p_n = p_0 R^n, n \geq 1,$$ (18)

where R is the non-negative minimal solution of the following quadratic matrix equation:

$$R^2 A_2 + R A_1 + A_0 = 0.$$

From (8)–(11) conclude that bound probabilities p_0 are determined from the following system of equations with the normalizing condition:

$$p_0(B + R A_2) = 0.$$

$$p_0(I - R)^{-1} e = 1,$$

where I indicate the identity matrix of dimension S+1.

4   Performance Measures

Main performance measures can be divided into two groups: stock-related metrics and queuing-related metrics. Stock-related metrics are the following:

- Average inventory level (S_{av})

$$S_{av} = \sum_{m=1}^{S} m \sum_{n=0}^{\infty} p(n, m) \qquad (19)$$

- Destruction rate of the stocks (DRS):

$$DRS = \kappa(1 - \sum_{n=0}^{\infty} p(n, 0)) \qquad (20)$$

- Average reorder rate (RR)

$$RR = \kappa \sum_{m=1}^{S} p(0, m) + (\mu + \kappa) \sum_{n=1}^{\infty} \sum_{m=1}^{S} p(n, m); \qquad (21)$$

Queuing-related metrics are the following:

- Loss rate (LR) of c-customers

$$LR = \lambda^{+} \varphi_2 \sum_{n=0}^{\infty} p(n, 0) + \lambda^{-}\left(1 - \sum_{m=0}^{S} p(0, m)\right) \qquad (22)$$

- Average length of the queue of c-customers (L_{av})

$$L_{av} = \sum_{n=1}^{\infty} n \sum_{m=0}^{S} p(n, m) \qquad (23)$$

5   Numerical Results

In Tables 1 through 6, we display the behavior of main system performance measures as well as Expected Total Cost (ETC) versus initial parameters. ETC is defined as follows:

$$ETC = K \cdot RR + c_h S_{av} + c_d DRS \cdot S_{av} + c_l LR + c_w L_{av}, \qquad (24)$$

where K is the fixed price of one order, c_h is the price of unit inventory holding per unit of time, c_d is the price of unit inventory destruction, c_l is the cost for a single c-customer loss, c_w is the price per unit time of queuing delay for a single c-customer.

In all our examples we take S = 50 and values of other parameters are shown in the table's titles. The coefficients in the expression for the functional in ETC (see (25)) were chosen as follows: K = 10, c_h = 20, c_l = 10, c_w = 20, c_d = 40.

Due to the limited volume of the paper, a detailed analysis of the results of numerical experiments is left to the reader. Here we briefly analyze the presented tables.

Common to all tables is the conclusion that all performance measures as well as ETC change smoothly. We register the following observations from these tables.

- An increase λ^+ results in a decrease(with very slow rate) in measures S_{av}, DRS and ETC; other measures are increasing versus λ^+ (see Table 1).
- Measures S_{av}, DRS and ETC are increasing(with very slow rate)ones versus λ^-; other measures are decrease (see Table 2).
- An increase μ results in a decrease in measures S_{av}, DRS (with very slow rate), L_{av}, ETC; other measures areal mos constants (see Table 3).
- S_{av}, ETC increase strongly compared to ν, and DRS and RR also increase, but very slowly; other measures are decreasing (see Table 4).
- An increase in κ leads to strong changes in all stock-related indicators and ETC, and only S_{av} decreases compared to κ; queue-related metrics are increasing at a moderate rate (see Table 5).
- All performance measures are almost constants versus φ_1, only RR and L_{av} are increased at a very slow rate (see Table 6).

Table 1. Effect of λ^+ on the performance measures; $\lambda^- = 2$, $\kappa = 3$, $\mu = 10$, $\nu = 3$, $\varphi_1 = 0.6$.

λ^+	S_{av}	DRS	RR	LR	L_{av}	ETC
5	24.3136	2.9395	5.9222	0.0202	0.7256	3418.97
5.2	24.2861	2.9394	6.1198	0.0202	0.7772	3418.14
5.4	24.2587	2.9393	6.3174	0.0202	0.8319	3417.37
5.6	24.2312	2.9393	6.5151	0.0202	0.8901	3416.66
5.8	24.2037	2.9392	6.7126	0.203	0.9522	3416.04
6	24.1763	2.9391	6.9102	0.0203	1.0185	3415.50
6.2	24.1489	2.9391	7.1078	0.0203	1.0894	3415.05
6.4	24.1214	2.9390	7.3054	0.0203	1.1655	3414.71
6.6	24.0941	2.9389	7.503	0.0204	1.2474	3414.48
6.8	24.0665	2.9389	7.7006	0.0204	1.3357	3414.38
7	24.0391	2.9388	7.8981	0.0204	1.4312	3414.42

Table 2. Effect of λ^- on the performance measures; $\lambda^+ = 6$, $\kappa = 3$, $\mu = 10$, $\nu = 3$, $\varphi_1 = 0.6$.

λ^-	S_{av}	DRS	RR	LR	L_{av}	ETC
1	24.1000	2.9389	7.4594	0.0204	1.2289	3414.52
1.2	24.1164	2.9390	7.3416	0.0203	1.1801	3414.66
1.4	24.1322	2.9390	7.2279	0.0203	1.1351	3414.83
1.6	24.1474	2.9391	7.1183	0.0203	1.0934	3415.03
1.8	24.1621	2.9391	7.0125	0.0203	1.0546	3415.26
2	24.1763	2.9391	6.9102	0.0203	1.0185	3415.51
2.2	24.1901	2.9392	6.8114	0.0203	0.9848	3415.76
2.4	24.2033	2.9392	6.7158	0.0203	0.9532	3416.03
2.6	24.2162	2.9392	6.6233	0.0203	0.9236	3416.32
2.8	24.2286	2.9393	6.5337	0.0202	0.8958	3416.61
3	24.2407	2.9393	6.4469	0.0202	0.8696	3416.91

Table 3. Effect of μ on the performance measures; $\lambda^+ = 6$, $\kappa = 3$, $\lambda^- = 2$, $\nu = 3$, φ_1 = 0.6.

μ	S_{av}	DRS	RR	LR	L_{av}	ETC
9	24.1883	2.9392	6.9088	0.0203	1.1761	3420.32
9.2	24.1851	2.9392	6.9092	0.0203	1.1323	3419.00
9.4	24.1822	2.9392	6.9096	0.0203	1.0916	3417.77
9.6	24.1791	2.9391	6.9099	0.0203	1.0538	3416.60
9.8	24.1763	2.9392	6.9102	0.203	1.0185	3415.50
10	24.1736	2.9391	6.9105	0.0203	0.9855	3414.46
10.2	24.1709	2.9391	6.9108	0.0203	0.9545	3413.47
10.4	24.1683	2.9391	6.9111	0.0203	0.9255	3412.53
10.6	24.1658	2.9391	6.9114	0.0203	0.8981	3411.63
10.8	24.1634	2.9391	6.9117	0.0203	0.8724	3410.78
11	24.1611	2.9391	6.9121	0.0203	0.8481	3409.97

Table 4. Effect of ν on the performance measures; $\lambda^+ = 6$, $\kappa = 3$, $\lambda^- = 2$, $\mu = 10$, $\varphi_1 = 0.6$.

ν	S_{av}	DRS	RR	LR	L_{av}	ETC
1	11.2953	2.8121	6.7197	0.0626	1.0631	1585.54
1.2	13.1308	2.8445	6.7687	0.0518	1.0509	1845.88
1.4	14.8016	2.8674	6.8032	0.0442	1.0427	2083.06
1.6	16.3284	2.8844	6.8287	0.0385	1.0367	2299.9
1.8	17.6057	2.8901	6.8456	0.0342	1.0301	2423.5
2	19.0176	2.9081	6.864	0.0306	1.0286	2682.03
2.2	20.2076	2.9166	6.8767	0.0278	1.0258	2851.19
2.4	21.3098	2.9337	6.8872	0.0254	1.0235	3007.88
2.6	22.3334	2.9296	6.8961	0.0235	1.0215	3153.43
2.8	23.2866	2.9347	6.9037	0.0218	1.0199	3288.97
3	24.1763	2.9391	6.9102	0.0203	1.0185	3415.50

Table 5. Effect of k on the performance measures; $\lambda^+ = 6$, $\lambda^- = 2$, $\nu = 3$, $\mu = 10$, φ_1 = 0.6.

κ	S_{av}	DRS	RR	LR	L_{av}	ETC
1	34.4452	1.2001	6.0726	0.0083	1.0074	2423.34
1.2	32.9607	1.3866	6.162	0.0096	1.0086	2569.23
1.4	31.4601	1.5923	6.2608	0.0112	1.0099	2715.87
1.6	30.2156	1.7779	6.3499	0.0123	1.0111	2836.97
1.8	28.9499	1.9825	6.4483	0.0137	1.0124	2959.58
2	27.8928	2.1671	6.5372	0.0157	1.0135	3061.49
2.2	26.8604	2.3609	6.6307	0.0163	1.0148	3160.55
2.4	25.9018	2.5542	6.724	0.0176	1.0161	3252.07
2.6	25.0093	2.7469	6.8172	0.0191	1.0173	3336.82
2.8	24.1763	2.9391	6.9102	0.0203	1.0185	3414.38
3	23.397	3.1309	7.0031	0.0216	1.0197	3488.68

Table 6. Effect of φ_1 on the performance measures; $\lambda^+ = 6$, $\lambda^- = 2$, $\kappa = 3$, $\nu = 3$, $\mu = 10$.

φ_1	S_{av}	DRS	RR	LR	L_{av}	ETC
0	24.1863	2.9392	6.8381	0.0203	0.9931	3415.72
0.1	24.1847	2.9392	6.8501	0.0203	0.9972	3415.68
0.2	24.1831	2.9392	6.8621	0.0203	1.0014	3415.64
0.3	24.1813	2.9392	6.8741	0.0203	1.0056	3415.60
0.4	24.1796	2.9392	6.8862	0.0203	1.0099	3415.57
0.5	24.178	2.9392	6.8982	0.0203	1.0142	3415.53
0.6	24.1763	2.9391	6.9102	0.0203	1.0185	3415.50
0.7	24.1746	2.9391	6.9223	0.0203	1.0229	3415.46
0.8	24.1732	2.9391	6.9343	0.0203	1.0273	3415.43
0.9	24.1713	2.939	6.9464	0.0203	1.0317	3415.40
1	24.1696	2.9391	6.9584	0.0203	1.0362	3415.37

6   Conclusions

In this paper, we considered single-server Markovian QIS with catastrophes in warehouses and negative customers under the base stock policy. The catastrophe only destroys the stocks of the system and does not force consumer customers out of the system. A mathematical model of the system under study is constructed in the form of a two-dimensional Markov chain, and the condition for its ergodicity is obtained. Formulas for calculating the main performance measures are proposed, and numerical examples show the impact of initial parameters on the behavior of performance measures.

Since the studied QIS is quite complex, here we have considered a simple Poisson/exponential model and developed an approach to its study. As directions for further research, one can indicate the study of QIS models with more complex distributions of flows of positive and negative customers and catastrophes (e.g., MAP flows) and service and lead times (e.g., PH distributions).

References

1. Abdul Reiyaz, M., Jeganathan, K.: Modeling of stochastic arrivals depending on base stock inventory system with a retrial queue. Inter. J. Appli. Comput. Math. **7**, 200 (2021)
2. Alizadeh, M., Eskandari, H., Sajadifar, S.M.: A modified (S-1, S) inventory system for deteriorating items with Poisson demand and non-zero lead time. Appli. Math. Model. **38**, 699–711 (2014)
3. Anbazhagan, N., Wang, J., Gomathi, D.: Base stock policy with retrial demands. Appli. Math. Model. **37**, 4464–4473 (2013)
4. Bijvank, M., Vis, Iris, F.A.: Lost-sales inventory theory: a review. Eur. J. Oper. Res. **215**, 1–13 (2011)
5. Chakravarthy, S.R.: Introduction to matrix-analytic methods in queues, vol. 1. John Wiley and Sons, Inc., London (2022a)

6. Chakravarthy, S.R., Introduction to matrix-analytic methods in queues, vol. 2. John Wiley and Sons Inc., London (2022b)
7. Dudin, A.N., Klimenok, V.I., Vishnevsky, V.M.: The Theory of Queuing Systems with Correlated Flows. Springer, Cham (2020). https://doi.org/10.1007/978-3-030-32072-0
8. Geetha Rani, M., Elango, C.: Markov process for service facility systems with perishable inventory and analysis of a single server queue with reneging. Stochastic Model. Inter. J. Comput. Appli. **44**(15), 18–23 (2012)
9. Gomathi, D., Jeganathan, K., Anbazhagan, N.: Two-commodity inventory system for base-stock policy with service facility. Glob. J. Sci. Front. Res. Math. Decis. Sci. **12**(1), 69–79 (2012)
10. He, Q.-M.: Fundamentals of Matrix-Analytic Methods. Springer, New York (2014). https://doi.org/10.1007/978-1-4614-7330-5
11. Kalpakam, S., Sapna, K.P.: (S-1, S) perishable inventory system with stochastic lead times. Math. Comput. Model. **21**(6), 95–104 (1995)
12. Kalpakam, S., Sapna, K.P.: A lost sales (S-1, S) perishable inventory system with renewal demand. Nav. Res. Logist. **43**, 129–142 (1996)
13. Kalpakam, S., Shanthi, S.: A perishable system with modified base stock policy and random supply quantity. Int. J. Comput. Math. Appl. **39**, 79–89 (2000)
14. Kalpakam, S., Shanthi, S.: A perishable inventory system with modified (S -1, S) policy and arbitrary processing times. Comput. Oper. Res. **28**, 453–471 (2013)
15. Kathiresan, J., Anbazhagan, N.: Base stock policy inventory system with multiple vacations and impatient customers. Int. J. Pure Appl. Math. **109**(10), 161–169 (2016)
16. Kathiresan, J., Anbazhagan, N.: An (S-1, S) inventory system with negative arrivals and multiple vacations. Int. J. Appl. Appl. Math. **14**(2), 672–686 (2019)
17. Krishnamoorthy, A., Manikandan, R., Lakshmy, B.: A survey on inventory models with positive service time. Opsearch. **48**, 153–169 (2011)
18. Krishnamoorthy, A., Shajin, D., Narayanan, W.: Inventory with Positive Service Time: A Survey. In: Anisimov, V., Limnios, N. (eds.) Advanced Trends in Queueing Theory. Mathematics and Statistics" Sciences, vol. 2, pp. 201–238. ISTE and Wiley, London, UK (2021)
19. Latouche, G., Ramaswami, V.: Introduction to matrix analytic methods in stochastic modeling. SIAM, Philadelphia (1999)
20. Melikov, A., Mirzayev, R.R., Nair S.S.: Numerical investigation of double source queuing-inventory systems with destructive customers. J. Comput. Syst. Sci. Inter. **61**(4), 581–598 (2022a)
21. Melikov, A.; Mirzayev, R.R., Nair S.S.: Double sources queuing-inventory system with hybrid replenishment policy. Mathematics, Article 2423 **10**(14), 2423 (2022b)
22. Melikov, A., Mirzayev, R.R., Sztrik, J.: Double sources QIS with finite waiting room and destructible stocks. Mathematics **11**(1), 226 (2023)
23. Melikov, A., Poladova, L., Sandhya, E., Sztrik, J.: Single server queuing-inventory systems with negative customers and catastrophes in the warehouse. Mathematics. **11**, 2380 (2023)
24. Melikov, A., Molchanov, A.: Stock optimization in transport/storage systems. Cybernetics. **28**(3), 484–487 (1992)
25. Neuts, M.F.: Matrix-geometric Solutions in Stochastic Models: An Algorithmic Approach. John Hopkins University Press, Baltimore (1981)
26. Otten, S.: Integrated models for performance analysis and optimization of queueing-inventory systems in logistic networks. Ph.D. thesis, Universität Hamburg, Department of Mathematics. Hamburg, Germany (2018)

27. Otten, S., Daduna, H.: Stability of queueing-inventory systems with customers of different priorities. Annals of Operations Research. https://doi.org/10.1007/s10479-022-05140-1

28. Schwarz, M., Daduna, H.: Queuing systems with inventory management with random lead times and with backordering. Math. Methods Operations Res. **64**(3), 383–414 (2006)

29. Schwarz, M., Sauer, C., Daduna, H., Kulik, R., Szekli, R.: M/M/1 queuing systems with inventory. Queuing Systems. Theory Appli. **54**(1), 55–78 (2006)

30. Sigman, K., Simchi-Levi, D.: Light traffic heuristic for an M/G/1 queue with limited inventory. Annals Operat. Res. **40**, 371–380 (1992)

A Machine-Learning Approach to Queue Length Estimation Using Tagged Customers Emission

Dmitry Efrosinin[1]([✉]) [ID], Vladimir Vishnevsky[2] [ID], and Natalia Stepanova[3] [ID]

[1] Johannes Kepler University Linz, Altenbergerstrasse 69, 4040 Linz, Austria
dmitry.efrosinin@jku.at
[2] V.A. Trapeznikov Institute of Control Sciences of Russian Academy of Sciences, Profsoyuznaya 65, 117997 Moscow, Russia
[3] Scientific and Production Company "INSET", Zvezdniy b-r 19-1, 129085 Moscow, Russia
http://www.jku.at, http://www.ipu.ru, http://www.incet.ru

Abstract. In this paper, we consider the problem of the queue length estimation if only some small number of a so-called tagged customers is observable. The problem is treated in terms of the queueing of vehicles behind a traffic light. A supervised machine learning, particularly an artificial neural network, is used to construct non-linear relationships between the feature and the target. For data generation we simulate an appropriate queueing system. We used an auxiliary Fourier series correction factor by training the neural network. As a result, the quality of the queue length estimation expressed in form of the empirical distribution function of an absolute error was considerably improved.

Keywords: Machine learning · neural network · queueing system · queue length · Fourier series

1 Introduction

Machine learning algorithms have found their active application in various fields of science and engineering, where it is necessary to solve complex classification and regression problems. There is a growing interest in the application of such methods to problems related to queueing systems. Such problems include both classical problems of system performance measures calculation, including evaluation of the corresponding distributions and their mean values, and the computation of optimal control policies in controlled systems. If queueing systems have a complex structure with phase-type or general non-Markovian distributions of the inter-arrival and service times, it is possible to combine simulation and machine learning methods to provide performance analysis and to solve optimization problems. Before we formulate the problem considered in this paper, we give a brief overview of recent work on the use of machine learning in the

V. M. Vishnevskiy et al. (Eds.): DCCN 2023, LNCS 14123, pp. 265–276, 2024.
https://doi.org/10.1007/978-3-031-50482-2_21

queueing theory. The first attempt to give a systematic introduction to application of machine learning by solving a queueing problems has been proposed in [11]. The number of busy servers in Markovian multi-server queueing system was predicted in [10] by means of an artificial neural network. The neural network approach was employed in [7] to estimate the mean performance measures and in [8] to calculate the stationary queue-length distribution in multi-server queueing system with generally distributed inter-arrival and service times. The prediction of the waiting time in queueing systems using machine learning was realized also in [6]. In [12], the complex priority multi-server queueing system with a marked Markovian arrival process and phase-type service time distribution was analyzed successfully by combining simulation and machine learning techniques. The polling systems with correlated arrivals were analyzed using a machine learning in [13]. The performance parameters of the closed queueing network by means of a neural network were evaluated in [5]. Artificial neural networks were used for simulation of the Markovian queueing systems in [9] In the following studies machine learning has been used for controlled systems. The combination of the Markov decision problem and the neural network to estimate the optimal threshold policy for different types of heterogeneous queueing models was studied in [2] and [3]. The problem of optimal scheduling in general multi-server queueing system by applying of neural network for decision making was presented in [4]. Based on the results obtained in the above-mentioned studies, it can be concluded that the application of machine learning in the queueing theory has great potential, especially when it concerns complex systems with arbitrary distributions, when it is difficult to obtain any analytical results or even approximations. However, this does not mean that this technology will replace classical methods of performance analysis and solving optimization problems. Here we are talking about complementing rather than replacing the classical approach.

In this paper, we consider extending the application of machine learning to problems of queueing theory. In many practical situations where the service process is described by a queueing system, information on the total number of customers in the system is often not available. Nevertheless, this characteristic needs to be estimated based on some observable information, such as the maximum number of customers of a certain type over a given period of time. We consider this problem in terms of the vehicle traffic control. In order to solve more global problems like city traffic optimization, or in particular the optimal traffic lights control, it is necessary to know the number of vehicles at intersections. As part of this task, we estimate the target value of the maximum total number of vehicles in the queue in front of the traffic lights. For more details on this subject the readers are referred to [1,14] and references therein. In the former paper, the queue length estimation at intersections is performed using the data sets from pair of advance and stop bar detectors that counts vehicles. In the latter paper, the hidden Markov model is used, where the queue length in each traffic signal cycle stands for a hidden state and the observed pattern of labeled vehicles is an emission. As an emission or a feature we use observable information about the number of tagged vehicles which can include taxis, public and municipal vehicles equipped with GPS-GLONASS trackers. The main idea consists in development of regression model using a machine learning approach to learn the complex non-linear relationship between the feature and target. The neural-network based regression was already used in queueing theory as a methodology for performance analysis of complicated queueing systems, see e.g. [12] or in controlled models to estimate the optimal control policy as was discussed in [3,4]. With this work we want to extend the application of machine learning algorithms to the problems of queueing theory.

Figure 1 illustrates a typical vehicle queueing process at traffic light. This illustration shows the situation of cars congregating in front of a traffic light when it is red and then passing them on the green light. Red cars are considered to be equipped with trackers which will refer to as tagged customers and yellow cars are all other cars, i.e. non-tagged customers. Note that we are modelling the operation of a single traffic light and for real practical purposes it is of course necessary to consider a set of such traffic lights in a particular area of the city. Note that the authors are not specialists in traffic control. This paper represents our subjective view of the problem, and it seems to us that it could be used by specialists who can easily adapt the proposed approach to their specific needs.

Fig. 1. The queue of vehicles by red-phase and passing at green-phase (Color figure online)

The rest of the paper is organized as follows. In Sect. 2 the queue length estimation problem is formulated. In Sect. 3 the simulation algorithm is described. The results of the queue length estimation by neural networks are given in Sect. 4. Section 5 is devoted to the Fourier series correction which considerably improve the estimation accuracy. Concluding remarks are summarized in Sect. 6.

2   Problem Statement and Target Queueing System

The aim is to estimate the hidden state of the maximal total queue length $q(n)$ at the traffic light in its nth cycle based on the observable emission (feature) which is given in form of maximal number of tagged vehicles $q_1(n)$ in this circle. For example, in Fig. 2, the emission $q_1 = 2$ corresponds to a target value $q = 7$.

Fig. 2. Observable $q_1(n)$ and hidden $q(n)$ queues

We will use supervised machine learning techniques to estimate the maximum number of vehicles. Therefore, we need data to train appropriate algorithms. In a real-world situation, mobile detectors can be used that temporarily set up at intersections and calculate the number of vehicles in front of a traffic light at each cycle over a period of time. From the data of such detectors it is also possible to estimate the inter-arrival and service time distributions or their parameters characterizing the corresponding queueing model. As we have no real data, the training samples are generated by a simulation of the appropriate queueing system. The vehicle queueing process is modelled by a type of $M(t)/G/1$ two class single server queuing system. The arriving vehicles to the traffic light form an inhomogeneous Poisson process

$$\mathbb{P}[N(t) = k] = \frac{e^{-\int_0^t \lambda(\tau)d\tau}}{k!} \left[\int_0^t \lambda(\tau)d\tau\right]^k, \tag{1}$$

where $N(t)$ is the number of arrived customers up to time t and

$$\lambda(t) = \alpha - \beta \sin\left(\frac{\pi t}{12} - \gamma\right), \lambda(t) \in [\alpha - \beta, \alpha + \beta]. \tag{2}$$

is a corresponding periodic (24h) arrival rate. The service time of a customer B, which is a passing time of a vehicle at a green-phase, is assumed to be uniformly distributed in the interval $[a, b]$, i.e. $B \sim \mathcal{U}(a, b)$. The service discipline is FIFO independently of the class of customers. The server has deterministic operation and vacation times which specify respectively the red- r and green g -phases of the traffic light. The arriving vehicle is tagged with a probability p. The schema of the queueing system is shown in Fig. 3. Here one can see a queue having FIFO service discipline, which is divided into two virtual parts according to the class of applications, tagged and non-tagged, where the corresponding ordinal numbers are indicated. The applications arrive with a rate $\lambda(t)$ and are divided into two streams with probabilities p and $1 - p$. The service is performed when the traffic light is green and when it is red, the service is blocked for a certain fixed time.

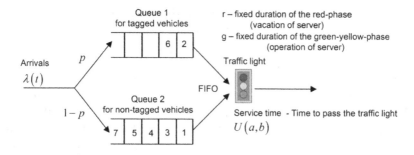

Fig. 3. Schema of the two-class queueing system

3   Simulation Results

Now we discuss shortly a simulation Algorithm 1 which is given in form of a pseudo-code. Using this algorithm we calculate the departure times of vehicles and herewith we determine the number of vehicles of each class in each traffic light cycle. The arrays **a** and **s** contain pre-generated arrival times of C customers of each type and their service times. Then until the end of service of the last customer, the time **d** of departure of the customers of a certain class is calculated. The expression with max means that the server should wait for the customer or the customer should wait for the server. Depending on where the customer's arrival time falls, in the green or red phase, the departure time of the customer from the system is corrected taking into account the traffic light cycle. After obtaining the array **d**, the times $(\mathbf{a}, \mathbf{d})$ are sorted in ascending order. This algorithm is very simple and computationally efficient in contrast to the standard approach, where the queue length is the system state and the arrays **a** and **d** are a list of events, where the arrival event is defined in **a** and the end-of-service event is continuously updated in **d**, which requires more time to obtain the result.

Algorithm 1. Queue Simulation

Require: $\mathbf{a} \in \mathbb{R}_+^C \times \{1, 2\}, \mathbf{s} \in \mathbb{R}_+^C, g, r \in \mathbb{R}_+$

1: Create scalars $b, n, p \in \mathbb{R}_+$ and vector $\mathbf{d} \in \mathbb{R}_+^C \times \{1, 2\}$
2: $i \leftarrow 1, b \leftarrow 0, n \leftarrow 1$
3: **while** $i \leq C$ **do**
4:     $p \leftarrow \max\{a_{ij}, b\} + s_i$
5:     **if** $p < ng + (n-1)r \,\|\, p \geq n(g+r)$ **then**
6:         $b \leftarrow p$
7:     **else if** $ng + (n-1)r \leq p < n(g+r)$ **then**
8:         $b \leftarrow n(g+r) + s_i$
9:         $n \leftarrow n+1$
10:    **end if**
11:    $d_{i1} \leftarrow b, d_{i2} \leftarrow j$
12:    $i \leftarrow i+1$
13: **end while**
14: Put $(\mathbf{a}, \mathbf{d})$ in ascending ordering

Example 1. Consider the queueing system with inhomogeneous Poisson arrival stream (1) where arrival rate (2) is of the form

$$\lambda(t) = 30 - 25 \sin\left(\frac{\pi t}{12} - 1\right),$$

i.e. the minimal rate is 5 and maximal rate is 55. The service time is uniformly distributed in time interval from 5 up to 7 s, or $B \in \mathcal{U}[5/3600, 7/3600]$. Traffic light cycle is equal to 10 min: $g + r$, where $r = 1/12, g = 1/12$. It is assumed further that 25% of vehicles are supplied by trackers, i.e. the probability that the arrived to the traffic light vehicle is tagged is equal to $p = 0.25$. The values of the maximum queue lengths of both tagged and all customers over time, i.e. in each successive cycle, are shown in Fig. 4. In Fig. 4(a), the process trajectory is shown for an interval of $N = 50$ cycles, i.e. just over 8 h; in Fig. 4(b), $N = 1700$ cycles or approximately 10 d. Blue indicates the maximum number of tagged customers and yellow indicates all customers. There is a pronounced periodicity and the presence of a large number of outliers.

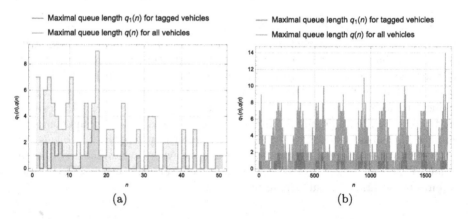

(a)   (b)

Fig. 4. Queue lengths $q_1(n)$ and $q(n)$ in the nth service cycle for 50 (a) and 1700 (b) cycles

4   Queue Length Estimation by Neural Networks

First, to show that the queue length estimation problem is not trivial, we give the results of a discrete classification using standard machine learning methods. We implemented gradient boosted trees which represents an ensemble of decision trees, logistic regression, hidden Markov model, naive Bayes, nearest neighbours, Neural Network with simple configuration, Random Forest and Support Vector Machines for the raw emission set $\{q_1(n) \rightarrow q(n)\}$. The empirical cumulative distribution functions (CDF) of the absolute error, i.e.

$$\hat{F}(t) = \frac{1}{N} \sum_{n=1}^{N} 1_{\{|q(n)-\hat{q}(n)|\leq t\}}, \tag{3}$$

and accuracies of the methods $\hat{F}(1)$ are illustrated in Fig. 5. Here $1_{\{A\}}$ denotes the indicator function which takes the value 1 if event A occurs and 0 otherwise. Based on presented results we see that the quality of standard classifiers implemented for the raw data is obviously unsatisfactory since for all methods $\hat{F}(1) < 0.4$. Thus, it is necessary to create a suitable method for the queue length estimation.

We propose to use the artificial neural network for regression which has the ability to learn the complex relationship between the features and target due to presence of activation function in each layer.

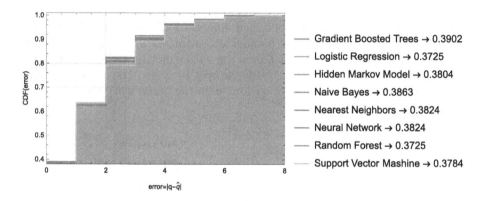

Fig. 5. CDF of the absolute error for the raw emission set

The main steps of the proposed queue length estimation algorithm using artificial neural networks (NN) are the following:

- Step 1: Set up training and test data by simulation.
- Step 2: Data pre-processing (denoising, smoothing, etc.).
- Step 3: Create the multi-layer NNs.
- Step 4: Use training data $(2/3)$ and learn NN.
- Step 5: Use test data $(1/3)$ to estimate the queue length.

The NN has 3–6 layers, regression type NN with a real-valued target and classifiers with discrete-valued classes. The input neurons correspond to $\{(n \mod 24), q_1(n)\}$. The output neuron corresponds to the value $q(n)$. The maximal queue lengths in 1700 subsequent traffic light cycles are generated in form of a sample,

$$s = \{\{(n \mod 24), q_1(n)\} \to q(n) | n \in N = [1, 1700] \cap \mathbb{N}\}, \, n \in N.$$

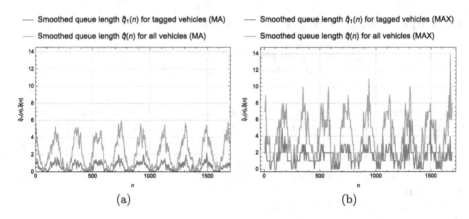

Fig. 6. Smoothed queue lengths by moving average (MA) (a) and by maximum (MAX) (b) over 10 cycles

In data pre-processing step we obtained smoothed queue lengths by moving average (MA) and by maximum number of customers in the queue over 10 cycles. Figure 6 shows pre-processed data, e.g. with a moving average filter and a maximum number of customers in the queue over 10 cycles. The moving average significantly reduces the number of customers in the queue while the maximum concentrates more on the outliers.

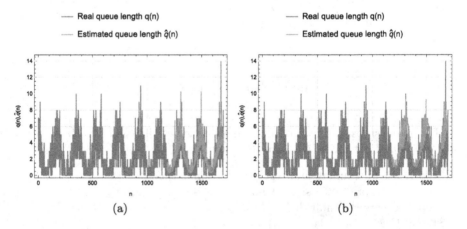

Fig. 7. Real and estimated queue length for raw (a) and smoothed by MAX (b) data

Figure 7 shows the raw data on the total number of customers in the system and the corresponding estimates obtained by the trained neural network over an interval of more than 3 d, based on the raw data and the data obtained after pre-processing with the MAX filter. In second case, the estimates are concentrated on the edges of the bursts of queue length values, which is maybe more preferable for us.

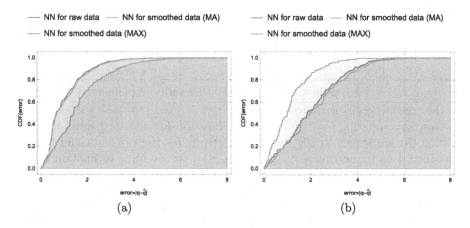

Fig. 8. CDF of absolute error for all $q(n)$ (a) and for $q(n) \geq 6$ (b)

The quality of the estimates obtained is compared in Fig. 8, which show again the empirical CDFs (3) of absolute estimation errors $|q(n) - \hat{q}(n)|$ for three estimators. As we can see, although in graphs the MAX filter looked more convincing than the estimate based on raw data, in Fig. 8(b) we get higher absolute estimation errors compared to the raw data or to the data treated by the moving average. The error probability $\hat{F}(1)$ is about 65% for raw and smoothed data, and only 30% for the MAX filter. On the other hand, small queue values may not be as interesting in terms of traffic management. And if we concentrate on estimating particularly large queue lengths, e.g. for $q(n) \geq 6$, in this case a neural network trained on data with a MAX filter will give lower absolute errors. The error probability $\hat{F}(1)$ in this case is respectively 60% for MAX and only 25% for the original and smoothed data.

Our further analyses are related to the desire to obtain even better estimates for the maximum queue length. Next, we show that the quality of estimates can be significantly improved by introducing as a feature a Fourier series based correction factor.

5   Fourier Series Correction

Given the periodicity of the data, we want to include an additional queue difference factor in our sample, which is estimated from the available data for network training. Denote by $f(n)$ the correction parameter for the cycle n, calculated from the smoothed (MA) values $\hat{q}_1(n)$ of the maximum number of tagged customers and the total number of customers $\hat{q}(n)$ as follows,

$$f(n) = \frac{\hat{q}(n) + 1}{\hat{q}_1(n) + 1}, \ n \in N. \tag{4}$$

We add 1 to the variables $\hat{q}_1(n)$ and $\hat{q}(n)$ to avoid the division by 0. We approximate this function, which is obviously also periodic with a period of 24 h, by a finite Fourier series.

$$f(n) \approx \hat{f}(n) = a_0 + \sum_{i=1}^{39} b_i \cos\left(\frac{i\pi n(g+r)}{12} + c_i\right), \; n \in N, \tag{5}$$

where $a_0 = \frac{1}{N_{24}} \sum_{j=1}^{N_{24}} a_{0j}$, $b_i = \frac{1}{N_{24}} \sum_{j=1}^{N_{24}} b_{ij}$, $c_0 = \frac{1}{N_{24}} \sum_{j=1}^{N_{24}} c_{ij}$. The parameters $\{a_{0j}, b_{ij}, c_{ij}\}$ are estimated from the Fourier series implemented separately for the jth day (24 h) withing the training set, $N_{24} = \lfloor \frac{2}{3} \frac{1700(g+r)}{24} \rfloor = 8$ days.

Graphs comparing the true correction function (4) and its Fourier approximation (5) are shown in Fig. 9(a,b) respectively for the raw data and for the smoothed data by moving average. In Fig. 9(a), we see some estimation of the cyclical trend of this time series. In Fig. 9(b), the approximation reflects the real changes and is quite suitable. We use it further as additional information to train machine learning algorithms. Estimation improvement by using a Fourier series correction (5) for the emission set $\{(q_1(n)+1)\hat{f}(n) - 1 \to q(n))\}$ is illustrated in Fig. 10. As we can see, the estimation accuracy increased for all standard classification methods. On a discrete domain, the maximum accuracy is now more than 67% using a neural network. If we build a neural network for the regression problem on a continuous domain of a target function, we are quite satisfied with the results. Figure 11 shows comparison of three empirical distribution functions constructed for absolute error in case of neural regression model with correction factor, Fourier series approximation and neural network without correction factor. The illustration speaks for itself. The use of an additional correction factor to account for differences in queue lengths according to periodic changes in process states in the regression neural network yields a 95% quality of estimate.

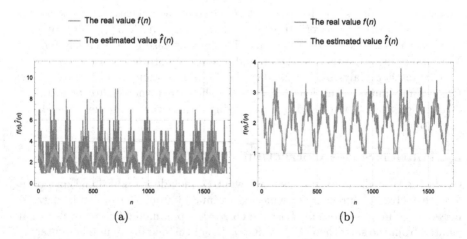

Fig. 9. The correction function $f(n)$ with an estimation $\hat{f}(n)$ for raw (a) and smoothed (MA) data (b)

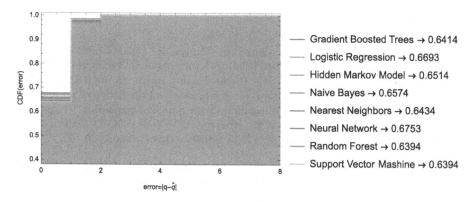

Fig. 10. CDF of the absolute error for the improved emission set

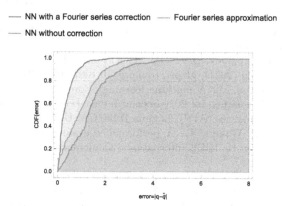

Fig. 11. CDF of the absolute error for the improved emission set

6 Conclusion

We have used machine-learning to estimate the hidden queue length based on emissions of the number of tagged customers in the system. The problem was exemplified by a traffic light queueing model. The neural network with a Fourier series correction illustrated a high quality of estimation with an overall accuracy about 95%. The results can be used further by solving traffic control problems, namely, by evaluation the optimal operation regime of traffic lights to keep the queue length below certain level, by determining the minimal number of tagged vehicles required to guarantee the appropriate quality of queue length estimation, etc.

Acknowledgements. The reported study was funded by RSF, project number 22-49-02023 (recipient V. Vishnevsky).

References

1. Amini, Z., Pedarsani, R., Skabardonis, A., Varaiya, P.: Queue-length estimation using real-time traffic data, In: 2016 IEEE 19th International Conference on Intelligent Transportation Systems (ITSC), Rio de Janeiro, Brazil, pp. 1476–1481 (2016)
2. Efrosinin, D., Rykov, V., Stepanova, N.: Evaluation and prediction of an optimal control in a processor sharing queueing system with heterogeneous servers. In: Vishnevskiy, V.M., Samouylov, K.E., Kozyrev, D.V. (eds.) DCCN 2020. LNCS, vol. 12563, pp. 450–462. Springer, Cham (2020). https://doi.org/10.1007/978-3-030-66471-8_34
3. Efrosinin, D., Stepanova, N.: Estimation of the optimal threshold policy in a queue with heterogeneous servers using a heuristic solution and artificial neural networks. Mathematics **9**, 1267 (2021)
4. Efrosinin, D., Vishnevsky, V., Stepanova, N.: Optimal scheduling in general multi-queue system by combining simulation and neural network techniques. Sensors **23**, 5479 (2023)
5. Gorbunova, A.V., Vishnevsky, V.: Evaluation of the performance parameters of a closed queuing network using artificial neural networks. In: Vishnevskiy, V.M., Samouylov, K.E., Kozyrev, D.V. (eds.) DCCN 2021. LNCS, vol. 13144, pp. 265–278. Springer, Cham (2021). https://doi.org/10.1007/978-3-030-92507-9_22
6. Kyritsis, A.I., Deriaz, M.: A machine learning approach to waiting time prediction in queueing scenarios. In: 2019 Second International Conference on Artificial Intelligence for Industries (AI4I), Laguna Hills, CA, USA, pp. 17–21 (2019)
7. Nii, S., Okuda, T, Wakita, T.: A performance evaluation of queueing systems by machine learning. In: IEEE International Conference on Consumer Electronics (ICCE-Taiwan) (2020)
8. Sherzer, E., Senderovich, A., Baron, O., Krass, D.: Can machines solve general queueing systems? arXiv preprint arXiv:2202.01729 (2022)
9. Sivakami, S.M., Senthil, K.K., Yamini, S., Palaniammal, S.: Artificial neural network simulation for Markovian queueing models. Indian J. Comput. Sci. Eng. **11**, 127–134 (2020)
10. Stintzing, J., Norrman, F.: Prediction of Queuing Behaviour through the Use of Artificial Neural Networks (2017). http://www.diva-portal.se/smash/get/diva2:1111289/FULLTEXT01.pdf
11. Vishnevsky, V., Gorbunova, A.V.: Application of machine learning methods to solving problems of queuing theory. In: Dudin, A., Nazarov, A., Moiseev, A. (eds) Information Technologies and Mathematical Modelling. Queueing Theory and Applications. ITMM 2021. CCIS, vol. 1605, pp. 304–316 (2022). https://doi.org/10.1007/978-3-031-09331-9_24
12. Vishnevsky, V., Klimenok, V., Sokolov, A., Larionov, A.: Performance evaluation of the priority multi-server system $MMAP/PH/M/N$ using machine learning methods. Mathematics **9**, 3236 (2021)
13. Vishnevsky, V., Semenova, O., Bui, D.T.: Using a machine learning approach for analysis of polling systems with correlated arrivals. In: Vishnevskiy, V.M., Samouylov, K.E., Kozyrev, D.V. (eds.) DCCN 2021. LNCS, vol. 13144, pp. 336–345. Springer, Cham (2021). https://doi.org/10.1007/978-3-030-92507-9_27
14. Zhao Y., Shen S., Liu H.X.: A hidden Markov model for the estimation of correlated queues in probe vehicle environments. Trans. Res. Part C: Emerging Technol. **128** (2021)

Analysis of Probabilistic Characteristics in the Integrated Access and Backhaul System

V. Feoktistov[1]([✉])[iD], D. Nikolaev[1][iD], Yu. Gaidamaka[1,2][iD], and K. Samouylov[1,2][iD]

[1] RUDN University, 6 Miklukho-Maklaya Street, Moscow 117198, Russian Federation
1032192939@pfur.ru, {1032201198,gaydamaka-yuv,samuylov-ke}@rudn.ru
[2] Federal Research Center "Computer Science and Control" of the Russian Academy of Sciences (FRC CSC RAS), 44-2 Vavilov Street, Moscow 119333, Russian Federation

Abstract. The presented research analyzes the probabilistic characteristics of the Integrated Access and Backhaul system developed by the 3GPP consortium for millimeter-wave wireless networks. This technology addresses the challenges of deploying mobile base stations including those on UAV in densely populated urban areas with high building density and the presence of both mobile and stationary radio signal blockers. The study encompasses the development of a model and scenario for IAB implementation, a GPSS simulator, and an analytical model of a system fragment represented as a polling-based queuing system. The achievable channel bandwidth, the average end-to-end delays, and the average number of packets on the individual routes are selected as the main performance metrics.

Keywords: mmWave · IAB · GPSS · Mobile networks · Integrated Access and Backhaul · Queuing system · Polling discipline

1 Introduction

The Integrated Access and Backhaul (IAB) technology in 5G networks is an innovative solution that combines the access and backhaul functions into a single network node using the same radio resources [1]. It reduces the need to install and maintain separate devices for interconnecting base stations. In 5G networks, the IAB technology utilizes wireless backhaul to reduce the reliance on fiber-optic connections and simplify installation [17]. IAB technology for 5G networks has proven itself well in the deployment of relay stations on UAVs, when the freedom of flight capabilities are used to find the optimal user and base station association, UAV 3D hovering locations and power allocations [14,18,19]. It allows 5G networks to be flexibly deployed, including mobile base stations that adapt to changing conditions in urban environments.

IAB technology reuses existing functions and interfaces defined for access in 5G networks to achieve its goals. Specifically, the IAB architecture employs

Mobile-Termination (MT), gNB-DU, gNB-CU, User Plane Function (UPF), Access and Mobility Management Function (AMF), and Session Management Function (SMF), along with corresponding interfaces, including NR Uu (between MT and gNB), F1, NG, X2, and N4. For this research, architecture 1a (Fig. 1), as described in 3GPP Technical Report 38.874 [1], has been selected.

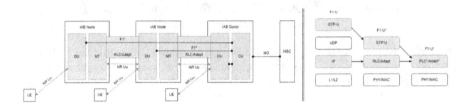

Fig. 1. The IAB architecture [1].

This architecture utilizes a CU/DU (Centralized Unit/Distributed Unit) split architecture and includes IAB-Nodes and IAB-Donor, which can be connected to multiple upper-level IAB-Nodes or DU of IAB-Donor. Each IAB-Node consists of a DU and an MT, where the MT connects the IAB-Node to an upper-level IAB-Node or an IAB-Donor, while the DU establishes RLC channels with user equipment (UE) and the MTs of lower-level IAB-Nodes. An IAB-Node may contain multiple DUs, but each DU has an F1-C connection only with one CU-CP IAB-Donor. The IAB-Donor also contains DUs to support UEs and the MTs of lower-level IAB-Nodes, and each DU in an IAB-Node is exclusively served by a single IAB-Donor.

The CU/DU split architecture protocol stack is discussed in 3GPP TS 38.401 [11], and the radio signal transmission and reception are covered in 3GPP TS 38.174 [12].

The Backhaul Adaptation Protocol (BAP) is employed for data routing within the IAB, which is part of the 3GPP TS 38.340 [13] specification. It aims to deliver packets to the destination node with multiple hops (multi-hopping). It operates over the RLC (Radio Link Control) layer and is designed to adapt data transmitted between IAB nodes for subsequent transmission over backhaul links. The protocol also allows configuring data transmission parameters, including service type, packet sizes, and timings and provides control over data flow management and error handling.

The BAP protocol acts similarly to the IP protocol in many aspects. IAB-Nodes resemble routers because they route and forward traffic (to other nodes) over uplink (UL) and downlink (DL) flows if a packet has not yet reached the final destination. The end-user equipment does not have a BAP layer. It means that the UE cannot directly interact with the BAP protocol and must rely on the IAB-Node to transmit data across the IAB network.

The main feature of IAB-systems is the half-duplex data transmission mode with mode separation for access and transport. In the transport mode, packets are transmitted between IAB-Nodes and the IAB-Donor, while in the access mode, between user devices and the connected base stations.

A half-duplex connection offers greater scalability than a full-duplex connection since it allows multiple base stations to share the same frequency band for uplink transmission without interference. It is achieved through a Time Division Duplexing (TDD) scheme, which switches the data transmission direction over time and allocates the available bandwidth between base stations and the core network using Time Division Multiplexing (TDM). It enables the deployment of more base stations within a single network and facilitates network scalability by adding additional time intervals to accommodate more base stations. In contrast, a full-duplex connection requires dedicated bandwidth for each communication direction, leading to increased bandwidth requirements as the number of base stations grows. This hampers network scalability as available bandwidth becomes a limiting factor.

Furthermore, a half-duplex connection offers several advantages over a full-duplex connection in terms of implementation and interference resilience [16]. It requires less complex and costly equipment, enabling faster and more cost-effective deployment of IAB technology. Additionally, a half-duplex connection reduces the likelihood of self-interference and cross-interference between different communication directions, enhancing network reliability and stability. These factors make IAB technology a more attractive solution for many operators.

The purpose of the paper is to develop a method for estimating the Quality of Service parameters for IAB technology in 5G wireless networks. The estimation is made in terms of key performance indicators such as an achievable channel bandwidth, average end-to-end delays while controlling the average number of packets in the IAB network. This study employs an approach that combines methods from queueing theory [21], simulation modeling, using software developed in the General Purpose Simulation System (GPSS) [15,20] for analysis of the above mentioned performance metrics of IAB network.

The paper is organized as follows. In Sect. 2 we formalize the description of data transmission processes in IAB network using the graph theory, considering the distinctive feature of the IAB technology, namely the half-duplex mode. Section 3 is devoted to analytical modeling of data transmission for an IAB-node using the theory of polling systems. Finally, in Sect. 4 we give the example of numerical experiment for illustration of the model possibilities from the point of view of several performance metrics.

2   IAB Network System Model

To formalize the description of the structure and processes in IAB network of an arbitrary topology we use the graph theory. We define the structure of the network in the form of a directed graph $< \mathcal{E}, \mathbf{G} >$ with a set of vertices $\mathcal{E}$ and an incidence matrix $\mathbf{G}$. According to [1] the IAB network consists of three types of elements: IAB-Donor, IAB-Node, User Equipment (UE). So we define three subsets of vertices.

- A subset of IAB-Donors, denoted as $\mathcal{E}_\mathcal{D} = \{E_0\}$, $|\mathcal{E}_\mathcal{D}| = 1$. This single IAB-Donor is connected to the external transport network via a wired connection.
- A subset of IAB-Nodes, denoted as $\mathcal{E}_\mathcal{N} = \{E_n | n = 1, ..., N\}$, where $|\mathcal{E}_\mathcal{N}| = N$ represents the number of IAB-Nodes in the system. These nodes can exchange packet data with the donor directly or through other IAB-Nodes via radio channels in half-duplex mode.
- Groups of user devices (i.e. UEs), denoted as $\mathcal{E}_\mathcal{U} = \{E_{N+n+1} | n = 0, ..., N\}$, $|\mathcal{E}_\mathcal{U}| = N + 1$. UEs can be associated either with IAB-Donor or with IAB-Node.

Let $\mathcal{E}_\mathcal{B} = \mathcal{E}_\mathcal{D} \cup \mathcal{E}_\mathcal{N}$ be the set of base stations in the IAB network, with a total of $|\mathcal{E}_\mathcal{B}| = N + 1$ stations, and $\mathcal{E} = \mathcal{E}_\mathcal{D} \cup \mathcal{E}_\mathcal{N} \cup \mathcal{E}_\mathcal{U}$ be the set of all network elements, where $|\mathcal{E}| = 2(N + 1)$.

Data from user devices can be transmitted to both the IAB-Donor and IAB-Nodes. Therefore, the number of user device groups coincides with the number of base stations in the network.

This study assumes that each user can only communicate with one base station via wireless links, and each IAB-Node can have direct connections with multiple base stations for downlink data transmission and only one base station for uplink transmission.

According to the adopted notations and assumptions, the combination of the IAB-Donor, IAB-Nodes, and groups of user devices form a tree-like topology of the IAB network. It is assumed that the network topology remains unchanged throughout the entire experiment. Therefore, the IAB network topology can be represented as a graph (Fig. 2) with an incidence matrix $\mathbf{G}$ (Fig. 3), which can be used to describe the characteristics of the channels.

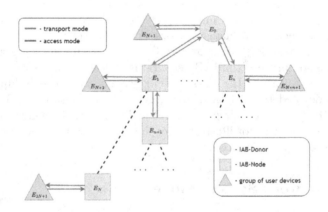

Fig. 2. Graph $< \mathcal{E}, \mathbf{G} >$ of the IAB network topology.

Here $i_{n_1, n_2} \in \{0, 1\}$, $n_1, n_2 = 0, ..., 2N + 1$ is the indicator, that equals 1 if there exists a direct radio channel from the network element E_{n_1} to element E_{n_2}, and 0 otherwise.

G	E_0	E_1	$\cdots$	E_N	E_{N+1}	$\cdots$	E_{2N+1}
E_0	0	$i_{0,1}$	$\cdots$	$i_{0,N}$	$i_{0,N+1}$	$\cdots$	0
E_1	$i_{1,0}$	0	$\cdots$	$i_{1,N}$	0	$\cdots$	0
$\vdots$	$\vdots$	$\vdots$	$\ddots$	$\vdots$	$\vdots$	$\ddots$	$\vdots$
E_N	$i_{N,0}$	$i_{N,1}$	$\cdots$	0	0	$\cdots$	$i_{N,2N+1}$
E_{N+1}	$i_{N+1,0}$	0	$\cdots$	0	0	$\cdots$	0
$\vdots$	$\vdots$	$\vdots$	$\ddots$	$\vdots$	$\vdots$	$\ddots$	$\vdots$
E_{2N+1}	0	0	$\cdots$	$i_{2N+1,N}$	0	$\cdots$	0

Fig. 3. The incidence matrix **G** for the graph $< \mathcal{E}, \mathbf{G} >$ of the IAB network.

To formalize the description of the data transmission process in IAB network, a model consisting of one IAB-Donor, one IAB-Node, and two groups of user devices UEs-1 and UEs-2 (Fig. 4 and 5) has been chosen.

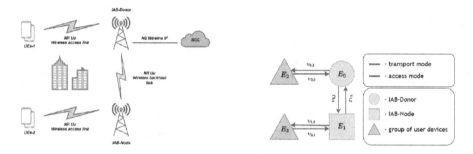

Fig. 4. Topology of the IAB network.   **Fig. 5.** Graph of the IAB network.

To ensure correct and uninterrupted system operation in half-duplex mode, a channel activity policy, i.e. a sequence of control actions $\mathcal{F} = \{\mathbf{F}_l | l = 1, ..., L\}$ is introduced. It consists of $|\mathcal{F}| = L$ incidence matrices that determine the allowed packet transmission through specific channels during certain time intervals. The switching time between control actions is referred to as a clock cycle. Note that the development of a channel activation policy for delay control in In-Band IAB systems is an actual problem, which can be formulated in general as a multi-criteria multi-parameter optimization problem with the constraints including the achievable channel bandwidth and the average end-to-end delay [22].

For the system model of the IAB network (Fig. 4), the set of vertices has the form $\mathcal{E} = \{E_0, E_1, E_2, E_3\}$, and the values of the elements of the incidence matrix **G** depend on the scenario of switching between transmission directions in the half-duplex mode and are given by the matrices **F**. In this paper the sequence of seven cyclic control actions $\mathcal{F} = \{\mathbf{F}_1, \mathbf{F}_2, ..., \mathbf{F}_7\}$, described in Table 1, and the corresponding matrices **F** are investigated:

Table 1. The sequence of control actions for research scenario.

Order of actions	Packets transmission direction
1	from the IAB-Donor to the IAB-Node
2	from the IAB-Donor to the UEs-1 group and from the IAB-Node to the UEs-2 group
3	from the UEs-1 group to the IAB-Donor and from the UEs-2 group to the IAB-Node
4	from the IAB-Node to the IAB-Donor
5	from the IAB-Donor to the UEs-1 group and from the IAB-Node to the UEs-2 group
6	from the IAB-Donor to the IAB-Node
7	from the IAB-Donor to the UEs-1 group and from the IAB-Node to the UEs-2 group

$$\mathbf{F}_1 = \mathbf{F}_6 = \begin{pmatrix} 0\,1\,0\,0 \\ 0\,0\,0\,0 \\ 0\,0\,0\,0 \\ 0\,0\,0\,0 \end{pmatrix}, \ \mathbf{F}_2 = \mathbf{F}_5 = \mathbf{F}_7 = \begin{pmatrix} 0\,0\,1\,0 \\ 0\,0\,0\,1 \\ 0\,0\,0\,0 \\ 0\,0\,0\,0 \end{pmatrix},$$

$$\mathbf{F}_3 = \begin{pmatrix} 0\,0\,0\,0 \\ 0\,0\,0\,0 \\ 1\,0\,0\,0 \\ 0\,1\,0\,0 \end{pmatrix}, \ \mathbf{F}_4 = \begin{pmatrix} 0\,0\,0\,0 \\ 1\,0\,0\,0 \\ 0\,0\,0\,0 \\ 0\,0\,0\,0 \end{pmatrix}.$$

The proposed formal description of the IAB network lies at the heart of the simulation model in GPSS presented in Sect. 4. The GPSS model also provides the parameter values for the queueing model, developed in Sect. 3 for the analysis of end-to-end delays in the IAB-Nodes on the rout.

3   IAB-Node Mathematical Model

The end-to-end delay in the IAB network can be estimated using the decomposition and aggregation approach as the sum of the delays in the network elements on the route. Within this approach, in Sect. 3, a single-server queueing model $M_2/M_2/1/(r_1, r_2)$ with two incoming flows and finite queues is constructed to estimate the packet delay in IAB-Node E_1 (Fig. 5) as the time spent by a request in the queueing system. The incoming flows corresponds to packet arrivals from parent BS E_0 (downlink on backhaul, queue Q_1) and from child UE E_3 (uplink on access, queue Q_2). The exhaustive polling discipline with a non-zero switching time between queues reflects the peculiarity of the half-duplex data transmission mode, while the server operates in a cycle with zero switching time from Q_1 to Q_2 and non-zero switching time from Q_2 to Q_1, as well as both queues are filled only during server's switching. Let λ_1 and λ_2 be the intensities of Poisson arrival flows to the first queue Q_1 and the second queue Q_2, μ_1 and μ_2 be the parameters of exponential service time for requests in Q_1 and Q_2, r_1, r_2 be the storage capacities and s be the mean value of exponential distributed switching time from Q_2 to Q_1.

The half-duplex data transmission entails the following:

$$\lambda_i = \begin{cases} \lambda_i, & q = 0 \text{ (switching)}, \\ 0, & q = 1,2 \text{ (serving queue } Q_1 \text{ or } Q_2), \end{cases} \tag{1}$$

where $i = 1, 2$.

The functioning the system can be described by a random process

$$\mathbf{X}(t) = \{(q(t), n_1(t), n_2(t)), t \geq 0\},$$

where $q(t) \in \{0, 1, 2\}$ is a server state ($q = 0$ is switching from Q_2 to Q_1, $q = 1, 2$ is service of Q_1 and Q_2 respectively), $n_1(t) \in \{0, 1, ..., r_1\}$ is a number of requests in Q_1, $n_2(t) \in \{0, 1, ..., r_2\}$ is a number of requests in Q_2 at time instant t.

Then the state space is represented as follows:

$$\begin{aligned} \mathbb{X} = \{&(0, n_1, n_2) : n_1 \in \{0, 1, \ldots, r_1\}, n_2 \in \{0, 1, \ldots, r_2\}, \\ &(1, n_1, n_2) : n_1 \in \{1, \ldots, r_1\}, n_2 \in \{0, 1, \ldots, r_2\}, \\ &(2, 0, n_2) : n_2 \in \{1, \ldots, r_2\}\} \end{aligned} \tag{2}$$

with the number of states

$$|\mathbb{X}| = (r_1 + 1)(r_2 + 1) + r_1(r_2 + 1) + r_2. \tag{3}$$

Let

$$p(q, n_1, n_2) = \lim_{t \to \infty} P\{\mathbf{X}(t) = (q(t), n_1(t), n_2(t))\}, (q, n_1, n_2) \in \mathbb{X}, \tag{4}$$

denote the stationary probabilities of $\mathbf{X}(t)$.

To find stationary distribution, we need to solve the following equilibrium system:

$$\begin{cases} \mu_1 p(1, 1, 0) + \mu_2 p(2, 0, 1) = (\lambda_1 + \lambda_2)p(0, 0, 0), \\ \lambda_1 p(0, i - 1, 0) = (u(r_1 - i)\lambda_1 + \lambda_2 + s^{-1})p(0, i, 0), i = 1, ..., r_1, \\ \lambda_2 p(0, 0, i - 1) = (\lambda_1 + u(r_2 - i)\lambda_2 + s^{-1})p(0, 0, i), i = 1, ..., r_2, \\ \lambda_1 p(0, i - 1, j) + \lambda_2 p(0, i, j - 1) = (u(r_1 - i)\lambda_1 + u(r_2 - j)\lambda_2 + s^{-1})p(0, i, j), \\ \quad i = 1, ..., r_1, j = 1, ..., r_2, \\ s^{-1}p(0, i, j) + u(r_1 - i)\mu_1 p(1, i + 1, j) = \mu_1 p(1, i, j), i = 1, ..., r_1, j = 0, ..., r_2, \\ s^{-1}p(0, 0, j) + \mu_1 p(1, 1, j) + u(r_2 - j)\mu_2 p(2, 0, j + 1) = \mu_2 p(2, 0, j), j = 1, ..., r_2, \end{cases} \tag{5}$$

where

$$u(x) = \begin{cases} 0, & x \leq 0, \\ 1, & x > 0. \end{cases}$$

Solving the system of equilibrium equations (5) with the normalization condition

$$\sum_{(q,n_1,n_2)\in X} p(q,n_1,n_2) = 1$$

we obtain the stationary probability distribution and then the essential metric to evaluate end-to-end delay in IAB system, i.e. an average waiting time of requests for both traffic flows. The formulas used to calculate some performance metrics of the system are listed below.

Blocking probability

$$P_{Block} = \sum_{n_2=0}^{r_2} p(0, r_1, n_2) + \sum_{n_1}^{r_1-1} p(0, n_1, r_2). \tag{6}$$

The average number of requests in the first

$$N_1 = \sum_{n_1=1}^{r_1} n_1 \sum_{q=0}^{2} \sum_{n_2=0}^{r_2} p(q, n_1, n_2) \tag{7}$$

and in the second queue

$$N_2 = \sum_{n_2=1}^{r_2} n_2 \sum_{q=0}^{2} \sum_{n_1=0}^{r_1} p(q, n_1, n_2). \tag{8}$$

According to the Little's Law the waiting time in the first

$$w_1 = \frac{N_1}{\lambda_1(1 - P_{Block})} \tag{9}$$

and in the second queue

$$w_2 = \frac{N_2}{\lambda_2(1 - P_{Block})}. \tag{10}$$

Knowing the waiting time (9) and (10) and parameters of service time distribution μ_1 and μ_2, we can easily obtain the sojourn time spent by a request in the system both for access and backhaul data transmission through the node E_1. The use of polling systems to simulate the process of data transmission in each node on the route over the IAB network allows to estimate the end-to-end delay as the sum of the sojourn times on the corresponding route.

Note that parameters μ_1 and μ_2 refer to the uplink and downlink data rates of the IAB-Node, i.e. the radio channel bandwidth between elements of the IAB network. For estimating these rates, considering the half-duplex mode of data transmission in the IAB network, the method from the 3GPP TS 38.306 [2] is proposed.

The supported maximum bandwidths for uplink and downlink flows for UEs and base stations are calculated using the following formula:

$$v = 10^{-6} \cdot \sum_{j=1}^{J} (\vartheta_{Layers}^{(j) \cdot Q_m^{(j)}} \cdot f^{(j)} \cdot R_{max} \cdot \frac{12 \cdot N_{PRB}^{BW^{(j)}, \mu}}{T_s^{\mu}} \cdot (1 - OH^{(j)})). \qquad (11)$$

The description of variables and their values used in the study are presented in Table 2.

Table 2. Data settings.

Parameter	Notation	Value
Frequency	ν	30 GHz
Bandwidth	$BW^{(j)}$	200 MHz
Number of aggregated carriers in the frequency band or band combinations	J	1
Number of multiplexed slots in UL	$\vartheta_{Layers,UL}^{(j)}$	4
Number of multiplexed slots in DL	$\vartheta_{Layers,DL}^{(j)}$	8
Number of multiplexed slots in transport mode	$\vartheta_{Layers,B}^{(j)}$	8
Modulation index in access mode	$Q_{m,B}^{(j)}$	6
Modulation index in transport mode	$Q_{m,A}^{(j)}$	8
Scaling factor	$f^{(j)}$	1
Error rate during encoding	R_{max}	948/1024
Numerology (3GPP TS 38.211 [3])	μ	3
Maximum number of resource blocks in the bandwidth	$N_{PRB}^{BW^{(j)}, \mu}$	132
The average duration of an OFDM slot in a μ subframe for numerology	T_s^{μ}	$\frac{10^{-3}}{14 \cdot 2^3}$
Transmission parameter in UL	$OH_{UL}^{(j)}$	0.1
Transmission parameter in DL	$OH_{DL}^{(j)}$	0.18
Transmission parameter in transit	$OH_B^{(j)}$	0.1

4   Numerical Experiment

For illustration of the model possibilities from the point of view of the achievable channel bandwidth, average end-to-end delays, and the average number of packets in IAB-Nodes, the simulation model based on the proposed in Sect. 2 formal description of the IAB network was developed. GPSS has been chosen as the simulation modeling language. It is based on mathematical principles of discrete event modeling, which involves representing a system or process as a series of discrete events occurring over time. The software includes various mathematical models and algorithms for modeling various aspects of system behavior, including queuing, resource utilization, and other factors that impact system performance. GPSS is closely related to queuing systems, which are mathematical models used to analyze the service processes of request flows. To simulate various models, GPSS implements the Monte Carlo method, which is a statistical method for approximate solutions using random numbers.

However, despite all the advantages of GPSS, it has limitations related to data storage, data processing, and visualization of simulation results. Therefore, data processing and visualization were performed using Python programming language version 3.7 along with libraries such as matplotlib [4], pandas [5], numpy [6], scipy [7], and PIL [8].

We begin the numerical experiment with model parametrization on close to real initial data with the following simulation model parameters:

- simulation time 10 s;
- duration of one cycle 1 ms;
- packet size distributed according to the probability density function (pdf) $f_s(x)$ for the IP packets length (refer to Fig. 6).

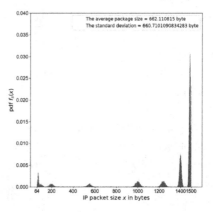

Fig. 6. Distribution of the IP packets length.

First, we obtain bandwidth for a channel BS-BS between base stations and for uplink and downlink channels UE-BS according to (11):

- channel bandwidth between base stations 9460.3 Mbps;
- downlink channel bandwidth for UEs 6464.5 Mbps;
- uplink channel bandwidth for UEs 3547.6 Mbps.

After that we obtain packet arrival intensity for UE downlink and packet generation intensity for UE uplink in assumption on Poisson arrival process and exponential service time:

– packet arrival intensity for UEs (downlink) 39000 packets/s;
– packet generation intensity for UEs (uplink) 1000 packets/s.

Finally, we obtain blocking probabilities for BS-BS and UE-BS channels:

– probability of successful packet delivery between base stations 0.95;
– probability of successful packet delivery between base stations and UEs 0.9.

As a result of the numerical experiment, four graphs were obtained. In Fig. 7, it can be observed that the average end-to-end delay obtained from the simulation model almost coincides with the average end-to-end delay obtained through Little's Law using the average number of packets in the network with relative error less than 1.5%. Moreover, as shown in Fig. 8, the average number of packets along routes and across the entire network has an almost linear dependence on the duration of a cycle.

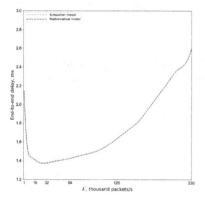
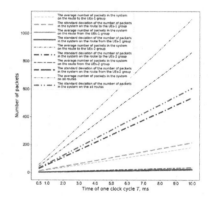

Fig. 7. Simulation and mathematical average end-to-end delay ($\lambda_{\text{UEs-1}} = \lambda_{\text{UEs-2}} = \lambda'$).

Fig. 8. The average number of packets on the network routes.

With an increase in the intensity of the packets arriving at the IAB-Donor from the external network to the groups of user devices ($\lambda_{\text{UEs-1}}$ is intensity for UEs-1, and $\lambda_{\text{UEs-2}}$ for UEs-2), the average number of packets across the entire network and along routes to groups of user devices increases exponentially, while the standard deviation has a more complex form (Fig. 9). It should be noted that the system under study remains in stationary mode up to the value of the incoming intensity from the outside on IAB-Donor for each user group up to $\lambda' \approx 240\,000$ packets/s, which corresponds to approximately 600 UEs per base station and 1200 UEs in the entire network in Fig. 5. Moreover up to $\lambda' \approx 240\,000$ packets/s the average end-to-end delays in the system will not exceed 4 ms, defined by the 3GPP standards for the provision of eMBB services [9,10]. This result should be considered when planning IAB networks with relay stations.

Continuing the analysis based on Fig. 10, we can conclude that only in the directions from the IAB-Donor to the UEs-1 and from the IAB-Node to the UEs-2, the queues reach the overloaded state at the mentioned intensity value λ'.

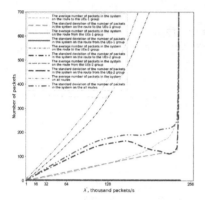

Fig. 9. The average number of packets on the network routes ($\lambda_{\text{UEs-1}} = \lambda_{\text{UEs-2}} = \lambda'$).

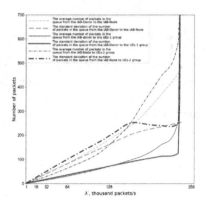

Fig. 10. The average number of packets in the three queues ($\lambda_{\text{UEs-1}} = \lambda_{\text{UEs-2}} = \lambda'$).

5   Conclusion

The paper proposes an approach to formalizing the description of the system model and scenarios for using the IAB in 5G networks, considering the half-duplex mode as a key feature of the IAB technology. The proposed approach formed the basis for the developed software tool in the GPSS language for analyzing end-to-end delay, which in IAB systems is the main indicator of the quality of service and a limitation on the use of systems with relay stations that add additional delay on the route. The analysis using the developed software tool includes preliminary parameterization of the system on a set of initial data close to real, subsequent study of the end-to-end delay and other key performance indicators, such as the achievable data transfer rate and the number of packets in the nodes on a route. Besides the software, a mathematical model of a polling queuing system is proposed to analyze the delay introduced by a terminal IAB-Node. Also, the data processing and visualization program were developed.

Numerical analysis of the studied network example with the proposed fixed control policy showed that the IAB technology can serve up to 1200 simultaneously connected user devices with moderate traffic load. End-to-end delays in the network remain relatively small until the congestion state, which ensures the required quality of service for users. The slight increase in end-to-end delay compared to traditional 5G networks is imperceptible to users and is compensated by increased network coverage at lower equipment deployment costs.

The further study may include modeling a route with several IAB-Nodes using the queuing network apparatus. Also, an interesting direction is the development of adapted traffic management methods.

Acknowledgments. The reported study was funded by RSF, project number 22-29-00694, https://rscf.ru/en/project/22-29-00694/.

References

1. 3GPP: Study on Integrated Access and Backhaul. Technical report 38.874 v16.0.0 (2018). https://portal.3gpp.org/desktopmodules/Specifications/SpecificationDetails.aspx?specificationId=3232
2. 3GPP: User Equipment (UE) radio access capabilities. Technical Specification 38.306 v17.4.0 (2023). https://portal.3gpp.org/desktopmodules/Specifications/SpecificationDetails.aspx?specificationId=3193
3. 3GPP: Physical channels and modulation. Technical Specification 38.211 v17.4.0 (2022). https://portal.3gpp.org/desktopmodules/Specifications/SpecificationDetails.aspx?specificationId=3213
4. Matplotlib: Matplotlib 3.7.1 documentation. https://matplotlib.org/stable/index.html
5. Pydata: Pandas documentation Version 2.0.1 (2023). https://pandas.pydata.org/docs/
6. NumPy: NumPy Reference. Release 1.23.0 (2022). https://numpy.org/doc/1.23/numpy-ref.pdf
7. SciPy: SciPy documentation (2023). https://docs.scipy.org/doc/scipy/
8. Readthedocs: Pillow (PIL Fork) 9.5.0 documentation. https://pillow.readthedocs.io/en/stable/index.html
9. 3GPP: Study on scenarios and requirements for next generation access technologies. Technical report 38.913 v17.0.0 (2022). https://portal.3gpp.org/desktopmodules/Specifications/SpecificationDetails.aspx?specificationId=2996
10. 5G Americas: Innovations in 5G Backhaul Technologies: IAB, HFC & Fiber, July 2020. https://www.5gamericas.org/wp-content/uploads/2020/06/Innovations-in-5G-Backhaul-Technologies-WP-PDF.pdf
11. 3GPP: Architecture description. Technical Specification 38.401 v17.3.0. NG-RAN (2022). https://portal.3gpp.org/desktopmodules/Specifications/SpecificationDetails.aspx?specificationId=3219
12. 3GPP: Integrated access and backhaul (IAB) radio transmission and reception. Technical Specification 38.174 v17.2.0 (2022). https://portal.3gpp.org/desktopmodules/Specifications/SpecificationDetails.aspx?specificationId=3665
13. 3GPP: Backhaul Adaptation Protocol (BAP) specification. Technical Specification 38.340 v17.4.0 (2023). https://portal.3gpp.org/desktopmodules/Specifications/SpecificationDetails.aspx?specificationId=3604
14. Fouda, A., Ibrahim, A.S., Guvenc, I., Ghosh, M.: UAV-based in-band integrated access and backhaul for 5G communications. In: IEEE 88th Vehicular Technology Conference (VTC-Fall), pp. 1–5 (2018). https://doi.org/10.1109/VTCFall.2018.8690860

15. Matyushenko, S.I., Pyatkina, D.A., Razumchik, R.V.: Modeling Queuing Systems in the GPSS World Environment: A Textbook, 2nd edn. Peoples' Friendship University of Russia, Moscow (2022)
16. Sadovaya, Y., et al.: Self-interference assessment and mitigation in 3GPP IAB deployments. In: ICC 2021 - IEEE International Conference on Communications, pp. 1–6 (2021). https://doi.org/10.1109/ICC42927.2021.9500769
17. Sadovaya, Y., et al.: Integrated access and backhaul in millimeter-wave cellular: benefits and challenges. IEEE Commun. Mag. **60**(9), 81–86 (2022). https://doi.org/10.1109/MCOM.004.2101082
18. Tafintsev, N., Moltchanov, D., Chiumento, A., Valkama, M., Andreev, S.: Airborne integrated access and backhaul systems: learning-aided modeling and optimization. IEEE Trans. Veh. Technol. (2023). https://doi.org/10.1109/TVT.2023.3293171
19. Tafintsev, N., et al.: Reinforcement learning for improved UAV-based integrated access and backhaul operation. In: IEEE International Conference on Communications Workshops (ICC Workshops), pp. 1–7. IEEE, June 2020. https://doi.org/10.1109/ICCWorkshops49005.2020.9145423
20. Thesen, A., Schmidt, J.: Computer Methods in Operations Research. Elsevier Science (2014)
21. Vishnevskiy, V., Semenova, O.: Polling Systems: Theory and Applications for Broadband Wireless Networks. LAP LAMBERT Academic Publishing, May 2012
22. Yarkina, N., Moltchanov, D., Koucheryavy, Y.: Counter Waves Link Activation Policy for Latency Control in In-Band IAB Systems. IEEE Commun. Lett. (2023)

Myopic Inventory Control with Returns in Case of Uncertainty: Adaptive Algorithms

A. Kozhan[1], V. Laptin[1], and A. Mandel[2(✉)]

[1] M.V. Lomonosov Moscow State University, Lenin's Mountings 1, Moscow, Russia
anton.kozhan@mail.ru , straqker@bk.ru
[2] V.A. Trapeznikov Institute of Control Sciences RAS, Profsoyuznaya 65, Moscow, Russia
almandel@yandex.ru

Abstract. A model of inventory control with returns is considered, when it is possible for consumers to return the products they have purchased in case of uncertainty when the probability distribution of a random variable of one-step demand is a priori unknown. It is shown that in this case, the optimal inventory management strategy is 4-parametric and adaptive algorithms are constructed that converge with probability 1 to the true values of these parameters.

Keywords: Inventory Control with Returns · Myopic Inventory Control Strategies · Case of Uncertainty · Adaptive Control Algorithms

1 Introduction

This paper continues the study of the inventory control system model with returns [1,2]. This time, we consider the case of the so-called "myopic" control strategies under conditions of uncertainty, when the distribution function of the demand at one step of the planning period is unknown. As stated in [2], the message for studying inventory management models with returns was that they provided an adequate mathematical tools for solving problems of channel switching process control in queuing systems [3–5]. The possibility of using the methods of the mathematical theory of inventory control for solving problems of the theory of queuing systems was noted by many authors, see, for example, books [6,7]. In contrast to [2], in this report, we study a "myopic" inventory control model, that is, an inventory control problem in which the minimum average cost at one (current) step of the planning period is used as an optimality criterion. The reason is that multi-step inventory control problems (like many problems in queuing theory) are usually solved under the assumption that the characteristics of the system are stationary in time. For the inventory control problem with returns, this means that the assumption of the invariability of demand characteristics is far from being always fulfilled. But if the probabilistic characteristics of demand change, then it is quite natural to use a one-step optimality criterion as an alternative. The second difference of this work from the

V. M. Vishnevskiy et al. (Eds.): DCCN 2023, LNCS 14123, pp. 291–303, 2024.
https://doi.org/10.1007/978-3-031-50482-2_23

system considered in [2] is that to solve the problem of inventory control under uncertainty, not Q-learning methods are used, but adaptive control theory algorithms developed in the works of Ya.Z. Tsypkin [8,9].

The concept of "myopic inventory management strategies with returns under uncertainty" is a concept related to managing inventory in conditions where returns from customers can occur with priory unknown probabilities, which, like the decisions made on their basis, can be refined with using newly arriving statistical data, and the criterion for choosing these solutions is the minimization of average costs at one step. In other words, the "myopia" of the strategy means that only a limited amount of information about the state at a given step is used in decision making. This does not take into account information that may become available in the future.

In this case, the "myopia" of the strategy means that when making decisions, managers can use only a limited amount of information (at this step) that is available at the time of the decision. This does not take into account all information that may become available in the future.

This is due to the fact that in conditions of uncertainty, when returns from customers can occur with a certain probability, but it is not known when and how many they will be, it is impossible to predict all possible future events and assess their consequences. Therefore, only the information that is currently available is used, and decisions that maximize the expected profit or minimize the expected losses are applied only at the next step.

This concept has a number of advantages that make it attractive compared to the farsighted control concept. Let's take a closer look at these benefits:

1. *More adaptive response.* In the context of inventory management, myopic decision making means that managers make decisions based on currently available information. They do not rely on long-term forecasts or plans, which may not be viable in the face of uncertainty. Myopic management allows you to quickly respond to changes in market trends, customer demand, or internal company factors such as supply or production issues. This contributes to more flexible and adaptive inventory management.

2. *Accounting for the Probability of Returns.* Returns of goods by customers can have a significant impact on inventory management. Myopic stock management with returns allows you to take into account the probability of returns when making decisions. Managers can take into account and account for the likelihood of returning goods when determining the optimal inventory level. This helps reduce return losses and optimize inventory levels for potential returns.

3. *Flexibility in Decision Making.* Myopic inventory management has more flexibility in decision making than a farsighted approach. Managers can base their decisions on up-to-date information that is available at the moment rather than relying on long-term forecasts or plans that may be out of date or untenable. This allows you to quickly adapt to changing market conditions, customer requirements or internal company factors. Myopic management allows decisions to be made based on the current situation, which can be critical in a rapidly changing environment.

4. *Reduce Costs and Increase Efficiency.* Myopic inventory management strategies also help reduce costs and improve the efficiency of a company's operations. By better predicting demand and managing inventory levels based on current information, myopic management helps avoid unnecessary overstocking or understocking.

5. *Reducing Excess Inventory.* Reducing inventory surplus helps to reduce storage, handling and disposal costs. It also helps reduce the risk of products becoming obsolete or damaged. On the other hand, optimal inventory management under conditions of uncertainty helps to avoid shortages of goods, which can lead to loss of customers and lost sales.

6. *The Efficiency of the Company's Operations.* Myopic inventory management also improves the efficiency of a company's operations. More accurate forecasting and inventory management based on current data allows you to more accurately plan production processes, orders from suppliers and distribution of goods. This avoids production downtime or lack of stock, improving the overall productivity and efficiency of the company.

As a result, myopic inventory management contributes to reducing the cost of warehousing and handling inventory, reducing the risks and losses from excess or shortage of inventory, as well as increasing the overall efficiency and productivity of the company.

Learning the concept of "myopic inventory management strategies with returns under uncertainty" is important for several reasons:

1. Inventory management is an important part of the operational management of a business. Optimal inventory management reduces inventory costs and improves customer service. The concept of "myopic strategies" helps to manage inventory more effectively in the face of uncertainty when returns are possible.

2. In conditions of uncertainty, when the probability of returning the goods is unknown, the use of "myopic strategies" may be the only way to make decisions. This allows you to quickly respond to changes in the environment and take into account the likelihood of returns when making decisions.

3. The concept of "myopic strategies" can be useful in a rapidly changing environment, when the expected outcomes of future events can vary greatly. The use of "myopic strategies" allows managers to make decisions faster and more flexibly that is especially relevant today.

The theoretical basis for inventory management with returns can later be used in the construction of channel switching strategies in queuing systems that are used in practice.

2 Inventory Control Model with Returns in Case of Uncertainty

A multistep model of inventory control is considered during the planning period $T = (0, N\tau)$, where N is a sufficiently large natural number, of one type of product with backlogging of the outstanding demand (so-called backorders). The demand at each of the steps of the process is described by a model of independent in the aggregate, identically distributed random variables $\{z(n), n = 1, 2, \ldots, N\}$ with the unknown distribution function $F(z)$. We will also assume that the delivery lag time is equal to 0. In this case, it is customary to associate the state of the inventory management system not with the stock on hand, but with a so-called inventory position. An inventory position (with a delivery time equal to 0) is defined [10] as the stock on hand minus the backordered demand.

Unlike the classical multi-step inventory control model, we will assume that the domain of the demand distribution function at one step $F(z)$ is the entire real axis: $-\infty < z < \infty$. Let also the mathematical expectation of this distribution be positive. The possibility of negative values of demand will be interpreted as the return by the consumer of the products purchased at the warehouse.

Let the criterion for the optimal functioning of the warehouse be the minimum of the one-step (current step) average costs during each step of the planning period $[0, T]$, where $T = N\tau$.

Formulation of the Problem. We will assume that at each of the decision-making moments, the warehouse has three possibilities: (1) to place the order of size $u > 0$, for which the warehouse pays the amount $A_1 + cu(A_1, c > 0)$, (2) not place an order at all and (3) return to suppliers $w > 0$ units of products, for which the warehouse pays a lump sum $A_2 > 0$. In addition to the supply costs, the costs of the warehouse for maintaining a positive on-hand stock are taken into account with a coefficient of unit costs for storage at one step h, and since all the back demand is fixed, the costs due to shortages are also taken into account with a specific coefficient equal to d.

So, we denote the average cost level at one step, which starts from the initial inventory level x and is followed by the submission of an order for replenishment of stocks in the amount of u, through $C(x, u)$. Then the problem under consideration is a minimization problem of the form

$$C^*(x) = \min_u C(x, u), \tag{1}$$

where $C^*(x)$ is the value of the minimum possible average cost per step, which starts from the initial stock level x.

Given that there are three options noted above in stock (reminder): (1) submit an order of size $u > 0$, (2) not place an order at all ($u = 0$) and (3) to return to suppliers a part of the products stored in the warehouse (we will consider this as the submission of an order of a negative size $u < 0$), we will write an expression for the costs for each of these options.

So, for $u > 0$ the costs at one step will be

$$C_{order}(x, u) = A_1 + cu + h \int_{-\infty}^{x+u} (x+u-z)dF(z) + d \int_{x+u}^{+\infty} (z-x-u)dF(z). \tag{2}$$

For $u = 0$, the costs at one step will be

$$C_0(x) = h \int_{-\infty}^{x} (x - z)dF(z) + d \int_{x}^{+\infty} (z - x)dF(z). \tag{3}$$

For $u < 0$, the costs at one step will be

$$C_{return}(x, u) = A_2 + h \int_{-\infty}^{x+u} (x + u - z)dF(z) + d \int_{x+u}^{+\infty} (z - x - u)dF(z). \tag{4}$$

Considering that the cost functions introduced above depend mainly on the sum $y = x + u$, we introduce new notation, writing for a given value of the initial stock x that when choosing a solution $y > x$, the costs at one step without taking into account the cost component that depends on x, and fixed costs A_1 will be

$$G_{order}(y) = cy + h \int_{-\infty}^{y} (y - z)dF(z) + d \int_{y}^{+\infty} (z - y)dF(z), \tag{5}$$

for $y = x$, the costs at one step will be

$$G_0(y) = C_0(x) = h \int_{-\infty}^{y} (y - z)dF(z) + d \int_{y}^{+\infty} (z - y)dF(z), \tag{6}$$

and for $y < x$ the costs at one step, excluding fixed costs A_2, will be

$$G_{return}(y) = h \int_{-\infty}^{y} (y - z)dF(z) + d \int_{y}^{+\infty} (z - y)dF(z), \tag{7}$$

Note that the functions $G_0(y)$ and $G_{return}(y)$ are actually the same. As a result, problem 1 can be rewritten as 8:

$$C^*(x) = -cx + \min \begin{cases} A_1 + \min_{y>x} G_{order}(y), \\ cx + G_0(x), \\ cx + A_2 + \min_{y<x} G_{return}(y), \end{cases} \tag{8}$$

Let' s find the minima in the first and third rows of the right side of formula 8, simultaneously making sure that the functions $G_{order}(y)$ and $G_{return}(y)$ are one-extremal downward convex functions. Indeed,

$$\frac{dG_{order}(y)}{dy} = c + hF(y) - d(1 - F(y)) = c - d - (h + d)F(y), \tag{9}$$

$$\frac{dG_{order}(y^*_{order})}{dy} = 0 \Rightarrow F(y^*_{order}) = \frac{d - c}{h + d} \tag{10}$$

and

$$\frac{dG^2_{order}(y)}{dy^2} = -(h + d)f(y) < 0, \tag{11}$$

$$\frac{dG_{return}(y)}{dy} = hF(y) - d(1 - F(y)) = -d - (h + d)F(y), \qquad (12)$$

$$\frac{dG_{return}(y^*_{return})}{dy} = 0 \Rightarrow F(y^*_{return}) = \frac{d}{h + d}, \qquad (13)$$

and

$$\frac{dG^2_{return}(y)}{dy^2} = -(h + d)f(y) < 0. \qquad (14)$$

In addition, from formulas 10 and 13 it can be seen that the points of minima of these functions y^*_{order} and y^*_{return} are ordered as follows: $y^*_{order} < y^*_{return}$. Let' s rename the values y^*_{order} and y^*_{return} as follows: $y^*_{order} = R_{myopic}$ and $y^*_{return} = S_{myopic}$.

It follows that the optimal ordering strategy is characterized by four parameters, the essence of which is clear from the following picture:

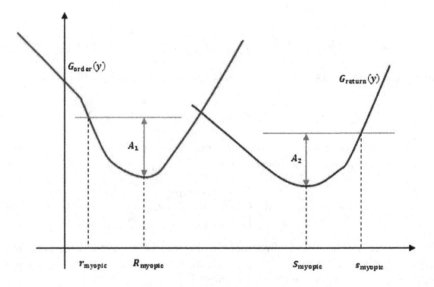

Fig. 1. Parameters of the optimal inventory control strategy.

In what follows, these strategy parameters will be abbreviated (in the order in which they are shown in Fig. 1) as follows: $r_{myopic} = r$, $R_{myopic} = R$, $S_{myopic} = S$ and $s_{myopic} = s$.

So (strictly it is proven shown in [1]) for the functions $G_{order}(y)$, $G_0(y)$ and $G_{order}(y)$ described by formulas 5–7 the optimal rule of inventory control is given by the formula:

$$u_n^*(x) = \begin{cases} R - x, & if\ x \le r, \\ 0, & if\ r < x < s, \\ x - S, & if\ x \ge s. \end{cases} \tag{15}$$

In this case, the expressions for the parameters look as follows.
Equation to determine R (minimizing point of function $G_{order}\ (y)$)

$$F(R) = \frac{d - c}{h + d}. \tag{16}$$

Equation to determine S (minimizing point of function $G_{return}\ (y)$)

$$F(S) = \frac{d}{h + d}. \tag{17}$$

Equation to determine r

$$A_1 + c(R - r) + h \int_{-\infty}^{R} (R - z)dF(z) + d \int_{R}^{+\infty} (z - R)dF(z) =$$
$$= h \int_{-\infty}^{r} (r - z)dF(z) + d \int_{r}^{+\infty} (z - r)dF(z). \tag{18}$$

Equation to determine s

$$A_2 + h \int_{-\infty}^{S} (S - z)dF(z) + d \int_{S}^{+\infty} (z - S)dF(z) =$$
$$= h \int_{-\infty}^{s} (s - z)dF(z) + d \int_{s}^{+\infty} (z - s)dF(z). \tag{19}$$

3 Adaptive Inventory Control Algorithms

In the context of control theory, the concept of uncertainty, as a rule, means that the control problem is solved in a probabilistic formulation under conditions when the probabilistic distributions describing the system are unknown.

In this connection, it is important to mention the following theorem, proved in [11] and called the Robbins-Monroe theorem.

Robbins-Monroe Theorem. Let a one-dimensional real equation of the form

$$M(\theta) = \alpha \tag{20}$$

is given, where $M(\theta)$ is a monotonically increasing function of the parameter θ. Moreover, the function $M(\theta)$ is the mathematical expectation of some random process $N(\xi, \theta)$, which depends on the parameter θ and is random through a random variable ξ, which takes in each experiment we observe independent values $\xi_1, \xi_2, \ldots, \xi_n, \ldots$, i.e.

$$M(\theta) = E\left[N(\xi,\ \theta)\right], \tag{21}$$

where E denotes the operator for calculating the mathematical expectation.

Then the iterative algorithm (subsequently called the Robbins-Monroe algorithm) of the form

$$\theta_{n+1} = \theta_n - \gamma_n\left[N\left(\xi_n,\ \theta\right) - \alpha\right], \tag{22}$$

under conditions[1]

$$\sum_{n=0}^{\infty} \gamma_n = \infty \quad and \quad \sum_{n=0}^{\infty} \gamma_T^2 < \infty, \tag{23}$$

with probability 1 converges to the solution of Eq. 20 (it is called the regression equation).

If we apply the Robbins-Monroe Algorithm 22 to solve Eqs. 16–19, then to solve inventory management problems with returns with a "myopic" optimality criterion under uncertainty.

Let's start with Eq. 16. The following calculations lead to the distribution function $F(x)$ in the form of expectation

$$F(x) = \int_{-\infty}^{x} f(z)dz = \int_{-\infty}^{\infty} \mathbf{1}(x - z)f(z)dz, \tag{24}$$

where $\mathbf{1}(\bullet)$ is the Heaviside function, and integration from $-\infty$ to $+\infty$ with a probability density $f(x)$ is an expectation operation, because the demand distribution is supported by a positive Euclidean semi-axis. From formula 28 it is clear that the role of the function $N(\xi,\ \theta)$ from the Robbins–Monroe theorem is played by the function $\mathbf{1}(x - z)$ under the integral sign. Hence we find that the algorithm of the form

$$R_{n+1} = R_n - \gamma_n\left[\mathbf{1}\left(R_n - \xi_n\right) - (d - c)/(h + d)\right], \tag{25}$$

with probability 1 converges to the true value of the desired answer R when choosing the coefficients γ_n that satisfy the Dvoretsky conditions 23.

Similarly, for Eq. 17 we find that the algorithm

$$S_{n+1} = S_n - \overline{\gamma}_n\left[\mathbf{1}\left(S_n - \xi_n\right) - d/(h + d)\right], \tag{26}$$

has the property that the values S_n with probability 1 converge to the true value of the desired answer S when choosing the coefficients $\overline{\gamma}_n$ satisfying the Dvoretsky conditions 23.

---

[1] These conditions, after the name of the mathematician who proposed them, are called the Dvoretsky conditions.

Now let' s move on to Eq. 18. Let' s write it in more detail:

$$A_1 + c(R - r) + hRF(R) - h\int_{-\infty}^{R} zdF(z) + d\int_{R}^{+\infty} zdF(z) - dR\left[1 - F(R)\right] -$$

$$-hrF(r) + h\int_{-\infty}^{r} zdF(z) + + dr\left[1 - F(r)\right] - d\int_{r}^{+\infty} zdF(z) = 0.$$

But since, by virtue of formula 16, $F(R) = \frac{d-c}{h+d}$, then, summing up like terms, we get as a result:

$$A_1 + c(R - r) + hR\frac{d - c}{h + d} - h\int_{-\infty}^{R} zdF(z) + d\int_{R}^{+\infty} zdF(z) -$$

$$-dR\left[1 - \frac{d - c}{h + d}\right] - hrF(r) + h\int_{-\infty}^{r} zdF(z) + dr\left[1 - F(r)\right] - d\int_{r}^{+\infty} zdF(z) = 0.$$

And, finally, Eq. 18 can be represented as

$$A_1 - (c - d)r - (h + d)\left[rF(r) + \int_{r}^{R} zdF(z)\right] = 0, \tag{27}$$

It can also be written that

$$\int_{r}^{R} zdF(z) = \int_{-\infty}^{+\infty} z\mathbf{1}(R - z)\mathbf{1}(z - r)dF(z). \tag{28}$$

Using formulas 24 and 28, we can write the following adaptive algorithm for estimating the parameter r from Eq. 18:

$$r_{n+1} = r_n +$$

$$+\gamma_n' \left[(c - d)r_n + (h + d)r_n\mathbf{1}(r_n - \xi_n) + (h + d)\xi_n\mathbf{1}(R_n - \xi_n)\mathbf{1}(\xi_n - r_n) - A_1\right], \tag{29}$$

in which the value of r_n with probability 1 converges to the true value of the desired answer r when choosing the coefficients γ_n that satisfy Dvoretsky' s conditions 23.

Similar calculations can be done for the transformation of Eq. 19), again using Eqs. 16 and 17:

$$A_2 + h\int_{-\infty}^{S}(S - z)dF(z) + d\int_{S}^{+\infty}(z - S)dF(z) =$$

$$= h\int_{-\infty}^{s}(s - z)dF(z) + d\int_{s}^{+\infty}(z - s)dF(z),$$

whence it follows that

$$A_2 + hSF(S) - h\int_{-\infty}^{S} zdF(z) + d\int_{S}^{+\infty} zdF(z) - dS\left[1 - F(S)\right] - hsF(s) +$$

$$+h\int_{-\infty}^{s} zdF(z) - d\int_{s}^{+\infty} zdF(z) + ds\left[1 - F(s)\right] = 0.$$

But since, by virtue of 17, $F(S) = \frac{d}{h+d}$, then, summing up similar terms, we obtain as a result

$$A_2 + hS\frac{d}{h+d} - h\int_{-\infty}^{S} zdF(z) + d\int_{S}^{+\infty} zdF(z) - dS\left[1 - \frac{d}{h+d}\right] - hsF(s) +$$

$$+h\int_{-\infty}^{s} zdF(z) - d\int_{s}^{+\infty} zdF(z) + ds\left[1 - F(s)\right] = 0.$$

As a result, by analogy with 29, we can write the following adaptive algorithm for estimating the parameter s:

$$s_{n+1} = s_n -$$

$$-\overline{\gamma}'_n\left[-s_nd + (h+d)s_n\mathbf{1}\left(s_n - \xi_n\right) + (h+d)\xi_n\mathbf{1}\left(s_n - \xi_n\right)\mathbf{1}\left(\xi_n - S_n\right) - A_2\right],$$
$$(30)$$

in which the values of s_n with probability 1 converge to the true values of the desired answers when choosing all the coefficients $\overline{\gamma}'_n$ that satisfy the Dvoretsky conditions 23).

4   Simulation

Below are the results of simulations performed using programs that were written in Python in Google Colab. The following libraries (modules) were used: SciPy, NumPy, Matplotlib, math, random. In this case, the following values of the introduced parameters were used for the case of an exponential law of demand distribution with the parameter λ at one step: $d = 20, c = 10, h = 2.6, A_1 = 5$ and $A_2 = 5$. At the same time, the true values of the strategy parameters were equal to $R = 0.584, S = 2.164, r = -0.232$ and $s = 2.651$ (Figs. 2, 3, 4 and 5).

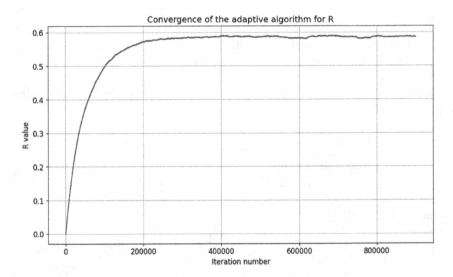

Fig. 2. Convergence of the adaptive algorithm for parameter R.

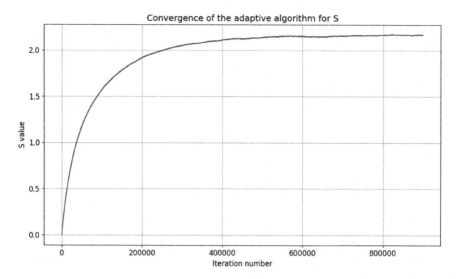

Fig. 3. Convergence of the adaptive algorithm for parameter S.

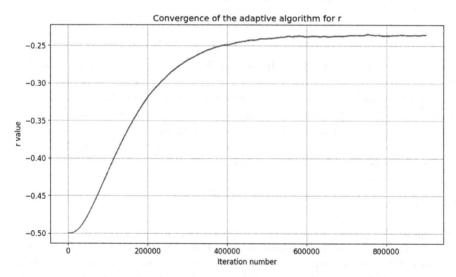

Fig. 4. Convergence of the adaptive algorithm for parameter r.

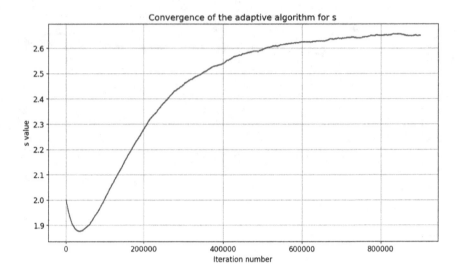

Fig. 5. Convergence of the adaptive algorithm for parameter s.

5   Summary

The model of inventory control with returns is considered under conditions when the minimum of total average costs at one step of the planning period is used as a criterion for the optimality of the selected order sizes. It is assumed that the probability distribution of demand at one step is a priori unknown. The optimal inventory management strategy is determined by setting four parameters. Stochastic approximation algorithms are proposed that converge to the true values of these parameters with probability 1.

References

1. Mandel, A., Granin, S.: Investigation of analogies between the problems of inventory control and the problems of the controllable queuing systems. In: Proceedings of the 11th International Conference Management of Large-Scale System Development (MLSD), pp. 1–4. IEEE (2018) https://doi.org/10.1109/MLSD.2018.8551852
2. Granin, S., Laptin, V., Mandel, A.: Inventory control with returns and controlled markov queueing systems. In: Vishnevskiy, V.M., Samouylov, K.E., Kozyrev, D.V. (eds.) Distributed Computer and Communication Networks: Control, Computation, Communications. DCCN 2022. Lecture Notes in Computer Science, vol. 13766. Springer, Cham (2023). https://doi.org/10.1007/978-3-031-23207-7_28
3. Mandel, A., Laptin, V.: Myopic channel switching strategies for stationary mode: threshold calculation algorithms. In: Vishnevskiy, V.M., Kozyrev, D.V. (eds.) DCCN 2018. CCIS, vol. 919, pp. 410–420. Springer, Cham (2018). https://doi.org/10.1007/978-3-319-99447-5_35
4. Mandel, A.S., Laptin, V.A.: Channel switching threshold strategies for multi-channel controllable queuing systems. In: Vishnevskiy, V.M., Samouylov, K.E.,

Kozyrev, D.V. (eds.) DCCN 2020. CCIS, vol. 1337, pp. 259–270. Springer, Cham (2020). https://doi.org/10.1007/978-3-030-66242-4_21

5. Laptin, V., Mandel, A.: Controlled markov queueing systems under uncertainty. In: Vishnevskiy, V.M., Samouylov, K.E., Kozyrev, D.V. (eds.) Distributed Computer and Communication Networks. DCCN 2022. Communications in Computer and Information Science, vol. 1748. Springer, Cham (2023). https://doi.org/10.1007/978-3-031-30648-8_20

6. Pervozvansky, A.: Mathematical Models of Inventory and Production Control. Science Publ. House, Moscow (1975). (In Russian)

7. Prabhu, N.U.: Stochastic Storage Processes: Queues, Insurance Risk, Dams, and Data Communication, p. 206. Springer (1997). https://doi.org/10.1007/978-1-4612-1742-8

8. Tsypkin, Y.Z.: Adaptation and learning in automatic systems. The main edition of physical and mathematical literature of the Nauka publishing house, p. 400 (1968). (In Russian)

9. Tsypkin, Y.Z.: Fundamentals of the theory of learning systems. The main edition of physical and mathematical literature of the Nauka publishing house, p. 252 (1968). (In Russian)

10. Hadley, G., Whitin, T.M.: Analysis of Inventory Systems. Prentice-Hall, Inc., Englewood Clifs, New Jersey (1963)

11. Robbins, H., Monro, S.: A stochastic approximation method. Ann. Math. Stat. **22**(1), 400–407 (1951)

Estimating the Distribution Parameter of Non-prolonging Random Dead Time Duration in Recurrent Semi-Synchronous Events Flow Through Maximum Likelihood

Anna Vetkina[(✉)] and Ludmila Nezhel'skaya

National Research Tomsk State University, 36 Lenina Avenue, Tomsk 634050, Russia
anyavetkina@gmail.com

Abstract. The paper explores the recurrent semi-synchronous events flow that is a common mathematical model of information flows of messages that operate in telecommunication and information-computing networks. This model belongs to the category of doubly stochastic events flows meaning that it is a stochastic process with both random arrival times and random intensity. General and special cases of this flow are presented. Functionality of the flow is considered under random non-prolonging dead time that has uniform distribution. Dead time refers to the time of event processing by the recording device. The aim of this work is to estimate a distribution parameter of the dead time duration applying maximum likelihood method. Results of statistical experiments are presented that demonstrate the high quality of the estimation.

Keywords: Semi-synchronous events flow · Random dead time · Maximum likelihood method · Parameter estimation

1 Introduction

Currently, the information message flows in different information and computational networks are best described by doubly stochastic events flows [1,2]. These types of flows are characterized by both the random time of occurrence of events and their random intensity. In general, doubly stochastic events flows are correlated flows [3] but under certain conditions they become recurrent.

Events flows that involve double randomness (stochasticity) can be divided into two classes: the first class consists of flows whose accompanying process (intensity) is a continuous random process [4,5]; the second class consists of flows whose accompanying process (intensity) is a piecewise-constant random process with a finite (arbitrary) number of states [6]. The latter are called MC-flows (Markov chain) or MAP-flows (Markovian Arrival Process) and are most commonly used in solving applied problems.

Depending on the way the flow intensity transitions from state to state, MC-flows are divided into:

1) synchronous events flows (flows whose state of the accompanying process changes at random times that are the times of occurrence of events) [7];

2) asynchronous events flows (flows whose transition from one state to another in the accompanying process occurs at random times and is independent of the times of occurrence of events) [8,9];

3) semi-synchronous events flows (flows whose accompanying process changes in part at the times of events occurrence and in part at random times not related to events occurrence) [10].

In practical information systems, it is often impossible to observe all events in a flow due to the dead time of recording devices [11]. This dead time occurs when recorded events cause other events to be lost, i.e. inaccessible for the observation. The dead time of recording devices can be categorized as non-prolonging or prolonging. Non-prolonging dead time means that during the duration of unobservability caused by processing of the recorded event, other occurred events do not prolong this duration. The dead time duration may be deterministic or random with a specific distribution law. The value and the nature of the duration of unobservability is influenced by various factors while for recorded devices this value is limited by an upper bound. Therefore, it is reasonable to consider the distribution of the dead time duration as uniform over some interval when considering it as a random variable and when we have no more knowledge about the nature of this duration.

The object of the present study is a recurrent semi-synchronous doubly stochastic events flow with an intensity that is a piecewise-constant random process, and functionality of the flow is considered under random uniformly distributed non-prolonging dead time. In order to identify lost events caused by a non-prolonging dead time factor, it is necessary to estimate the value of the duration of the dead time. The maximum likelihood (ML) method is a widely used approach for determining unknown parameters in various fields of study. In this work, 1) an algorithm to find the ML estimations for the parameter of the dead time distribution is applied to the observed semi-synchronous events flows; 2) for the investigation of obtained estimations, a numerical study of constructed likelihood function in the domain of the unknown variable is conducted. Numerical results implemented on a simulation model of the observed flow demonstrate that this method produces accurate ML estimates.

2   Mathematical Model

This research focuses on a semi-synchronous doubly stochastic events flow that accompanying process (intensity) is a piecewise-constant random process $\lambda(t)$ with two states S_1 and S_2. It is assumed that the i-th state of the process S_i occurs when $\lambda(t) = \lambda_i$, and during that time a Poisson events flow with intensity λ_i occurs, $i = 1, 2$. The transition from the state S_1 of the process $\lambda(t)$ to the state S_2 is possible only at the time of occurrence of an event (synchronicity property of the flow), and this transition occurs with probability p (the process $\lambda(t)$ remains in the first state with probability $1 - p$). The transition from the state S_2 to the state S_1 can occur at any time that is not related to the occurrence of an event (asynchronicity property of the flow). The duration of the process $\lambda(t)$ being in the second state is a random variable distributed according to an exponential law $F(t) = 1 - e^{-\alpha_2 t}$, $t \geq 0$, where α_2 is the intensity of the transition from the state S_2 to the state S_1. Hence, we have a semi-synchronous doubly stochastic events flow. Under these assumptions, $\lambda(t)$ is a hidden Markov process ($\lambda(t)$ is a fundamentally unobservable process; only the moments of the time of events occurrences in the flow are observable).

After each recorded event at time t_k, a period of the dead time of random duration occurs, which is generated by this event, so that other events in the flow that occur during this dead time period are not observable and do not cause its prolongation. It is assumed that the random duration of the dead time is uniformly distributed with probability density $p(T) = 1/T^*$, where T is the value of the duration of the dead time, $0 \leq T \leq T^*$, T^* is an unknown parameter of the uniform distribution.

In this work we consider the semi-synchronous events flow when $p = 1$, i.e. a flow that instantly transitions to the second state at each occurrence of an event in the first state. When this restriction is satisfied, the original semi-synchronous events flow, operating under conditions of deterministic dead time, becomes a recurrent flow:

$$p(\tau_1, \tau_2 \mid T) = p(\tau_1 \mid T)p(\tau_2 \mid T), \ \tau_1 \geq T, \tau_2 \geq T$$

where T is deterministic duration of the dead time, τ_1, τ_2 are independent random variables of the duration values between adjacent events in the observed flow, $p(\tau_i \mid T)$, $i = 1, 2$, is the probability density of the interval duration between flow events, $p(\tau_1, \tau_2 \mid T)$ is the joint probability density [14].

A possible case for that flow is shown in Fig. 1, where S_1 and S_2 are states of the random process $\lambda(t)$; the time axis $(0, t)$ is the axis of moments of occurrence of observable events at times $t_1, t_2, \ldots$; the time axis $(0, t^{(i)})$, $i = 1, 2$, is the axis of moments of occurrence of events in the i-th state of the $\lambda(t)$ process, on which the values of the duration of dead times generated by observable events in the flow are also indicated; white circles denote observable events, black ones denote unobservable (lost) events, the hatching denotes the period of dead time; the trajectory of the $\lambda(t)$ process is attached to the time axis $(0, t^{(1)})$.

Note that entering the non-prolonging random dead time into the mathematical model of a recurrent flow with $p = 1$ keeps the observed flow in the class of recurrent events flows. Also note that the stationary mode of operation of the flow is considered, i.e. we assume that the flow operates infinitely, and its probabilistic characteristics do not depend on time.

For the semi-synchronous events flow with $p = 1$ we consider two cases of the ratio of parameters of this flow: a) general case, when $\lambda_1 - \lambda_2 - \alpha_2 \neq 0$; b) special case, when $\lambda_1 - \lambda_2 - \alpha_2 = 0$.

The objectives of this study are:

1. Using the maximum likelihood method, to estimate the parameter of the uniform distribution of non-prolonging dead time duration T^*, based on a

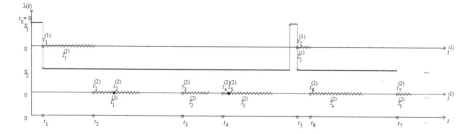

Fig. 1. Formation of the observed recurrent events flow.

sample of moments of observed events occurrence $t_1, t_2, \ldots, t_n$ on the time interval $(0, T_m)$, where T_m is the observation time of the flow $(t_n < T_m)$.

2. To investigate the estimation $\hat{T}^*$ for the general and special cases of the considered flow. For that, to conduct statistical experiments to determine the properties of the obtained estimates.

3   Approximate ML Estimate of the Parameter T^*

The maximum likelihood method [12,13] is used to estimate the unknown parameter T^* of the uniform distribution of non-extendable dead time duration in the observed events flow. This method consists in maximizing a likelihood function to estimate the parameter so that, under the assumed statistical model, the observed data is the most probable.

Denote $\tau_k = t_{k+1} - t_k$, $k = 1, 2, \ldots$, are values of the duration of the interval between neighboring events in the observed flow $(\tau_k \geq 0)$. As we consider the stationary mode of operation of the flow, then a probability density function of the duration of the k-th interval is $p(\tau_k) = p(\tau)$, $\tau \geq 0$, for each k, i.e. the moment of the occurrence of the event is $\tau = 0$. Let we have n values of time intervals between observed events: $\tau_1, \tau_2, \ldots, \tau_n$ that are measured on the time interval $(t_0, t]$. Then we have the form of the likelihood function:

$$L(T^* \mid \tau_1, \tau_2 \ldots, \tau_n) = \prod_{k=1}^{n} p(\tau_k \mid T^*), \ T^* > 0, \tag{1}$$

where $p(\tau_k \mid T^*)$ is the probability density function of the duration of the interval between neighboring events in the observed flow with $\tau = \tau_k$ (τ_k is a measurement); T^* is a variable value ($T^* > 0$).

Range of the random variable T of the dead time duration is an union of two ranges: when $0 \leq \tau < T^*$ and when $\tau \geq T^*$, then the general expression for the density $p(\tau \mid T^*)$ has the following form:

$$p(\tau \mid T^*) = \begin{cases} p_1(\tau \mid T^*), & 0 \leq T \leq \tau, \ 0 \leq \tau < T^*, \\ p_2(\tau \mid T^*), & 0 \leq T \leq \tau, \ \tau \geq T^*. \end{cases}$$

a) For the general case, we have:

$$p_1\left(\tau \mid T^*\right) = \frac{1}{T^*}\left\{1 - e^{-\lambda_1\tau} - e^{-(\lambda_1+\alpha_2)\tau} + e^{-(\lambda_2+\alpha_2)\tau}\right\},\ 0 \le \tau < T^*,$$

$$p_2\left(\tau \mid T^*\right) = \frac{1}{T^*}\left\{e^{-\lambda_1\tau}\left[-1 + C_1\,e^{\lambda_1 T^*} + C_2\,e^{-\alpha_2 T^*}\right]\right.$$

$$\left. + e^{-(\lambda_2+\alpha_2)\tau}\left[-1 + C_1 e^{-(\lambda_1-\lambda_2)T^*} + C_2 e^{(\lambda_2+\alpha_2)T^*}\right]\right\},\ \tau \ge T^*,$$

where

$$C_1 = \frac{-\alpha_2\left(\lambda_2+\alpha_2\right)}{\left(\lambda_1+\alpha_2\right)\left(\lambda_1-\lambda_2-\alpha_2\right)},$$

$$C_2 = \frac{\lambda_1\left(\lambda_1-\lambda_2\right)}{\left(\lambda_1+\alpha_2\right)\left(\lambda_1-\lambda_2-\alpha_2\right)},$$

$$C_1 + C_2 = 1,$$

$$\lambda_1 - \lambda_2 - \alpha_2 \ne 0.$$

b) For the special case, we have:

$$p_1\left(\tau \mid T^*\right) = \frac{1}{T^*}\left\{1 - 2e^{-\lambda_1\tau} + e^{-(\lambda_1+\alpha_2)\tau}\right\},\ 0 \le \tau < T^*,$$

$$p_2\left(\tau \mid T^*\right) = \frac{1}{T^*}e^{-\lambda_1\tau}\left\{-2 + \left[1 + \frac{\lambda_1\alpha_2}{\lambda_1+\alpha_2}\left(\tau - T^*\right)\right]e^{\lambda_1 T^*}\right.$$

$$\left. + \left[1 - \frac{\lambda_1\alpha_2}{\lambda_1+\alpha_2}\left(\tau - T^*\right)\right]e^{-\alpha_2 T^*}\right\},\ \tau \ge T^*,$$

where $\lambda_1 - \lambda_2 - \alpha_2 = 0$.

Then we can rewrite the expression (1) in the form:

$$L(T^* \mid \tau_1, \tau_2 \dots, \tau_n) = L_1(T^* \mid \tau_1, \tau_2 \dots, \tau_n)L_2(T^* \mid \tau_1, \tau_2 \dots, \tau_n),$$

$$L_1(T^* \mid \tau_1, \tau_2 \dots, \tau_n) = \prod_{k:0 \le \tau_k < T^*} p_1(\tau_k \mid T^*), \tag{2}$$

$$L_2(T^* \mid \tau_1, \tau_2 \dots, \tau_n) = \prod_{k:\tau_k \ge T^*} p_2(\tau_k \mid T^*).$$

Since the values τ_k, $k = \overline{1,\,n}$, are independent, we can order them: $0 < \tau^{(1)} < \dots < \tau^{(n)} < \infty$. Then for the expression (2) we have:

$$L(T^* \mid \tau^{(1)}, \tau^{(2)}, \dots, \tau^{(n)}) = \prod_{k=1}^{i} p_1(\tau^{(k)}|T^*) \prod_{k=i+1}^{n} p_2(\tau^{(k)}|T^*), \tag{3}$$

$$\tau^{(i)} \le T^* < \tau^{(i+1)},\ i = \overline{1,\,n},$$

assuming $\tau^{(0)} = 0$, $\tau^{(n+1)} = \infty$.

Previously in the paper [15] the following theorem for the recurrent events flow with $p = 1$ was proven.

Theorem 1. *The function* $p\left(\tau^{(k)}|T^*\right)$ *of the variable* T^* $(T^* > 0)$ *in both general and special cases reaches its global maximum at a point* $T^* = \tau^{(k)}$.

The result of the theorem indicates that the likelihood function (2) is an increasing function at the interval $\left[0, \tau^{(1)}\right]$ and it reaches its local maximum at a point $T^* = \tau^{(1)}$, and it is an decreasing function at the interval $\left[\tau^{(n)}, \infty\right)$ and reaches its local maximum at a point $T^* = \tau^{(n)}$. Therefore, to find the global maximum of the likelihood function (2), it is necessary to investigate the interval $\left[\tau^{(1)}, \tau^{(n)}\right]$ of change of the variable T^* $(\tau^{(1)} \leq T^* \leq \tau^{(n)})$.

Since the function $p\left(\tau^{(k)}|T^*\right)$ $(T^* > 0)$ reaches its global maximum at the point $T^* = \tau^{(k)}$, $k = \overline{1, n}$, than the point $T^* = \tau^{(k)}$, $k = \overline{1, n}$, is considered as a point of a local maximum of the likelihood function (2), and the algorithm for finding the value of an approximate MP estimate of the parameter T^* is as follows: 1) the values of the likelihood function (2) are calculated at the points $T^* = \tau^{(k)}$, $k = \overline{1, n}$; 2) the maximum value of the function (2) is found on the set of these points; 3) the value $\hat{T}^*$ that provide the maximum value of the function (2) at the previous step of the algorithm is selected as the value of the approximate ML estimate of the parameter T^*.

It is not analytically possible to find out explicitly the behavior of the likelihood function (2) over the entire segment $\left[\tau^{(1)}, \tau^{(n)}\right]$ of variable T^* change. Instead, we can numerically investigate the behavior of the likelihood function at some intermediate points of the segment $\left[\tau^{(1)}, \tau^{(n)}\right]$ by dividing it evenly into several parts with a given sampling step ΔT^*.

The present study focuses on examining and improving the approximations derived from the maximum likelihood approach explained earlier by calculating all the values of the likelihood function on the segment $\left[\tau^{(1)}, \tau^{(n)}\right]$ of variable T^* change with a given sampling step ΔT^* and comparing its maximum value with the maximum found from the values of the likelihood function at the points $T^* = \tau^{(k)}$, $k = \overline{1, n}$. Therefore, we find out the behavior of the likelihood function (2) over the entire interval of variable T^* change (with a given accuracy of calculations), and not only at the points where the global maximum of the function $p\left(\tau^{(k)}|T^*\right)$ is reached.

For that, numerous experiments were provided on a simulation model of the observed flow in both general and special cases. Approximate and refined ML estimations of the parameter T^* were obtained numerically according to the two algorithms presented above. Results of the experiments indicate the high accuracy of the ML estimates $\hat{T}^*$ calculated by methodology of estimating the unknown parameter of the model by the maximum likelihood method, since they received insignificant deviations from the approximate estimates and refined ones.

4   Numerical Results

Statistical experiments were provided to establish the stationary mode and to identify the characteristics of the obtained estimates.

4.1   Establishment of a Stationary Mode

First statistical experiment is an establishment of a stationary mode. A case study was performed on a simulation model of the recurrent semi-synchronous events flow with $p = 1$.

$N = 100$ implementations of the observed flow were obtained for general and special cases of the flow with defined parameter of uniform distribution $T^* = 1$ time units and modeling time $T_m = 100, 200, \ldots, 2000$ time units. For each value of the modeling time, 100 estimates of the parameter T^* were obtained by the maximum likelihood method. Each i-th solution is the value of the estimate $\hat{T}_i^*$, $i = 1, \ldots, N$, of the parameter T^*.

Based on the data obtained, the sample mean of the estimates $\hat{M}(\hat{\boldsymbol{T}}^*) = \frac{1}{N} \sum_{i=1}^{N} \hat{T}_i^*$ and their sample variation $\hat{V}(\hat{\boldsymbol{T}}^*) = \frac{1}{N} \sum_{i=1}^{N} (\hat{T}_i^* - T^*)^2$ were calculated, where T^* is the parameter value known from the simulation model.

General Case. The recurrent semi-synchronous events flow in general case is considered with parameters $\lambda_1 = 2$, $\lambda_2 = 1$, $\alpha_2 = 0.2$, $p = 1$. A Table 1 shows results of the experiment. The first row of the table shows the modeling time T_m (time of flow observation) ($T_m = 100, 200, \ldots, 2000$ time units); the second and third rows of the table show the sample mean $\hat{M}(\hat{\boldsymbol{T}}^*)$ and sample variation $\hat{V}(\hat{\boldsymbol{T}}^*)$.

Table 1. Numerical results of the first statistical experiment for the general case.

T_m	100	200	300	...	1800	1900	2000
$\hat{M}(\hat{\boldsymbol{T}}^*)$	0.9629	0.9811	0.9911	...	1.0004	1.0007	1.0014
$\hat{V}(\hat{\boldsymbol{T}}^*)$	0.0327	0.0171	0.0112	...	0.0013	0.0013	0.0011

Figures 2 and 3 show plots of the sample mean $\hat{M}(\hat{\boldsymbol{T}}^*)$ and sample variation $\hat{V}(\hat{\boldsymbol{T}}^*)$ respectively versus the modeling time T_m constructed according to data in the Table 1.

It follows from the analysis of results of the first statistical experiment that for a recurrent semi-synchronous events flow in general case the stationary mode is established at $T_m \geq 1300$ time units because the sample mean $\hat{M}(\hat{\boldsymbol{T}}^*)$ is setting near a constant value $T^* = 1$ at these values of the modeling time.

Special Case. Defined parameters for the special case of the recurrent flow are $\lambda_1 = 2$, $\lambda_2 = 1.8$, $\alpha_2 = 0.2$, $p = 1$. The results of the first statistical experiment for the special case are shown in a Table 2 and in Figs. 4, 5. The structure of the Table 2 is identical to the structure of the Table 1.

Similarly, for the special case of the observed flow, it follows that the stationary mode is established at $T_m \geq 1300$ time units, since the sample mean $\hat{M}(\hat{\boldsymbol{T}}^*)$ tends to a constant value at $T_m \geq 1300$ time units.

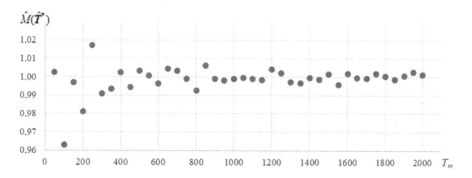

Fig. 2. Sample mean of the ML estimates in general case.

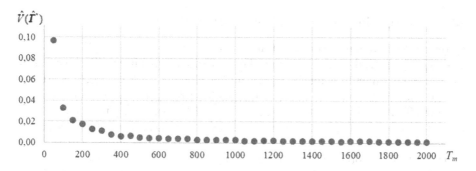

Fig. 3. Sample variance of the ML estimates in general case.

Table 2. Numerical results of the first statistical experiment for the special case.

T_m	100	200	300	...	1800	1900	2000
$\hat{M}(\hat{T}^*)$	0.9849	1.0001	1.0091	...	0.9987	1.0011	1.0005
$\hat{V}(\hat{T}^*)$	0.0110	0.0056	0.0035	...	0.0004	0.0004	0.0004

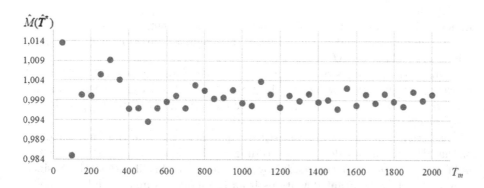

Fig. 4. Sample mean of the ML estimates in special case.

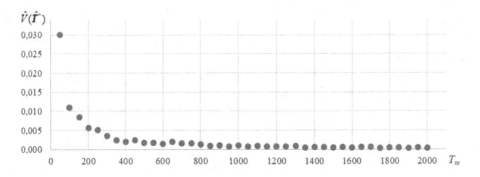

Fig. 5. Sample variance of the ML estimates in general case.

4.2　Investigation of the Quality of ML Estimates

Second statistical experiment is an investigation of the quality of ML estimates. The experiment is designed to refine the approximate maximum likelihood estimate of the parameter using the algorithm described in the Sect. 3 of this article.

The modeling time was fixed at $T_m = 1300$ time units, which corresponds, as follows from Tables 1, 2, to the time of establishment of the stationary mode; the simulation model parameters were defined the same as at the first statistical experiment.

By setting the precision parameter (sampling step) $\Delta T^* = 0.05$ time units, at each time step $T^* = \Delta T_i^* = i \, \Delta T^*$, $i \in \mathbb{Z}$, the value of the likelihood function (3) is calculated, and the maximum of the obtained values is found, and the point providing the maximum is taken as the desired refined estimate (tabulated) $\hat{T}_{Tab}^*$. Additionally, according to the algorithm for finding an approximate ML estimate, the values of the likelihood function (3) at the points $T^* = \tau^{(k)}$, $k = \overline{1, \, n}$, are calculated, and the point that yields the maximum within a specific range of variable values T^* is taken as an approximate ML estimate $\hat{T}_{ML}^*$.

Results of the second statistical experiment for general and special cases are shown in Figs. 6 and 7 respectively as a plot showing the dependence of the likelihood function $L(T^* \mid \tau^{(1)}, \tau^{(2)}, \ldots, \tau^{(n)}$ of the parameter T^*. Squares indicate the values of the likelihood function at the points $T^* = \tau^{(k)}$, $k = \overline{1, \, n}$, $n = 15$, and circles indicate the values of the likelihood function at intermediate points $T^* = \Delta T_i^* = i \, \Delta T^*$, $i \in \mathbb{Z}$.

Table 3 shows values of the obtained estimates for the general and special cases, where the first row shows the type of considered case, the second and third rows show the values of approximate and refined estimates $\hat{T}_{ML}^*$ and $\hat{T}_{Tab}^*$ respectively.

The results of the second statistical experiment indicate the high accuracy of the ML estimates calculated according to the algorithm for estimating the unknown parameter of the model by the maximum likelihood method, since they received insignificant deviations from the model estimates and intermediate ones. The values of the likelihood function (3) calculated at intermediate

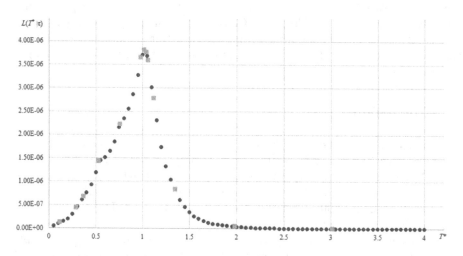

Fig. 6. Likelihood function in general case.

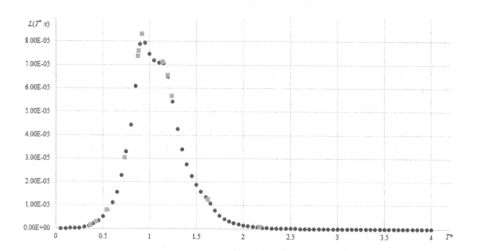

Fig. 7. Likelihood function in special case.

Table 3. Numerical results of the second statistical experiment.

case	General	Special
$\hat{T}^*_{ML}$	1.0213	1.0500
$\hat{T}^*_{Tab}$	0.9214	0.9500

points of the variable change interval are close to its values at model points, and strongly depend on the sampling step. Consequently, in order to achieve satisfactory results of estimating an unknown parameter T^*, it is required to maximize the likelihood function (3) only on a set of model points, which reduces computational costs.

5 Conclusion

In this paper, we considered a recurrent semi-synchronous doubly stochastic events flow with $p = 1$ with uniformly distributed random dead time in the general case, when $\lambda_1 - \lambda_2 - \alpha_2 \neq 0$ and in the special case, when $\lambda_1 - \lambda_2 - \alpha_2 = 0$. An explicit form of the likelihood function was given for estimating the parameter T^* of a uniform distribution of the duration of dead time. The procedure for constructing the ML estimates $\hat{T}^*$ was described.

The results of numerical calculations were presented, indicating an acceptable quality of estimation due to the small mean variation of the estimates and the insignificant difference between approximate and refined ML estimates, that approved the possibility of effective use of the maximum likelihood method for estimating an unknown parameter in the recurrent semi-synchronous events flow.

References

1. Cox, D.R.: The analysis of non-Markovian stochastic processes by the inclusion of supplementary variables. Proc. Camb. Philos. Soc. **51**(3), 433–441 (1955)
2. Kingman, Y.F.C.: On doubly stochastic Poisson process. Proc. Camb. Philos. Soc. **60**(4), 923–930 (1964)
3. Vishnevsky, V.M., Dudin, A.N., Klimenok, V.I.: Stochastic Systems with Correlated Flows. Theory and Applications in Telecommunication Networks. Technosphere, Moscow (2018)
4. Basharin, G.P., Kokotushkin, V.A., Naumov, V.A.: On the equivalent substitutions method for computing fragments of communication networks. Part 1. Proc. USSR Acad. Sci. Tech. Cybern. **6**, 92–99 (1979)
5. Neuts, M.F.: A versatile Markov point process. J. Appl. Probab. **16**, 764–779 (1979)
6. Lucantoni, D.M., Neuts, M.F.: Some steady-state distributions for the MAP/SM/1 queue. Commun. Stat. Stoch. Models **10**, 575–598 (1994)
7. Gortsev, A.M., Nezhelskaya, L.A.: Parameter estimation of synchronous twice-stochastic flow of events using the method of moments. Tomsk State Univ. J. **S1–1**, 24–29 (2002)
8. Nezhel'skaya, L.: Probability density function for modulated MAP event flows with unextendable dead time. In: Dudin, A., Nazarov, A., Yakupov, R. (eds.) ITMM 2015. CCIS, vol. 564, pp. 141–151. Springer, Cham (2015). https://doi.org/10. 1007/978-3-319-25861-4_12
9. Gortsev, A.M., Zuevich, V.L.: Optimal estimation of states of the asynchronous doubly stochastic flow of events with arbitrary number of the states. Tomsk State Univ. J. Control Comput. Sci. **2**(11), 44–65 (2010)

10. Kalyagin, A.A., Nezhel'skaya, L.A.: The comparison of maximum likelihood method and moments method by estimation of dead time in a generalized semysynchronous flow of events. Tomsk State Univ. J. Control Comput. Sci. 3(32), 23–32 (2015)
11. Apanasovich, V.V., Kolyada, A.A., Chernyavsky, A.F.: Statistical analysis of random flows in a physical experiment. University, Minsk (1988)
12. Malinkovsky, Y.V.: Probability Theory and Mathematical Statistics (Part 2. Mathematical Statistics). Francisk Skorina Gomel State University, Gomel (2004)
13. Nezhelskaya, L.A., Pershina, A.A.: Estimation of the uniform distribution parameter of unextendable dead time duration in a generalized recurrent asynchronous flow of events by the maximum likelihood method. Tomsk State Univ. J. Control Comput. Sci. 55, 53–64 (2021)
14. Nezhelskaya, L.A.: Estimation of states and parameters of doubly stochastic event flows. Dissertation for the degree of Doctor of Physical and Mathematical Sciences, Tomsk, p. 341 (2016)
15. Gortsev A.M., Vetkina A.V.: An application of the maximum likelihood estimation for the parameter of uniform distribution of the duration of unextendable dead time in recurrent alternating semi-synchronous flow. Tomsk State Univ. J. Control Comput. Sci. 62, 36–49 (2023)

Stochastic Modelling for Energy Efficiency in LTE-A and LTE-5G Networks

S. Dharmaraja[1]([✉]), Anisha Aggarwal[1], and Priyanka Kalita[2]

[1] Department of Mathematics, IIT Delhi, New Delhi 110016, India
dharmar@iitd.ac.in

[2] Department of Statistics, Bhattadev University, Pathsala, Assam 781325, India

Abstract. Maximizing energy efficiency in User Equipment (UE) is a critical concern due to the limited power sources available in these devices. This holds true for both LTE-A and LTE-5G networks. To assess energy efficiency in these networks, we've developed three stochastic models: the Markov model, semi-Markov model, Markov regenerative process (MRGP) model. The Markov and semi-Markov model focuses on the discontinuous reception (DRX) mechanism, and the MRGP model pertains to the Modified DRX mechanism. We've derived explicit expressions for steady-state system size probabilities across all these models. Additionally, we've computed energy savings in UE based on the DRX mechanism states and performed sensitivity analyses for both the Markov and semi-Markov models. In the case of the MRGP model, we've determined the optimal system's energy savings and throughput by optimizing parameters like maximum short sleep and short sleep duration. To evaluate energy efficiency in LTE-5G networks, we've explored the trade-off between energy savings, average energy consumption, and throughput. Our analysis of the DRX mechanism provides valuable insights into its performance and lays the groundwork for potential enhancements to existing DRX mechanisms. We believe that these models have the potential to be extended for studying energy savings in hardware and other system components, contributing to overall energy efficiency improvements.

Keywords: Markov model · Semi-Markov process model · Markov regenerative process model · Energy · DRX mechanism

1 Introduction

Long-Term Evolution (LTE) stands as a cornerstone in wireless communication technology (WCT), while LTE Advanced (LTE-A) represents the next-generation standard, boasting advancements beyond LTE networks. LTE-A brings forth a novel cellular network paradigm characterized by high peak data rates, exceptional spectral efficiency, rapid round-trip times, reduced latency, prolonged battery life, and enhanced bandwidth frequency flexibility. Nonetheless, a prevalent challenge in wireless technology remains the heightened power consumption associated with achieving faster data rates in LTE-A networks. In response to this challenge, we turn our attention to the Discontinuous Reception mechanism (DRX), an energy-saving mechanism designed for User Equipment (UE) within LTE-A networks.

Efficient reduction of power consumption in UE is accomplished by implementing a sleep mode functionality. To bolster energy efficiency, the LTE-A standard has introduced the DRX mechanism, permitting UEs to deactivate components when not actively transmitting or receiving data traffic. Numerous approaches have been proposed to optimize the DRX scheme, as seen in references such as [3,16], which prioritize Quality of Service (QoS) and energy efficiency considerations. Moreover, the work in [17] presents an uncomplicated yet highly effective application-aware DRX mechanism, tailored to enhance LTE-A network performance. Additionally, [19] introduces a semi-Markov model to assess DRX performance metrics under specific conditions involving an exponential packet arrival rate. Notably, Arunsundar et al. [20] conducted an extensive analysis of the DRX mechanism, considering the impact of long and short cycles on buffering, average latency, DRX-sleep, and energy consumption.

Despite the variety of models explored in existing literature by different researchers, a stochastic model with a particular focus on UE energy savings at the packet level has been notably absent. This paper introduces two distinct stochastic models aimed at evaluating UE performance in LTE-A networks. Our primary objective is to demonstrate that energy savings in UE can be maximized through Markov modeling of the DRX mechanism, as opposed to semi-Markov modeling.

In a related study [21], Philip and Malarkodi introduced a Semi-Markov (SM) Model to assess HD-DRX performance in 5G communication scenarios involving 4G base stations. They employed beam search techniques to enable dual connectivity for User Equipment (UE), connecting to both LTE eNB (4G base station) and NR NodeB (5G New RAN base station). Sallam et al. [22] utilized a Semi-Markov (SM) Process to determine UE state probabilities within the HD-DRX system, leveraging these probabilities to calculate power-saving factors and average delays. Additionally, in the context of mmWave-enabled 5G communication and directional air interfaces, a novel D-DRX mechanism was explored in [1]. This mechanism emphasized the significance of beam searching states following each sleep cycle to enhance power—saving. Maheshwari et al. [1] further proposed three distinct D-DRX variants-Integrated D-DRX (ID-DRX), Standalone D-DRX (SD-DRX), and Cooperative D-DRX (CD-DRX)—and obtained probabilistic results for UE delay and power-saving across these mechanisms. Furthermore, within the realm of wireless technology, artificial intelligence (AI) plays a crucial role, as discussed by Memon et al. [2]. They introduced an AI-based DRX mechanism (AI-DRX) aimed at enhancing energy efficiency in LTE-5G connected devices, with a particular focus on dynamic short and long sleep cycles in DRX through the application of an AI-DRX algorithm for multiple beam communications.

While Arunsundar et al. [20] previously introduced a Semi-Markov Model (SMM) to derive probabilistic outcomes related to UE power-saving, energy consumption, and delay within the context of the DRX mechanism for LTE-A and LTE-5G networks, the existing literature lacks research that specifically addresses the optimization of energy-saving and energy consumption while considering system throughput and employs a stochastic model utilizing the Markov regenerative process. Therefore, this study aims to bridge this gap by developing a stochastic model for the MD-DRX mechanism in LTE-5G networks using the Markov regenerative process.

The paper is arranged as follows: In Sect. 2, we delve into the various stochastic models, including the Markov model, semi-Markov model, and Markov regenerative process model. Section 3 is dedicated to the analysis of performance measures across all the models. In Sect. 4, we present the numerical results pertaining to the proposed models. Finally, in Sect. 5, we provide concluding remarks.

2   Stochastic Modelling

2.1   Markov Model

In this section, we will begin by providing a detailed explanation of the UE's Discontinuous Reception (DRX) mechanism, which serves as a foundation for comprehending the proposed stochastic models. The DRX mechanism is composed of three distinct periods: working, idle, and sleep periods, as documented in [3,4]. Within the sleep period, two categories of sleep modes exist, specifically referred to as short sleep and long sleep. To illustrate the structure of the DRX mechanism, please refer to Fig. 1. It's important to note that energy consumption levels vary across these different DRX periods, and these energy values have been sourced from existing literature [5,6]. Detailed energy consumption values for each period are presented in Table 1.

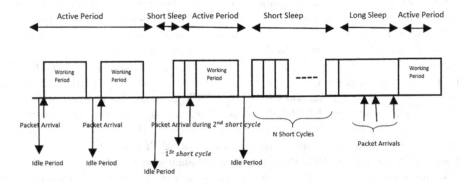

Fig. 1. Structure of DRX Mechanism

Table 1. Energy consumption values of UE in different periods of the DRX mechanism

Period	Energy Consumption (mW/ms)
Idle period	255.5
Working period	500
Short sleep	11
Long sleep	0

In this paper, we initially explore two models: Markov model and the semi-Markov model, which are founded on the sojourn time distributions within sleep periods. In the Markov model, the sojourn time within each state follows an exponential distribution. In contrast, the semi-Markov model diverges from this by having non-exponential sojourn times within sleep periods.

Description of Proposed Model. We introduce a stochastic model that accounts for the various states experienced by the UE in the DRX Mechanism. This stochastic model characterizes each state's duration using independent exponential distributions, ensuring that the overall stochastic process adheres to a continuous-time Markov chain. In the following section, we provide a comprehensive explanation of our proposed Markov model, outlining its four distinct states:

- Idle period: This state represents the condition in which the UE is active and awaiting incoming packets to deliver its services.
- Working period: This state signifies that the UE is actively providing services to the packets.
- Short cycle: This state involves the UE powering down most of its components. If no data packet arrives, this short cycle repeats a set number of times (e.g., N), and it terminates if a data packet does arrive.
- Long sleep: This state is entered when no data packet arrives upon the expiration of the short cycle.

Consider the continuous-time Markov chain denoted by $\{(N(t), S(t)), t \geq 0\}$. Here, $\{N(t), t \geq 0\}$ signifies the count of packets in the system at any given time t, while $\{S(t), t \geq 0\}$ indicates the system's state at time t, with its state space defined as $\Omega = \{(i,j); i = 0, 1, \ldots; j = 0, 1, \ldots, N, N+1\}$.

The system's states are defined as follows:

- State $(0, 0)$ represents an idle period.
- States $(i, 0)$, where $i = 1, 2, \ldots$, represent working periods where the system serves the ith packet.
- States $(0, j)$, where $j = 1, 2, \ldots, N$, correspond to short cycles.
- State $(i, N+1)$ denotes a long sleep state with i packets. During both idle and working periods, the UE actively serves packets.

The idle period before entering the first short cycle follows an exponential distribution with parameter β. Packet arrivals in the system adhere to a Poisson process with parameter λ, and packet service times are exponentially distributed with parameter μ. Short cycle durations and long sleep periods are exponentially distributed with parameters α and ν, respectively.

Initially, the system is in an idle period, waiting for data packets to arrive for service. If a packet arrives during an idle period, the system transitions to state $(1,0)$ and starts serving packets. If another packet arrives before the current one is serviced, the system moves to state $(i+1,0)$, where i packets are already in the system. After each service, the system transitions from state $(i,0)$ to state $(i-1,0)$. In the absence of packet arrivals during an idle period, the system transitions to state $(0,1)$ and continues moving to states $(0,j)$ if no packet arrivals occur in state $(0,j-1)$, where $j=1,2,\ldots,N$. If a packet arrives in any state $(0,j)$, where $j=1,2,\ldots,N$, the system transitions back to state $(1,0)$.

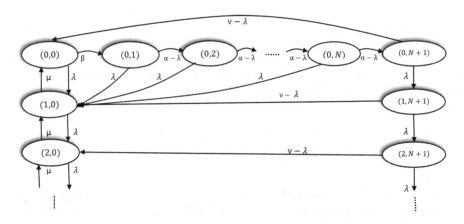

Fig. 2. State transition diagram for the proposed Markov model

If no packet arrives in state $(0,N)$, the system transitions to state $(0,N+1)$. In the event of a packet arrival during this long sleep state, the system moves to state $(i+1,N+1)$, with i packets already in the system. Upon completing the long sleep, if i packets arrive, the system transitions to state $(i,0)$. The state transition diagram for the proposed Markov model is illustrated in Fig. 2. The steady state probabilities of the system are obtained from forward Kolmogorrov equation. It is known that, $\lim_{t\to\infty} P_{i,j}(t) = \pi_{i,j}$ and $\lim_{t\to\infty} P'_{i,j}(t) = 0$. Hence, Eqs. (1)–(7) becomes

$$0 = -(\beta+\lambda)\pi_{0,0} + \mu\pi_{1,0} + (\nu-\lambda)\pi_{0,N+1},$$
$$0 = -(\mu+\lambda)\pi_{1,0} + \lambda\pi_{0,0} + \mu\pi_{2,0} + \lambda(\pi_{0,1}+\pi_{0,2}+\ldots+\pi_{0,N}) + (\nu-\lambda)\pi_{1,N+1},$$
$$0 = -(\mu+\lambda)\pi_{i,0} + \lambda\pi_{i-1,0} + \mu\pi_{i+1,0} + (\nu-\lambda)\pi_{i,N+1}, \quad i=2,3,\ldots$$
$$0 = -\alpha\pi_{0,1} + \beta\pi_{0,0},$$

$$0 = -\alpha\pi_{0,j} + (\alpha - \lambda)\pi_{0,j-1}, \, j = 2, 3, \ldots, N$$
$$0 = -\nu\pi_{0,N+1} + (\alpha - \lambda)\pi_{0,N},$$
$$0 = -\nu\pi_{i,N+1} + \lambda\pi_{i-1,N+1}, \, i = 1, 2, \ldots.$$

The stability conditions for the existence of these steady state probabilities are $\lambda \leq \alpha$, $\lambda < \mu$ and $\lambda < \nu$.

Theorem 1. *Steady-state probabilities of UE being in different states, i.e., $(0,0)$, $(0,j)$, $(i, N+1)$ and $(n,0)$ for $n = 1, 2, \ldots; 1 \leq j \leq N; i \geq 0$ are given by*

$$\pi_{0,j} = \frac{\beta(\alpha - \lambda)^{j-1}}{\alpha^j}\pi_{0,0}, j = 1, 2, \ldots, N.$$

$$\pi_{i,N+1} = \frac{\beta\lambda^i(\alpha - \lambda)^N}{\nu^{i+1}\alpha^N}\pi_{0,0}, i = 0, 1, \ldots$$

$$\pi_{n,0} = A_n\pi_{0,0}, n \geq 1$$

where

$$A_n = \left[\frac{\mu b^{n+1}}{(\mu-\lambda)} + \frac{\beta b^{n-1}((\alpha-\lambda)^N - \alpha^N)}{\alpha^N(\mu-\lambda)}\right]\left[b^{n+1} - b^{n-1}\right]$$
$$+ \frac{\beta b^n(\alpha-\lambda)^N}{(\mu-\lambda)\alpha^N}\left[\frac{(\nu-\lambda)b^{n+2}(1-(\frac{\lambda}{\nu b^2})^{n+1})}{(\nu b^2 - \lambda)}\right.$$
$$\left. - b^n\left(1 - \left(\frac{\lambda}{\nu}\right)^{n+1}\right) + \left(\frac{\lambda}{\nu}\right)^{n+1}\left(b^{-n} - b^n\right)\right].$$

$$\pi_{0,0} = \frac{1}{1 + \sum_{i=1}^{\infty} A_i + \sum_{j=1}^{N} \frac{\beta(\alpha-\lambda)^{j-1}}{\alpha^j} + \sum_{i=0}^{\infty} \frac{\beta\lambda^i(\alpha-\lambda)^N}{\nu^{i+1}\alpha^N}}.$$

For the proof of above theorem readers can refer to [24]. Using the probability $\pi_{0,0}$, we can compute all the probabilities $\pi_{0,j}$, $\pi_{i,N+1}$; $1 \leq j \leq N; i \geq 0$ and $\pi_{n,0}$ for $n = 1, 2, \ldots$.

2.2   Semi-Markov Process

The state space in this proposed semi-Markov model aligns with that of the proposed Markov model. However, the key difference lies in the sojourn time within states related to the sleep period, which follows a non-exponential distribution.

Let I_0 represent a state where the system lies in an idle period with no packets, signifying that the system is awaiting packets. W_k denotes a state where the UE is in a working period with k packets, corresponding to the $(k, 0)^{th}$ state. S_j signifies a state where the UE is in the jth short cycle, corresponding to the $(0, j)^{th}$ state. L_i represents a state where the UE is in a long sleep mode with i packets to serve, indicating the $(i, N+1)^{th}$ state. Here, $k = 1, 2, \ldots, i = 0, 1, \ldots,$ and $j = 1, 2, \ldots, N$.

Let $F_{m,n}(t)$ denote the Cumulative Distribution Function (CDF) associated with the transition from state m to state n, where $m, n \in \Omega$, with $\Omega = \{I_0, S_j, W_k, L_i, i = 0, 1, \ldots; k = 1, 2, \ldots; j = 1, 2, \ldots, N\}$. Several papers have discussed the sojourn time distribution for various states within the DRX mechanism [5,14]. Therefore, the sojourn time distribution for various states in the semi-Markov model is presented in Table 2. The transitions between states are subject to various influencing factors, resulting in a stochastic behavior. Within this dynamic process, the distribution of transitions between states is directly impacted by the real-time conditions of DRX mechanisms. Consequently, the sojourn time in certain states adheres to deterministic distribution patterns, with time epochs exhibiting the Markov property. As a result, this stochastic process can be conceptualized as a Semi-Markov Process (SMP) [13]. The state transition diagram for the proposed semi-Markov model is depicted in Fig. 3. The state space of this model is given as $\Omega = \{I_0, W_k, S_j, L_i, k = 1, 2, \ldots; i = 0, 1, \ldots; j = 1, 2, \ldots, N\}$. The steady-state probabilities $\pi_m \forall\, m \in S$ exist under the stability conditions, $\lambda < \mu$, $\lambda \leq \alpha$ and $\lambda < \nu$.

Table 2. Distributions of sojourn times in semi-Markov model

CDF	Distribution	Parameter
$F_{I_0, S_1}(t)$	Deterministic	$\frac{1}{\beta}$
$F_{S_j, S_{j+1}}(t)$	Deterministic	$\frac{1}{\alpha}$
$F_{S_N, L_0}(t)$	Deterministic	$\frac{1}{\alpha}$
$F_{W_k, W_{k+1}}(t)$	exponential	λ
$F_{W_1, I_0}(t)$	exponential	μ
$F_{W_k, W_{k-1}}(t)$	exponential	μ
$F_{S_j, W_1}(t)$	exponential	λ
$F_{L_i, L_{i+1}}(t)$	exponential	λ
$F_{L_0, I_0}(t)$	Deterministic	$\frac{1}{\nu}$
$F_{L_i, W_i}(t)$	Deterministic	$\frac{1}{\nu}$

Theorem 2. *The steady-state probabilities for the semi-Markov model are given as:*

$$\pi_{I_0} = \frac{1 - e^{-\frac{\lambda}{\beta}}}{\lambda M},$$

$$\pi_{W_1} = \frac{1 - e^{-\left(\frac{\lambda}{\beta} + \frac{\lambda N}{\alpha} + \frac{\lambda}{\nu}\right)}}{\mu M},$$

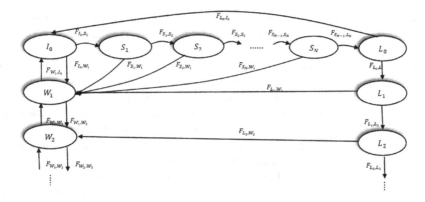

Fig. 3. State transition diagram for the proposed semi-Markov model

$$\pi_{W_k} = \frac{1}{\mu M} \left[\left(\frac{\lambda}{\mu}\right)^{k-1} \left(1 - e^{-\left(\frac{\lambda}{\beta} + \frac{\lambda N}{\alpha} + \frac{\lambda}{\nu}\right)}\right) + \right.$$
$$\left. + \frac{\left(\frac{\lambda}{\mu}\right)\left(1 - e^{-\frac{\lambda}{\nu}}\right)^2}{\left(\frac{\lambda}{\mu} - 1 + e^{-\frac{\lambda}{\nu}}\right)} e^{-\left(\frac{\lambda}{\beta} + \frac{\lambda N}{\alpha}\right)} \left[\left(\frac{\lambda}{\mu}\right)^{k-2} - \left(1 - e^{-\frac{\lambda}{\nu}}\right)^{k-2} \right] \right], \ k = 2, 3, \ldots$$

$$\pi_{S_j} = \frac{e^{-\frac{\lambda}{\beta}} \left(e^{-\frac{\lambda}{\alpha}}\right)^{j-1} \left(1 - e^{-\frac{\lambda}{\alpha}}\right)}{\lambda M}, \ j = 1, 2, \ldots, N$$

$$\pi_{L_i} = \frac{e^{-\left(\frac{\lambda}{\beta} + \frac{\lambda N}{\alpha}\right)} \left(1 - e^{-\frac{\lambda}{\nu}}\right)^{i+1}}{\lambda M}, \ i = 0, 1, \ldots$$

For the proof of above theorem readers can refer to [24].

2.3 Markov Generative Process Model

In LTE networks, the management of any DRX mechanism is overseen by the RRC layer. A device or User Equipment (UE) in LTE can operate in either the $RRC_{CONNECTED}$ mode or the RRC_{Idle} mode. In RRC_{Idle} mode, the UE's sole activity is monitoring paging signals during the paging cycle, during which it neither receives nor transmits data packets. Conversely, all data exchange between the eNB and UE occurs within the $RRC_{CONNECTED}$ mode. Consequently, this work primarily focuses on the utilization of the MD-DRX mechanism in the $RRC_{CONNECTED}$ mode, as illustrated in Fig. 4. The DRX mechanism is designed to conserve power by intermittently monitoring the physical downlink control channel (PDCCH). Here, we elucidate the MD-DRX mechanism employed by UEs in LTE-5G networks. The proposed MD-DRX mechanism encompasses the following operational states: inactive, beam searching, active, ON, long sleep, and short sleep. The short sleep state comprises a maximum of M short sleep periods. These states are mentioned as:

- **Active state:** In the active state, the UE receives/transmits data packets, representing a state of availability where it actively provides service to each packet.
- **Inactive state:** Following the active state, the UE transitions into an inactive state for a randomly determined duration. If a packet arrives before this inactive duration expires, the UE will wait until it ends, then move into the ON period. Conversely, if no packets are available in the buffer when the inactive duration concludes, the UE switches to the short sleep state.
- **Short sleep state:** During the short sleep state, the UE powers down most of its energy hungry components and can go through a maximum of M short sleep cycles. After each short sleep if the system finds empty, the UE can have up to M short sleep cycles in this state. However, if no data packets arrive until the Mth short sleep, the UE switches to the beam searching state. Conversely, if data packets do arrive during the short sleep, the UE will wait until the duration concludes before transitioning to the ON period state.
- **Beam searching state:** After completing a maximum of M short sleep and long sleep cycles, the UE enters this state. During the beam searching state, the UE conducts measurements on available beams to identify the optimal beam pair for communication. While engaged in this beam alignment process, if the UE receives a data packet, it will remain in this state until the duration expires before transitioning to the ON period state. However, if no suitable beam pair or data packets are detected during the beam searching state, the UE switches to the long sleep state.
- **Long sleep state:** In the long sleep state, the UE can receive data packets but cannot serve them. When the long sleep duration ends, if the system's buffer is empty, the UE transitions to the beam searching state. However, if there is at least one data packet in the buffer when the long sleep duration is completed, the UE enters the ON period.
- **ON period state:** When the inactive state, short sleep state, beam searching state, or long sleep state ends, and the UE's buffer contains at least one data packet, the UE transitions to the ON period. In this state, the UE briefly prepares to start serving packets before entering the active state.

We assume that the active state's packet service time follows a deterministic distribution characterized by the parameter $\frac{1}{\mu}$. Additionally, we consider the inter-arrival times of packets to be independent and following an exponential distribution with the parameter λ.

Description of the Proposed Model. The system state bivariate stochastic process for the MD-DRX mechanism is represented as $\{(N(t), S(t)); t \geq 0\}$, where $\{N(t); t \geq 0\}$ indicates the number of packets in the system at time t, and $\{S(t); t \geq 0\}$ denotes the state of the system at time t with state space Ω. The state space Ω encompasses various combinations of the number of packets and system states. Consider $\Omega \equiv \varsigma_{(i,j)} \in \{0, 1, \ldots\} \times \{0, 1, 2, 3, 4, 5, 6, \ldots, M + 4\}$. Here, $i \in \{0, 1, 2, \ldots\}$ symbolizes the number of

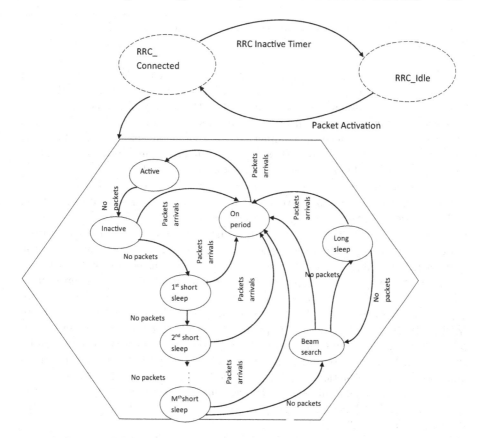

Fig. 4. LTE-DRX in RRC connected for the proposed model as in current 5G system

packets in the system (if any packets is being served, incorporating the one also) and $j \in 0, 1, 2, 3, 4, 5, 6, \ldots, M + 4$ symbolizes the states of the system i.e., whether the system lies in Beam searching state (B), or in inactive system (I), or in active state (A), or in long sleep state (L), or in ON state (0), or 1st short sleep, or in 2nd short sleep, $\ldots$, or in M^{th} short sleep (which are in between short sleep (S)). Each state in Ω is identified as follows:

- $A_{i,2}$ represents the system lies in the active state with i number of packets.
- $I_{i,1}$ symbolizes the system in the inactive state with i number of packets.
- $B_{i,0}$ represents the system in the beam searching state with i number of packets.
- $L_{i,3}$ symbolizes the system in the long sleep state with i number of packets.
- $O_{i,4}$ represents the system in the ON state with i number of packets.
- $S_{i,j}$ represents the system in the jth short sleep state with i number of packets.

Hence, Ω consists of these state combinations, i.e., Hence, $\Omega = \{B_{i,0}; i = 0, 1, \ldots, I_{i,1}; i = 0, 1, \ldots, A_{i,2}; i = 1, 2, \ldots, L_{i,3}; i = 0, 1, \ldots, O_{i,4}; i = 1, 2, \ldots, S_{i,j}; i = 0, 1, \ldots, j = 5, 6, \ldots, M + 4\}$.

The process $\{(N(t), S(t)); t \geq 0\}$ is not a continuous-time Markov chain (CTMC) because the sojourn times in each state are not exponentially distributed. Additionally, it is not a Semi-Markov Process (SMP) since the number of packets can change between two sequential service completion points due to new packet arrivals before the completion of previous packet services. Moreover, the transitions between states (beam search, inactive, active, long sleep, ON, short sleep) do not happen immediately upon new arrivals.

Due to the time remaining when new arrivals occur during the time durations in the beam search state, inactive state, active state, long sleep state, ON state, and short sleep state (including the 1st short sleep, 2nd short sleep, and so on, up to the Mth short sleep), all the state completion points become regeneration points. We define a sequence of epochs, denoted as $\{\tau_n, n \geq 0\}$, at which the process $\{(N(t), S(t)); t \geq 0\}$ is observed. These τ_n epochs correspond to the termination of the beam search period, inactive period, service, long sleep period, ON period, and short sleep period. At these regeneration epochs, we denote ζ_n, which represents the state of the system. These elements in the state space Ω are referred to as regeneration epochs because they occur at the completion points of states, including the service completion point, inactive period completion epoch, long sleep period completion epoch, ON period completion epoch, and short sleep completion epoch. As a result, we can assert that the sequence $\{\zeta_n, \tau_n\}$ constitutes an embedded Markov renewal sequence. Consequently, the stochastic process $\{(N(t), S(t)); t \geq 0\}$ is classified as a Markov Regenerative Process (MRGP) [12].

Packet arrivals in the system are modeled as a Poisson process with the parameter λ. The inter-arrival times of packets are independent and follow exponential distributions with the parameter λ. The service time, inactive time, beam searching duration, short sleep duration, long sleep duration, and ON duration all follow deterministic distributions with parameters $\frac{1}{\mu}$, $\frac{1}{\nu}$, $\frac{1}{\xi}$, $\frac{1}{\beta}$, $\frac{1}{\gamma}$, and $\frac{1}{\alpha}$, respectively. The state transition diagram for the proposed model is depicted in Fig. 5. Table 3 provides explanations for all the cumulative distribution functions (CDFs) along with their corresponding parameters.

Theorem 3. *The steady-state probabilities for the Markov regenerative process model are given as:*

$$\phi_{B_{0,0}} = \frac{1}{G}\left[(e^{-\frac{\lambda}{\beta}})^{M-1}e^{-\lambda(\frac{1}{\mu}+\frac{1}{\nu})}(1 - e^{-\frac{\lambda}{\beta}}) + \frac{e^{-\lambda(\frac{1}{\xi}+\frac{1}{\nu})}(e^{-\frac{\lambda}{\beta}})^M}{(1 - e^{-\lambda(\frac{1}{\gamma}+\frac{1}{\xi})})}(1 - e^{-\frac{\lambda}{\gamma}})e^{-\frac{\lambda}{\mu}}\right],$$

$$\phi_{I_{0,1}} = \frac{1}{G}(1 - e^{-\frac{\lambda}{\mu}}),$$

$$\phi_L = \frac{1}{G}\frac{e^{-\lambda(\frac{1}{\mu}+\frac{1}{\nu})}(e^{-\frac{\lambda}{\beta}})^M}{(1 - e^{-\lambda(\frac{1}{\gamma}+\frac{1}{\xi})})}(1 - e^{-\frac{\lambda}{\xi}}),$$

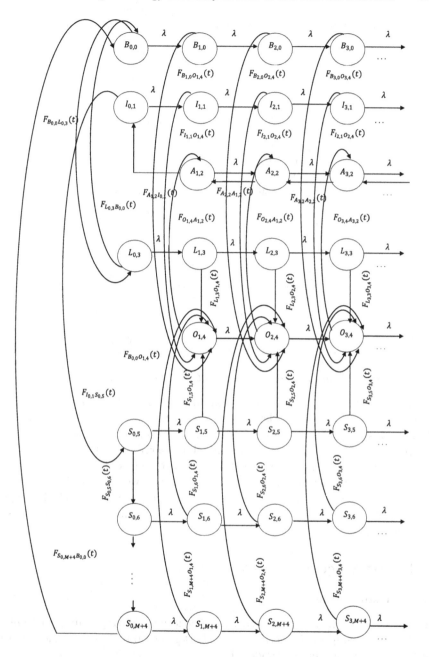

Fig. 5. State transition diagram for the proposed MRGP model

$$\phi_{O_{i,4}} = \frac{1}{G}\left[(1-e^{\frac{\lambda}{\nu}})^i e^{-\frac{\lambda}{\mu}}(1-e^{-\frac{\lambda}{\nu}}) + (e^{-\frac{\lambda}{\beta}})^{j-5}(1-e^{-\frac{\lambda}{\beta}})^i e^{-\lambda(\frac{1}{\nu}+\frac{1}{\mu})}(1-e^{-\frac{\lambda}{\beta}})\right.$$

$$\left. + (1-e^{-\frac{\lambda}{\zeta}})^i \frac{e^{-\frac{\lambda}{\nu}}(e^{-\frac{\lambda}{\beta}})^M}{(1-e^{-\lambda(\frac{1}{\zeta}+\frac{1}{\gamma})})}e^{-\frac{\lambda}{\mu}}(1-e^{-\frac{\lambda}{\zeta}})\right]; i = 0,1,\ldots, j = 5,6,\ldots, M+4,$$

Table 3. Cumulative distribution functions with parameters

CDF	parameter
$F_{B_{i,j}O_{i,j+4}}(t); i = 1, 2, ldots, j = 0$	$\frac{1}{\zeta}$
$F_{B_{0,0}L_{0,3}}(t)$	$\frac{1}{\zeta}$
$F_{I_{0,1}S_{0,5}}(t)$	$\frac{1}{\nu}$
$F_{I_{i,j}O_{i,j+3}}(t);$ $i = 1, 2, \ldots, j = 1$	$\frac{1}{\nu}$
$F_{A_{1,2}I_{0,1}}(t)$	$\frac{1}{\mu}$
$F_{S_{i,j}O_{i,j-k}}(t);$ $i = 1, 2, \ldots, j - k = 4$ $j = 5, 6, \ldots, M + 4, k = 1, 2, \ldots, M$	$\frac{1}{\beta}$
$F_{A_{i,2}A_{i-1,2}}(t); i = 2, 3, \ldots,$	$\frac{1}{\mu}$
$F_{L_{0,3}B_{0,0}}(t)$	$\frac{1}{\gamma}$
$F_{L_{i,j}O_{i,j+1}}(t); i = 1, 2, \ldots, j = 3$	$\frac{1}{\gamma}$
$F_{O_{i,j}A_{i,j-2}}(t)$	$\frac{1}{\alpha}$
$F_{S_{0,j}S_{0,j+1}}(t); j = 5, 6, \ldots, M + 4$	$\frac{1}{\beta}$
$F_{S_{0,M+4}B_{0,0}}(t)$	$\frac{1}{\beta}$

$$\phi_S = \frac{1}{G}\left[e^{-\frac{\lambda}{\mu}}\{(1 - e^{-\frac{\lambda}{\nu}}) + e^{-\frac{\lambda}{\nu}}(1 - (e^{-\frac{\lambda}{\beta}})^M)\}\right],$$

for the value of G and the steady state probability $\phi_{A_{i,2}}$, refer [23].

In this context, the system state probabilities ϕ_n for all states n in the state space Ω exist when $\lambda < \mu$. Please note: Given that we assume packet arrivals follow a Poisson process, service times follow a deterministic distribution, and the system operates as an infinite capacity single-server model, it conforms to an $M/D/1$ queue. Referring to the steady-state probabilities of the Erlang's Model C ($M/D/1$) queue, as mentioned in [10], page 261, we can conclude that the condition for system stability requires that the traffic intensity ρ (which is equal to $\frac{\lambda}{\mu}$) should be less than 1. Therefore, the stability condition for the proposed model is $\lambda < \mu$.

3   Performance Measures

3.1   Markov Model

We use the notations A_M, S_M, and L_M to denote the steady-state probabilities of the system being in the active state, short sleep state, and long sleep state, respectively, for the Markov model. Then,

$$A_M = \sum_{n=0}^{\infty} \pi_{n,0} = \left\{1 - \left[\frac{\lambda}{(\mu-\lambda)} + \frac{\beta((\alpha-\lambda)^N - \alpha^N)}{\alpha^N(\mu-\lambda)}\right]\right.$$
$$\left. + \frac{\beta(\alpha-\lambda)^N}{(\mu-\lambda)\alpha^N}\left[\frac{(\nu-\lambda)b^4}{(\nu b^2-\lambda)(1-b^2)} - \frac{\lambda^2}{\nu(\nu b^2-\lambda)} - \frac{b^2}{1-b^2} + \frac{\lambda^2}{\nu(\nu-\lambda)}\right]\right\}\pi_{0,0},$$

$$S_M = \sum_{j=1}^{N} \pi_{0,j} = \frac{\beta(\alpha^N - (\alpha - \lambda)^N)}{\lambda \alpha^N} \pi_{0,0},$$

$$L_M = \sum_{n=0}^{\infty} \pi_{n,N+1} = \frac{\beta(\alpha - \lambda)^N}{(\nu - \lambda)\alpha^N} \pi_{0,0}.$$

Let $E[N_M]$ denote the mean number of packets in the system, and let $E[N_{A_M}]$ and $E[N_{S_M}]$ denote the mean number of packets in the active mode and sleep mode, respectively. Then,

$$E[N_M] = E[N_{A_M}] + E[N_{S_M}],$$

where

$$E[N_{A_M}] = \sum_{n=0}^{\infty} n\pi_{n,0} = \left\{ \frac{\mu}{(\lambda - \mu)} \left[\frac{\lambda}{(\mu - \lambda)} \right.\right.$$

$$\left. + \frac{\beta((\alpha - \lambda)^N - \alpha^N)}{\alpha^N(\mu - \lambda)} \right] + \frac{\beta(\alpha - \lambda)^N}{(\mu - \lambda)\alpha^N} \left[\frac{(\nu - \lambda)b^4}{(\nu b^2 - \lambda)(1 - b^2)^2} \right.$$

$$\left.\left. - \frac{\lambda^2}{(\nu - \lambda)(\nu b^2 - \lambda)} - \frac{b^2}{(1 - b^2)^2} + \frac{\lambda^2}{(\nu - \lambda)^2} \right] \right\} \pi_{0,0},$$

$$E[N_{S_M}] = \sum_{n=0}^{\infty} n\pi_{n,N+1} = \frac{\beta\lambda(\alpha - \lambda)^N}{(\nu - \lambda)^2 \alpha^N} \pi_{0,0}.$$

The energy-saving factor is determined by calculating the percentage of time the UE remains in the sleep state, which involves comparing the time spent in a sleep period to the overall time spent in all states. To compute this energy-saving factor, we utilize steady-state probabilities denoted as $\pi_{i,j}$ for values of $i = 0, 1, 2, ...$ and $j = 1, 2, ..., N$. The formula for calculating the energy-saving factor E_S can be found in [9,11]:

$$E_S = \sum_{j=1}^{N} \pi_{0,j} + \sum_{i=0}^{\infty} \pi_{i,N+1}$$

$$= \left\{ \frac{\beta(\alpha^N - (\alpha - \lambda)^N)}{\lambda \alpha^N} + \frac{\beta(\alpha - \lambda)^N}{(\nu - \lambda)\alpha^N} \right\} \pi_{0,0}.$$

3.2   Semi-Markov Model

Let A_{SM}, S_{SM}, and L_{SM} represent the steady-state probabilities of being in the active state, short sleep, and long sleep, respectively, for the semi-Markov model in the long run. Then

$$A_{SM} = \pi_{I_0} + \sum_{k=1}^{\infty} \pi_{W_k} = \frac{1}{M}\left[\frac{\mu}{\lambda(\mu - \lambda)} - \frac{e^{-\left(\frac{\lambda}{\beta} + \frac{\lambda N}{\alpha} + \frac{\lambda}{\nu}\right)}}{\mu - \lambda} - \frac{e^{-\frac{\lambda}{\beta}}}{\lambda} \right.$$

$$\left. - \frac{\lambda}{\mu}\frac{e^{-\left(\frac{\lambda N}{\alpha} + \frac{\lambda}{\beta}\right)}\left(1 - e^{-\frac{\lambda}{\nu}}\right)^2}{\left(\frac{\lambda}{\mu} - 1 + e^{-\frac{\lambda}{\nu}}\right)}\left[\frac{e^{\frac{\lambda}{\mu}}}{\mu} - \frac{1}{\mu - \lambda} \right] \right],$$

$$S_{SM} = \sum_{j=1}^{N} \pi_{S_j} = \frac{1}{M}\left[\frac{e^{-\frac{\lambda}{\beta}}\left(1 - e^{-\frac{\lambda N}{\alpha}}\right)}{\lambda}\right],$$

$$L_{SM} = \sum_{i=0}^{\infty} \pi_{L_i} = \frac{1}{M}\left[\frac{e^{-\left(\frac{\lambda N}{\alpha} + \frac{\lambda}{\beta}\right)}\left(e^{\frac{\lambda}{\nu}} - 1\right)}{\lambda}\right].$$

Let $E[N_{SM}]$ denote the mean number of packets in the system, and let $E[N_{A_{SM}}]$ and $E[N_{S_{SM}}]$ represent the mean number of packets in active mode and sleep mode, respectively. Then,

$$E[N_{SM}] = E[N_{A_{SM}}] + E[N_{S_{SM}}],$$

where

$$E[N_{A_{SM}}] = \sum_{n=1}^{\infty} n\pi_{W_n} = \frac{1}{M}\left[\frac{\mu\left(1 - e^{-\left(\frac{\lambda}{\beta} + \frac{\lambda N}{\alpha} + \frac{\lambda}{\nu}\right)}\right)}{(\mu - \lambda)^2}\right.$$
$$\left. + \frac{\lambda}{\mu^2}\frac{e^{-\left(\frac{\lambda N}{\alpha} + \frac{\lambda}{\beta}\right)}\left(1 - e^{-\frac{\lambda}{\nu}}\right)}{\left(\frac{\lambda}{\mu} - 1 + e^{-\frac{\lambda}{\nu}}\right)}\left[1 - e^{\frac{2\lambda}{\nu}} - \frac{\left(1 - e^{-\frac{\lambda}{\nu}}\right)\left(2 - \frac{\lambda}{\mu}\right)}{\left(1 - \frac{\lambda}{\mu}\right)^2}\right]\right],$$

$$E[N_{S_{SM}}] = \sum_{n=0}^{\infty} n\pi_{L_n} = \frac{1}{M}\left[\frac{e^{-\left(\frac{\lambda N}{\alpha} + \frac{\lambda}{\beta}\right)}e^{\frac{2\lambda}{\nu}}\left(1 - e^{\frac{\lambda}{\nu}}\right)^2}{\lambda}\right].$$

The energy saving factor is represented as the percentage of time the UE spends in the sleep state, which is the ratio of the time the UE spends in a sleep period to the total time across all the states. Therefore, using steady-state probabilities π_m for $m \in \Omega$, the energy saving factor E_S is computed as:

$$E_S = \sum_{j=1}^{N}\pi_{S_j} + \sum_{i=0}^{\infty}\pi_{L_i}$$
$$= \frac{1}{M}\left[\frac{e^{-\frac{\lambda}{\beta}}\left(1 - e^{-\frac{\lambda N}{\alpha}}\right)}{\lambda} + \frac{e^{-\left(\frac{\lambda N}{\alpha} + \frac{\lambda}{\beta}\right)}\left(e^{\frac{\lambda}{\nu}} - 1\right)}{\lambda}\right].$$

3.3   Markov Regenrative Process Model

Let P_L, P_B, P_I, P_S, P_A, and P_O represent the steady-state probabilities for being in the long sleep state, beam searching state, inactive state, short sleep state, active state, and ON state, respectively.

$$P_L = \frac{1}{G}\frac{e^{-\lambda\left(\frac{1}{\mu} + \frac{1}{\nu}\right)}\left(e^{-\frac{\lambda}{\beta}}\right)^M}{\left(1 - e^{-\lambda\left(\frac{1}{\gamma} + \frac{1}{\zeta}\right)}\right)}\left(1 - e^{-\frac{\lambda}{\zeta}}\right),$$

$$P_B = \frac{1}{G}\left[\left(e^{-\frac{\lambda}{\beta}}\right)^{M-1}e^{-\lambda\left(\frac{1}{\mu} + \frac{1}{\nu}\right)}\left(1 - e^{-\frac{\lambda}{\beta}}\right) + \frac{e^{-\lambda\left(\frac{1}{\zeta} + \frac{1}{\nu}\right)}\left(e^{-\frac{\lambda}{\beta}}\right)^M}{\left(1 - e^{-\lambda\left(\frac{1}{\gamma} + \frac{1}{\zeta}\right)}\right)}\left(1 - e^{-\frac{\lambda}{\gamma}}\right)e^{-\frac{\lambda}{\mu}}\right],$$

$$P_I = \frac{1}{G}\left(1 - e^{-\frac{\lambda}{\mu}}\right),$$

$$P_S = \frac{1}{G}\left[e^{-\frac{\lambda}{\mu}}\{(1-e^{-\frac{\lambda}{\nu}})+e^{-\frac{\lambda}{\nu}}(1-(e^{-\frac{\lambda}{\beta}})^M)\}\right],$$

$$P_O = \frac{1}{G}bigl[e^{\frac{\lambda}{\nu}}e^{-\frac{\lambda}{\mu}}(1-e^{-\frac{\lambda}{\nu}})+e^{\frac{\lambda}{\beta}}\frac{(1-(e^{-\frac{\lambda}{\beta}})^M)}{1-e^{-\frac{\lambda}{\beta}}}e^{-\lambda(\frac{1}{\mu}+\frac{1}{\nu})}(1-e^{-\frac{\lambda}{\beta}})$$

$$+\frac{e^{-\lambda(\frac{1}{\nu}+\frac{1}{\mu})}e^{\frac{\lambda}{\xi}}(e^{-\frac{\lambda}{\beta}})^M}{(1-e^{-\lambda(\frac{1}{\xi}+\frac{1}{\gamma})})}(1-e^{-\frac{\lambda}{\xi}})],$$

$$P_A = \sum_{i=6}^{\infty}\phi_{A_{i,2}}$$

Now, we calculate the energy-saving factor (ES), defined as the percentage of time the device (UE) spends in the sleep state, including short sleep and long sleep periods.

$$ES = (P_L + P_S) \times 100\%.$$

The average energy consumption is the power consumed by the UE's transceiver, and it varies in different states [7]. Let EC_L, EC_S, EC_A, EC_I, and EC_B, and EC_O represent the energy consumed during the long sleep state, short sleep state, active state, inactivity timer, beam searching state, and ON state, respectively.

Hence, we obtain the average energy consumption by UE

$$E_{cons} = P_L EC_L + P_S EC_S + P_A EC_A + P_I EC_I + P_B EC_B + P_O EC_O.$$

The energy consumption in the MD-DRX mechanism in LTE-5G is demonstrated in Fig. 6.

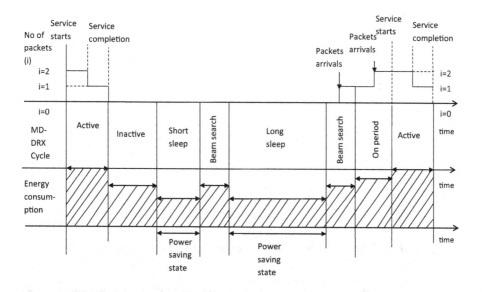

Fig. 6. Energy consumption in MD-DRX mechanism

The throughput of the system is a crucial factor in determining the optimal energy consumption and energy-saving strategies. It represents the efficiency of the system in serving customers. To calculate the system's throughput, we consider the fraction of time the system is available to customers, multiplied by the servicing rate in the case of an infinite capacity model [8]. Therefore, the system's throughput is defined as follows:

$$\text{Throughput} = P_A \times \mu,$$

where μ is the service rate.

4   Numerical Illustration

To create the graphs, we've established fixed parameter values that adhere to the stability conditions of our proposed models, which are $\frac{\lambda}{\mu} < 1$, $\frac{\lambda}{\nu} < 1$, and $\alpha \geq \lambda$, with $\beta > 0$ and $\lambda > 0$. The specific parameter values are as follows: $\alpha = 1$, $\beta = 1.2$, $\lambda = 0.5$, $\mu = 1.5$, and $\nu = 1.8$. In order to present the results of the Markov model, we've labeled it as MM with varying parameter values, and for the Semi-Markov model, we've denoted it as SMM with the corresponding parameter variations. The energy-saving factor is plotted against the values of N in Fig. 7, keeping all other parameter values constant. Notably, in the case of the Semi-Markov model, the power-saving factor is lower compared to the Markov model, as the steady-state probability of the UE being in the active state is higher. However, this factor stabilizes as the value of N increases.

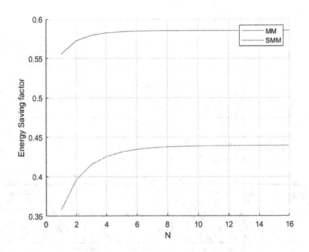

Fig. 7. Energy Saving factor varying number of short cycles

The percentage difference in measurements between both models is calculated as the measured value reaches its steady-state value. To facilitate comparison, we've used the following parameter values for both models: $\alpha = 1$, $\beta = 1.2$, $\lambda = 0.5$, $\mu = 1.5$, and $\nu = 1.8$. The comparison between the energy-saving factors for both models is presented in Table 3. Our observations from Table 4 reveal that the energy-saving factor approaches its steady-state value for the Markov model when $N \geq 12$ and for the semi-Markov model when $N \geq 16$. Furthermore, we've noted that the energy-saving factor is at least 33.19% higher for the Markov model compared to the semi-Markov model. Hence, the maximization of energy savings in the UE can be achieved through the Markov modeling of the DRX mechanism.

Table 4. Values for energy saving factor by varying number of short cycles

N	Markov model (MM)	Semi-Markov model (SMM)	% Difference between MM and SMM
8	0.5852	0.4377	33.69
9	0.5853	0.4384	33.50
10	0.5853	0.4388	33.38
11	0.5853	0.4391	33.29
12	0.5854	0.4393	33.25
13	0.5854	0.4393	33.25
14	0.5854	0.4394	33.22
15	0.5854	0.4394	33.22
16	0.5854	0.4395	33.19
17	0.5854	0.4395	33.19
18	0.5854	0.4395	33.19

Using parameter values of $\nu = 0.02$, $\mu = 0.05$, $\zeta = 0.01$, $\alpha = 1$, $\gamma = 0.0125$, and $M^* = 6$, we have generated a plot illustrating the trade-off between the energy-saving factor and throughput. This plot varies the arrival rate from 0.001 to 0.004 for the MRGP model. Key observations from Fig. 8 include:

- As the arrival rate (λ) ranges from 0.001 to 0.004, it's observed that the energy-saving factor (ES) decreases, while the throughput increases, which aligns with the expected behavior.
- As the short sleep duration ($\frac{1}{\beta}$) increases, it is observed that the energy-saving factor (ES) also increases, while the throughput decreases. This behavior can be attributed to the fact that during longer short sleep durations, fewer components of the UE remain active, resulting in increased energy savings for the UE.

Using the same parameters as in Fig. 8, we have plotted the trade-off between average energy consumption and throughput, varying the arrival rate from 0.001

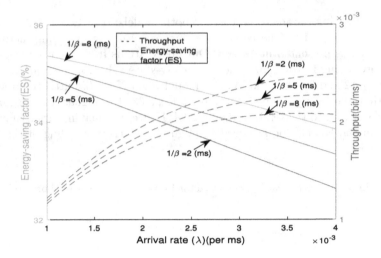

Fig. 8. Trade-off between Energy-saving factor and throughput as arrival rate varies

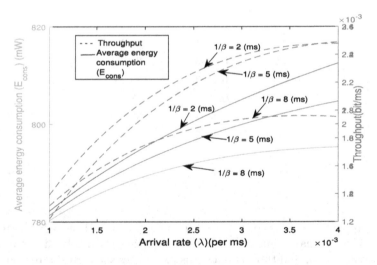

Fig. 9. Trade-off between average energy consumption and throughput as arrival rate varies

to 0.004 for the MRGP model. The energy consumption values in different states are as follows: $EC_L = 11.4$, $EC_S = 11.4$, $EC_A = 1680.2$, $EC_I = 1060$, $EC_B = 1060$, $EC_O = 1280.04$ [7]. From Fig. 9, the following observations can be made:

- As the arrival rate (λ) varies from 0.001 to 0.004, both the average energy consumption (E_{cons}) and throughput increase as expected.
- As the short sleep duration $\left(\frac{1}{\beta}\right)$ increases, the average energy consumption (E_{cons}) decreases, resulting in a decrease in throughput. This occurs because, in a longer short sleep duration, only a few components of the UE are active,

while most of the components remain inactive. As a result, the longer short sleep duration reduces the average energy consumption (E_{cons}) of the UE, but it also decreases the throughput.

5   Conclusion and Future Work

In this study, we delve into the exploration of stochastic models designed to investigate User Equipment (UE) energy-saving mechanisms, with a particular emphasis on the DRX (Discontinuous Reception) mechanism. Our research introduces a variety of continuous-time models grounded in different assumptions and DRX states. We conduct analyses centered on energy savings within the UE, making use of steady-state probabilities and expected sojourn times as key metrics. Our results indicate that the Markov model outperforms the semi-Markov model in terms of overall performance.

Furthermore, we present an MRGP (Markov Regenerative Process) model tailored specifically for MD-DRX, providing insights into the intricate trade-offs between energy-saving, energy consumption, and throughput. We carry out a comparative evaluation of our proposed MRGP model against existing models, underlining its superior accuracy and capability for real-time simulations.

Our future research endeavors will focus on potential improvements by exploring different state sojourn time distributions and extending the modeling framework to encompass LTE-A, 5G, and NR networks. Additionally, we plan to develop a validation testbed for the DRX mechanism and conduct comparative studies with other LTE-DRX mechanisms, with the ultimate goal of integrating our findings into the evolving landscape of LTE-6G networks.

Acknowledgement. One of the authors S. Dharmaraja thanks Bharti Airtel Limited, India, for financial support in this research work.

References

1. Maheshwari, M.K., Agiwal, M., Saxena, N., Roy, A.: Directional discontinuous reception (DDRX) for mmWave enabled 5G communications. IEEE Trans. Mob. Comput. **18**(10), 2330–2343 (2018)
2. Memon, M.L., Maheshwari, M.K., Saxena, N., Roy, A., Shin, D.R.: Artificial intelligence-based discontinuous reception for energy saving in 5G networks. Electronics **8**(7), 1–19 (2019)
3. Gautam, A., Choudhury, G., Dharmaraja, S.: Performance analysis of DRX mechanism using batch arrival vacation queueing system with N-policy in LTE-A networks. Annals Telecommun. **75**, 353–367 (2020)
4. Gautam, A., Dharmaraja, S.: Selection of DRX scheme for voice Traffic in LTE-A networks: Markov modeling and performance analysis. J. Indust. Manag. Optimiz. **15**, 739–756 (2019)
5. Tseng, C., Wang, H., Kuo, F., Ting, K., Chen, H., Chen, G.: Delay and power consumption in LTE/LTE-A DRX mechanism with mixed short and long cycles. Proc. IEEE Trans. Veh. Technol. **65**(3), 1721–1734 (2015)

6. Kolding, T.E., Wigard, J., Dalsgaard, L.: Balancing power saving and single user experience with discontinuous reception in LTE. In: IEEE International Symposium on Wireless Communication Systems, pp. 713–717 (2008)

7. Maheshwari, M.K., Agiwal, M., Rashid Masud, A.: Analytical modeling for signaling-based DRX in 5G communication. Trans. Emerg. Telecommun. Technol. **32**(1), 1–18 (2021)

8. Thomasian, A.: Vacationing server model for M/G/1 queues for rebuild processing in raid5 and threshold scheduling for readers and writers. Inf. Process. Lett. **135**, 41–46 (2018)

9. Maheshwari, M.K., Agiwal, M., Masud, A.R.: Analytical modeling of DRX with flexible TTI for 5G communications. Trans. Emerg. Telecommun. Technol. **29**(2), e3275 (2018)

10. Medhi, J.: Stochastic Models in Queueing Theory. Elsevier (2002)

11. Maheshwari, M.K., Agiwal, M., Masud, A.R.: Analytical modeling for signaling-based DRX in 5G communication. Trans. Emerg. Telecommun. Technol. (2021)

12. Kulkarni, V.G.: Modelling and Analysis of Stochastic Systems. Chapman & Hall, London (1995)

13. Castaneda, L.B., Arunachalam, V., Dharmaraja, S.: Introduction to Probability and Stochastic Processes with Applications. Wiley, New Jersey (2012)

14. Mihov, Y.Y., Kassev, K.M., Tsankov, B.P.: Analysis and performance evaluation of the DRX mechanism for power saving in LTE. In: IEEE Convention of Electrical and Electronics Engineers in Israel, Eilat, pp. 520–524 (2010)

15. Jayadia, R., Lai, Y-C., Chen, L-C.: An interleaved-sleeping-listening scheduler for power saving in mobile stations. Comput. Electric. Eng. **67**, 278–290 (2018)

16. Wang, C., Li, C.M., Ting, K.C.: Energy-efficient and QoS-aware discontinuous reception using a multi-cycle mechanism in 3GPP LTE/LTE-advanced. Telecommun. Syst. **64**, 599–615 (2017)

17. Yin, F.: An application aware discontinuous reception mechanism in LTE-advanced with carrier aggregation consideration. Ann. Telecommun. **67**, 147–159 (2012)

18. Wu, J., Park, J.: Analysis of discontinuous reception (DRX) on energy efficiency and transmission delay with bursty packet data traffic. Annals Telecommun. **76**, 429–446 (2021)

19. Chavarria-Reyes, E., Fadel, E., Almasri, S., Malik, M.G.A.: Reducing the energy consumption at the user equipment through multi-stream crosscarrier-aware discontinuous reception (DRX) in LTE-advanced systems. J. Netw. Comput. Appl. **64**, 43–61 (2016)

20. Arunsundar, B., Sakthivel, P., Natarajan, E.: Analysis of energy consumption and latency in advanced wireless networks through DRX mechanism. J. Supercomput. **76**, 3765–3787 (2020)

21. Philip, N.R., Malarkodi, B.: Extended hybrid directional DRX with auxiliary active cycles for light traffic in 5G networks. Trans. Emerg. Telecommun. Technol. **30**(1), 1–14 (2019)

22. Sallam, M.M., Nafea, H.B., Zaki, F.W.: Comparative study of power saving and delay in LTE DRX, directional-DRX and hybrid-directional DRX. Wirel. Pers. Commun. **98**(4), 3299–3317 (2018)

23. Kalita, P., Dharmaraja, S.: Stochastic modelling for energy efficiency in modified discontinuous reception (MD-DRX) for LTE-5G networks. Int. J. Commun. Syst. **36**(6), e5434 (2023). 33 pages

24. Dharmaraja, S., Aggarwal, A., Sudhesh, R.: Analysis of energy saving in user equipment in LTE-A using stochastic modelling. Telecommun. Syst. **80**, 123–140 (2022)

Analyzing Reliability Metrics
of All-Optical Switches

E. A. Barabanova[1][(✉)] ⓘ, K. A. Vytovtov[1] ⓘ, and A. N. Fedorovskaya[2] ⓘ

[1] V. A. Trapeznikov Institute of Control Sciences of RAS, Profsoyuznaya Street 65,
08544 Moscow, Russia
`elizavetaalexb@yandex.ru`
[2] Astrakhan State Technical University, Tatishchev Street 16, 414056 Astrakhan,
Russia

Abstract. The reliability of all-optical switches based on new approach is analyzed. The proposed approach allows to take into account
three main criteria influencing on switch functioning such as switch
architecture, switch technology and method of switching control. The
description of all-optical switch schemes are presented. The mathematical expressions of the reliability functions and the mean time to failure
of well-known all-optical switches have been obtained. The models allow
comparing the reliability of all-optical switches with different number
of inputs and based on well-known architectures and switching fabric
technologies. The numerical results of reliability functions calculation
for Close, Banyan, Dual and Crossbar switches based on electro-optical
basic elements are presented. A comparison of reliability functions of
Dual and Banyan switches with different types of basic elements has
been considered.

Keywords: all-optical switches · a reliability function · a reliability
block diagram · mean time to failure

1 Introduction

All-optical networks are the perspective new generation networks with low
latency and high bandwidth. The main element of an all-optical network is an
all-optical switch which is characterized by such parameters as switching speed,
insertion loss, crosstalk, scheme complexity, throughput and reliability [1–6].

Today, all-optical switching is practically poorly implemented. One of the
reasons is the poor knowledge of the basic parameters of all-optical switches and
the lack of adequate mathematical models for calculating these parameters.

To date, there are a number of approaches that allow calculating the reliability metrics of switching systems such as reliability function and mean time

The reported study was funded by Russian Science Foundation, project number 23-29-
00795, https://rscf.ru/en/project/23-29-00795/.

to failure based on analysis of their architecture [7,8]. The reliability metrics of multistage switching systems have been calculated in [7]. The 3D model of the all-optical MEMS-switch and his operating mode block diagram is used for investigating reliability of the optical switch in [8]. To evaluate the reliability of various optical network-on-chip architectures, a simulator called "Reliability Assessment of Photonic Network-on-Chip" is proposed [9]. A high-level Python-based user simulator is proposed and used to assess the reliability of optical networks on a chip [10]. A new architecture of optical cross-connection is proposed to improve the reliability parameters of photonic networks. The proposed architecture can provide port scalability and failure resiliency at the same time [12]. Traditional plasmonic waveguides have high losses due to the presence of a metal layer in nanostructures. To reduce light propagation losses, a new photonic switch in a hybrid plasmonic waveguide is proposed [13]. To reduce the insertion loss and crosstalk of photonic devices, a thermally tuned symmetric optical switch (TTSOS) is proposed and its thermal effects are analyzed [6]. TTSOS improves the reliability and performance of multi-stage optical networks through temperature modification.

A separate problem in reliability theory is the study of non-stationary reliability characteristics of systems. The reliability of all-optical switches in transient mode has been investigated in [4]. The authors have analyzed the reliability function depending on the number of redundant switching elements.

It should be noted that for the accurate evaluating the all-optical switch reliability metrics beside of the switch architecture it must be taking into account the other important features such a method of switching scheme controlling and the type of its basic switching elements.

In this paper the new mathematical models for reliability analysis of all-optical switches are presented. The authors propose the approach taking into account the impact of several characteristics of all-optical switches on their reliability metrics at once. Such characteristics are a switch architecture, a switching control method, and a switch technology.

2   The Reliability Block Diagrams of All-Optical Switches

First of all, to calculating and analyzing the reliability of all-optical switches, it is necessary to develop their reliability block diagrams. In this work we consider well-known switching schemes that can be used for constructing all-optical switches such as Close [3,14,15], Banyan [11], Dual [5] and crossbar switches [7]. The switch schemes and corresponded them reliability block diagrams are presented in Figs. 1, 2, 3, 4, 5, 6 and 7.

The reliability block diagram of Close scheme consists of basic switching elements of three stages and control device connected in series. The m basic switching elements of central stage are connected in parallel. They are necessary for providing strictly non-blocking switching function but also can be considered as redundancy blocks. The minimum number of such elements is determined by using following rule: $m = 2n - 1$, where $n = \sqrt{N/2}$ is the optimal number of inputs of the first stage switching blocks [3].

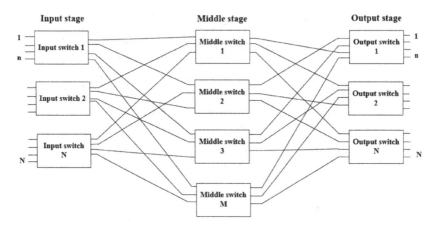

Fig. 1. The Close scheme with centralized control

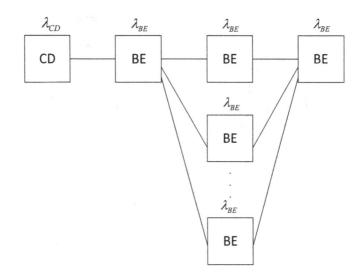

Fig. 2. The reliability block diagram of a Close switch

As Banyan (Fig. 3) and Dual (Fig. 4) schemes are decentralized control ones their reliability block diagrams do not include the control device block and contain only M basic switching elements connected in series, where M is equal to the number of switch stages (Fig. 5). For the Banyan scheme there is the following relation between the number of switch stages M and the number of switch inputs N: $M = \log_2 N$ [3]. For Dual one this relationship has the form: $M = 0,5 \log_2 N$.

The scheme and the reliability block diagram of crossbar switch are presented in Fig. 6 and Fig. 7 accordingly. The reliability block diagram consists of $2N - 1$ basic elements which are necessary for the switching function in the case of the

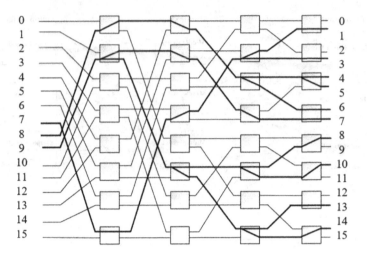

Fig. 3. The Banyan scheme with decentralized control

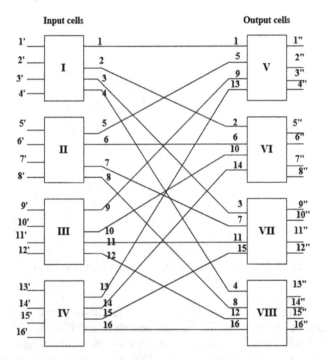

Fig. 4. The Dual switch with decentralized control

worst case. The worst case is the case when N input can be connected with the N output.

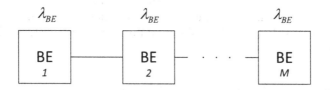

Fig. 5. The reliability block diagram of a Banyan and a Dual switches

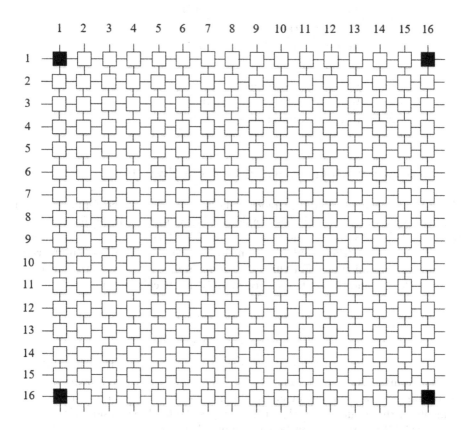

Fig. 6. The Crossbar scheme with centralized control

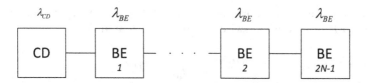

Fig. 7. The reliability block diagram of a crossbar switch

After analyzing recent work on calculating the failure rate of optical switching elements based on the most commonly used technologies [16,17], a comparative Table 1 was compiled. The values of failure rates of Table 1 were used in subsequent numerical calculations.

Table 1. Failure rates of basic optical switching elements based on different technologies

Type of switch	Failure rate $\lambda, 1/h$
Electro-optical	10^{-9}
Acousto-optical	10^{-8}
Thermo-optical	10^{-7}
Electromechanical(MEMS)	10^{-11}
Liquid crystal	10^{-8}
Electronic control device	10^{-7}

3   Reliability Metrics of All-Optical Switches

Using the developed reliability block diagrams as well as the failure rates of different types of optical switches (Table 1) it is possible to calculate a reliability function $R(t)$ and a mean time to failure $MTTF$ (Table 2).

Let us consider the simplest exponential failure model when failure rates are constant for all times. For this case the reliability function of the switching element $R_{BE}(t)$ and the control device $R_{CD}(t)$ can be calculated using formulas:

$$R_{BE}(t) = exp(-\lambda_{BE}t) \tag{1}$$

$$R_{CD}(t) = exp(-\lambda_{CD}t) \tag{2}$$

where λ_{BE} and λ_{CD} are failure rates of the base element and the control device correspondingly. As the reliability block diagram of the Close scheme includes m redundancy switching elements in the intermediate stage and two switching elements in the input and in the output stages the final expression of the Close switch reliability function can be written as following

$$R(t) = (1 - (1 - exp(-\lambda_{BE}t))^m) \cdot exp(-(2\lambda_{BE} + \lambda_{CD})t) \tag{3}$$

Using the well-known formula from the reliability theory [1] the mean time to failure $MTTF$ of all-optical switches can be calculated as

$$MTTF = \int_0^\infty R(t)\,dt \tag{4}$$

Analogously the reliability functions of the other all-optical switching systems have been obtained and presented in Table 2.

Table 2. The reliability metrics of all-optical switches

Switch Arhitecter	$R(t)$	$MTTF$
Close	$(1 - (1 - exp(-\lambda_{BE}t))^{\sqrt{2N}-1}) \times$ $\times exp(-(2\lambda_{BE} + \lambda_{CD})t)$	$\frac{1}{2\lambda_{BE}+\lambda_{CD}}$ $- \int_0^\infty (1 - exp(-\lambda_{BE}t))^{\sqrt{2N}-1} \times$ $\times exp(-(\lambda_{BE} + \lambda_{CD})t)dt$
Dual	$exp(-0.5\lambda_{BE} \cdot \log_2 N \cdot t)$	$\frac{1}{0.5 \log_2 N \cdot \lambda_{BE}}$
Banyan	$exp(-\lambda_{BE} \cdot \log_2 N \cdot t)$	$\frac{1}{log_2 N \cdot \lambda_{BE}}$
Crossbar	$exp(-((2N-1) \cdot \lambda_{BE} + \lambda_{CD})t)$	$\frac{1}{(2N-1) \cdot (\lambda_{BE}+\lambda_{CD})}$

At the next stage of reliability calculation, the failure rates of switching and control elements must be found. These rates depend on fabric technology and reliability metrics of electron processors [14]. The results of numerical calculation of the reliability functions for the four described above switch architectures based on electro-optical basic switching elements are presented in Fig. 8.

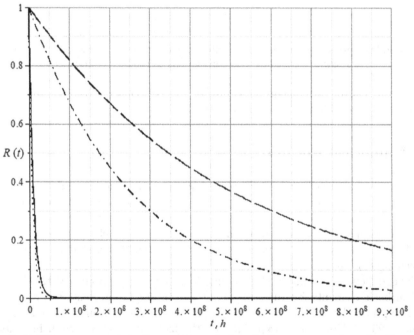

Dotted line corresponds to the crossbar switch; solid line corresponds to the Close switch; dash-dotted line corresponds to the Banyan switch; longdash line corresponds to the Dual switch

Fig. 8. Comparison of reliability functions of all-optical switches based on electro-optical basic elements

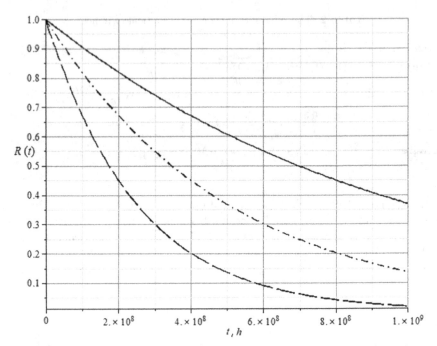

Solid line corresponds to the 4 × 4-switch; dash-dotted line corresponds to the
16 × 16-switch; longdash line corresponds to the 256 × 256-switch

Fig. 9. Comparison of reliability functions for Dual switches based on electro-optical
basic elements for different numbers of inputs

4   Numerical Results

The proposed approach and the developed mathematical models make it possible
to compare the reliability of all-optical switches based on different architectures
and different switch-fabric technologies. Comparison of reliability functions of
16 × 16-all-optical switches based on electro-optical basic elements is presented
in Fig. 8. In this case the numerical results have been shown that more reliable
switch is Dual one.

The dependencies of reliability functions for Dual switches with electro-
optical basic elements for different numbers of inputs are presented in Fig. 9.

The graph shows that the most reliable switch is 4 × 4, and the most unreliable is 256 × 256. This is explained by the fact that such a switch has more
intermediate stages, which increases complexity and, accordingly, reduces reliability.

Figure 10 shows the dependencies of reliability functions of 16 × 16 Dual
switches with different types of basic elements. As it can be seen from the graphs
the most reliable are MEMS-switches.

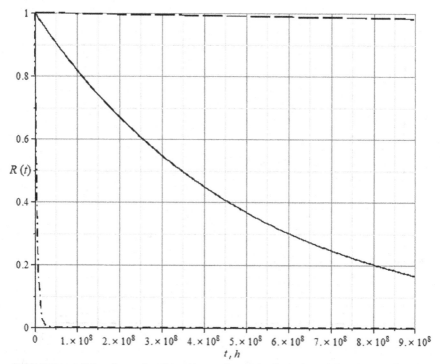

Solid line corresponds to the electro-optical switch; dash-dotted line corresponds to
the thermo-optical switch; longdash line corresponds to the MEMS-switch

Fig. 10. Comparison of reliability functions for Dual switches with different types of
basic elements

Comparison of reliability functions of 16 × 16-Dual and 16 × 16-Banyan
switches with different types of basic elements is presented in Fig. 11. The analysis of graphs has been shown that if the Banyan switch will be built on 2D
MEMS-elements [18], which are more reliable than electro-optical ones based on
$LiNbO_3$ elements [17], then it will have higher reliability metrics.

Thus it is possible to obtain such a combination of architecture/type of basic
elements, in which a switch built according to a less reliable architecture but
using more reliable basic elements will have higher reliability metrics.

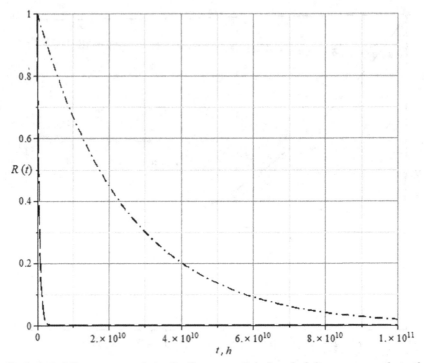

Dash-dotted line corresponds to the Banyan switch; longdash line corresponds to the Dual switch

Fig. 11. Comparison of reliability functions for Dual and Banyan switches with different types of basic elements

5   Conclusion

The main aim of this work is investigations of the all-optical switch reliability as well as comparison of all-optical switches reliability metrics. The proposed approach is based on the derivation of mathematical expressions of reliability metrics taking into account the reliability block diagrams, the control method, the operation switch algorithms and the type of basic elements. This approach can be used to obtain mathematical models of the reliability of the other switching circuits of all-optical switches, such as Benes, Spanker schemes, and others. The obtained mathematical models allow comparing a reliability of different types of all-optical switches that can accelerate practical realization of all-optical switching.

References

1. Birolini, A.: Reliability Engineering. Theory and Practice, 3rd.edn. Springer, Berlin, Heidelberg (1999). https://doi.org/10.1007/978-3-662-03792-8
2. Maier, M.: Optical Switching Networks, 1st.edn. Cambridge University Press, Great Britain (2008)
3. Clos, C.: A study of non-blocking Switching Networks. The Bell System Technical Journal, vol. 32, pp. 406–424 (1953)
4. Barabanova, E., Vytovtov, K., Vishnevsky, V., Dvorkovich, A., Shurshev, V.: Investigating the reliability of all-optical switches in transient mode. In: Journal of Physics Conference Series: 5th International Scientific Conference on Information, Control, and Communication Technologies (ICCT-2021), vol. 2091, 012039. Astrakhan, Russian Federation (2021). https://doi.org/10.1088/1742-6596/2091/1/012039
5. Barabanova, E., Vytovtov, K., Vishnevsky, V., Podlazov, V..: High-capacity strictly non-blocking optical switches based on new dual principle. In: Journal of Physics Conference Series: 5th International Scientific Conference on Information, Control, and Communication Technologies (ICCT-2021), vol. 2091, 012040. Astrakhan, Russian Federation (2021). https://doi.org/10.1088/1742-6596/2091/1/012040
6. Dehghani, F., Mohammadi, S., Barekatain, B., Abdollahi, M.: Power loss analysis in thermally-tuned nanophotonic switch for on-chip interconnect. Nano Commun. Netw. **26**, 100323 (2020). https://doi.org/10.1016/j.nancom.2020.100323
7. Bistouni, F., Jahanshahi, M.: Scalable crossbar network: a non-blocking interconnection network for large-scale systems. J. Supercomput. **71**, 697–728. Netherlands (2015). https://doi.org/10.1007/s11227-014-1319-2
8. Hasanov, M., Agayev, N., Atayev, N., Fataliyev, V.: A new generation of controlled optic switch. T-Comm. **15**(3), 64–68. Russia (2021). https://doi.org/10.36724/2072-8735-2021-15-3-64-68
9. Abdollahi, M., Baharloo, M., Shokouhinia, F., Ebrahimi, M.: RAP-NoC: reliability assessment of photonic network-on-chips, a simulator. In: Proceedings of the Eight Annual ACM International Conference on Nanoscale Computing and Communication, pp 1–7. Italy (2021). https://doi.org/10.1145/3477206.3477455
10. Baharloo, M., Abdollahi, M., Baniasadi, A.: System-level reliability assessment of optical network on chip. Microprocessors and Microsyst. **99**(3), 104843. Netherlands (2023). https://doi.org/10.1016/j.micpro.2023.104843
11. Zulfin, M., Pinem, M., Fauzi, R., Razali, M.: Reducing crosspoints on multistage switching by using batcher banyan switches. In: Proceedings of The 2nd International Conference On Advance And Scientific Innovation, pp. 484–491. Banda Aceh, Indonesia (2019). https://doi.org/10.4108/eai.18-7-2019.2288554
12. Yamakami, S., Mori, Y., Hasegawa, H., Sato, K.: Highly reliable and large-scale subsystem modular optical cross-connect. OSA Optics Express **25**(15), 17982–17994. USA (2017). https://doi.org/10.1364/OE.25.017982
13. Kumar Sahu, S., Singh, M.: High-performance all-optical hybrid plasmonic switch using zn-doped cadmium oxide. IEEE Trans. Plasma Sci. **51**(2), 605–612. USA (2023). https://doi.org/10.1109/TPS.2023.3239429
14. Kutuzov, D., Osovsky, A., Starov, D., Stukach, O., Maltseva, N., Surkov, D.: Crossbar switch arbitration with traffic control for NoC. In: 2022 International Siberian Conference on Control and Communications, SIBCON 2022 - Proceedings (2020). https://ieeexplore.ieee.org/document/10002976

15. Osovsky, A., Stukach, O., Kutuzov, D., Popov, I.: Simulation of a crossbar switch node by differential-Taylor transformation with risetime estimation. In: Moscow Workshop on Electronic and Networking Technologies, MWENT 2022 - Proceedings (2022). https://ieeexplore.ieee.org/document/9802180
16. Ma, X., Kuo. G.: Optical switching technology comparison: optical mems vs. other technologies. IEEE Optical Commun. **41**(11), pp. 16–23. USA (2003). https://doi.org/10.1109/MCOM.2003.1244924
17. Berghmans, F., Eve, S., Held, M.: Introduction to reliability of optical components and fiber optic sensors. In: Optical Waveguide Sensing and Imaging. Part of the NATO Science for Peace and Security Series Book Series, pp. 73–100 (2008)
18. Dobbelaere, P., Falta, K., Fan, L., Gloeckner, S., Patra, S.: Digital MEMS for Optical Switching. IEEE Commun. Mag., 88–95 (2002)

Investigating Transient Behavior of All-Optical Switch

Konstantin Vytovtov[1], Elizaveta Barabanova[1(✉)], Georgii Vytovtov[2], and Nikolai Antonov[3]

[1] V. A. Trapeznikov Institute of Control Sciences, RAS, 65 Profsoyuznaya Street, Moscow, Russia
elizavetaalexb@yandex.ru

[2] Astrakhan State Technical University, 16 Tatishchev Street, Astrakhan, Russia

[3] MIREA - Russian Technological University, 78 Vernadsky Avenue, Moscow, Russia

Abstract. The article considers a transient behavior of all-optical switch described using a single-channel finite buffer queuing system with Poisson input flow and exponential service time distribution. To solve the Kolmogorov equations system and find the probabilities of states in transient mode the Laplace transform is used. Analytical expressions for determining the probability of losses and the throughput of all-optical switch at a given point of time are found. The performance metrics of all-optical switch with a buffer size of two packets are analyzed tn the numerical section. The values of probability of losses, the throughput, and the transient time for different network loads in transient and stationary modes are found.

Keywords: Optical switch · Transient mode · Queueing systems · Laplace transform · State probabilities · Throughput

1 Introduction

With the development of next-generation optical networks, there has been an increased interest in their modeling. In most cases, mathematical tools from the theory of queuing systems [1,2] are employed for this purpose. Depending on the parameters of optical network devices, various types of queuing systems can be used to describe their operation [3]. Performance metrics of queuing systems, such as buffer size, number of requests in the system, loss probability, and waiting time of requests in the queue, can be described both in steady-state mode [1–3] and in transient mode [4–11]. In the steady-state, also known as the stationary mode, the system's characteristics do not depend on time. The transient mode is achieved after a certain period from the start of the system's operation or

The reported study was funded by Russian Science Foundation, project number 23-29-00795, https://rscf.ru/en/project/23-29-00795/.

V. M. Vishnevskiy et al. (Eds.): DCCN 2023, LNCS 14123, pp. 349–360, 2024.
https://doi.org/10.1007/978-3-031-50482-2_27

its restart and continues for some time before transitioning to the steady-state mode. Such modes also occur when there is a sudden increase in the flow of requests in the system or an abrupt failure of a servicing device.

It should be noted that a significant portion of research on queuing systems and the search for their performance metrics are conducted in the steady-state mode [1,3]. However, as the capacity of next-generation communication networks increases and the transition to all-optical systems occurs [2], the duration of the transient mode becomes comparable to the duration of an optical packet. Therefore developing a model for calculating packet loss probability and other performance metrics of all-optical switching devices in non-stationary conditions will allow communication operators to formulate accurate and precise requirements for the technical specifications of switching equipment, leading to improved a quality of provided services.

The critical problem of communication systems design is investigating a transient time. Evaluating the duration of transition mode based on input flow parameters and servicing device characteristics enables the optimization of system operation during the transition mode, and minimizing its impact on overall system performance.

The analytical expressions of the queuing systems transient characteristics is essential for analyzing operation of queuing systems and are necessary for solving synthesis problems for such systems.

This paper investigates the model of the all-optical switch with a finite buffer and analyzes its characteristics in the transient mode using the analytical method based on Laplace transforms. In Sect. 2 the state diagram of the system is presented, and the statement of the problem is described. The analytical method for solving Kolmogorov equations which describe the operation of the considered system in the transition mode is used in Sect. 3. The mathematical expressions for calculation of the performance metrics in transient mode are obtained in Sect. 4. The numerical results for the $M/M/1/2$ system are presented in Sect. 5.

2   Statement of the Problem

In this study, the model of all-optical switch in the transient mode is considered. The model represents a queuing system $M/M/1/N$, where the input flow is a simple arrival process with intensity λ, and the service time of requests has an exponential distribution with parameter μ. When the device is busy, a request enters the buffer if there is available space. If all N places in the buffer are occupied, the request exits the system unprocessed. The state diagram of the analyzed queuing system is presented in Fig. 1.

In Fig. 1 the state S_0 is the initial state at which the system has no requests. The state S_1 is the state when one request is processed by the server and the buffer is idle. In the state S_2 one request is processed by the server and one request is in the buffer; S_n is the state when one request is processed by the server and n requests are in the buffer; S_{n+1} is the state when one request is processed by the server, the buffer is full, and a newly received request is lost.

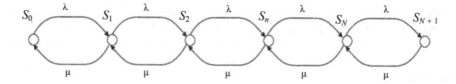

Fig. 1. Graph of states of $M/M/1/N$ system

The Kolmogorov equations describing the operation of this system have the form

$$
\begin{cases}
\dfrac{dP_0(t)}{dt} = -\lambda P_0(t) + \mu P_1(t), n = 0 \\[2ex]
\dfrac{dP_n(t)}{dt} = \lambda P_{n-1}(t) - (\lambda + \mu)P_n(t) + (n+1)\mu P_{n+1}(t),\ 1 \le n \le N \\[2ex]
\dfrac{dP_{N+1}(t)}{dt} = \lambda P_N(t) - \mu P_{N+1}(t)
\end{cases}
\tag{1}
$$

where $P_n(t)$ is the probability of the system being in state n at time t, corresponding to the probability of having n packets in the system. Thus, $P_0(t)$ is the probability of having no requests in the system, $P_{n-1}(t)$ is the probability of having $N-1$ packets in the system, and $P_{n+1}(t)$ is the probability of the buffer being fully occupied, resulting in the loss of the next arriving packet, which corresponds to the packet loss probability.

The development of an analytical model for all-optical switch with a limited buffer is required to analyze the state probabilities of its operation and its performance metrics in both steady-state and transient modes.

3   The Analytical Method

3.1   Probability Translation Matrix

The so-called probability transformation matrix [10] is the general solution of the system (1). In this section we propose the construction of the probability transformation matrix by using the Laplace transform. The advantage of this approach compared to the classical approach [10, 11] is the fact that the proposed approach does not involve finding the exponent of the coefficient matrix of (1). This fact greatly simplifies the solution. Of course, for the system of a large dimension, the roots of an algebraic equation of the Nth order can be found numerically only. However, all other expressions are analytical.

To find the final expression of the probability translation matrix, in the first step it is necessary to do the transformation from the functions of the time in (1) to their images in the complex domain by using the direct Laplace transform. Then we must solve the obtained system of the algebraic equations with respect to the probability images for some given initial conditions. Next, we need to set independent initial conditions for finding images of the probability translation matrix elements in the complex region, similar to [10,11]. Finally, it is necessary to transform from the images of probability translation matrix elements to the functions of time.

To solve the system of equations (1), we will use the Laplace transform. This approach allows us to transform the system of differential equations into the system of linear algebraic equations, that significantly simplify for solution. To begin let's writes the system (1) in matrix form

$$\frac{\vec{P}(t)}{dt} = \mathbf{B}\vec{P}(t) \tag{2}$$

where $\vec{P}(t)$ is the vector of the state probability, $\mathbf{B}$ is the 4×4 coefficient matrix of the system (1). In accordance to the method let us apply the Laplace transform to the system (2)

$$\int\limits_0^\infty \exp(-st)\frac{\vec{P}(t)}{dt}dt = \int\limits_0^\infty \exp(-st)\mathbf{B}\vec{P}(t)dt \tag{3}$$

Utilizing the property of Laplace transformation we can write

$$\begin{aligned} \frac{\vec{P}(t)}{dt} &\div s\vec{P}(s) - \vec{P}(0) \\ \mathbf{B}\vec{P}(t) &\div \mathbf{B}\vec{P}(s) \end{aligned} \tag{4}$$

where s is a complex variable; $\vec{P}(s)$ is Laplace domain representation of the vector of state probabilities $\vec{P}(t)$, $\vec{P}(0)$ is vector of initial conditions. Now (3) can be written as the algebraic equation

$$\vec{P}(s) - \vec{P}(0) = \mathbf{B}\vec{P}(s) \tag{5}$$

Thus, we obtain a non-homogeneous system of linear algebraic equations

$$(\mathbf{B} - s\mathbf{I})\vec{P}(s) = -\vec{P}(0) \tag{6}$$

where $\mathbf{I}$ is the unit diagonal matrix. Let us introduce the notation $\mathbf{A} = \mathbf{B} - s\mathbf{I}$, and write

$$\mathbf{A}\vec{P}(s) = -\vec{P}(0) \tag{7}$$

instead (6). Solving the system (7) using Cramer's method, we can obtain the vector of the state probabilities $\vec{P}(s)$ in complex domain, where the k-th element of this vector is

$$P_k(s) = \frac{\Delta_k(s)}{\Delta(s)} \tag{8}$$

where $\Delta(s)$ is the determinant of the matrix $\mathbf{A}$, $\Delta_k(s)$ is the determinant of the matrix $\mathbf{A}_k$ that obtained by replacing the k-th column of matrix $\mathbf{A}$ with the vector of initial condition $\overrightarrow{P}(s)$.

Now we must use the independent initial conditions for the construction of the translation matrix [10,11] in the complex region. It is known [10,11] that the initial conditions $\overrightarrow{P}(0) = \{1,0,0,...,0\}$ allow us to find the first column of the translation matrix, the initial conditions $\overrightarrow{P}(0) = \{0,1,0,...,0\}$ give us the second column of the matrix and so on. As the result, the elements of the probability translation matrix are

$$L_{i,j}(s) = (-1)^{i+j}\frac{\Delta_{i,j}(s)}{\Delta(s)} \tag{9}$$

where $\Delta(s)$ is the determinant of the matrix $\mathbf{A}$, $\Delta_{ji}(s)$ is the minor of the element A_{ji} of the matrix $\mathbf{A}$.

Now let us consider the inverse Laplace transform. First of all, we must note that the image of the element (9) of the probability transformation matrix is the proper fraction

$$L_{i,j}(s) = (-1)^{i+j}\frac{a_m s^m + a_{m-1}s^{m-1} + ... + a_1 s + a_0}{b_N s^N + b_{N-1}s^{N-1} + ... + b_1 s + b_0} \tag{10}$$

Moreover, $m < N$, since the numerator is the determinant of the algebraic complement of the matrix element whose determinant is in the denominator. Then the fraction in (10) can be factorized

$$L_{i,j}(s) = (-1)^{i+j}\frac{\Delta_{i,j}(s)}{\Delta(s)}$$

$$= (-1)^{i+j}\left(A_0\frac{1}{s-s_0} + A_1\frac{1}{s-s_1} + ... + A_{N-1}\frac{1}{s-s_{N-1}}\right) \tag{11}$$

$$= (-1)^{i+j}\sum_{i=0}^{N-1}A_i\frac{1}{s-s_i}$$

Let's multiply both sides of the expression by the factor $s - s_0$:

$$(-1)^{i+j}\frac{\Delta_{i,j}(s)}{\Delta(s)}(s-s_0)$$

$$= A_0 + (s-s_0)(-1)^{i+j}\sum_{i=0}^{N-1}A_i\frac{1}{s-s_i} \tag{12}$$

The right side of (12) for $x \to x_1$ is equal to A_1, since $s - s_1 \to 0$. The left side represents the uncertainty 0/0 because the factor $s - s_1$ is present in both the numerator and the denominator. Let us reveal this uncertainty by using

L'Hopital's rule and obtain the left-hand side in the form

$$(-1)^{i+j} \lim_{s\to s_1} \frac{\Delta_{i,j}(s)}{\Delta(s)}(s-s_0) = (-1)^{i+j} \lim_{s\to s_1} \frac{\Delta_{i,j}(s)+(s-s_1)\frac{d\Delta_{i,j}(s)}{ds}}{\frac{d\Delta(s)}{ds}} \tag{13}$$

$$= (-1)^{i+j} \frac{\Delta_{i,j}(s_1)}{\left.\frac{d\Delta(s)}{ds}\right|_{s=s_1}}$$

Then the coefficient A_1 can be written as

$$A_1(s) = (-1)^{i+j} \frac{\Delta_{i,j}(s_1)}{\left.\frac{d\Delta(s)}{ds}\right|_{s=s_1}} \tag{14}$$

Similarly, we find the kth coefficient in (11) as

$$A_k(s) = (-1)^{i+j} \frac{\Delta_{i,j}(s_k)}{\left.\frac{d\Delta(s)}{ds}\right|_{s=s_k}} \tag{15}$$

Thus, expression (11) can be written in the form

$$L_{i,j}(s) = \sum_{k=0}^{N-1}\left((-1)^{i+j}\frac{\Delta_{i,j}(s_k)}{\left.\frac{d\Delta(s)}{ds}\right|_{s=s_k}} \cdot \frac{1}{s-s_k}\right) \tag{16}$$

Applying the inverse Laplace transform to (16) and carrying out mathematical transformations, we obtain

$$L_{i,j}(t) = \frac{1}{2\pi i}\int_{-\infty}^{\infty}\sum_{k=0}^{N-1}\left((-1)^{i+j}\frac{\Delta_{i,j}(s_k)}{\left.\frac{d\Delta(s)}{ds}\right|_{s=s_k}} \cdot \frac{1}{s-s_k}\right)\exp(st)ds$$

$$= \sum_{k=0}^{N-1}\left((-1)^{i+j}\frac{\Delta_{i,j}(s_k)}{\left.\frac{d\Delta(s)}{ds}\right|_{s=s_k}} \cdot \exp(s_k t)\right) \tag{17}$$

Thus, the analytical expression for the elements of the probability translation matrix is obtained here. It is important to note that this matrix allows us to find the state probabilities of the queuing system at any time under any initial conditions.

4   Performance Metrics of All-Optical Switch in Transient Mode

Let's examine the key performance metrics of all- optical switch in the transient mode.

4.1   Packet Loss Probability

The packet loss probability is equal to the probability of the system being in state when the servicing device is busy, and the switch buffer is completely filled. In this state, an incoming packet is discarded. Thus, the expression for the packet loss probability is given by

$$P_{loss}(t) = \sum_{i=1}^{n} \frac{\Delta_n(s_i)}{\frac{d\Delta(s_i)}{ds}} e^{s_i t} \tag{18}$$

4.2   Throughput of All-Optical Switch in Transient Mode

As the optical switch processes incoming packets in all system states except $P_{loss}(t)$, either by immediately transmitting them to the servicing device, as in state $P_0(t)$, or by temporarily holding them in the buffer when the switch is in one of the k-th $P_k(t)$ states $k = \overline{1, n-1}$, and since the sum of the probabilities of all states equals one, the throughput capacity of the optical switch in the transient mode is determined by $A(t) = [1 - P_{loss}]\lambda$, where λ is the arrival rate of packets. The final expression for the throughput capacity of the optical switch in the transient mode, taking into account (6), is given by

$$A(t) = \left[1 - \sum_{i=1}^{n} \frac{\Delta_n(s_i)}{\frac{d\Delta(s_i)}{ds}} e^{s_i t}\right] \lambda$$

4.3   Transient Time

Transient time refers to the period from the start of the transient mode until it reaches a steady-state, i.e., a state where its characteristics can be considered constant and independent of time. To determine the transient time, it is initially necessary to calculate the time constant using the formula

$$\tau = max\{\tau_i, i = \overline{1, n+2}\} = max\left\{\frac{1}{|\alpha_i|, i = \overline{1, n+2}}\right\} \tag{19}$$

where α is the real part of the complex variable Δ_s.

Expression (7) signifies that out of all the roots of the polynomial $\Delta(s)$ it is necessary to select a nonzero root of the equation $\Delta(s) = 0$, where $s_i = \alpha_i + j\beta_i$ with the smallest real part α_i, $j\beta_i$ - imaginary part. The smallest α_i will determine the largest value of the time constant [10]. Having the time constant, the transient time can be determined using the formula $t_{tr} = (3 \div 5)\tau$.

5   Numerical Results

Let's consider a numerical calculation of the performance metrics of an optical switch using the example of a queuing system $M/M/1/2$ with a buffer size of 2 and a total number of states equal to 4. It should be noted that the chosen buffer size is determined by the characteristics of building optical delay lines, which are not intended for storing a large volume of information. The system of Kolmogorov equations describing the operation of this system takes the form

$$
\begin{cases}
\dfrac{dP_0(t)}{dt} = -\lambda P_0(t) + \mu P_1(t), n = 0 \\[2mm]
\dfrac{dP_n(t)}{dt} = \lambda P_{n-1}(t) - (\lambda + \mu)P_n(t) + (n+1)\mu P_{n+1}(t), 1 \le n \le 2 \quad (20) \\[2mm]
\dfrac{dP_3(t)}{dt} = \lambda P_2(t) - \mu P_3(t)
\end{cases}
$$

Algorithm for finding numerical values of state probabilities of all-optical switch consists of the following steps:

1. Finding the determinant of the coefficient matrix for the system (20): $\Delta(s) = 4s^4 + 9(\lambda + \mu)s^3 + 2(3\lambda^2 + 4\lambda\mu + 3\mu^2)s^2 + (\lambda^3 + \lambda^2\mu + \lambda\mu^2 + \mu^3)s$
2. Finding the roots of the equation: $\Delta(s) = 4s^4 + 9(\lambda + \mu)s^3 + 2(3\lambda^2 + 4\lambda\mu + 3\mu^2)s^2 + (\lambda^3 + \lambda^2\mu + \lambda\mu^2 + \mu^3)s = 0$
   From the analysis of this equation it follows that one of the roots is zero, and to find the remaining three, it is necessary to solve the equation:

$$
4s^3 + 9(\lambda + \mu)s^2 + 2(3\lambda^2 + 4\lambda\mu + 3\mu^2)s + (\lambda^3 + \lambda^2\mu + \lambda\mu^2 + \mu^3) = 0 \quad (21)
$$

3. Determining the polynomials Δ_k, where $k = \overline{0,3}$.
4. Transitioning from Laplace domain representations to the original domain using formula (17).

To obtain numerical values, consider an all-optical network with a bandwidth of 1 Gbps and a packet length of 1500 bytes. Thus, the arrival rate of packets λ corresponding to these network parameters is 89479 packets/s. Let's analyze the temporal dependence of the state probabilities, throughput, and transient time for the cases when $\rho > 1$, $\rho = 1$, $\rho < 1$, where $\rho = \lambda/\mu$ is the system load. Let's consider the first case when $\lambda < \mu$, $\lambda = 89479$ packets/s, $\mu = 134218.5$ packets/s, i.e., $\rho = 0.67$. The initial conditions, indicating that the system's buffer is completely empty at time $t = 0$, are defined as follows: $(P_0(0), P_1(0), P_2(0), P_3(0)) = (1, 0, 0, 0)$. Figure 2 shows the dependence of state probabilities on time for this case.

In this example the following values have been obtained: $\alpha_{min} = 68715$, the time constant $\tau = 1.46 \cdot 10^{-5}$ s, and the transient time $t_{tr} = 7.3 \cdot 10^{-5}$ s. It should be noted that in the steady-state mode the state probabilities have the

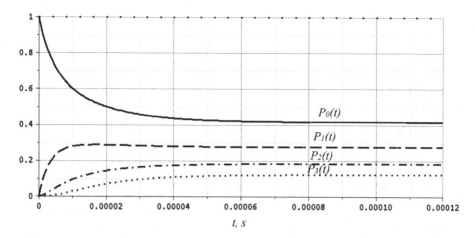

Fig. 2. Dependence of the states probabilities of the $M/M/1/2$ system on time, $\lambda < \mu$

following values $\pi_0 = 0.42$; $\pi_1 = 0.28$; $\pi_2 = 0.19$; $\pi_3 = 0.12$. The obtained results confirm the ones obtained using the well-known formula:

$$\pi_i = \rho^i \frac{1-\rho}{1-\rho^{n+2}} [1]. \tag{22}$$

Let's consider the second case, when $\lambda = \mu = 89479$ packets/s, i.e. $\rho = 1$ (Fig. 3). For this case, $\alpha_{min} = 52417$, the time constant $\tau = 1,9 \cdot 10^{-5}$ s, and the transient time $t_{tr} = 9,5 \cdot 10^{-5}$ s. In the steady-state mode the state probabilities are equal to $\pi_0 = \pi_1 = \pi_2 = \pi_3 = 0,25$, which also coincides with the known results [1].

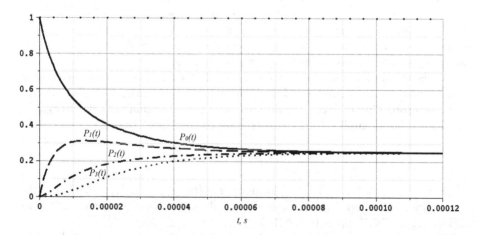

Fig. 3. Dependence of the states probabilities of the $M/M/1/2$ system on time, $\lambda = \mu$

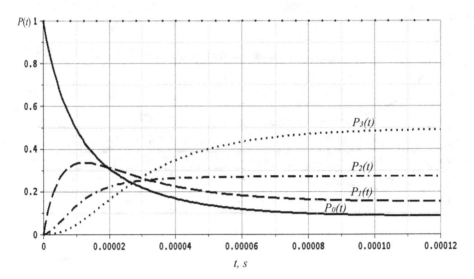

Fig. 4. Dependence of the states probabilities of the $M/M/1/2$ system on time, $\lambda > \mu$

Let's consider the third case when $\lambda > \mu$, $\rho > 1$, $\lambda = 89479$ packets/s, $\mu = 49479$ packets/s, i.e., $\rho = 1.81$ (Fig. 4). In the steady-state mode, the state probabilities are as follows: $\pi_0 = 0.08$; $\pi_1 = 0.15$; $\pi_2 = 0.27$; $\pi_3 = 0.49$, these values coincide with the ones calculated using the formula [1], confirming the correctness of the obtained results. According to the calculations, $\alpha_{min} = 44859$, the time constant $\tau_{min} = 2.2 \cdot 10^{-5}$ s, and the transient time $t_{tr} = 1.1 \cdot 10^{-4}$ s. Thus, as the service intensity of the optical switch μ decreases, the transient time increases.

Investigating the dependencies of the throughput capacity of the optical switch $A(t)$ in the transient mode for the three cases are described. In Fig. 5 $A_1(t)$ represents the throughput capacity of the optical switch for the case $\rho < 1$; $A_2(t)$ represents the throughput capacity of the optical switch for the case $\rho = 1$; $A_3(t)$ represents the throughput capacity of the optical switch for the case $\rho > 1$.

In the steady-state mode the throughput of all-optical switch for the cases of $\rho < 1$ and $\rho > 1$ is calculated using the formula $A = (1 - \pi_{loss}) \cdot \lambda$. Considering that the probability of losses can be calculated using formula (22), where $\pi_{loss} = \pi_3$ is the loss probability in the steady-state mode [1] we obtain:

$$A = (1 - \rho^3 \frac{1 - \rho}{1 - \rho^4}) \cdot \lambda \tag{23}$$

For the case of $\rho = 1$, the throughput $A(t) = (1 - \pi_{loss}) \cdot \lambda$. Thus, the throughput of the optical switch in the steady-state mode for three cases is as follows: $A_1 = 78357$ pps; $A_2 = 67109$ pps; $A_3 = 45322$ pps.

The analysis of the dependencies presented in Fig. 5 demonstrates how the throughput of the optical switch decreases when transitioning to the steady-state mode under different network loads. The higher system load ρ corresponds to the greater the difference between the throughput value at the initial time and its value in the steady-state mode.

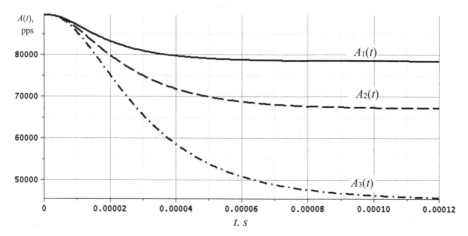

Fig. 5. Dependence of the $M/M/1/2$ system throughput on time at different ρ

6   Conclusion

This study has conducted an analysis of the transient behavior of all-optical switch, represented as a model of a single-line queuing system with a limited buffer, Poisson input flow, and exponential service time distribution. The Laplace transform technique was utilized to solve the Kolmogorov equations and determine the state probabilities of the system in the transient mode. The numerical example considered a switch model with a buffer designed to store two packets and input flow rates corresponding to the real optical network. Both normal network operation and overload scenarios were examined.

The proposed analytical model enables the investigation of transient performance metrics of all-optical network switch, including throughput, loss probability, transient time, and can be employed in the design of optical switching devices with predefined performance characteristics.

References

1. Dudin, A.N., Klimenok, V.I., Vishnevsky, V.M.: The Theory of Queuing Systems with Correlated Flows. Springer, Cham (2020). https://doi.org/10.1007/978-3-030-32072-0
2. Barabanova, E., Vytovtov, K., Vishnevsky, V., Khafizov, I.: Analysis of functioning photonic switches in next-generation networks using queueing theory and simulation modeling. In: Vishnevskiy, V.M., Samouylov, K.E., Kozyrev, D.V. (eds.) Distributed Computer and Communication Networks, DCCN 2022. CCIS, vol. 1748, pp. 356–369. Springer, Cham (2023). https://doi.org/10.1007/978-3-031-30648-8_28

3. Tomar, R.S., Shrivastav, R.K.: Three phases of service for a single server queueing system subject to server breakdown and Bernoulli vacation. Int. J. Math. Trends Technol. (IJMTT) **66**(5), 124–136 (2020)

4. Singla, N., Garg, P.C.: Transient and numerical solution of a feedback queueing system with correlated departures. Am. J. Numer. Anal. **2**(1), 20–28 (2014)

5. Sah, S.S., Ghimire, R.P.: Transient analysis of queueing model. J. Inst. Eng. **11**(1), 165–171 (2015)

6. Kempa, W.M., Paprocka, I.: Transient behavior of a queueing model with hyper-exponentially distributed processing times and finite buffer capacity. Sensors **22**(24), 9909 (2022). https://doi.org/10.3390/s22249909

7. Sah, S.S., Ghimire, R.P.: Transient analysis of queueing model. J. Inst. Eng. **11**(1), 165–171 (2015). https://doi.org/10.3126/jie.v11i1.14711

8. Kumar, R., Soodan, B.S.: Transient numerical analysis of a queueing model with correlated reneging, balking and feedback. Reliab. Theor. Appl. **4**(55), 46–54 (2019)

9. Kaczynski, W.H., Leemis, L.M., Drew, J.H.: Transient queueing analysis. INFORMS J. Comput. **24**(1), 10–28 (2012)

10. Vishnevsky, V.M., Vytovtov, K.A., Barabanova, E.A., Semenova, O.V.: Transient behavior of the MAP/M/1/N queuing system. Mathematics **9**(20), 2559 (2021)

11. Vytovtov, K.A., Barabanova, E.A.: An analytical method for the analysis of inhomogeneous continuous Markov processes with piecewise constant transition intensities. Autom. Remote. Control. **82**(12), 2111–2123 (2021). https://doi.org/10.1134/S0005117921120043

Approbation of Asymptotic Method for Queue with an Unlimited Number of Servers and State-Dependent Service Rate

Anatoly Nazarov[ID], Ivan Lapatin[ID], and Olga Lizyura[(✉)][ID]

Institute of Applied Mathematics and Computer Science, National Research Tomsk State University, 36 Lenina ave., Tomsk 634050, Russia
oliztsu@mail.ru

Abstract. The article discusses a model of the operation of a computing processor in the form of a queuing system with an unlimited number of servers and degradation of service rate. Degradation of service rate is the functional dependence of the service rate on the total number of requests (customers) in the system. Such function allows us to take into account the decrease in processor performance as the load on it increases. As a result of modification and application of the asymptotic analysis method, a Gaussian approximation of the probability distribution of the number of customers in the system was obtained, which corresponds to the distribution of simultaneously executed tasks by the processor.

Keywords: Competition · Concurrency · Degradation function · Queuing theory · Asymptotic analysis method

1 Introduction

Physical computing machines can vary in scale from a personal computer with several processor cores to huge servers and data centers. For all these cases, there is a phenomenon of performance degradation as the load on them grows, i.e. the number of simultaneously performed tasks increases. Coverage of this problem in the literature is primarily related to the study of the operation of cloud nodes [2,9] and the operation of different numbers of virtual machines on them. Modeling the service rate of such systems should include a dependence on the number of clients working with the cloud node. In this work, we propose to consider the model of operation not of a cloud node, where virtual machines can change their operating modes, but of a model of a separate processor that performs tasks sent to it and among them there is a competition for computing resources.

This study was supported by the Tomsk State University Development Programme (Priority-2030).

Simulation of the computers operation can be carried out using methods of queuing theory. This makes it possible to take into account the stochastic nature of the object under study, since the moments of launching virtual machines in the cloud, starting the next process on a personal computer occur at random times, and the duration of the virtual machine lifetime or the time it takes to complete a task is a non-deterministic value. In addition, the results of such studies allow one to make probabilistic conclusions (for example, to find quantiles) based on the characteristics being studied.

A significant complication of the model for describing the operation of computers is taking into account the dependence of the service rate on the total number of requests in the system. The most popular type of state-dependent service rate models is step function. This modeling approach is used in the works of [4,5,13]. Another method is to use simulations of service rates, arrival rates, and waiting times (before service) as dependent random variables [15].

Degradation of service due to the increasing load is common problem for computing systems. The competition for resources shared between customers leads to the situation when service time of one request is grower then the same time required for operation in empty system. In literature, authors propose various methods of avoiding such phenomena using scheduling algorithms and taks migration [8,11,12,16]. Few papers are devoted to the investigation of the nature of degradation [3,6] or methods of its detection [7,10].

The problem of performance degradation in computing systems is a pressing one, since it causes unpredictability of performance measures. Analysis of performance characteristics and their prediction is one of the main problems in computing services. Accurate performance evaluation allows maintaining the required level of service quality and provide effective load balancing [1,17].

We propose to use the so-called service rate degradation function, which assigns a service rate reduction coefficient to each value of the number of requests in the system. Using this approach, it is possible to simulate various situations of sensitivity of the service rate to the number of requests in the system.

Note that determining the specific type of degradation function is a separate task and is not a goal of this research.

To simulate the operation of the processor, we will use a queuing system with an unlimited number of servers. The incoming flow describes some aggregated task stream that arrives at the processor. We will identify each busy server with the task being performed. The more servers are busy, the more tasks are performed simultaneously. To model the reduction in service rate, we will use the degradation function, which actually changes the service rate depending on the number of tasks in the system. Thus, the random nature of the execution time of requests is generated by the different volume of incoming tasks and the variable rate of their execution. The study will be carried out using the method of asymptotic analysis.

In this paper, we propose a modification of the asymptotic analysis method for queues with an unlimited number of servers and state-dependet service rate.

We demonstrate the effectiveness and efficiency of the asymptotic analysis on a basic model with a Poisson input flow.

The rest of the paper is organized as follows. In Sect. 2, we describe the model and derive some preliminary equations. Section 3 is dedicated to the modification of asymptotic analysis method for investigation of system with state-dependent service rate. In Sect. 4 we obtain the prelimit distribution for the number of busy servers in the system. Section 5 shows the numerical example based on the obtained results. In Sect. 6, we discuss the directions of future studies and give concluding remarks.

2   Mathematical Model

Let us consider a queueing model with unlimited of servers (Fig. 1).

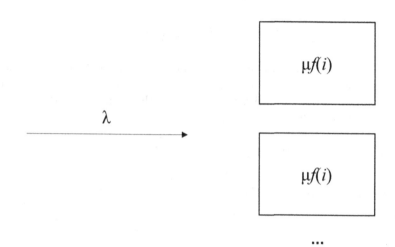

Fig. 1. Mathematical model of the processor

The input of the system is stationary Poisson process with rate λ. Each incoming request immediately starts the service. Service rates are state-dependent and equal $\mu f(i)$, where i is the current number of busy servers and μ is the initial service rate (without competition). Here $f(i)$ represents degradation function, the values of which are a dimensionless coefficient of reduction in service intensity depending on the number of requests i in the system. The more requests are in the system at the same time, the lower the intensity of their service. Thus, the degradation function is monotonically nonincreasing function.

We introduce process $i(t)$ as the number of busy servers (incomplete requests) in the system and its probability distribution $P(i, t) = P\{i(t) = i\}$, which satisfies the system of Kolmogorov differential equations

$$\frac{\partial P(i,t)}{\partial t} = -(\lambda + i\mu f(i))P(i,t) + (i+1)\mu f(i+1)P(i+1,t) + \lambda P(i-1,t).$$

Considering the last system in stationary regime, we obtain the main system of equations

$$- (\lambda + i\mu f(i))P(i) + (i+1)\mu f(i+1)P(i+1) + \lambda P(i-1) = 0. \tag{1}$$

In order to solve system (1), we use asymptotic analysis method under high rate of the input flow limiting condition.

3   Asymptotic Analysis Method

3.1   First Step: Average Number

The condition of high rate of the input flow is given by infinitely large parameter $T \to \infty$ and the following equalities

$$\lambda = \lambda_1 T, \quad f(i) = f_1\left(\frac{i}{T}\right).$$

We will call T as the parameter of high input rate. Here λ_1 is a fixed value. In practice, the intensity $\lambda_1 T$ is considered as a uniform value with sufficiently large but finite values [14]. Taking into account the notation made, we rewrite the system (1) in the form

$$-\left(\lambda_1 + \frac{i}{T}\mu f_1\left(\frac{i}{T}\right)\right)P(i) + \mu\frac{i+1}{T}f_1\left(\frac{i+1}{T}\right)P(i+1) + \lambda_1 P(i-1) = 0. \tag{2}$$

Denoting $\varepsilon = \frac{1}{T}$, we make the following substitutions in system (2)

$$\frac{i}{T} = i\varepsilon = x, \quad P(i) = P_1(x, \varepsilon),$$

which yields

$$-(\lambda_1 + x\mu f_1(x))\, P_1(x,\varepsilon) + (x+\varepsilon)\mu f_1(x+\varepsilon)P_1(x+\varepsilon,\varepsilon) + \lambda_1 P_1(x-\varepsilon,\varepsilon) = 0. \tag{3}$$

After that, we use the following Taylor decompositions:

$$\lambda_1 P_1(x-\varepsilon,\varepsilon) = \lambda_1 P_1(x,\varepsilon) - \varepsilon\frac{\partial \lambda_1 P_1(x,\varepsilon)}{\partial x} + O(\varepsilon^2),$$
$$(x+\varepsilon)\mu f_1(x+\varepsilon)P_1(x+\varepsilon,\varepsilon) = x\mu f_1(x)P_1(x,\varepsilon) + \varepsilon\frac{\partial x f_1(x)P_1(x,\varepsilon)}{\partial x} + O(\varepsilon^2).$$

Substituting them into (3), we obtain

$$\varepsilon\frac{\partial x\mu f_1(x)P_1(x,\varepsilon)}{\partial x} - \varepsilon\frac{\partial \lambda_1 P_1(x,\varepsilon)}{\partial x} = O(\varepsilon^2).$$

We divide the equation by ε and take the limit by $\varepsilon \to 0$, which yields

$$\frac{d}{dx}\{P_1(x)x\mu f_1(x)\} - \frac{d}{dx}\{P_1(x)\lambda_1\} = 0,$$

where $P_1(x) = \lim_{\varepsilon \to 0} P_1(x, \varepsilon)$. Reorganizing the equation, we obtain degenerate Fokker-Planck equation

$$-\frac{d}{dx}\{P_1(x)(\lambda_1 - x\mu f_1(x))\} = 0,$$

the drift coefficient of which produces the equation for the mean number of occupied servers in the system

$$\lambda_1 - x\mu f_1(x) = 0. \tag{4}$$

We denote the root of the equation as $x = \kappa_1$, which is asymptotical average number of simultaneously processed requests.

3.2   Second Step: Variance

For the next step of analysis, we consider (2) again and make the following substitutions

$$\frac{1}{T} = \varepsilon^2, \ \frac{i}{T} = i\varepsilon^2 = x + \varepsilon y, \ P(i) = P_2(y, \varepsilon),$$

then we have

$$- (\lambda_1 + (x + \varepsilon y)\mu f_1 (x + \varepsilon y)) P_2(y, \varepsilon)$$
$$+ (x + \varepsilon(y + \varepsilon))\mu f_1 (x + \varepsilon(y + \varepsilon)) P_2(y + \varepsilon, \varepsilon) + \lambda_1 P_2(y - \varepsilon, \varepsilon) = 0. \tag{5}$$

Let us use Tailor decompositions up to the $O(\varepsilon^3)$:

$$\lambda_1 P_2(y - \varepsilon, \varepsilon) = \lambda_1 P_2(y, \varepsilon) - \varepsilon \frac{\partial}{\partial y} (\lambda_1 P_2(y, \varepsilon)) + \frac{\varepsilon^2}{2} \frac{\partial^2}{\partial y^2} (\lambda_1 P_2(y, \varepsilon)) + O(\varepsilon^3),$$
$$(x + \varepsilon(y + \varepsilon))\mu f_1 (x + \varepsilon(y + \varepsilon)) P_2(y + \varepsilon, \varepsilon)$$
$$= (x + \varepsilon y)\mu f_1 (x + \varepsilon y) P_2(y, \varepsilon) + \varepsilon \frac{\partial}{\partial y} ((x + \varepsilon y)\mu f_1 (x + \varepsilon y) P_2(y, \varepsilon))$$
$$+ \frac{\varepsilon^2}{2} \frac{\partial^2}{\partial y^2} ((x + \varepsilon y)\mu f_1 (x + \varepsilon y) P_2(y, \varepsilon)) + O(\varepsilon^3).$$

Substituting these decompositions into Eq. (5), we obtain

$$\varepsilon \frac{\partial}{\partial y} \{(x + \varepsilon y)\mu f_1 (x + \varepsilon y) P_2(y, \varepsilon)\}$$
$$+ \frac{\varepsilon^2}{2} \frac{\partial^2}{\partial y^2} ((x + \varepsilon y)\mu f_1 (x + \varepsilon y) P_2(y, \varepsilon)) - \varepsilon \frac{\partial}{\partial y} (\lambda_1 P_2(y, \varepsilon))$$
$$+ \frac{\varepsilon^2}{2} \frac{\partial^2}{\partial y^2} (\lambda_1 P_2(y, \varepsilon)) = O(\varepsilon^3).$$

We also use similar decomposition for degradation function

$$(x + \varepsilon y)\mu f_1 (x + \varepsilon y) = x\mu f_1(x) + \varepsilon y \frac{d}{dx} x\mu f_1(x) + O(\varepsilon^2),$$

which yields

$$\varepsilon \frac{\partial}{\partial y} \{(x\mu f_1(x) + \varepsilon y \frac{d}{dx} x\mu f_1(x)) P_2(y, \varepsilon)\}$$
$$+ \frac{\varepsilon^2}{2} \frac{\partial^2}{\partial y^2} ((x\mu f_1(x)) P_2(y, \varepsilon)) - \varepsilon \frac{\partial}{\partial y} (\lambda_1 P_2(y, \varepsilon))$$
$$+ \frac{\varepsilon^2}{2} \frac{\partial^2}{\partial y^2} (\lambda_1 P_2(y, \varepsilon)) = O(\varepsilon^3).$$

Let us reorganize the equation as follows:

$$-\varepsilon \frac{\partial}{\partial y} \left\{ \left(\lambda_1 - x\mu f_1(x) - \varepsilon y \frac{d}{dx} x\mu f_1(x) \right) P_2(y, \varepsilon) \right\}$$
$$+ \frac{\varepsilon^2}{2} \frac{\partial^2}{\partial y^2} \left((\lambda_1 + x\mu f_1(x)) P_2(y, \varepsilon) \right) = O(\varepsilon^3).$$

Taking (4) into account, we have

$$\varepsilon^2 \frac{\partial}{\partial y} \left\{ y \frac{d}{dx} (x\mu f_1(x)) P_2(y, \varepsilon) \right\} + \frac{\varepsilon^2}{2} \frac{\partial^2}{\partial y^2} (2\lambda_1 P_2(y, \varepsilon)) = O(\varepsilon^3).$$

Then we divide the equation by ε^2 and take the limit by $\varepsilon \to 0$:

$$\frac{d}{dy} \left\{ y \frac{d}{dx} (x\mu f_1(x)) P_2(y) \right\} + \frac{1}{2} \frac{d^2}{dy^2} (2\lambda_1 P_2(y)) = 0.$$

We denote

$$a = \frac{d}{dx} (x\mu f_1(x)),$$

and lower the degree of the integral

$$y a P_2(y) + \lambda_1 \frac{d}{dy} \{P_2(y)\} = 0.$$

Thus, we have an ordinary differential equation

$$\frac{d}{dy} P_2(y) = -\frac{a}{\lambda_1} y P_2(y),$$

the solution of which is as follows:

$$P_2(y) = C \exp \left\{ -\frac{y^2}{2} \frac{a}{\lambda_1} \right\}. \tag{6}$$

Here C is the integration constant, which we obtain using the normalization condition

$$C = \frac{1}{\sqrt{2\pi \frac{\lambda_1}{a}}}.$$

3.3   Results of Asymptotic Analysis

In Sect. 3.1, we derived the equation for the mean number of processed requests (4). In Sect. 3.2, we shown that the distribution of the process $i(t)$ is asymptotically Gaussian (6). Taking into account the obtained results, we propose the following approximation for the prelimit probability distribution of $i(t)$:

$$p_{discrete}(i) = C_1 \exp \left\{ -\frac{(i - \kappa_1 T)^2}{2\kappa_2 T} \right\}.$$

Here C_1 is the normalization constant, κ_1 is a unique solution of the equation

$$\lambda_1 - \kappa_1 f_1(\kappa_1) = 0. \tag{7}$$

κ_2 is given by formula

$$\kappa_2 = \lambda_1 \left(\mu \frac{\partial}{\partial x} (x f_1(x)) \right)^{-1}. \tag{8}$$

Thus, we can approximate the prelimit distribution $p(i)$ using Gaussian distribution with mean number of busy servers $\kappa_1 T$ and variance $\kappa_2 T$.

4   Prelimit Distribution for the Number of Busy Servers

In this paper, we have chosen a base model to illustrate the quality of the Gaussian approximation. For the proposed model, we can obtain formulas for calculating the prelimit distribution

$$P(i) = P(0) \prod_{n=1}^{i} \frac{\lambda}{n\mu f(i)}, i \geq 1, \tag{9}$$

$$P(0) = \left\{ 1 + \sum_{i=1}^{\infty} \prod_{n=1}^{i} \frac{\lambda}{n\mu f(i)} \right\}^{-1} \tag{10}$$

This allows to avoid using simulation to show the effectiveness and efficiency of asymptotic results.

5   Numerical Example

Let us consider the following values of the model parameters:

- $\lambda = \lambda_1 T = 50$ as the average number of tasks that are sent to the processor per unit of time, that is, the intensity of task arrivals;
- $\mu = 10$ as the inverse value of the average query execution time in the absence of other tasks on the processor;
- we define the service rate degradation function as a non-negative decreasing function

$$f(i) = \frac{1}{\sqrt{1+i}}. \tag{11}$$

The form of the degradation function is presented in Fig. 2.

Figure 3 shows the Gaussian approximation of the probability distribution of the number of busy servers in the system under study, taking into account the degradation function of the form (11), specified by the average number of occupied devices in the system $\kappa_1 T = 5$ and variance $\kappa_2 T = 5$.

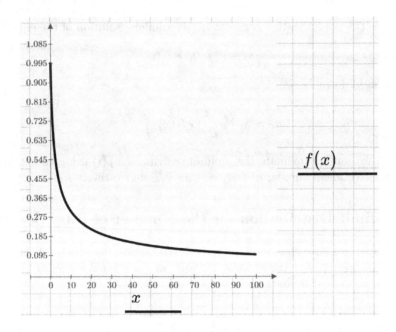

Fig. 2. Service rate degradation function

For estimating an accuracy of the obtained Gaussian approximation, we consider an examples of the queueing system with Poisson arrivals and unlimited number of servers. Service rate degradation is (11), the initial service rate (without competition) $\mu = 10$. In Figs. 4, 5 and 6, we show examples of the comparison of the asymptotic and the exact distributions for different values of λ. We compare the exact and the asymptotic distributions of the number of busy servers (incomplete requests) in the system for making conclusions about applicability area of the asymptotic result. We use Kolmogorov distance as a measure of the difference between the distributions:

$$\Delta = \max_{i \geq 0} \left| \sum_{k=0}^{i} [P(k) - P_{discrete}(k)] \right|, \tag{12}$$

where $P(k)$ is the exact probability distribution (9)-(10) and $P_{discrete}(k)$ is the asymptotic distribution. We have found that the error decreases when the average number of tasks grows.

Obtaining the result in the form of an estimate of the probability distribution allows us to draw conclusions, for example, about the average number of simultaneously executed tasks on the processor or the number of tasks that will not be exceeded with a certain probability q. By substituting the obtained values into the degradation function, we can estimate what task performance was on average or below what level performance did not fall with probability q.

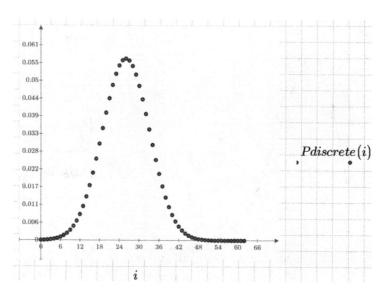

Fig. 3. Gaussian approximation of the number of simultaneously performed request in the system

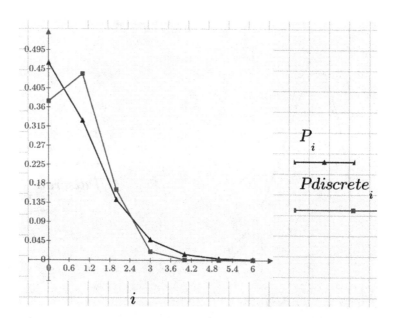

Fig. 4. Asymptotic and Numerical Probability Distribution of Busy Servers in the System, $\lambda = 5, \Delta = 0.09$

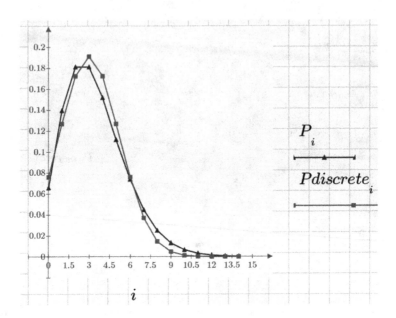

Fig. 5. Asymptotic and Numerical Probability Distribution of Busy Servers in the System, $\lambda = 15, \Delta = 0.037$

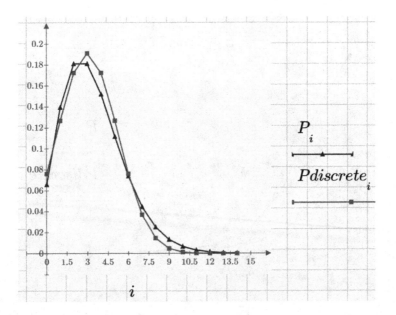

Fig. 6. Asymptotic and Numerical Probability Distribution of Busy Servers in the System, $\lambda = 25, \Delta = 0.029$

6   Conclusion

The paper considers a queuing system with an unlimited number of servers and degradation of service rate depending on the number of requests in the system. Such a model may be useful for addressing certain systems that, in a growing number of clients, are causing performance degradation on a per-client basis. In this example, we considered the model of processor operation in conditions of competition between tasks for a computing resource. As a result of the study, a Gaussian approximation of the probability distribution of the number of operating servers in the system was obtained, taking into account the degradation of service.

In future, we plan to adapt the method of studying systems with state-dependent parameters for models with a more complex structure.

References

1. Aslanpour, M.S., Gill, S.S., Toosi, A.N.: Performance evaluation metrics for cloud, fog and edge computing: a review, taxonomy, benchmarks and standards for future research. Internet of Things **12**, 100273 (2020)
2. Bermejo, B., Juiz, C.: A general method for evaluating the overhead when consolidating servers: performance degradation in virtual machines and containers. J. Supercomput. **78**(9), 11345–11372 (2022)
3. Bhatele, A., Mohror, K., Langer, S.H., Isaacs, K.E.: There goes the neighborhood: performance degradation due to nearby jobs. In: Proceedings of the International Conference on High Performance Computing, Networking, Storage and Analysis, pp. 1–12 (2013)
4. Bruneo, D.: A stochastic model to investigate data center performance and QoS in IAAS cloud computing systems. IEEE Trans. Parallel Distrib. Syst. **25**(3), 560–569 (2013)
5. Choudhary, A., Chakravarthy, S.R., Sharma, D.C.: Analysis of map/ph/1 queueing system with degrading service rate and phase type vacation. Mathematics **9**(19), 2387 (2021)
6. Ekanayake, J., Fox, G.: High performance parallel computing with clouds and cloud technologies. In: Avresky, D.R., Diaz, M., Bode, A., Ciciani, B., Dekel, E. (eds.) CloudComp 2009. LNICSSTE, vol. 34, pp. 20–38. Springer, Heidelberg (2010). https://doi.org/10.1007/978-3-642-12636-9_2
7. Hao, J., Zhang, B., Yue, K., Wu, H., Zhang, J.: Measuring performance degradation of virtual machines based on the Bayesian network with hidden variables. Int. J. Commun Syst **31**(13), e3732 (2018)
8. Houssein, E.H., Gad, A.G., Wazery, Y.M., Suganthan, P.N.: Task scheduling in cloud computing based on meta-heuristics: review, taxonomy, open challenges, and future trends. Swarm Evol. Comput. **62**, 100841 (2021)
9. Huber, N., von Quast, M., Brosig, F., Hauck, M., Kounev, S.: A method for experimental analysis and modeling of virtualization performance overhead. In: Ivanov, I., van Sinderen, M., Shishkov, B. (eds.) CLOSER 2011. SSRISE, pp. 353–370. Springer, New York (2012). https://doi.org/10.1007/978-1-4614-2326-3_19
10. Ibidunmoye, O., Metsch, T., Elmroth, E.: Real-time detection of performance anomalies for cloud services. In: 2016 IEEE/ACM 24th International Symposium on Quality of Service (IWQoS), pp. 1–2. IEEE (2016)

11. Kishor, A., Chakarbarty, C.: Task offloading in fog computing for using smart ant colony optimization. Wireless Pers. Commun., 1–22 (2021)
12. Kumar, M., Sharma, S.C., Goel, A., Singh, S.P.: A comprehensive survey for scheduling techniques in cloud computing. J. Netw. Comput. Appl. **143**, 1–33 (2019)
13. Liu, X., Li, S., Tong, W.: A queuing model considering resources sharing for cloud service performance. J. Supercomput. **71**, 4042–4055 (2015)
14. Moiseev, A., Nazarov, A.: Infinite-linear queueing systems and networks. NTL Publications, Tomsk (2015)
15. Morozov, E.: A general multi-server state-dependent queueing system (2010)
16. Rejiba, Z., Masip-Bruin, X., Marín-Tordera, E.: A survey on mobility-induced service migration in the fog, edge, and related computing paradigms. ACM Comput. Surv. (CSUR) **52**(5), 1–33 (2019)
17. Sefati, S., Mousavinasab, M., Zareh Farkhady, R.: Load balancing in cloud computing environment using the grey wolf optimization algorithm based on the reliability: performance evaluation. J. Supercomput. **78**(1), 18–42 (2022)

Computer and Communication Networks

Risk Management in the Design of Computer Network Topology

Alexander Shiroky$^{(\boxtimes)}$ (iD)

V. A. Trapeznikov Institute of Control Sciences of Russian Academy of Sciences,
65 Profsoyuznaya Street, Moscow, Russia
shiroky@ipu.ru

Abstract. The problems of risk management are considered from different points of view. Most often, they are investigated as part of organizational management. The results obtained within this approach are widely used, but their validity is debated. On the other hand, risk management can be considered as a system of mathematical models. In this form, it is discussed in the mechanism design section of control theory. The results obtained with this approach have rigorous substantiation, but also require a large number of conditions to be met, which is often hard to achieve in practice. Therefore, the convergence of these approaches is an important practical problem. The proposed paper attempts to put a formal solution to building a risk-minimizing topology of a computer network into the risk management framework. When choosing the optimal set of countermeasures to protect a computer network, the possibility of modifying the network topology is usually not considered. At the same time, topology rebuilding is a natural process for some classes of networks (in particular, wireless mesh networks). This study considers the problem of designing the topology of computer networks that minimizes risk. The proposed solution is based on the results obtained previously in research on the influence of the structure of complex systems on their integral risk. The article considers a particular case of the exact solution of the problem for a peer-to-peer distributed network of the 'bus' topology. In addition, it proposes an algorithm to design a risk-minimizing topology in attack scenarios of a simple chain structure.

Keywords: Computer networks · Topology design · Risk management

1 Introduction

Risk management is generally understood as coordinated activities to direct and control an organization with regard to risk [1]. To solve problems in this area, one usually builds a framework defining the rules for identifying risks, analyzing and evaluating them, as well as an algorithm for selecting control actions to minimize risks. The content of such a framework strongly depends on the subject area (in particular, for risk management in the information security domain, there is an ISO/IEC 27005:2022 [2] standard based on ISO 31000:2018). Note that the

V. M. Vishnevskiy et al. (Eds.): DCCN 2023, LNCS 14123, pp. 375–386, 2024.
https://doi.org/10.1007/978-3-031-50482-2_29

mentioned set of components is minimal and can be expanded by the rules of risk monitoring and logging, reporting forms, and other sections, depending on the specifics of the organization.

The choice of a specific model or method for each of the typical problems within the risk management framework is usually not formalized and attributed to the competence of the risk manager. At the same time, the list of suitable methods to choose from can be quite long. In particular, IT specialists often use statistical analysis to identify cybersecurity risks [3–5], as well as evolutionary algorithms [6–8], and data mining and machine learning methods [9–11].

Qualitative assessment of cybersecurity risks is typically performed with the approaches described in the Cybersecurity Framework by the US National Institute of Standards and Technology [12], the ISO / IEC 27005 standard [2], and, more rarely, using OCTAVE (Operationally Critical Threat, Asset and Vulnerability Evaluation [13,14]).

For quantitative risk assessment cybersecurity specialists use models based on fault and attack trees [15–17], Markov models [18,19], Bayesian decision network [20,21], decision diagrams [22], and Petri nets [23,24].

Finally, to choose the risk-minimizing control actions, one can use game theory methods [25–28]. The values of their parameters can be obtained using optimization methods, including linear and non-linear programming [29,30], bilevel and trilevel optimization [31,32], multi-objective optimization [33,34], and stochastic optimization [35,36].

However, risk management in general can also be considered as a mathematical problem. Suppose we have a system described with a finite set of parameters and there is a set in their space that we treat as system's desired operation mode. Then the problem of risk management consists in minimizing the distance between the current state of the system and the desired set. Let us show that such a problem is equal to the problem of effective system control under uncertainty. Detailed discussion of this equality can be found, for example, in [37].

Let U be the set of control actions, Q and Θ be the sets of states of the system and the external environment, respectively. Suppose that the state of the system $q \in Q$ is determined by the state of the external environment and the applied control actions, that is, $q = q(u, \theta), u \in U, \theta \in \Theta$. Let us introduce the functional $K = K(u, q, \theta) = K\left(q\left(u, \theta\right)\right)$ as an *efficiency criterion* of the control u in the state of the system q and the state of the external environment θ. The problem is to find such control actions $u^* \in U$ that

$$K(u^*, q, \theta) = \max_{u \in U} K(u, q, \theta). \tag{1}$$

Let us assume that there is a set (perhaps not the only one) of arguments (u^*, q^*, θ^*) in which the efficiency criterion takes the maximum possible value (a perfect situation):

$$K(u^*, q^*, \theta^*) = \max_{u \in U} \max_{q \in Q} \max_{\theta \in \Theta} K(u, q, \theta). \tag{2}$$

The value
$$\varphi(u, q, \theta) = K(u^*, q^*, \theta^*) - K(u, q, \theta) \tag{3}$$

equal to the difference between the maximum and the current (for some u, q and θ) value of the efficiency criterion is called the *loss function*. It is easy to see that

$$\min_{u \in U} \varphi(u, q, \theta) = K(u^*, q^*, \theta^*) - \max_{u \in U} K(u, q, \theta). \tag{4}$$

Consequently, the problem of maximizing the efficiency criterion is equal to the problem of minimizing the loss function.

Full awareness is rare in practice and is mainly typical for control problems of technical systems. More often, the control system does not have complete and/or reliable information about the states of both the controlled system and the external environment. In other words, there is some kind of uncertainty. To solve the control problem under uncertainty, one needs to resolve the latter in some way. For example, the solution could use the most likely states or the least favorable states. More ways to resolve the uncertainty are discussed in [38].

Let us choose a control action $u \in U$ and set the uncertainty-resolving operator $\mathfrak{I}$ such that the result of its application to the efficiency criterion depends only on the chosen control u:

$$\mathfrak{K}(u) = \mathfrak{I}K(u, \cdot, \cdot). \tag{5}$$

Then we can write down the problem of finding a control u^* that maximizes the value of the efficiency criterion under uncertainty like we do at (2):

$$\mathfrak{K}(u^*) = \max_{u \in U} \mathfrak{K}(u). \tag{6}$$

Let us define a real-valued risk function $\rho(u) = \mathfrak{K}(u^*) - \mathfrak{K}(u)$. The values of $\rho(u)$ are interpreted as the risk associated with the execution of the chosen control action u. Now we can set the problem of minimizing the risk:

$$\rho(u^*) = \min_{u \in U} \rho(u). \tag{7}$$

It is obvious that

$$u^* = \operatorname{Arg\,max}_{u \in U} \mathfrak{K}(u) = \operatorname{Arg\,min}_{u \in U} \rho(u). \tag{8}$$

Thus, when managing a complex system under uncertainty with a risk function determined by (7), the problem of maximizing the efficiency criterion (6) is equal to the problem of minimizing risk as shown in (8).

The foregoing considerations explain why risk management as a system of mathematical models is discussed in control theory, more specifically in the mechanism design section.

The formal definition of risk as a measure of deviation between the state of the system and the target set agrees well with the definitions of cybersecurity risk in ISO/IEC 27005:2022 and the NIST Cybersecurity Framework. This article is an attempt to apply the principles of risk management in complex networks considering their structure (obtained as an abstract mathematical result) within the general risk management framework on the simple example of constructing the risk-minimizing computer network topology. Namely, it considers the influence on the integral risk of the position of the nodes relative to the perimeter in a network of a 'bus' topology. The result obtained can be used both directly (for example, when designing a wireless mesh network) and in scenario analysis of possible attacks. In addition, the paper proposes the algorithm for designing the risk-minimizing topology of an arbitrary network under attack by the simple chain structure.

2   General Problem Statement

Consider a complex system consisting of a finite set of elements (objects, of an arbitrary nature so far): $S = \{s_1, ..., s_i, ..., s_n\}, i \in N = \{1, ..., n\}$. We assume that the elements $s_i \in S, i \in N$ of the system are autonomous so that they cannot influence each other's states.

Suppose that there are two subjects (also of an arbitrary nature for the time being), which we will call player A (otherwise, *the Attacker*) and player D (otherwise, *the Defender*). These two subjects have different intentions towards the state of the system S.

We assume that the player D has a certain amount of resource $X \geq 0$, which he can arbitrarily distribute among the elements of the system S: $x = (x_1, ..., , x_n), x_i \geq 0, i \in N, \sum_{i=1}^{n} x_i \leq X$.

Similarly, we assume that the player A also has a certain amount of resource $Y \geq 0$, which he can arbitrarily distribute among the elements of the system S: $y = (y_1, ..., y_n), y_i \geq 0, i \in N, \sum_{i=1}^{n} y_i \leq Y$.

In the framework of the model described above, we will consider the 'resource' as any measurable and arbitrarily divisible asset represented by a non-negative real number. It could be financial, labor, time, production, and other resources/costs depending on the context.

We will call *the local risk* some local characteristic of an element $s_i \in S$, depending on the amount of resources allocated by the players D and A. The local risk characteristic represents possible losses (damage) due to changing the state of the element.

In turn, we will call *the integral risk* some overall characteristic of the entire system S, depending on the amount of resources allocated by players D and A to all its elements, and associated with possible losses (damage) due to a state change of each element.

If the elements of the system do not affect each other, the local risk of any element will depend on the amount of resources allocated to it by the players D and A. We define the local risk function for each element as $\rho_i(x_i, y_i) : \mathbb{R}_+^0 \times \mathbb{R}_+^0 \rightarrow \mathbb{R}_+^0$, where $\mathbb{R}_+^0$ is a set of non-negative real numbers.

Considering the described model, we will further assume that the local risk functions $\rho_i(\cdot, \cdot), i \in N$ have the following properties:

1. Risk non-negativity:

$$\forall i \in N, x_i, y_i \geq 0 : \rho_i(x_i, y_i) \geq 0. \tag{9}$$

2. Risk monotonicity:

$$\forall i \in N : \frac{\partial \rho_i(x_i, y_i)}{\partial x_i} \leq 0, \frac{\partial \rho_i(x_i, y_i)}{\partial y_i} \geq 0. \tag{10}$$

3. Risk finiteness:

$$\forall i \in N, x_i, y_i \geq 0 \, \exists \rho_i^x = \text{const}, \rho_i^y = \text{const} : \rho_i^x \leq \rho_i(x_i, y_i) \leq \rho_i^y. \tag{11}$$

The non-negativity of the risk means that the potential damage associated with an occurrence of the local risk for any element $s_i \in S$ cannot be negative. We assume that there is always a positive *residual risk* in the common case, regardless of the measures taken to reduce it. There are only separate exceptional cases where a risk can be reduced to zero.

The monotonicity of the risk means that the *additional* allocation of resources to any element $s_i \in S$ by the Defender should *not increase* the local risk for any element of the system S. On the other hand, the *additional* resource allocation by the Attacker should *not decrease* the local risk for any element of the system S.

The finiteness of the risk means that the Defender cannot reduce the residual risk of any element $s_i \in S$ to zero, and, on the other hand, there is a final positive *marginal risk* for any element $s_i \in S$ (regardless of the amount of resources spent by the Attacker).

Let the structure $W = \langle G(S, E), T \rangle$ be a graph with a set of elements S as its vertices, a set of edges E, and a specific subset of vertices $T \subseteq S$, which we will call the perimeter of the system S.

We assume that the player A attacks the elements of the system along the path $c = \langle u, v \rangle, u \in T, v \in S$ and transits from some vertex (or element) $s_i \in c$ to an adjacent vertex $s_j \in c$ only if his attack on the element s_i was successful.

Let $x = (x_1, ..., x_n), y = (y_1, ..., y_n)$ be some valid resource distributions among the elements of the system S for players D and A, respectively. Consider the local risk functions for each element $s_i \in S$ as follows:

$$\rho_i(x, y) = u_i(x, y) \cdot p_i(x, y). \tag{12}$$

Here $u_i(x, y) : \mathbb{R}_+^n \times \mathbb{R}_+^n \to \mathbb{R}_+^0$ is a function describing the dependence of the expected damage in the case of a successful attack depending on the resource distributions x and y, and $p_i(x, y) : \mathbb{R}_+^n \times \mathbb{R}_+^n \to (0, 1]$ is the probability of a successful attack depending on the distributions x and y.

The following tuple defines the basic risk management model for complex systems with the structure and the perimeter:

$$\langle S = \{s_i\}_{i \in N}, T, E, D, A, X, Y, \{\rho_i(\cdot, \cdot)\}_{i \in N}, \rho(\cdot, \cdot) \rangle \tag{13}$$

If the structure $W = \langle G(S, E), T \rangle$ is fixed, then:

– The Defender's goal is to allocate the available resource X among the elements of the system S to reduce the value of the integral risk function $\rho(x, y)$ to a minimum;

– On the contrary, the Attacker's goal is to allocate the available resource Y among the elements of the system S to increase the value of the integral risk function $\rho(x, y)$ to a maximum.

Let $\mathscr{X}(X)$ be the set of valid allocations of resource X among the elements of the system S by the player D, and $\mathscr{Y}(Y)$—the set of valid allocations of resource Y among the elements of the system S by the player A:

$$\mathscr{X}(X) = \left\{ (x_1, ..., x_n) \in \mathbb{R}^n : x_i \geq 0, i \in N, \sum_{i=1}^{n} x_i \leq X \right\}, \tag{14}$$

$$\mathscr{Y}(Y) = \left\{ (y_1, ..., y_n) \in \mathbb{R}^n : y_i \geq 0, i \in N, \sum_{i=1}^{n} y_i \leq Y \right\}. \tag{15}$$

Then the player's D problem ("Defender's problem") is to find the resource allocation $x^* \in \mathscr{X}$ that minimizes the integral risk, and can formally be written as:

$$x^* = \operatorname*{Arg\,min}_{x \in \mathscr{X}} \rho(x, y) = \operatorname*{arg\,min}_{x \in \mathscr{X}} \sum_{i=1}^{n} \rho_i(x, y). \tag{16}$$

Similarly, the player's A problem ("Attacker's problem") is to find the resource allocation $y^* \in \mathscr{Y}$ that maximizes the integral risk, and can be written as:

$$y^* = \operatorname*{Arg\,min}_{y \in \mathscr{Y}} \rho(x, y) = \operatorname*{arg\,min}_{y \in \mathscr{Y}} \sum_{i=1}^{n} \rho_i(x, y). \tag{17}$$

If the structure of $W = \langle G(S, E), T \rangle$ can be altered (for example, by modifying the sets of E and/or T), then before solving the problem (16), the Defender can further reduce risks by solving the problem of building the structure minimizing risks.

3 Minimizing the Risk for a Peer-to-Peer Distributed Network with the 'Bus' Topology

First, let us consider a particular problem of choosing the node placement that minimizes the integral risk in a scenario of a sequential attack starting from some predetermined node, which we will call the perimeter.

Suppose that we have a set of $S = \{s_1, ..., s_n\}$ nodes, each of which can act as a router. For each element of the set s_i, we know the probability p_i of a successful attack on it. Let us assume that we know the values of u_i damage to the protected system in case of a successful attack on the node. We will consider scenarios in which the player A sequentially attacks the nodes of the protected network, starting from some predetermined perimeter node and moving further along the neighbors. The Attacker visits each node exactly once.

Let us write down the formal problem setting with modified designations and definitions introduced in [39].

Definition 1. *Let the graph be* $G\left(V = \{v_1, ..., v_n\}, E = \{(v_i, v_{i+1})\}_{i=1}^{n-1}\right), n \in$ $\mathbb{N}$ *and the perimeter* $T = \{v_1\}$. *Then we will say that the tuple* $W_n = \langle G(V, E), T \rangle$ *sets the attack scenario of length* n.

Note that the graph G is a simple chain.

Definition 2. *Consider the one-to-one mapping* $M^{-1} : S \rightarrow V, S = \{s_1, ..., s_n\}, n \in \mathbb{N} : \forall i \leq n \, \exists! j \leq n : v_j = M^{-1}(s_i)$ *as the placement of elements* S *in the* W_n *scenario. The corresponding inverse mapping* $M : V \rightarrow S$ *will be called the projection of the scenario* W_n *onto the set of nodes* S.

In the future, we will omit the subscript in the designation of the scenario for convenience.

For an arbitrary given placement $M^{-1} : S \rightarrow V \setminus \{v_{n+1}, ..., v_n\}$, we can calculate the value of the integral risk

$$\rho\left(S, W, M^{-1}\right) = \sum_{i=1}^{n} \rho_{M(v_i)}, \tag{18}$$

where $\rho_{M(v_i)}$ is the local risk value for the element $M(v_i)$. Now let us state the problem of minimizing the integral risk. The problem consists in finding a set of placements $\mathbf{M}_{min}^{-1}$, for each of the elements of which the minimum value of the integral risk ρ_{min} is achieved:

$$\mathbf{M}_{min}^{-1} = \operatorname*{Arg\,min}_{M^{-1}} \rho\left(S, W, M^{-1}\right) : \rho_{min} = \sum_{i=1}^{n} \rho_{M(v_i)} \forall M^{-1} \in \mathbf{M}_{min}^{-1}. \tag{19}$$

Definition 3. *Let us say that nodes* $s_i, s_j \in S, i, j \in N, i \neq j$ *are loosely ordered in ascending (descending) order of local risk and write* $s_i \preceq s_j (s_i \succeq s_j)$ *if, under a given attack scenario* W, *for any placements* M^{-1}, K^{-1}, *and for any such indices* $p, q, k, l, \ p < q, \ k > l$, *that* $s_i = M(v_p) = K(v_k), \ s_j = M(v_q) = K(v_l)$, *the following inequality holds* $\rho\left(S, W, M^{-1}\right) \leq \rho\left(S, W, K^{-1}\right) \left(\rho\left(S, W, M^{-1}\right) \geq \rho\left(S, W, K^{-1}\right)\right)$.

The following statements were proved in [39].

Proposition 1. *Let* $N = \{1, ..., n\}, S = \{s_1, ..., s_n\}$. *Then* $\forall i \in N \setminus \{n\} \ s_i \preceq s_{i+1} \iff \frac{u_i}{u_{i+1}} \leq \frac{p_{i+1}(1-p_i)}{p_i(1-p_{i+1})}; \ s_i \succeq s_{i+1} \iff \frac{u_i}{u_{i+1}} \geq \frac{p_{i+1}(1-p_i)}{p_i(1-p_{i+1})}.$

Proposition 2. *Let* $N = \{1, ..., n\}, S = \{s_1, ..., s_n\}$. *Then* $\forall i, j, k \in N : i < j < k \; s_i \preceq s_j \preceq s_k \implies s_i \preceq s_k$.

These statements define a transitive criterion for ordering nodes in an attack scenario and allow us to solve the problem (19) for any scenario considered.

4   An Algorithm for Designing the Risk-Minimizing Topology of a Computer Network

Consider a computer network with n nodes specifying the set $S = \{s_1, ..., s_n\}$. Initially, we will assume that all elements are available to the attacker, that is, the starting network model is a complete graph $G(V, E), V = S, E = \cup_{i \neq j}(s_i, s_j), 1 \leq i, j \leq n$, and the possible attack scenarios are routes in it. Here, we will consider scenarios that are simple paths.

Let us assume that the Attacker wants to inflict maximum damage on the considering network. Then he should successfully attack all its nodes without any exceptions. Attacker will solve the inverse of the problem (19) to find the desired route:

$$\mathbf{M}_{max}^{-1} = \operatorname*{Arg\,max}_{M^{-1}} \rho\left(S, W, M^{-1}\right) : \rho_{max} = \sum_{i=1}^{n} \rho_{M(v_i)} \forall M^{-1} \in \mathbf{M}_{max}^{-1}, \quad (20)$$

where $\mathbf{M}_{max}^{-1}$ is a set of placements, for each of the elements of which the maximum value of the integral risk ρ_{max} is achieved.

Taking into account the result described in the previous paragraph, the solution is trivial and consists in choosing a simple path $\left(v_1^A, v_2^A, ..., v_n^A\right)$, including all vertices of the model graph G ordered such that $v_i \succeq v_{i+1} \forall i < n$.

The Defender's problem, in turn, is to direct the Attacker along the least "advantageous" path for him. This path is also easily constructed and, as in the previous case, is a simple path $\left(v_1^D, v_2^D, ..., v_n^D\right)$, including all vertices of the graph G ordered such that $v_i \preceq v_{i+1} \forall i < n$.

Note that in the case when the following relation is fulfilled

$$\frac{1 - p_i}{u_i p_i} = \frac{1 - p_j}{u_j p_j} \iff i = j, i, j \in \{1, ..., , n\}, \quad (21)$$

both problems have a unique solution, and $v_i^D = v_{n-i+1}^A$. In other words, Attacker and Defender tend to implement opposite paths.

Now we can construct the algorithm for solving the Defender's problem under condition (21).

1. Provide a single gateway to the external network (set the perimeter) at node v_1^D (otherwise, v_n^A).
2. Assign the node v_1^D as the current one (put i equal to 1).
3. Sequentially remove the edges connecting the current node v_i^D with the nodes $v_n^D, v_{n-1}^D, ..., v_{i+1}^D$.

4. If $i < n$, then assign the current node $v_{(}i + 1)^D$ (put i equal to $i + 1$).
5. Go to Step 3.

Note that under the following condition

$$\frac{1 - p_i}{u_i p_i} = \frac{1 - p_j}{u_j p_j} \; \forall i, j \in \{1, ...,, n\} \tag{22}$$

nodes become permutation-neutral in the sense of the Proposition 1. In this case, if the Defender knows that he will have some limited resource which he may use to reduce the probabilities of a successful attack of certain nodes, then he can use alternate order criterion. Namely, the Defender will bring closer to the perimeter the nodes that increase their attack resistance with resource allocation more responsively. This problem will be considered in further studies.

5   Conclusion

This study considers the problem of designing the risk-minimizing topology of a computer network. The result obtained is the rule for the placement of network nodes depending on their distance from the network perimeter. Also, the study proposes an algorithm for designing the risk minimizing topology for computer networks under attack scenarios of a simple chain structure.

The formal result obtained is presented in the form of a simple rule and algorithm, which allows to include it into the risk management framework as one of the possible control actions. At the same time, there are easily formulated criteria for its applicability, allowing the risk manager to strictly justify its application or, conversely, rejection in each specific case.

The further development of the study is to examine the possibility of formulating similar rules to construct a network topology that minimizes risk in more complex scenarios. Another promising problem is dynamic topology optimization under attacks of various profiles. Such rules can also be adopted for a risk management framework—therefore, solving these problems will contribute both to expanding the scope of mathematical modeling in risk management and to increasing the validity of decision-making by a risk manager.

References

1. ISO 31000:2018. Risk management—Guidelines. https://www.iso.org/obp/ui/en/#iso:std:iso:31000:ed-2:v1:en. Accessed 12 July 2023
2. ISO/IEC 27005:2022(en). Information security, cybersecurity and privacy protection–Guidance on managing information security risks. https://www.iso.org/obp/ui/en/#iso:std:iso-iec:27005:ed-4:v1:en. Accessed 12 July 2023
3. Pacheco, J., Benitez, V., Félix, L.: Anomaly behavior analysis for IoT network nodes. In: Proceedings of the 3rd International Conference on Future Networks and Distributed Systems, pp. 1–6. ACM Press, Paris (2019)

4. Kavallieratos, G., Spathoulas, G., Katsikas, S.: Cyber risk propagation and optimal selection of cybersecurity controls for complex cyberphysical systems. Sensors **21**(5), e1691 (2021)

5. Naqash, T., Shah, S.H., Islam, M.N.U.: Statistical analysis based intrusion detection system for ultra-high-speed software defined network. Int. J. Parallel Prog. **50**(1), 89–114 (2022)

6. Jinarajadasa, G.M., Liyanage, S.R.: Evolutionary algorithms for enhancing mobile ad hoc network security. In: Bhatt, C., Wu, Yu., Harous, S., Villari, M. (eds.) Security Issues in Fog Computing from 5G to 6G: Architectures, Applications and Solutions, pp. 15–30. Springer, Cham (2022). https://doi.org/10.1007/978-3-031-08254-2_2

7. Li, J., Zhao, Z., Li, R., Zhang, H.: AI-based two-stage intrusion detection for software defined IoT networks. IEEE Internet Things J. **6**(2), 2093–2102 (2019)

8. Thakkar, A., Lohiya, R.: Role of swarm and evolutionary algorithms for intrusion detection system: a survey. Swarm Evol. Comput. **53**, e100631 (2020)

9. Hasan, Z., Jishkariani, M.: Machine learning and data mining methods for cyber security: a survey. Mesopotam. J. Cybersecur. **2022**, 47–56 (2022)

10. Subasi, A., et al.: Intrusion detection in smart grid using data mining techniques. In: Proceedings of the 2018 21st Saudi Computer Society National Computer Conference (NCC), pp. 1–6. IEEE, Riyadh (2018)

11. Roopak, M., Tian, G.Y., Chambers, J.: Deep learning models for cyber security in IoT networks. In: Proceedings of the 2019 9th Annual Computing and Communication Workshop and Conference (CCWC), pp. 452–457. IEEE, Las Vegas (2019)

12. NIST Cybersecurity Framework (2018). https://www.nist.gov/cyberframework. Accessed 13 July 2023

13. Caralli, R.A., Stevens, J.F., Young, L.R., Wilson, W.R.: Introducing Octave Allegro: Improving the Information Security Risk assessment process. Carnegie Mellon University, Hansom AFB (2007)

14. Awad, A.I., Shokry, M., Khalaf, A.A., Abd-Ellah, M.K.: Assessment of potential security risks in advanced metering infrastructure using the OCTAVE Allegro approach. Comput. Electr. Eng. **108**, e108667 (2023)

15. Barrère, M., Hankin, C.: Fault tree analysis: identifying maximum probability minimal Cut sets with MaxSAT. In: 2020 50th Annual IEEE-IFIP International Conference on Dependable Systems and Networks – Supplemental Volume (DSN-S), pp. 53–54. IEEE, Valencia (2020)

16. Yadav, V., Youngblood, R.W., Blanc, K.L.L., Perschon, J., Pitcher, R.: Fault-tree based pevention analysis of cyber-attack scenarios for PRA applications. In: Proceedings of the 2019 Annual Reliability and Maintainability Symposium (RAMS), pp. 1–7. IEEE, Orlando (2019)

17. Tantawy, A., Abdelwahed, S., Erradi, A., Shaban, K.: Model-based risk assessment for cyber physical systems security. Comput. Secur. **96**, e101864 (2020)

18. Zegeye, W.: Quantitative risk assessment tied to HMM based intrusion detection system. In: Proceedings of the 56th Annual International Telemetering Conference (ITC 2021), pp. 104–113. International Foundation for Telemetering, Las Vegas (2021)

19. Hoffmann, R.: Markov model of cyber attack life cycle triggered by software vulnerability. Int. J. Electron. Telecommun. **67**, 35–41 (2021)

20. Giang, V.T.H., Tuan, N.M.: Application of Bayesian network in risk assessment for website deployment scenarios. J. Sci. Technol. Inf. Secur. **2**(14), 3–17 (2021)

21. Du, H., Liu, D.F., Holsopple, J., Yang, S.J.: Toward ensemble characterization and projection of multistage cyber attacks. In: Proceedings of the 19th International Conference on Computer Communications and Networks (ICCCN), pp. 1–8. IEEE, Zurich (2010). https://doi.org/10.1109/ICCCN.2010.5560087
22. Manikas, T.W., Thornton, M.A., Feinstein, D.Y.: Modeling system threat probabilities using mixed-radix multiple-valued logic decision diagrams. Multip. Value Logic Soft Comput. **24**(1–4), 135–149 (2015)
23. Pasandideh, S., Gomes, L., Maló, P.: Improving attack trees analysis using petri net modeling of cyber-attacks. In: Proceedings of the 2019 IEEE 28th International Symposium on Industrial Electronics (ISIE), pp. 1644–1649. IEEE, Vancouver (2019)
24. Berger, S., van Dun, C., Häckel, B.: IT availability risks in smart factory networks-analyzing the effects of IT threats on production processes using petri nets. Inf. Syst. Front. **2022**, 1–20 (2022)
25. Graf, J., Batchelor, W., Harper, S., Marlow, R., Carlisle, E., Athanas, P.: A practical application of game theory to optimize selection of hardware trojan detection strategies. J. Hardw. Syst. Secur. **4**, 98–119 (2020)
26. Hu, H., Liu, Y., Chen, C., Zhang, H., Liu, Y.: Optimal decision making approach for cyber security defense using evolutionary game. IEEE Trans. Netw. Serv. Manage. **17**(3), 1683–1700 (2020)
27. Sokri, A.: Optimal resource allocation in cyber-security: a game theoretic approach. Procedia Comput. Sci. **134**, 283–288 (2018)
28. Zhang, Y., Malacaria, P.: Bayesian Stackelberg games for cyber-security decision support. Decis. Support Syst. **148**, e113599 (2021)
29. Patterson, I., Nutaro, J., Allgood, G., Kuruganti, T., Fugate, D.: Optimizing investments in cyber-security for critical infrastructure. In: Proceedings of the 8th Annual Cyber Security and Information Intelligence Research Workshop (CSIIRW 2013), pp. 1–4. ACM Press, New York (2013). https://doi.org/10.1145/2459976.2459999
30. Wang, C., Hou, Y.: Reliability-based updating strategies of cyber infrastructures. In: Proceedings of the 2015 IEEE Power and Energy Society General Meeting, pp. 1–5. IEEE, Denver (2015). https://doi.org/10.1109/PESGM.2015.7286403
31. Khanna, K., Panigrahi, B.K., Joshi, A.: Bi-level modelling of false data injection attacks on security constrained optimal power flow. IET Gen. Transmiss. Distrib. **11**(14), 3586–3593 (2017)
32. Zheng, K., Albert, L.A.: Interdiction models for delaying adversarial attacks against critical information technology infrastructure. Nav. Res. Logist. **66**(5), 411–429 (2019)
33. Khouzani, M.H., Liu, Z., Malacaria, P.: Scalable min-max multi-objective cyber-security optimisation over probabilistic attack graphs. Eur. J. Oper. Res. **278**(3), 894–903 (2019)
34. Reilly, J., Martin, S., Payer, M., Bayen, A.M.: Creating complex congestion patterns via multi-objective optimal freeway traffic control with application to cyber-security. Transp. Res. Part B: Methodol. **91**, 366–382 (2016)
35. Heyman, D.P., Sobel, M.J.: Stochastic Models in Operations Research: Stochastic Optimization, vol. 2. Dover Publications, Mineola (2004)
36. Zhang, Y., Wang, L., Sun, W.: Trust system design optimization in smart grid network infrastructure. IEEE Trans. Smart Grid **4**(1), 184–195 (2013)
37. Shiroky, A.A., Kalashnikov, A.O.: Natural computing with application to risk management in complex systems. Control Sciences **2021**(4), 2–17 (2021)

38. McManus, H., Hastings, D.: A framework for understanding uncertainty and its mitigation and exploitation in complex systems. IEEE Eng. Manage. Rev. **34**(3), e81 (2006)
39. Shiroky, A.A., Kalashnikov, A.O.: Mathematical problems of managing the risks of complex systems under targeted attacks with known structures. Mathematics **9**(19), e2468 (2021)

Age of Information Performance of Ultra Reliable Low Latency Service in 5G New Radio Networks

Elena Zhbankova[1]([✉]) [iD], Varvara Manaeva[1], Ekaterina Markova[1] [iD],
and Yuliya Gaidamaka[1,2] [iD]

[1] RUDN University, 6 Miklukho-Maklaya Street, Moscow 117198, Russian Federation
{zhbankova-ea,1032201197,markova-ev,gaydamaka-yuv}@rudn.ru
[2] Federal Research Center "Computer Science and Control" of the Russian Academy of Sciences (FRC CSC RAS), 44-2 Vavilov Street, Moscow 119333, Russian Federation

Abstract. Ultra-reliable low latency (URLLC) service is one of the cornerstone services that needs to be supported in fifth-generation (5G) cellular systems. For a class of applications utilizing URLLC service – periodic state updates, a vital metric of interest is Age of Information (AoI) characterizing the timeliness of updates received. The analysis of latency-related metrics in 5G cellular systems is however complicated by the orthogonal-frequency division multiple access (OFDMA) specifics resulting in batch service of packets. In this paper, we formalize the solve a queueing model capturing the specifics of URLLC service over OFDMA-enabled systems by explicitly accounting for batch service. For this model, we derive the mean peak AoI (PAoI). Our numerical results demonstrate, that the PAoI behavior is mainly dictated by the loss performance required by the URLLC service. Specifically, the difference between PAoI is minimal often less than 10%. Thus, we conclude that when the loss guarantees are satisfied, the mean PAoI is insensitive to the choice of system parameters.

Keywords: AoI · Age of Information · Peak Age of Information · 5G · URLLC · OFDMA

1 Introduction

The fifth generation (5G) systems needs to support a plethora of services with different quality of service requirements (QoS) including enhanced mobile broadband (eMBB), massive machine type communications (mMTC), ultra-reliable low latency service (URLLC) as well as different intermediate services. The recently standardized New Radio (NR) interface incorporates various advanced link level capabilities such as flexible frame structure and numerologies, different operational bands, network slicing for traffic isolation, etc. However, aside from

This research was funded by the Russian Science Foundation grant number 22-79-10053 (https://rscf.ru/en/project/22-79-10053/).

eMBB service, the performance of mMTC and URLLC applications running over 5G NR radio interface is still loosely addressed in the literature.

A critical use-case for URLLC service in 5G systems is the exchange of state update information between end devices (ED) and the control center [17,18]. Such use-case finds its application in many mission-critical fields such as telemedicine, control of production lines in industrial environments, automotive networks, video surveillance systems, energy grids. The main metric of interest for such services is the timeliness of the remote system updates available at the receiver.

Conventionally, the performance of time-constrained applications has been evaluated by utilizing latency as the main metric. However, this metric depends on the traffic load in the network that in turn is a function of the update interarrival time at the sources. Recently, a new measure of timeliness for state update services, the so-called Age of Information (AoI) has been proposed [8]. AoI quantifies how fresh the information available at the receiver with respect to the last update generated at the source. The metric is an explicit function of the update interarrival times and presumes that only timely received updates can reflect the current state of the system. AoI allows to describe the detailed behavior of the URLLC service operating over the 5G NR systems and can be considered as a new measure of QoS.

In spite of the significant interest AoI and peak AoI (PAoI) metrics attracted over the last few years, the models developed so far does not account for specifics of URLLC service over 5G systems. First of all, most of the models proposed so far assumed a single ED as an input. Furthermore, orthogonal-frequency division multiple access scheme specifics of 5G interface organization has not been addressed in the existing literature. The rationale is that this access scheme naturally leads to queueing formulations with batch service process. Such queueing systems have been loosely studied in the literature. We aim to fill these gap.

The aim of this paper is to characterize URLLC service performance operating over the 5G cellular network interface with orthogonal-frequency division multiple access (OFDMA) channel organization. To this aim, we first develop a system model that accounts for the specifics of URLLC service organization in 5G systems. Then, we proceed formalizing this model as a queueing system in discrete time and further provide its continuous approximation. We solve the latter for the mean PAoI.

The main contributions of our study are

- System model for the service process of URLLC service in 5G NR systems with OFDMA channel access;
- Mathematical formalization of the service process of URLLC EDs as a queueing model with multiple input flows and batch service time in both discrete and continuous time;
- Numerical results showing that the impact of system parameters such as service rate and buffer size on the PAoI is negligible in the operational regime of the system, where packet loss guarantees of URLLC service are satisfied.

The rest of the paper is organized as follows. We start with the overview of the related work in Sect. 2. In Sect. 3, we formalize our system and specify two models in discrete and continuous time. We then proceed solving the latter in Sect. 4 for PAoI as the metric of interest. The conclusions are drawn in the last section.

2   Related Work

Nowadays, new applications are being actively designed that require nodes to periodically share their time-critical state information with neighboring nodes or a base station (BS). For example, applications in automotive networks, where vehicles share their data with each other to improve road safety, or applications in manufacturing industry, where sensors transmit important information and help avoid machine breakdowns.

The concept of AoI was introduced in 2011 in [8] to quantify the recency of existing knowledge about the state of a remote system. AoI is defined as the time elapsed at the destination since the generation of the last successfully received packet containing information about the state of the source system. The first set of studies investigating AoI characteristics appeared already in 2012 spawning the wave of interest in the research community. Unlike latency (delay), AoI has an interesting property: it increases at both low and high packet generation rates in queueing systems with First Come First Served (FCFS) service discipline.

The AoI has been proposed as a metric that quantifies the novelty of information updates in a communication system. Therefore, researchers have often tried to understand and optimize various AoI statistics for classic queueing systems serving as an abstraction of a network. The authors of [19] studied the AoI distribution for the $GI/GI/1/1$ and $GI/GI/1/2$ systems with non-preemptive service procedure and calculated the upper bounds on AoI using the computer simulations. Next, they analyzed single-source, single-server queueing systems and derived analytical expressions for PAoI in $D/GI/1/1$ and $M/GI/1/1$ systems.

In [9], the authors introduced the mean AoI metric and considered the $M/M/1$, $M/D/1$, and $D/M/1$ queueing systems with FCFS discipline. The authors derived closed-form expressions for the mean AoI in these systems and showed that the lowest mean AoI in systems with FCFS service discipline is achieved if a new packet is generated exactly when the service of the previous packet is completed. The further studies in [4] and [5] considered queueing systems with one or more independent sources that can discard some of the generated packets to improve AoI. The authors considered systems with Poisson arrivals, FCFS service discipline, and exponentially distributed service times with two packet discarding policies modeled by: $M/M/1/1$ and $M/M/1/2$ queueing systems.

An important property of status update systems is the memoryless property of the observed stochastic process. If the received data has this property, then only the last status update is important and the recipient is not required to keep track of a history of updates. This is specifically useful in applications that deal with time-critical information. Therefore, if a packet arrives at its destination after a new packet is generated at the source, then it contains no useful information. Thus, researchers started to investigate systems with Last Come First Served (LCFS) queueing discipline. To this end, the study in [10] for the first time allowed newly generated status updates to outperform old updates. The idea was to use LCFS, whereby the most recent updates take precedence in transmission and replace queued updates with newer ones from the same source. This buffer management mechanism improves the performance of the system with respect to the AoI metric. In [22], $M/M/1$ system with FCFS and LCFS service disciplines were considered. For such systems, the authors obtained and characterized the areas of acceptable AoI.

Aside from the queueing-theoretic studies, applied studies of the AoI in various systems started to appear. As an example, the study in [14] explores AoI in IoT-based control systems operating over the wireless channel. Specifically, the authors characterized AoI in distributed Carrier Sense Multiple Access With Collision Avoidance (CSMA/CA) [6] and centralized Time Division Multiple Access (TDMA) networks. Their analytical results have been confirmed using the simulation studies. The authors demonstrated that different types of access schemes are best suited for different system parameters. Specifically, CSMA and TDMA networks have different attributes that make them useful depending on the performance and specific aspects of the system being designed.

The AoI metric is of special importance for URLLC services that are intended to carry the state update information with strict reliability and timing constraints. However, to the best of the authors knowledge there have been no studies estimating the AoI metric in 5G NR systems with multiple sources, where OFDMA access scheme is utilized.

3   System Model

Consider the system shown in Fig. 1. The 5G NR base station (BS) serves N URLLC service end-devises (ED). We assume that URLLC traffic is completely isolated from other types of traffic at the BS. This can be achieved by using the Network Slicing (NS) function based on resource reservation [12, 21]. The length of one subframe, which is the transmission time interval (TTI), is $T = 10^{-6}$ s, the bandwidth available to URLLC traffic is W MHz. We assume the OFDMA access scheme with the size of one resource block (RB) of R_W MHz. Overall, there are K RBs available at each subframe.

We assume that each URLLC ED generates a packet of length L bits in each TTI with probability a, independently of other EDs. In URLLC applications the value L is often significantly smaller than the size of the RB. Therefore, we assume that one packet is transmitted in exactly one RB, regardless of the type of modulation and coding scheme (MSC) utilized. Finally, we assume that all N URLLC EDs are characterized by the same service priority. By considering these two facts one may observe that the utilized service discipline at the BS is first come first served (FCFS). The size of the buffer at BS is limited to R packets.

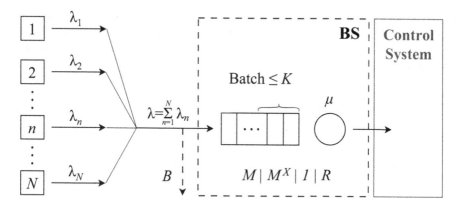

Fig. 1. The considered 5G BS serving URLLC end devices.

Due to the use of OFDMA channel organization the service process of packets is batch in nature. Specifically, we assume that at most K packets can be transmitted per TTI, where $K = \lfloor W/R_W \rfloor$. Here, K can be calculated using the assumptions about the blockage and micro-mobility of user equipment according to the methodology proposed in [11]. We also take into account the requirements for the reliability of URLLC traffic delivery. Specifically, we assume that the probability of erroneous reception of a packet is small which is achieved via repetition coding [7].

That considered technical system can be represented as a queueing system of the form $D^X/D^X/1/R$, where X indicates the batch arrivals and service [3] with binomial distribution of an arrival batch length $B(N, a)$ and at most K customers in a batch under service. Such a model is a special case of more general systems considered in [1,15].

The problem with discrete queueing analysis is that (i) there might be no closed-form solution for stationary probabilities of the number of packets in the system, (ii) the delay distribution is often expressed in terms of infinite sums. Thus, in our paper, we approximate such a discrete system by a continuous system of the form $M/M^X/1/R$ in a way described in [2]. In our case the arrival rate $\lambda = E[A]/T$, where $E[A] = Na$ is the mean arrival batch length, T is the frame duration, the batch service rate is $\mu = 1/T$ and each batch has no more than K packets.

4   Performance Evaluation

We consider an $M/M^X/1/R$ queueing system of capacity R with a single server. The arrival process is a homogeneous Poisson with intensity λ. Packets are served in batches of no more than K packets depending on how many packets are available in the queue at the instant when service starts. The batch service time distribution is exponential with intensity μ. The service starts at the instant the server is released if there are requests in the queue or at the instant the first request arrives if the server is not busy. We consider the FCFS service discipline.

Let $X(t)$ be the Markov process of the packets number in the queue over the state space

$$\mathcal{X} = \{0, 1, \ldots, R\}, \tag{1}$$

with stationary state probabilities

$$q_i = \lim_{t \to \infty} P\{X(t) = i\}, \ i \in \mathcal{X}. \tag{2}$$

The transition intensity diagram of the considered systems is shown in Fig. 2.

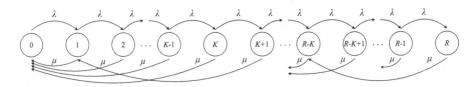

Fig. 2. Transition intensity diagram for $M/M^X/1/R$ queueing system.

Since the queueing system $M/M^X/1/R$ is stable for finite state space $\mathcal{X}$, one may utilize this diagram to define the system of equilibrium equations:

$$\begin{cases} \lambda q_0 = \mu \sum_{i=1}^{K} q_i, \\ (\lambda + \mu) q_i = \lambda q_{i-1} + \mu q_{i+K}, i = \overline{1, R - K}, \\ (\lambda + \mu) q_i = \lambda q_{i-1}, i = \overline{R - K + 1, R - 1}, \\ \mu q_R = \lambda q_{R-1} \end{cases} \tag{3}$$

with normalized condition $\sum_{i=0}^{R} q_i = 1$.

The defined system can be solved numerically using conventional approaches, e.g. Gauss-Zeidel method. Once the stationary distribution is found, the performance measures immediately follow:

– The mean number of packets in the queue

$$Q = \sum_{i=1}^{R} i q_i, \tag{4}$$

– The packet loss rate (PLR)

$$B = q_R, \tag{5}$$

– The mean waiting time

$$\bar{w} = Q/\lambda(1 - B), \tag{6}$$

– The mean PAoI

$$PAoI = \lambda^{-1} + \bar{w} + \mu^{-1}. \tag{7}$$

5 Numerical Results

In this section, we elaborate on our numerical results. To this aim, we assume that the 5G system utilizes NS to provide a certain fixed amount of resources to URLLC service [13,20]. Note that these resources can be converted into the link capacity measured in packets per second by utilizing the approach described in [16].

In what follows, we take the following approach. First, we determine the operational regime of the system by estimating the required amount of resources that need to be allocated for a URLLC slice, such that the packet loss rate (PLR) is less than 10^{-5} to satisfy the requirements of URLLC service, i.e. $B \leq 10^{-5}$. Then, we proceed assessing the delay and PAoI performance of the system.

5.1 Required Amount of Resources

In order to investigate PAoI performance in operational URLLC conditions, we first need to determine the operational regime of the system, where reliability guarantees are satisfied. To this aim, we start with Fig. 3 showing the PLR as a function of the arrival packet intensity λ and for different batch sizes of K packets and buffer size of just 20 packets. As one may observe, for the considered range of arrival intensity the PLR varies from 10^{-8} to 10^{-3}. Recalling that the target PLR for URLLC service is 10^{-6}, we see that for the selected buffer and batch sizes λ of up to around 7000 packets/s can be supported.

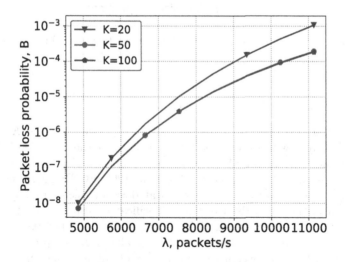

Fig. 3. PLR as a function of the packet arrival intensity.

In addition to λ, the operational regime of the system is also affected by the buffer and batch sizes. Figure 4 shows the dependence of PLR on these parameters for two distinctive values of λ, i.e. 5×10^3 and 10^4. By analyzing the presented data, we observe that for the selected values of λ the impact of the buffer size is much more profound as compared to the size of the batch. Indeed, even for $\lambda = 10^3$ packets/s increasing the batch size from 50 packets to 150–200 packets allows to improve the PLR from 10^{-2} to approximately 10^{-6}. Increasing the amount of allocated resources by increasing the batch size from 20 packets to 30 packets leads to the gradual decrease in the PLR and then all the curves plateaus as seen in Fig. 5.

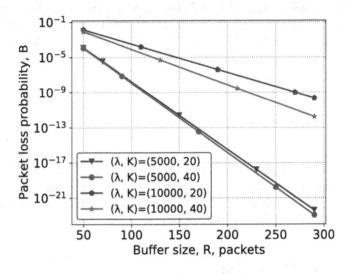

Fig. 4. PLR as a function of R and K: as a function of the buffer space.

5.2   Delay and PAoI Performance

Having identified the operational regime of the service, including the K and R for considered λ, we now proceed assessing the behavior of the mean PAoI. We start with Fig. 6 showing the mean PAoI as a function of the arrival packet intensity λ for two selected sets of (R, K), i.e. $(200, 20)$ and $(200, 40)$. By analyzing the presented data, we see that the amount of allocated resources captured by K affects the behavior of the mean PAoI metrics. Specifically, for $(R, K) = (200, 20)$ the optimal value of λ minimizing the PAoI is around 6×10^3. However, by improving the batch size from 20 to 40 packets, this value increases proportionally to around 10^4 packets/s. Nevertheless, we note that the difference between PAoI corresponding to optimal and non-optimal values of λ is minimal often less than 10% of the optimal value.

We now proceed quantifying the impact of the buffer and batch size on the mean PAoI illustrated in Fig. 7. Here, we see that for the low arrival packets intensity of 5×10^3, the impact of the buffer size is negligible. When the λ

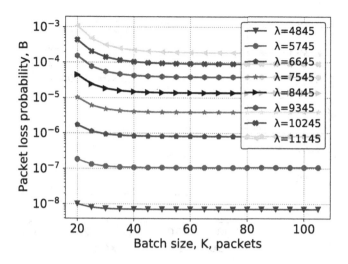

Fig. 5. PLR as a function of R and K: as a function of the batch size.

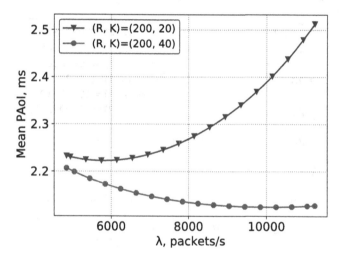

Fig. 6. Mean PAoI as a function of the packet arrival intensity.

increases there is a noticeable jump for small values of the buffer size. However, even for the couple $(\lambda, K) = (10^4, 20)$ it is bounded by approximately 0.1 ms which is around 5% of the absolute value. The impact of the amount of allocated resources captured by K in Fig. 8 is more profound. However, even for $\lambda = 10^4$ packets/s, the mean PAoI decreases by at most 0.3 ms which is slightly more than 10%.

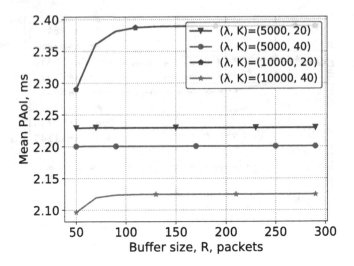

Fig. 7. Mean PAoI as a function of R and K: as a function of the buffer space.

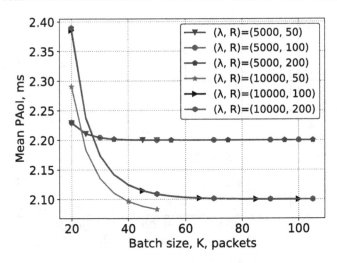

Fig. 8. Mean PAoI as a function of R and K: as a function of the batch size.

6   Conclusion

In this paper, we investigated the PAoI performance of URLLC services in 5G NR systems. To this aim, we first proposed a system model of a 5G NR BS serving multiple URLLC UEs. Then, we proceeded formalizing the queueing models in both discrete and continuous times by taking into account the specifics of the OFDMA-based access. For the latter model, we derived the mean PAoI time.

Our numerical results for a range of close to real values of system parameters show that for the operational regime of the system, where the PLR is kept around

10^{-6} as required by URLLC service, the impact of the buffer size and the amount of allocated resources on mean PAoI is negligible. Furthermore, the difference between PAoI corresponding to optimal and non-optimal values of λ is minimal often less than 10%. In general, we can conclude that when loss guarantees are satisfied the mean PAoI is insensitive to the choice of system parameters.

References

1. Alfa, A.S.: Queueing Theory for Telecommunications: Discrete Time Modelling of a Single Node System. Springer, Boston (2010). https://doi.org/10.1007/978-1-4419-7314-6
2. Basharin G.P., E.V.: Issledovanie odnolineynoy sistemy s zayavkami neskol'kikh tipov v diskretnom vremeni. Probl. Inf. Transmiss. **20**(1), 95–104 (1984)
3. Chaudhry, M., Templeton, J.G.: First Course in Bulk Queues. Wiley (1983)
4. Costa, M., Codreanu, M., Ephremides, A.: Age of Information with packet management. In: 2014 IEEE International Symposium on Information Theory, pp. 1583–1587. IEEE (2014)
5. Costa, M., Codreanu, M., Ephremides, A.: On the age of information in status update systems with packet management. IEEE Trans. Inf. Theory **62**(4), 1897–1910 (2016)
6. Cuozzo, G., et al.: Enabling URLLC in 5G NR IIoT networks: a full-stack end-to-end analysis. In: 2022 Joint European Conference on Networks and Communications and 6G Summit (EuCNC/6G Summit), pp. 333–338. IEEE (2022)
7. Ivanova, D., Markova, E., Moltchanov, D., Pirmagomedov, R., Koucheryavy, Y., Samouylov, K.: Performance of priority-based traffic coexistence strategies in 5G mmWave industrial deployments. IEEE Access **10**, 9241–9256 (2022)
8. Kaul, S., Gruteser, M., Rai, V., Kenney, J.: Minimizing age of information in vehicular networks. In: 2011 8th Annual IEEE Communications Society Conference on Sensor, Mesh and Ad Hoc Communications and Networks, pp. 350–358. IEEE (2011)
9. Kaul, S., Yates, R., Gruteser, M.: Real-time status: how often should one update? In: 2012 Proceedings IEEE INFOCOM, pp. 2731–2735. IEEE (2012)
10. Kaul, S.K., Yates, R.D., Gruteser, M.: Status updates through queues. In: 2012 46th Annual Conference on Information Sciences and Systems (CISS), pp. 1–6. IEEE (2012)
11. Khayrov, E.M., Prosvirov, V.A., Platonova, A.: Traffic arrival model for millimeter wave 5G NR systems. In: Vishnevskiy, V.M., Samouylov, K.E., Kozyrev, D.V. (eds.) Distributed Computer and Communication Networks: Control, Computation, Communications, DCCN 2022. LNCS, vol. 13766, pp. 161–175. Springer, Cham (2022). https://doi.org/10.1007/978-3-031-23207-7_13
12. Kochetkova, I., Vlaskina, A., Burtseva, S., Savich, V., Hosek, J.: Analyzing the effectiveness of dynamic network slicing procedure in 5G network by queuing and simulation models. In: Galinina, O., Andreev, S., Balandin, S., Koucheryavy, Y. (eds.) Internet of Things, Smart Spaces, and Next Generation Networks and Systems. NEW2AN ruSMART 2020. LNCS, vol. 12525, pp. 71–85. Springer, Cham (2020). https://doi.org/10.1007/978-3-030-65726-0_7
13. Koucheryavy, Y., Lisovskaya, E., Moltchanov, D., Kovalchukov, R., Samuylov, A.: Quantifying the millimeter wave new radio base stations density for network slicing with prescribed SLAs. Comput. Commun. **174**, 13–27 (2021)

14. Mena, J.P., Núñez, F.: Age of Information in IoT-based networked control systems: a MAC perspective. Automatica **147**, 110652 (2023)
15. Moltchanov, D., Koucheryavy, Y., Harju, J.: Loss performance model for wireless channels with autocorrelated arrivals and losses. Comput. Commun. **29**(13–14), 2646–2660 (2006)
16. Moltchanov, D., Sopin, E., Begishev, V., Samuylov, A., Koucheryavy, Y., Samouylov, K.: A tutorial on mathematical modeling of 5G/6G millimeter wave and terahertz cellular systems. IEEE Commun. Surv. Tutor. **24**(2), 1072–1116 (2022)
17. Okano, M.T.: IoT and industry 4.0: the industrial new revolution. In: International Conference on Management and Information Systems, vol. 25, pp. 75–82 (2017)
18. Schulz, P., et al.: Latency critical IoT applications in 5G: perspective on the design of radio interface and network architecture. IEEE Commun. Mag. **55**(2), 70–78 (2017)
19. Yang, H.H., Song, M., Xu, C., Wang, X., Quek, T.Q.: Locally adaptive status updating for optimizing age of information in poisson networks. IEEE Trans. Mobile Comput. 1–13 (2022)
20. Yarkina, N., Correia, L.M., Moltchanov, D., Gaidamaka, Y., Samouylov, K.: Multi-tenant resource sharing with equitable-priority-based performance isolation of slices for 5G cellular systems. Comput. Commun. **188**, 39–51 (2022)
21. Yarkina, N., Gaydamaka, A., Moltchanov, D., Koucheryavy, Y.: Performance assessment of an ITU-T compliant machine learning enhancements for 5 G RAN network slicing. IEEE Trans. Mobile Comput. 1–17 (2022)
22. Yates, R.D.: The age of information in networks: moments, distributions, and sampling. IEEE Trans. Inf. Theory **66**(9), 5712–5728 (2020)

On the Classification of Cytological Images of Leukocytes Using Depthwise Separable Convolutional Neural Networks

E. Yu. Shchetinin[1] , A. G. Glushkova[2] , A. V. Demidova[3] ,
and L. A. Sevastianov[3,4(✉)]

[1] Financial University under the Government of the Russian Federation, 49
Leningradsky Prospekt, 125993 Moscow, Russia
riviera-molto@mail.ru
[2] Endeavor, Chiswick Park, 566 Chiswick High Road, London W4 5HR, UK
[3] RUDN University, 6, Miklukho-Maklaya Street, Moscow 117198, Russia
demidova-av@rudn.ru , sevastianov-la@rudn.ru
[4] Joint Institute for Nuclear Research, 6, Joliot-Curie Street, Dubna, Moscow Region
141980, Russia

Abstract. Blood cell analysis is the most important diagnostic process in medical practice. In particular, the detection of white blood cells (WBCs) is necessary for the diagnosis of many diseases. Manual screening of blood smears is labor-intensive and subjective, and can lead to inconsistencies and errors. However, automated blood cell detection can improve the accuracy and efficiency of the screening process. In this paper, a computer-based approach to the classification and detection of white blood cells in cytological images of blood cells using deep learning methods is implemented. A model for the classification of blood images, LeucoCyteNetv2, is proposed, which includes depthwise separable convolutional layers, Separable-Conv2D, in its architecture. The developed model classifies leukocytes with an accuracy of 98.86%, which made it possible to confirm its use as an auxiliary tool for hematological blood analysis. The proposed model represents a good alternative for automated diagnostics to support hematologists in the clinical laboratory in the evaluation of leukocytes in blood from blood smear images.

Keywords: leukocytes · classification · deep learning · depthwise separable convolutional neural networks

1 Introduction

Leukocytes, also known as white blood cells (WBC), play an important role in protecting the human body against foreign invaders, including bacteria and viruses, and the dangerous diseases they cause [24]. In addition, the presence of

This paper has been supported by the RUDN University Strategic Academic Leadership Program (recipients L. A. Sevastianov, A. V. Demidova).

leukocytes is also used as an indicator for the detection and diagnosis of diseases such as leukemia, anemia, cancer and others. Reliable detection of malignant leukocytes is a key step in the diagnosis of hematological malignancies, such as acute myeloid leukemia [3,6].

Blood consists of three different components: white blood cells, red blood cells and platelets. White blood cells consist of five different cell types. There are 0–1% basophils, 2–10% monocytes, 1–5% eosinophils, 20–45% lymphocytes and 50–70% neutrophils. Basophils are a type of WBC that helps prevent and treat wound infections. They contain essential substances that help relieve allergies and control blood clotting. Monocytes remove microorganisms, unknown agents and dead cells. Eosinophils are white blood cells that control the inflammation that causes asthma and allergies. Eosinophils fight and prevent parasitic and bacterial infections. Neutrophils are the most common type of white blood cell in the body. This type of cell protects against infections caused by toxic substances, fights viruses, bacteria, fungi and even cancer cells. Lymphocytes fight and prevent infections [4]. Examples of WBC images are shown in Fig. 1.

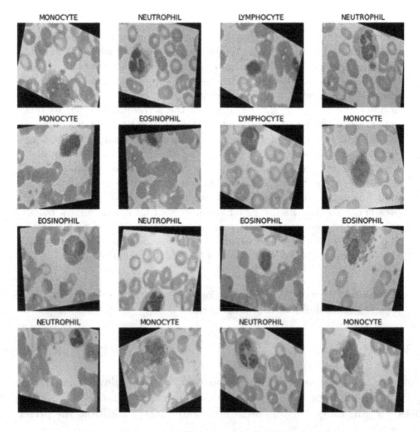

Fig. 1. Images of different classes of blood cells

The main clinical methods for leukocyte classification include flow cytometry and visual inspection. Microscopic morphological examination of blood cells is typically performed by trained hematologist experts. However, this process is laborious, time-consuming, and challenging to automate. Additionally, the low quality of the original medical images can significantly affect the accuracy of the analysis and the conclusions drawn.

Manual leukocyte counting is based on microscopic observation of a blood smear by an analyst who can differentiate subtypes mainly on the basis of morphological characteristics of the cell nucleus and cytosol. However, this process is highly dependent on the analyst's time and experience, which leads to errors if the analyst is not sufficiently trained. In addition, because this hematological evaluation is often in high demand in clinical laboratories, this represents an increased workload affecting productivity. Thus, the development of automated diagnostic tools for diagnostic assistance in the laboratory is relevant.

Such automated leukocyte classification allows for faster and more reproducible results, reducing error and variability of results. On the other hand, due to the complexity inherent in the observation of medical examinations, their automation is still a challenge for researchers. Achieving a level of accuracy comparable to that of a professional in the field is critical. Automation is subject to error or bias in the recognition and classification of these images. They will directly impact the diagnosis, increasing the cost of treatment and negatively impacting patient recovery and survival.

The main component of an automated system for WBC classification is an algorithm for cell detection and segmentation. Based on image processing analysis, it identifies different elements of interest, taking into account various aspects related to cell morphology: size, shape, texture, nucleus, etc. As a rule, cell segmentation is a difficult task in tissue samples. However, in cell smears this task is relatively simpler, given the dark-stained nucleus of leukocytes. The difficulty lies mainly in determining cell boundaries, separating overlapping cells, and removing noise and artifacts in the image during image acquisition.

2 State of Research on Leukemia Detection and Classification

The problem of white blood cell digital image classification using deep learning methods has been the subject of a significant number of publications. In [2] it was proposed to use deep convolutional network models VGG-16, Inceptionv3 to extract features of different WBC classes, achieving an average classification accuracy of 96.2%. In [19], a deep learning model utilizing DenseNet121 was implemented for the classification of various types of WBCs. The DenseNet121 model was optimized using data normalization and augmentation techniques, resulting in a classification accuracy of 98.84%.

In a study [25], a two-module weighted optimized deformable convolutional neural network method was proposed for WBC classification. The method was compared with classical deep learning models VGG16, VGG19, Inception-V3,

ResNet-50, as well as support vector machine (SVM), multilayer perceptron (MLP), decision tree (DT) and random forest (RF) to check the effectiveness. The authors of [5] optimized the loss function to improve WBC CNN classification using regularization and weighted loss function. Jiang et al [11] constructed a new CNN model called WBCNet, which can fully and entail the features of the WBC image by combining a batch normalization algorithm, a residual convolution architecture and an improved activation function. Similarly, Khan et al [12] presented a new model called MLANet-FS, which combined AlexNet with a feature selection strategy for WBC type identification.

Another approach is the application of hybrid deep learning method as proposed, for example, in [7]. It presents an approach that combines AlexNet and GoogleNet models to extract features from WBC images. These features are then combined and classified using the SVM method. Similarly, Özyurt [15] proposed ensemble of convolutional networks for WBC detection, where pre-trained AlexNet, VGG-16, GoogleNet and ResNet deep neural network models were used as feature extractors. The extracted features were then combined and the ELM algorithm was used for further classification. Patil et al. [16] presented a deep hybrid model based on convolutional and recurrent neural networks (RNN) with canonical correlation analysis to extract overlapping and multiple kernels from blood cell images. In [22] the blood cell image were preprocessed and segmented using the SegNet deep learning model. Then, the features were extracted using the EfficientNet neural network. Finally, the WBCs were classified into four different types using the XGBoost classifier. The evaluation results showed that the proposed approach achieved an accuracy of 99.02%.

3   Our Contribution

In this paper we have used popular convolutional neural network models VGG16, Res-Net50, DensNet201, InceptionResnetv2, MobileNetv3, and also developed a custom deep learning model. The significance of its creation is due to the fact that most studies on WBC classification have been performed only for binary classification of WBC or leukemia detection, while a small existing portion of research on multiclass WBC classification have shown low accuracy. However, none of these methods have an architecture similar to that proposed in our study, so we believe that the proposed LeucoCyteNetv2 model is novel and can make an effective contribution to the problem of leukocyte classification.

The goal of this research is to develop a computer system for the classification of leukocytes of four types of blood cells: eosinophils (Eosinophil, E.P.), lymphocytes (Lymphocyte, L.C.), monocytes (Monocyte, M.C.), neutrophils (Neutrophil, N.P.). The contribution of this paper to research on WBC classification using deep learning can be summarized as follows:

- We proposed a deep neural network model, LeucoCyteNetv2, for blood cell classification, utilizing depthwise separable convolutional layers (SeparableConv2D) in its architecture.

– Performance evaluation of the LeucoCyteNetv2 model using an extended set of cytological images of WBC obtained by applying different digital transformations to them was performed, and the average classification accuracy of 98.86% was achieved.
– A comparative analysis of the performance indicators of the model with other models is performed, and it is shown that the proposed model LeucoCyteNetv2 model exceeds or is comparable in accuracy to other most popular convolutional network-based architectures in image classification and computer vision when applied to the WBC classification tasks.

4 Data Description and Preprocessing

The dataset used in this study is based on a public dataset [14]. The dataset consists of 9 663 WBC images divided into four classes: {'Lymphocyte, L.C.', 'Eosinophil, E.P.', 'Monocyte, M.C.', 'Neutrophil, N.P.'}. Initially, the images were presented in RGB format with pixel values ranging from 0 to 255 and had the same size of $(320 \times 240 \times 3)$ pixels. During the image processing the normalization procedure was applied to scale the pixel values to the range of 0 to 1. Also augmentation techniques were employed to increase the number of images and balance the classes. In particular, transformations such as rotation, brightness and scaling were implemented. After the class balancing, the total number of images reached 12 444. Thus, the dataset consists of four classes of images: eosinophils, lymphocytes, monocytes, and neutrophils, which have a total of 3 110, 3 103, 3 098 and 3123 images, respectively. The images were resized from their original (320×240) to (120×120) pixels, which was necessary for more convenient use in computer experiments with neural networks. The expanded set of images was further divided into training, verification and test samples with image sizes of 9 716, 1 215 and 1 215 respectively.

5 Computer Experiments

In this paper we carried out computer experiments on the classification of digital images of blood cells using the data described above and deep-learning models. The following neural network architectures were implemented: VGG16 [21], ResNet50 [10], InceptionResNetv2 [23], Dense201 [9], MobileNetv2 [18]. These models were chosen due to their great popularity among researchers in various fields of artificial intelligence which use computer vision. We also proposed our own model of deep neural network that uses SeparableConv2D depth-separable convolutional layers in its architecture [1].

The SeparableConv2D convolutional layer is a variation of the classical convolutional layer designed for faster computation. It performs a spatial depth convolution followed by a point convolution that blends the resulting output channels at the output. Such layers are computationally more efficient as they require less computer resources, thus providing more accurate and faster results of computer experiments.

Let us describe the architecture of the proposed LeucoCyteNetv2 model of WBC image classification. It consists of an input layer which receives the input image with dimensions (120,120,3) and then passes it to the first convolution block. This, in turn, consists of the following layers: convolutional layer Conv2D(16,(3,3)) and convolutional layer Conv2D(16,(3,3)) each with activation function 'ReLu', layer MaxPooling2D(pool_size = (2,2)). This is followed by a second convolutional block consisting of two identical separable convolutional layers Separable2D(32,(3,3), activation = 'relu', padding = 'same'). Next, layers BatchNormalization() and MaxPooling2D(pool_size = (2,2)) are included.

The third layer contains two separable convolution layers Separable2D(64,(3,3), activation = 'relu', padding = 'same'). Next, layer BatchNormalization() and MaxPooling2D(pool_size = (2,2)) are included. The fourth layer contains two separable convolution layers Separable2D(128,(3,3), activation = 'relu', padding = 'same'). Then, the layers BatchNormalization() and MaxPooling2D(pool_size = (2,2)) are included. Finally, thinning layer Dropout(0.2) is included into this block. The fifth layer contains two Separable2D(256,(3,3), activation = 'relu', padding = 'same') layers. In the end, BatchNormalization(), MaxPooling2D(pool_size = (2,2)), and Dropout(0.2) are included.

The architecture of the LeucoCyteNetv2 model is concluded with the alignment layer Flatten(), the full-bound layers Dense(units = 512, activation = 'relu'), Dense(units = 128, activation = 'relu'), Dense(units = 64, activation = 'relu'), and the output layer Dense(units = 4, activation = 'softmax') with softmax activation function and 4 output nodes. Thus, the model is compiled with loss function 'sparse_categorical_crossentropy' and optimizer function 'Adam'. During training, the number of epochs was 100 and the training rate was 1.E-06. A visualization of the architecture of the proposed Leuco-CyteNetv2 neural network model is shown in Fig. 2. We have run our models with Kaggle and Google Colab web-environment with GPU, and the models were implemented with Keras and TensorFlow [8].

The deep neural network models VGG16, ResNet50, Inception-ResNetv2, Dense201, MobileNetv2 were pre-trained on the ImageNet-1k image set [13] and then tuned on the digital blood cell image set described above. At this stage, all models were evaluated using the four metrics: accuracy, precision, recall and F1-score [17,20].

$$accuracy = \frac{TP + TN}{TP + FP + TN + FN}, \tag{1}$$

$$recall = \frac{TP}{TP + FN}, \tag{2}$$

$$precision = \frac{TP}{TP + FP}, \tag{3}$$

$$F1 - score = \frac{2 \times (precision - recall)}{(precision + recall)}, \tag{4}$$

where (TP) – true positive – indicates the number of precisely identified blood cells, while the number of erroneously recognized blood cell types is shown by

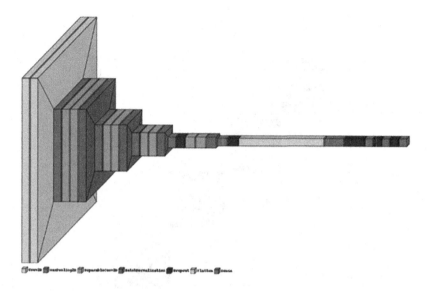

Fig. 2. Block diagram of LeucoCyteNetv2 neural network architecture

false negative (FN). The number of cells correctly identified as the wrong blood cell type is called true negative (TN), and the number of cells incorrectly identified as the wrong blood cell type is called false positive (FP). Values (1)–(4) are calculated from the error matrix for each class, which is shown in Fig. 3. We can also see the accuracy of the classification of the individual types of leukocytes. For eosinophils it is 97.4%, for lymphocytes – 99.66%, for monocytes – 98.7%, for neutrophils – 98.03%. Calculated accuracy, precision, recall, and F1-score values are shown in Table 1.

Table 1. Metrics calculation results for classification of blood cell images.

Model	Accuracy, %	Precision, %	Recall, %	f1-score, %
VGG16	93	90	87.51	88.84
ResNet50	89	91.13	88.52	89.1
Inception_ResNetv2	95	89.74	90.0	94.11
DenseNet201	9	88.87	93.42	91
LeucoCyteNetv2	98.86	99.56	98.89	99.22
MobileNet_v3	92	87.58	90.16	88.93

Figure 4 shows the results of WBC image class prediction from the test set built using the LeucoCyteNetv2 model. It includes real image class, prediction of image class, value of loss function, probability of the given image belonging to the specified class.

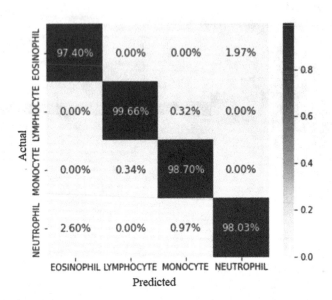

Fig. 3. Error matrix built on the test set of WBC images using LeucoCyteNetv2 model

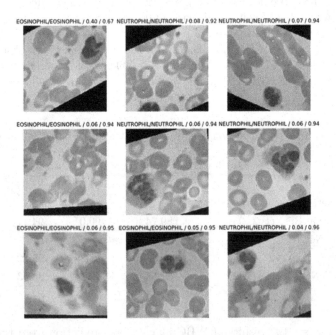

Fig. 4. Prediction of WBC image class from the test set

The results of this paper were also compared with the results obtained in the research of other authors. They are presented in Table 2, from which we can see that the results obtained in this paper either exceed or are comparable with other papers' results. In particular, the research [5] obtained an accuracy of 99.1%, but the classification was carried out only for two classes of images.

Table 2. Results of comparative analysis of blood cell classification, where '–' means that the author did not cite the value of this indicator in the publication.

Paper	Accuracy, %	F1-score, %	Model
A. Acevedo et al. [2]	95.70	–	VGG16, Inceptionv3
S. Sharma et al. [19]	98.84	–	Dense121
X. Yao et al. [25]	95.70	95.00	Deformable CNN
J. Basnet et al. [5]	98.80	97.70	DCNN
M. Jiang et al [11]	83.00	–	AlexNet, GoogleNet, WBCNet
A. Khan et al. [12]	99.10	99.00	DCNN
A. Cinar et al [7]	99.00	99.00	Ensemble of CNN models and SVM
F. Ozyurt et al. [15]	95.29	95.00	Ensemble of CNN models with ELM classifier
A.M. Patil et al. [16]	95.90	95.80	Xception
SivaRao BBS [22]	99.02	98.86	EfficientNet+XGB
Our proposal	98.84		LeucoCyteNetv2

6   Conclusion

This paper investigated the problem of classifying blood cells based on the analysis of their histological images using deep learning methods. An image set containing 12 444 cytological images of blood cells with 3 120 images of eosinophils, 3 103 leukocytes, 3 098 monocytes and 3 123 images of neutrophils was used as a study data set. The proposed LeucoCyteNetv2 deep learning model, which uses SeparableConv2D depthwise separable convolutional layers in its architecture, provided the best or comparable results to the other deep learning models discussed in this article. It achieved accuracy of 98.86%, precision of 99.56%, recall of 98.89% and F1-score of 99.22%.

The high accuracy of the developed model indicates the possibility to use artificial neural networks as an auxiliary tool for leukocyte classification in hematological blood analysis. The proposed model can be used for rapid pre-screening and quantitative decision-making by cytology experts and can increase its efficiency when combined with additional quantitative methods used in the diagnosis of hematological malignancies, such as flow cytometry or molecular genetics.

References

1. Keras, SeparableConv2D layer (2023). https://keras.io/api/layers/convolution_layers/separable_convolution2d
2. Acevedo, A., Alférez, S., Merino, A., Puigví, L., Rodellar, J.: Recognition of peripheral blood cell images using convolutional neural networks. Comput. Methods Programs Biomed. **180**, 105020 (2019). https://doi.org/10.1016/j.cmpb.2019.105020
3. Ahmad, Z., et al.: Immunology in Medical Biotechnology. In: Anwar, M., Rather, R.A., Farooq, Z. (eds.) Fundamentals and Advances in Medical Biotechnology, pp. 179–207. Springer, Heidelberg (2022). https://doi.org/10.1007/978-3-030-98554-7_6
4. Al-Dulaimi, K.A.K., Banks, J., Chandran, V., Tomeo-Reyes, I., Thanh, K.N.: Classification of white blood cell types from microscope images: techniques and challenges. In: Mendez-Vilas, A., Torres-Hergueta, E. (eds.) Microscopy science: Last approaches on educational programs and applied research (Microscopy Book Series, 8), pp. 17–25. Formatex Research Center, Spain (2018). https://eprints.qut.edu.au/121783/
5. Basnet, J., Alsadoon, A., Prasad, P.W.C., Aloussi, S.A., Alsadoon, O.H.: A novel solution of using deep learning for white blood cells classification: enhanced loss function with regularization and weighted loss (elfrwl). Neural Process. Lett. **52**, 1517–1553 (2020). https://doi.org/10.1007/s11063-020-10321-9
6. Bonilla, M., Menell, J.: Chapter 13-disorders of white blood cells. In: Lanzkowsky's Manual of Pediatric Hematology and Oncology. Elsevier, Amsterdam (2016)
7. Çinar, A., Arslan Tuncer, S.: Classification of lymphocytes, monocytes, eosinophils, and neutrophils on white blood cells using hybrid alexnet-googlenet-svm. SN Applied Sciences **3**, 503 (2021). https://doi.org/10.1007/s42452-021-04485-9
8. Chollet, F.: Deep Learning with Python. Manning Publishing, Shelter Island (2018)
9. Goodfellow, I., Bengio, Y., Courville, A.: Deep Learning. MIT Press, Cambridge (2016)
10. He, K., Zhang, X., Ren, S., Sun, J.: Deep residual learning for image recognition (2016). https://doi.org/10.48550/arXiv.1512.03385. https://arxiv.org/abs/1512.03385
11. Jiang, M., Cheng, L., Qin, F., Du, L., Zhang, M.: White blood cells classification with deep convolutional neural networks. Int. J. Pattern Recogn. Artif. Intell. **32**, 1857006 (2018). https://doi.org/10.1142/S0218001418570069
12. Khan, A., Eker, A., Chefranov, A., Demirel, H.: White blood cell type identification using multi-layer convolutional features with an extreme-learning machine. Biomed. Signal Process. Control **69**, 102932 (2021). https://doi.org/10.1016/j.bspc.2021.102932
13. Krizhevsky, A., Sutskever, I., Hinton, G.E.: Imagenet classification with deep convolutional neural networks. Commun. ACM **60**(6), 84–90 (2017). https://doi.org/10.1145/3065386
14. Mooney, P.: Blood cell images (2018). https://www.kaggle.com/paultimothymooney/blood-cells. Accessed 10 June 2023
15. Ozyurt, F.: A fused CNN model for WBC detection with feature selection and extreme learning machine. Soft. Comput. **24**(11), 8163–8172 (2020). https://doi.org/10.1007/s00500-019-04383-8
16. Patil, A.M., Patil, M.D., Birajdar, G.K.: White blood cells image classification using deep learning with canonical correlation analysis. Innov. Res. BioMed. Eng. **42**(5), 378–389 (2021). https://doi.org/10.1016/j.irbm.2020.08.005

17. Raschka, S., Vahid, M.: Python Machine Learning, 3rd edn. Packt Publishing, Birmingham (2019)
18. Sandler, M., Howard, A., Zhu, M., Zhmoginov, A., Chen, L.C.: Mobilenetv 2: inverted residuals and linear bottlenecks. In: 2018 IEEE/CVF Conference on Computer Vision and Pattern Recognition, pp. 4510–4520 (2018). https://doi.org/10.1109/CVPR.2018.00474
19. Sharma, S.: Deep learning model for the automatic classification of white blood cells. Comput. Intell. Neurosci. **2022**, 7384131 (2022). https://doi.org/10.1155/2022/7384131
20. Shchetinin, E.Y., Glushkova, A.G.: Arrhythmia detection using resampling and deep learning methods on unbalanced data. Comput. Opt. **46**(6), 980–987 (2022). https://doi.org/10.18287/2412-6179-CO-1112
21. Simonyan, K., Zisserman, A.J.: Very deep convolutional networks for large-scale image recognition (2014). https://doi.org/10.48550/arXiv.1409.1556. https://arxiv.org/abs/1409.1556
22. SivaRao, B.S.S., Rao, B.S.: Efficientnet - xgboost: an effective white-blood-cell segmentation and classification framework. Nano Biomed. Eng. (2023). https://doi.org/10.26599/NBE.2023.9290014
23. Szegedy, C., Vanhoucke, V., Ioffe, S., Shlens, J., Wojna, Z.: Rethinking the inception architecture for computer vision (2015). https://doi.org/10.48550/arXiv.1512.00567. https://arxiv.org/abs/1512.00567
24. Tkachuk, D.C., Hirschmann, J.V.: Wintrobe's Atlas of Clinical Hematology. Lippincott Williams & Wilkins, Philadelphia (2017)
25. Yao, X., Sun, K., Bu, X., Zhao, C., Jin, Y.: Classification of white blood cells using weighted optimized deformable convolutional neural networks. Artif. Cells Nanomed. Biotechnol. **49**(1), 147–155 (2021). https://doi.org/10.1080/21691401.2021.1879823

Utilization of Machine Learning Algorithms to Identify User Applications

Svetlana Dugaeva[1] , Vyacheslav Begishev[1(✉)] , and Nikita Stepanov[2]

[1] Peoples' Friendship University of Russia (RUDN University), 6 Miklukho -Maklaya Str., Moscow 117198, Russia
{dugaeva-sa,begishev-vo}@rudn.ru
[2] Saint-Petersburg State University of Aerospace Instrumentation (GUAP University), Bolshaya Morskaya Str., Saint Petersburg 190000, Russia
stepanov.nikita@guap.ru

Abstract. Beamtracking is a critical functionality in modern millimeter wave (mmWave) 5G New Radio (NR) systems and is expected to become even more critical in future 6G systems operating in terahertz (THz) frequency band. To enable uninterrupted connectivity base stations (BS) need to invoke this procedure periodically. Due to the use of massive antenna arrays in 6G THz systems, the amount of resources consumed by beamtracking will be extremely large making the time interval between sweeping beam configurations a very critical parameter. One of the phenomena affecting the choice of this interval is a user equipment (UE) micromobility – quick displacements and rotations of UE in the hands of a user happening even when the latter is in a stationary position. In this paper, by utilizing machine learning (ML) algorithms, we propose a procedure for the detection of the beam center at the BS side for applications characterized by different types of micromobility. We demonstrate that one can safely differentiate between applications characterized by low as well as distinctively different micromobility speeds. All the considered classifieds including the tree, random forest, and neural network perform qualitatively similarly. For applications having fast and similar micromobility speeds such as VR and gaming, the classification accuracy stays at around 85–90%. However, this loss in accuracy does not affect the ultimate goal of the remote application detection algorithm – understanding how often the beam alignment procedure must be invoked at UE and BS.

Keywords: 6G · terahertz · cellular systems · micromobility · application detection

Sections 2 and 4 were written by Vyacheslav Begishev under the support of the Russian Science Foundation, project No. 21-79-10139. This publication has been supported by the RUDN University Scientific Projects Grant System, project No. 021928-2-074 (recipients Svetlana Dugaeva, Sections 3 and 5).

V. M. Vishnevskiy et al. (Eds.): DCCN 2023, LNCS 14123, pp. 410–422, 2024.
https://doi.org/10.1007/978-3-031-50482-2_32

1 Introduction

Enabling the capacity boost at the access interface of future cellular systems is only feasible by providing more bandwidth. To this aim, 5G New Radio systems specified operation in the so-called millimeter frequency band at carriers around 28, 38, and 72 GHz, where multiple channels each 400 MHz wide are available [20]. Future 6G terahertz (THz) systems are expected to push the frequency bands even higher to the 100–300 GHz range with bandwidth expected to the on the order of 2 GHz per single channel [12].

To compensate for small antenna aperture at mmWave/THz frequencies massive antenna arrays operating in beamforming mode need to be used at base stations (BS) in 5G/6G systems [9,16]. Such an antenna creates extremely directional radiation patterns towards user equipment (UE) allowing for high gains in transmit and receive direction and compensating for the limited emitted power of a single array element [6]. As a side effect, these arrays posse very directional main lobe whose half-power beamwidth (HPBW) may reach a fraction of a degree [3]. As a result, as opposed to 4G and the previous generation of cellular systems beamtracking procedure keeping BS and UE synchronized at all times needs to be utilized.

There are two critical phenomena affecting continuous connectivity between BS and UE. The first one is blockage by small moving objects and large stationary objects that have been deeply studied in the past [5,8]. To efficiently avoid this problem, 3GPP has recently proposed a multiconnectivity solution [1] that allows for fairly well avoid the blockage events [2,7,11,17,18]. In addition to blockage, the micromobility phenomenon referring to quick displacements and rotations in the hand of a user has been recently identified [4,10,14]. Although, similarly to the blockage, this effect may lead to the temporal loss of connectivity and its characteristics are somewhat documented in the literature, the measures against this effect has not been developed so far. Specifically, the authors in [15] characterized the channel capacity in the presence of micromobility showing that for sub-seconds HPBWs on-demand beamtracking outperforms the current regular beamtracking approach utilized currently in 5G NR systems. In [13], it has been shown that the array switching time needs to be drastically decreased (by three orders of magnitude) if one wants to completely remove the connectivity losses occurring as a result of multiconnecitity.

Having much larger antenna arrays 6G THz systems would require extremely large resources for periodic beamtracking procedure in the presence of micromobility. To decrease the resource usage of may attempt to optimize the interval between beamtracking time instants. However, as demonstrated in [19] different applications utilized by modern smartphones are characterized by principally different statistical characteristics including time to outage. Thus, based on the knowledge of the application type utilized at the UE may provide BS with appropriate information on how often the beamtracking procedure needs to be invoked. Unfortunately, 3GPP stack does not currently provide the signaling information for notifying the BS about the currently utilized application.

By utilizing the measurements reported in [19] in this paper we propose a method for remote detection of the type of application utilized at UE. To this aim, we utilize the information about the trajectory of the center of the beam that is indirectly available at the BS when the antenna arrays at UE and BS are both known. We then apply ML approaches including trees, random forests, and neural networks to differentiate between 4 types of applications including video watching, virtual reality (VR) gaming, racing gaming, and phone calling.

The main contributions of our paper are:

- the non-intrusive procedure that does not rely upon signaling information to determine the type of applications utilized at the UE for further adaptation of beamtracking interval at the BS;
- numerical results showing that: (i) the accuracy of differentiating between applications having low speeds or distinctively different speeds stays at 100% (ii) all the considered algorithms including the tree, random forest, and neural network perform qualitatively similar.

The rest of the paper is organized as follows. We start in Sect. 2 reporting the summary of the results of the micromobility measurements and modeling campaign. Then, in Sect. 3, we introduce the proposed approach. In Sect. 4, we report our numerical results. The conclusions are drawn in the last section.

2   Micromobility Measurements, Properties, and Modeling

In this paper, we utilize the micromobility measurement results and model reported in [19]. In this section, we will briefly recall the measurement setup, statistical characteristics, and the utilized modeling framework.

2.1   Measurements Setup

We consider a point-to-point communication between a base station (BS) and a UE, as shown in Fig. 1a. Both UE and BS are assumed to operate utilizing arrays creating directional antenna radiation patterns. In theproposed THz system, the BS appears to be firmly fixed while the user associated with UE is in a stationary position. However, even in stationary conditions, UE in hands of a user is expected to perform small displacements and rotations that are modulated by the perceived content. Specifically, these are small displacements along Ox and Oy axes and rotations over the vertical and transverse axes. These types of movements may cause the loss of connectivity between UE and BS. Note that the micro changes along Oz-axis, as well as rotations along the longitudinal axis do not affect the state of connection.

To capture the micromobility of the UE, it is sufficient to characterize the motion pattern of the UE's beam center. The measurements performed in [19] were conducted by utilizing a laser pointer rigidly connected to a smartphone. A laser pointer with a beam diameter of 3 mm, an output power of 5 mW and a

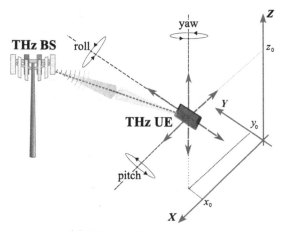

(a) The considered scenario

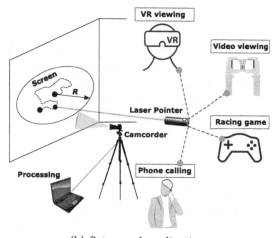

(b) Setup and applications

Fig. 1. The considered scenario, setup, and applications

long wavelength of light of 650 ± 10 nm was used. The laser spot detection was facilitated by utilizing 1280×720 30 frames-per-second video camera, as shown in Fig. 1b. Four different applications were considered: (i) watching video, (ii) phone calling, (iii) watching virtual reality (VR) video, and (iv) playing a racing game. We carried out our experiments at a distance of 2 m from the UE to the screen. The observed area is 1.5×2.5 m.

To obtain statistical data 10 independent experiments were carried out for each considered application. The duration of each experiment is set to 10 s. The original camera resolution was 30 frames per second, resulting in a source trace for each application containing approximately 3300 sample points.

Table 1. Summary of micromobility characteristics and modeling [19]

Characteristics	Video	Phone	VR	Racing
Radial symmetry	Yes	No	No	No
Velocity	3–10 m/s	7 m/s	9–13 m/s	9–5 m/s
Drift to the origin	0.17–0.11	0.17–0.3	0.17	0.17
Axes dependence	Negligible	Moderate	Negligible	Strong
Correlation coefficient	0.0	−0.2	0.0	−0.4
X-axis velocity	Increases, 1–6 m/s	3–6 m/s	6–9 m/s	7–4 m/s
Y-axis velocity	Increases, 2–8 m/s	3–5 m/s	5–8 m/s	3–2 m/s
X-axis drift	0.17–0.05	0.17–0.13	0.17	0.13–0.21
Y-axis drift	0.17–0.21	0.17–0.25	0.17	0.19–0.14
Markov modeling	Yes	Yes	Limited	No

2.2   Statistical Results

A summary of the statistics and simulation results is in Table 1. As one may observe, none of the applications are characterized by radial symmetry. The speed and drift of random mobility patterns are inherently distance dependent. Thus, in order to accurately capture the stochastic properties of UE micromobility models, these properties require complex modeling approaches involving 2D random walks with independent increments.

2.3   Modeling

The authors in [19] also proceeded offering 2D Markov models capturing micromobility patterns of applications. They utilized a direct approach by segmenting the screen into N by N grid for different values of N ranging from 50 to 200. Then they proceeded to parameterize the Markov model, determining the transition probabilities p_{ij}, $i, j = 1, 2, \ldots, N^2$ based on statistical data. They further demonstrated that these models are suitable for applications with low and purely random dynamics, such as video viewing, VR viewing, and phone calling and may not fully resemble the properties of the more complex patterns when the application directly controls the user's behavior, as in the case of racing game. Still in the latter case such a model can still serve as a first-order approximation. In what follows, we will utilize these models built by utilizing the empirical data reported in [19] to generate traces of micromobility patterns.

3   The Proposed Approach

3.1   Proposal

The core of the proposed idea is to differentiate between different types of applications having distinctively different micromobility speeds, and thus requiring different frequencies of beam alignments, remotely, without modifying the current signaling interface between BS and UE. To this aim, we propose BS to constantly monitor the current received signal strength. By knowing this information and antenna radiation patterns at both UE and BS sides one may deduce the location of the beam center. BS may track the current trajectory of the beam center and utilize the pre-trained ML algorithms to classify the type of applications.

To evaluate the proposed approach, we utilize the two-dimensional Markov models reported in [19]. We utilize these models to generate random traces of beam center movement of different applications. These traces are then utilized to train the ML models. Finally, to assess the accuracy of the proposal, we provide the classification of applications based on the remaining sets of generated traces.

3.2 Description of the Utilized Algorithms

In our study we consider three types of ML algorithms for remote application detection. These are decision trees, random forests, and neural networks. Below, we provide their short description and discuss parameterization.

Decision Tree. The decision tree algorithm works by recursively splitting the dataset into subsets based on the values of the features, with the aim of maximizing the information gain at each split. Information gain is a measure of the reduction in entropy or disorder in the dataset, and it is calculated as

$$I(D, F) = H(D) - H(D|F), \tag{1}$$

where $I(D, F)$ is the information gain, D is the dataset, F is the feature being considered, $H(D)$ is the entropy of the dataset, and $H(D|F)$ is the conditional entropy of the dataset given the feature F.

The entropy of a dataset is calculated as

$$H(D) = -\sum (p_i \log_2(p_i)), \tag{2}$$

where p_i is the proportion of instances in the dataset that belong to class i.

The decision tree algorithm split the dataset based on the features with the highest information gain until a stopping criterion is met, such as a maximum tree depth or a minimum number of instances in each leaf node.

Random Forest. Random forest is an ensemble learning algorithm that constructs a multitude of decision trees at training time and outputs the class that is the mode of the classes (classification) or mean prediction (regression) of the individual trees. The basic idea behind the random forest is to combine multiple decision trees in determining the final output, with each decision tree being constructed on a different random subset of the training data.

Assuming a classification problem with a dataset of n examples with m features, the algorithm works as follows: (i) randomly select k features from m total features, where $k << m$, creating the so-called "feature subspace" for the given decision tree, (ii) use the selected features to build a decision tree, where each node in the tree splits the data based on the value of one of the k randomly selected features, (iii) repeat steps (i) and (ii) t times to generate a total of t decision trees, (iv) to make a prediction for a new example, classify the example using each of the t decision trees and take the majority vote as the final prediction.

In random forests, the predicted output of each decision tree is aggregated to determine the final prediction. Let Y denote the class label of an input example, and let T be the number of decision trees in the random forest. Then, the output of the random forest, denoted as $\hat{Y}$, is computed as

$$\hat{Y} = M(Y_1, Y_2, \ldots, Y_t), \tag{3}$$

where M is the statistical mode.

Neural Network. The most basic form of a neural network is a feedforward neural network, which consists of an input layer, one or more hidden layers, and an output layer. Each layer contains a set of neurons, and each neuron is connected to all neurons in the adjacent layers by a set of weights.

The output of a neuron in a neural network is determined by the weighted sum of its inputs, passed through an activation function. The activation function is a nonlinear function that introduces non-linearity into the model, allowing it to learn complex patterns in the data. The most commonly used activation functions are the sigmoid function and the rectified linear unit (ReLU) function. Specifically, the output of a neuron in layer j can be computed as

$$z_j = \sum (w_{i,j} * x_i) + b_j, \, a_j = g(z_j), \tag{4}$$

where $w_{i,j}$ is the weight of the connection between neuron i in layer $j - 1$ and neuron j in layer j, x_i is the input value to neuron i, b_j is the bias term for neuron j, $g(\cdot)$ is the activation function, z_j is the weighted sum of inputs to neuron j, and a_j is the output value of neuron j.

The output of the neural network is obtained by applying the feedforward process to the input layer and propagating the computed values through each layer until the output layer is reached. During training, the weights and biases of the neural network are adjusted to minimize the difference between the predicted output and the actual output, using a loss function. The most commonly used loss function for regression problems is the mean squared error (MSE) function, which measures the average squared difference between the predicted and actual outputs. The weights and biases are updated using an optimization algorithm, such as gradient descent, which iteratively adjusts the weights in the direction of the steepest descent of the loss function. This process is called backpropagation, and it allows the neural network to learn the optimal weights and biases that minimize the loss function.

3.3   Implementation

To generate traces, the initial data must be presented in the following format: (i) the starting point (x and y-coordinates), (ii) followed by the transition point

(x, y), and (iii) the probability of transitioning from the initial position to the next. Next, we calculate the trajectory of the UE, generating 10^4 states for each trace by applying the conventional procedure for generating trajectories of the Markov process. We generated 10^5 traces for each application.

The following attributes have been selected for classification: (i) mean distance from the center, (ii) mean speed of movement, (iii) total distance traveled, (iv) mean distance from the center on the Ox axis, (v) mean distance from the center on the Oy axis. Once the set of attributes for each trace was calculated, the next step was to use them for classification. For this purpose, Matlab was used, specifically its built-in *Classification Learner (CL)* function. To classify the data, the attributes have been provided to the CL function, where each feature was marked with the scenario of use to which it belonged.

The classification process involved selecting the size of the test sample to train the models, followed by the selection of the necessary classification algorithms. The following decision tree, random forests, and neural network algorithms have been considered. The models were then trained and the classification was carried out.

To determine the weight of the contribution of each feature to the accuracy of classification, the *Feature Selection (FS)* function was used. Note that the model is trained on the selected sample size, and the classification accuracy is calculated on the entire data, including test data. Finally, as we also want to understand how useful each feature was for classification, the weight of the contribution of each of the attributes to the accuracy of classification was calculated.

4   Numerical Results

In this section, we report our numerical results for the classification of applications at the UE. We first consider three types of applications' classification: (i) video and gaming (ii) voice and video (iii) VR and gaming. Then, we proceed to report the classification results for all four considered applications.

Overall, 10^4 traces were generated for each application. A fraction of those was utilized to train the considered models, while the rest – for assessing the performance. In all the cases, we utilized the prediction (classification) accuracy metric defined as the number of correctly predicted applications to the overall number of test samples. In addition, we also report the attributes' contribution to the decision making.

4.1   Pairwise Classification

Recalling that the ultimate goal of identifying the type of applications utilized at UE is to understand how often beam alignment procedure needs to be invoked, we start with the pairwise classification of the types of applications including: (i) video and gaming (ii) voice and video (iii) VR and gaming. Note that, based on Table 1, voice and video applications are characterized by similar statistical properties including the speed of motion of the beam center. VR and gaming

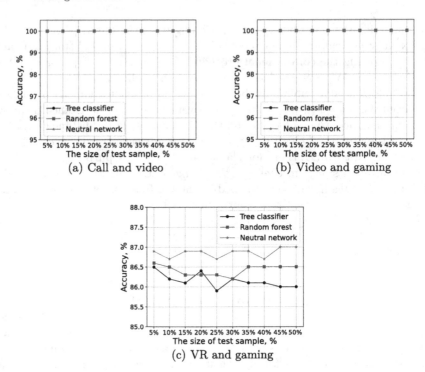

Fig. 2. The classification accuracy for pairwise applications classification

applications also have qualitatively comparable statistics with the exception that gaming applications are characterized by much higher speed over the Ox axis. Thus, we may expect that it might be difficult to distinguish between voice and video applications, while the classification should work much better for video and gaming applications.

The classification accuracy metric is reported in Fig. 2 for all the considered pairs of applications and three utilized algorithms as a function of the percentage of generated traces utilized for testing. By analyzing the presented data we see that for the classification of call and video as well as video and gaming applications the accuracy remains perfect at 100% even when the size of the training sample decreases. However, for VR and gaming applications the results are not perfect leading to approximately 13–15% of erroneously predicted cases. Out of all the considered models, the neural network shows the most promising results slightly outperforming tree and forest classifiers by approximately 0.5–1%. However, this difference is not relevant in practical applications implying that the choice of the classifier can be solely based on implementation complexity. Finally, we also notice that the tree classifier is the only one showing slight degradation in performance as compared to the other two algorithms when the training sample size decreases.

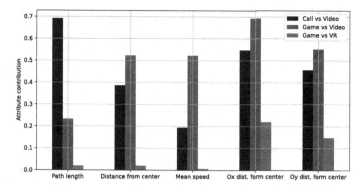

Fig. 3. The attributes contribution to the pairwise applications classification

The abovementioned results of pairwise applications classification can be explained by analyzing attributes' contribution to the decision making shown in Fig. 3 for all the pairwisely classified applications. Here, we see that for call and video as well as video and gaming applications, all the attributes play significant roles in the final decision. However, gaming and VR applications appear to be quite similar in terms of three attributes – length traversed by a beam center, distance from the center, and mean speed. The main contributions in this case come from different distances over Cartesian axes. However, by noticing that all the considered algorithms perfectly distinguish between different applications having low micromobility speeds and applications having distinctively different speeds, we observe that failure to perfectly distinguishing between applications characterized by high micromobility speeds does not affect the ultimate goal consisting in understanding how often the beam alignment procedure needs to be invoked.

4.2   Joint Classification

Having studied the pairwise applications classification, we now proceed to report the results of classifying all the applications altogether. To this aim, Fig. 4 shows the classification accuracy for all the considered algorithms as a function of the fraction of traces utilized for testing. Here, we again see that the neural network is characterized by negligibly better accuracy while the tree classifiers are more sensitive to the size of the training set. By comparing these results to those reported for pairwise comparison we may deduce that the loss in classification accuracy is mainly attributed to the presence of 50% of traces with high and comparable micromobility speed – VR and gaming applications. This is implicitly confirmed by the contribution of the micromobility speed to the overall decision making shown in Fig. 5. Here, as the overall collection of traces becomes quite heterogeneous, logically, the distance to the center plays the most significant role.

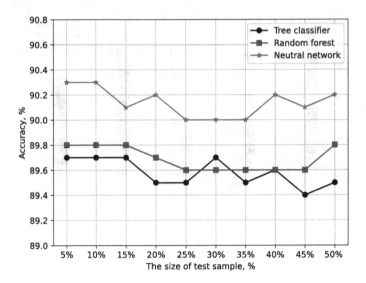

Fig. 4. The classification accuracy for all the applications

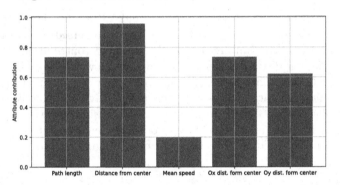

Fig. 5. The attributes contribution for all applications classification

5   Conclusions

Motivated by the need to optimize regular beamtracking interval at mmWave/THz BSs in 5G/6G systems utilizing massive antenna arrays, in this paper we proposed a non-intrusive procedure for remote detection of the application running at UE. Specifically, by utilizing the trajectory of the center of HPBW available at BS, we proposed to use ML techniques to discriminate the type of application. The proposed approach can be used to optimize the time interval between beamtracking time instants such that continuous connectivity between UE and BS is ensured consuming the minimum amount of resources.

Our numerical results demonstrate that all the considered classifiers, including tree, random forest and neural network allow us to perfectly distinguish between applications having low micromobility speed and applications having distinctively different speeds. However, the accuracy of classification of those applications having fast and similar speeds such as VR and gaming is not perfect staying at approximately 85–90%. Nevertheless, as the main goal of the proposed classification is to understand how often the beam alignment procedures must be invoked, the abovementioned observation does not lead to significant challenges. Finally, we note that all the considered classifiers including the tree, random forest, and neural network perform qualitatively similary with the tree being more sensitive to the training sample size.

References

1. 3GPP: NR; Multi-connectivity; stage 2 (Release 16). 3GPP TS 37.340 V16.0.0, 3GPP (2019)
2. Agiwal, M., Kwon, H., Park, S., Jin, H.: A survey on 4g–5g dual connectivity: road to 5g implementation. IEEE Access **9**, 16193–16210 (2021)
3. Akyildiz, I.F., Jornet, J.M.: Realizing ultra-massive mimo (1024× 1024) communication in the (0.06–10) terahertz band. Nano Commun. Netw. **8**, 46–54 (2016)
4. Al-Ogaili, F., Shubair, R.M.: Millimeter-wave mobile communications for 5g: challenges and opportunities. In: 2016 IEEE International Symposium on Antennas and Propagation (APSURSI), pp. 1003–1004. IEEE (2016)
5. Bai, T., Vaze, R., Heath, R.W.: Analysis of blockage effects on urban cellular networks. IEEE Trans. Wirel. Commun. **13**(9), 5070–5083 (2014)
6. Balanis, C.A.: Antenna Theory: Analysis and Design. John wiley & sons, Hoboken (2015)
7. Begishev, V., et al.: Joint use of guard capacity and multiconnectivity for improved session continuity in millimeter-wave 5G NR systems. IEEE Trans. Veh. Technol. **70**(3), 2657–2672 (2021)
8. Gapeyenko, M., et al.: Analysis of human-body blockage in urban millimeter-wave cellular communications. In: 2016 IEEE International Conference on Communications (ICC), pp. 1–7. IEEE (2016)
9. Giordani, M., Polese, M., Roy, A., Castor, D., Zorzi, M.: A tutorial on beam management for 3g pp nr at mmwave frequencies. IEEE Commun. Surv. Tutor. **21**(1), 173–196 (2018)
10. Ichkov, A., Gehring, I., Mähönen, P., Simić, L.: Millimeter-wave beam misalignment effects of small-and large-scale user mobility based on urban measurements. In: Proceedings of the 5th ACM Workshop on Millimeter-Wave and Terahertz Networks and Sensing Systems, pp. 13–18 (2021)
11. Kulkarni, M.N., Singh, S., Andrews, J.G.: Coverage and rate trends in dense urban mmwave cellular networks. In: 2014 IEEE Global Communications Conference, pp. 3809–3814. IEEE (2014)
12. Matthaiou, M., Yurduseven, O., Ngo, H.Q., Morales-Jimenez, D., Cotton, S.L., Fusco, V.F.: The road to 6G: ten physical layer challenges for communications engineers. IEEE Commun. Mag. **59**(1), 64–69 (2021)
13. Moltchanov, D., Gaidamaka, Y., Ostrikova, D., Beschastnyi, V., Koucheryavy, Y., Samouylov, K.: Ergodic outage and capacity of terahertz systems under micromobility and blockage impairments. IEEE Trans. Wirel. Commun. **21**(5), 3024–3039 (2021)

14. Petrov, V., Moltchanov, D., Koucheryavy, Y., Jornet, J.M.: The effect of small-scale mobility on terahertz band communications. In: Proceedings of the 5th ACM International Conference on Nanoscale Computing and Communication, pp. 1–2 (2018)
15. Petrov, V., Moltchanov, D., Koucheryavy, Y., Jornet, J.M.: Capacity and outage of terahertz communications with user micro-mobility and beam misalignment. IEEE Trans. Veh. Technol. **69**(6), 6822–6827 (2020)
16. Petrov, V., Pyattaev, A., Moltchanov, D., Koucheryavy, Y.: Terahertz band communications: applications, research challenges, and standardization activities. In: 2016 8th International Congress on Ultra Modern Telecommunications and Control Systems and Workshops (ICUMT), pp. 183–190. IEEE (2016)
17. Polese, M., Giordani, M., Mezzavilla, M., Rangan, S., Zorzi, M.: Improved handover through dual connectivity in 5g mmwave mobile networks. IEEE J. Sel. Areas Commun. **35**(9), 2069–2084 (2017)
18. Sopin, E., Moltchanov, D., Daraseliya, A., Koucheryavy, Y., Gaidamaka, Y.: User association and multi-connectivity strategies in joint terahertz and millimeter wave 6G systems. IEEE Trans. Veh. Technol. **71**(12), 12765–12781 (2022)
19. Stepanov, N., Moltchanov, D., Begishev, V., Turlikov, A., Koucheryavy, Y.: Statistical analysis and modeling of user micromobility for THz cellular communications. IEEE Trans. Veh. Technol. **71**(1), 725–738 (2022). https://doi.org/10.1109/TVT.2021.3124870
20. Zaidi, A., Athley, F., Medbo, J., Gustavsson, U., Durisi, G., Chen, X.: 5G Physical Layer: Principles, Models and Technology Components. Academic Press, Cambridge (2018)

Blockage Attenuation and Duration Over Reflected Propagation Paths in Indoor Terahertz Deployments

Anatoliy Prikhodko[1,2] , Abdukodir Khakimov[3] , Evgeny Mokrov[3] ,
Vyacheslav Begishev[3(✉)] , Alexander Shurakov[1,2] ,
and Gregory Gol'tsman[1,2]

[1] Moscow Pedagogical State University, Moscow, Russia
anprihodko@hse.ru, {alexander,goltsman}@rplab.ru
[2] National Research University Higher School of Economics, Moscow, Russia
[3] Peoples' Friendship University of Russia (RUDN University), Moscow, Russia
{khakimov-aa,mokrov-ev,begishev-vo}@rudn.ru

Abstract. The future 6G cellular systems are expected to utilize the lower part of the terahertz frequency band, 100–300 GHz. As a result of high path losses, the coverage of such systems will be limited to a few tens of meters making them suitable for indoor environments. As compared to outdoor deployments, indoor usage of THz systems is characterized by the need to operate over shorter distances using the reflected propagation paths. This paper aims to characterize the impact of blockage of reflected propagation paths in typical scenarios. Specifically, we carry out a detailed measurements campaign at 156 GHz and report reflection losses, blockage losses over the reflected path as well as blockage duration, signal fall and rise times. Our results show that signal polarization has a profound impact on the reflection losses with E-plane horizontally oriented signal losses being at least 8 dB higher as compared to H-plane signal horizontal orientation. Furthermore, the reflection material types do not affect the mean blockage attenuation over the reflected paths. Generally, the presence of a reflector neither quantitatively nor qualitative changes the mean attenuation induced by a blockage phenomenon.

Keywords: 6G · terahertz · reflections · blockage · attenuation · duration · signal fall · rise times

1 Introduction

Seeking for the capacity boost at the access interface in cellular systems, ITU-R and 3GPP utilize millimeter wave (mmWave) bands, 30–100 GHz for 5G New

This study was conducted as a part of strategic project "Digital Transformation: Technologies, Effectiveness, Efficiency" of Higher School of Economics development programme granted by Ministry of science and higher education of Russia "Priority-2030" grant as a part of "Science and Universities" national project. Support from the Basic Research Program of the National Research University Higher School of Economics is gratefully acknowledged.

Radio (NR) systems [11,14]. The next step in the evolution of such systems is the utilization of the lower part of the terahertz (THz) frequency band, where large parts of the spectrum are still not regulated and tens of gigahertz of bandwidth can be allocated to 6G systems [5,9].

To partially compensate for the reduction in the effective antenna aperture that reduces with the increase of the carrier frequency, similarly to 5G NR mmWave systems, 6G sub-THz systems will heavily rely upon the use of antenna arrays at both base station (BS) and user equipment (UE) operating in beamforming mode [10]. Nevertheless, the coverage of such systems will still be limited to tens or hundreds of meters making them a suitable choice for indoor areas, where most of the traffic demands originate. The landscape of applications in the indoor environment is rather large including conventional 4k/8k video watching, virtual/augmented reality (AR/VR) gaming, and forthcoming applications such as collective VR gaming, holographic communications [6], etc.

Indoor deployments of 6G sub-THz systems are characterized by several propagation specifics. First of all, the link distances are on average smaller as compared to those outdoors. Secondly, due to rather small heights of BS, human body blockage is more likely to occur. Finally, as a result of the complex geometry of indoor premises, communications over reflected paths are expected to be much more common. Specifically, short distances have been recently shown to lead to much smaller human body blockage attenuation at 156 GHz varying in the range of $8 - 13$ dB [12] as compared to 15–35 dB losses over larger distances and at lower frequencies, e.g., as reported in [4,7].

The aim of this paper is to characterize reflected propagation paths in indoor environments in the sub-THz frequency band. Specifically, by carrying out a large-scale measurement campaign at carrier frequency of 156 GHz, we characterize reflection losses of different materials and blockage attenuation of reflected propagation paths. In addition to attenuation, we also investigate time-related metrics such as blockage duration as well as signal rise and fall times. The main findings of our paper acquired empirically are:

- the orientation of the antenna polarization plane has a profound impact on the reflection losses with horizontally oriented polarization plane losses being at least 7 dB higher as compared to vertically orientated polarization plane;
- the reflection material types do not affect the mean blockage attenuation over the reflected paths;
- the presence of reflector neither quantitatively nor qualitative changes the mean attenuation induced by a blockage phenomenon;
- blockage, fall and rise times for drywall are characterized by slightly smaller mean values as compared to concrete for vertical polarization plane.

The paper is organized as follows. First, in Sect. 2, we overview related studies reported in the literature. Next, in Sect. 3, we outline experimental setup. The main results of the conducted experiments are reported further in Sect. 4. And conclusions are provided in the last section.

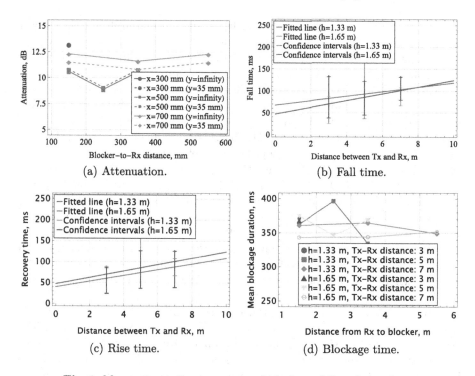

Fig. 1. Mean attenuation, mean signal blockage, fall and rise times

2   Related Work

In this section, we outline related works. We start by briefly reminding the results for non-reflected path propagation and blockage. They are recapitulated for comparison purposes in our study. Then, we discuss results similar to those reported in our study which are related to reflection losses and blockage of reflected propagation paths in the mmWave and THz bands.

As one of the paper goals is to compare blockage statistics over reflected paths to that over primary line-of-sight (LoS) paths, we briefly introduce the latter as reported earlier in [12]. To this aim, Fig. 1 provides mean attenuation, mean signal blockage, fall and rise times for different Tx-to-Rx distances, x, LoS heights, h, and Rx-to-blocker distances. By analyzing the presented results, one may observe that the mean attenuation varies between 8–13 dB and is generally independent of blocker-to-Rx and Tx-to-Rx distances. Furthermore, both fall and rise times increase as a function of the distance potentially making it more feasible to detect blockage events timely. The absolute difference between the reported times is insignificant and lies within 2–4% of the nominal value (e.g., 60 ms for $x = 3$ m, 80 ms for $x = 5$ m and 100 ms for $x = 7$ m for fall times). It is worth noting that both fall and rise times have almost identical nominal values, and, in general, the rise time is 7–10% smaller than the fall time.

The attenuation caused by reflections from different materials was the subject of several studies. In [3], the authors reported the measurement results of the received signal reflected from aluminum, glass, plastic, hardboard and concrete using THz time-domain spectroscopy (THz-TDS) equipment for different angles of incidence. The time duration of utilized pulses was chosen such that the energy is mainly concentrated in the 0.1–4 THz band. The observed losses were in the range of 10–60 dB depending on the angle of incidence and type of the material. Specifically, in the 0.1–0.3 THz band, aluminum demonstrated the least attenuation of around 25–35 dB for the angle of incidence of $\pi/4$. The rest of the materials provided higher attenuation.

The authors in [1,8] reported the results of reflections from typical vehicle materials at 300 GHz for different configurations including front and rear reflections, side-lane and under-vehicle reflections. The front and rear reflections were reported to result in 24–42 dB and 15–30 dB attenuations, respectively, while side-lane reflections led to additional 16–20 dB losses. The authors also proposed to model the under-vehicle reflection losses of the asphalt by utilizing the $\alpha d^{-\beta}$ function, where d is the separation distances, while α and β are some coefficients tabulated in Table 1 in their manuscript.

Similar studies have been performed in the mmWave band. In [2], specifically, the authors investigated the impact of polarization properties of the indoor 38 GHz channel after single bounce reflection for different angles of incidence and observation and two types of materials: aluminum and concrete. They highlighted that, when the polarization of Tx and Rx antennas coincide, much smaller losses are experienced. The difference can reach 15–20 dB and is maximized for peculiar reflections. Concrete results in 8–12 dB higher losses as compared to aluminum.

3   Experimental Setup

In this section, we introduce our measurement setup and present details on the acquisition of experimental data.

A schematic of the measurement setup employed for blockage studies is presented in Fig. 2a. We use a 156 GHz constant waveform source with amplitude modulation at 25 kHz. It provides a 6° wide beam incident on the material under test (MUT) at a constant angle of 70°. The reflected beam is received by a low-barrier diode detector equipped with the same optics as the source. A lock-in amplifier is used to readout the detector response voltage. When a blocker walks across the reflected beam at the midpoint between MUT and the detector, the voltage-vs-time series is registered by a digital signal oscilloscope (DSO). The measurement covers a time frame of 4 s with a resolution of 100 μs. Referring to Fig. 2a, list of the employed measurement equipment includes the following items:

- microwave synthesizer: Hittite HMC-T2220;
- frequency multiplier: RPG Tx-134-158-20;
- ultrafast switch: ELVA-1 VCVA-06;

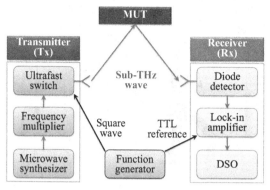

(a) Measurement equipment.

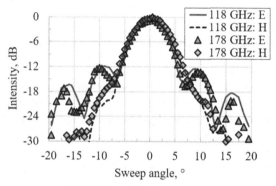

(b) Tx/Rx radiation pattern.

Fig. 2. Experimental setup

- function generator: SRS DS345;
- diode detector: DOK WR-06;
- lock-in amplifier: SRS SR844;
- DSO: R&S RTO1012.

The source-to-detector (Tx-to-Rx) optical path of 3 m is chosen in all the measurements. The source provides 52 mW of power, and the setup ensures a signal-to-noise ratio of up to 3×10^4 at the detector output. Measured response voltages are further converted into power levels at the detector input via its responsivity, which is equal to 500 V/W at 156 GHz. MUT is successively presented by concrete, drywall and class samples with thicknesses of 50, 12.5 and 6 mm, respectively. The sample linear dimensions of 0.5 m × 0.5 m are chosen to overlap a 156 GHz beam upon reflection. The sample is installed in a wooden frame to set its center at 1.65 m above the floor which corresponds to the LoS height between the source and the detector. Spurious reflections from the wall behind the sample are geometrically filtered out by the positioning of the measurement equipment. Input optics of the source and the detector rely on a pyra-

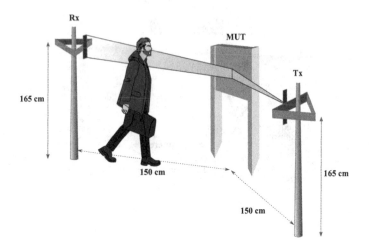

Fig. 3. Illustration of the considered scenario

midal horn known as a wide-band antenna with different side lobe levels in H- and E-planes, see Fig. 2b. We make use of this feature and conduct a series of measurements for two orientations of the horn antennas, when either their H- or E-planes are horizontally oriented, i.e., coinciding with the plane of incidence of the transmitted 156 GHz beam.

The schematic illustration of the scenario is shown in Fig. 3.

4   Measurements Results

In this section, we report our results. We start with visual illustrations of the blockage phenomenon over the reflected paths. Then, we characterize the mean attenuation caused by reflection for typical types of wall materials such as drywall, concrete and glass. Further, we investigate blockage attenuation over reflected paths. Finally, we report mean and cumulative distribution function (CDF) of blockage duration, signal fall and rise times for reflected paths.

4.1   Time Series

We start with a time series representation of the blockage over the reflected paths, as demonstrated in Fig. 4, for all the considered types of materials and orientations of the Tx/Rx antennas. Here, for comparison purposes, we also demonstrate blockage over the LoS path ("direct trace" in Fig. 4).

By visually inspecting the presented data, one may notice that the received signal level in no blockage condition is significantly lower as compared to direct LoS propagation. This is attributed to the reflection losses considered in details below. Furthermore, we observe significant difference between signal levels and blockage profiles for horizontally and vertically oriented E-planes of Tx/Rx

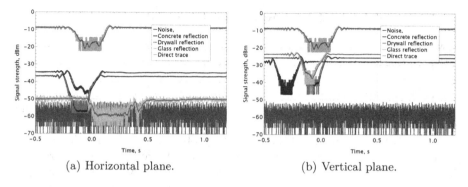

(a) Horizontal plane.   (b) Vertical plane.

Fig. 4. Comparison of reflection and direct traces

antennas (the H-planes are respectively orthogonal). The former condition is referred to as horizontal plane (HP), and the latter – as vertical plane (VP). Specifically, we see that the signal strength for the HP reflection from glass is barely higher than the noise level. And for the VP reflection case, it is comparable to other materials. Finally, visual inspection does not allow clearly highlight any difference between attenuation- and time-related blockage profiles for different materials requiring detailed statistical analysis.

4.2 The Impact of Reflection

We start our analysis with the characterization of the reflection losses by comparing the reflected signal against the LoS signal studied earlier in [12]. In order to determine reflection losses, we utilize the propagation model, $L(x)$, with the coefficients y_1 and y_2 empirically derived from the signal difference between Tx and Rx at different distances.

$$L(x) = y_1 \log_{10} x + 20 \log_{10} f_c + y_2 + I_B L_B(x, d), \tag{1}$$

In the case of LoS propagation for the carrier frequency of 156 GHz, the coefficients are $y_1 = -22.04$ and $y_2 = -251.704$ [12].

By utilizing (1), we subtract the model's values from the path losses observed in non-LoS conditions obtaining the reflection losses. These losses are reported in Table 1 for different types of materials and polarizations. By analyzing the presented data, one may deduce that the impact of the polarization plane orientation is critical. For the VP condition, all the materials behave similarly leading to 14–19 dB losses with glass having 4 dB gain on top of concrete. The HP condition, however, is characterized by at least 7–13 dB stronger attenuations. Specifically, there are 7 dB higher losses for concrete and 11 dB higher losses for drywall. In the HP condition, glass attenuates the signal on reflection by approximately 42 dB making the reflected path signal strength comparable to noise, see Fig. 4. We specifically note that attenuation of at least 30 dB may lead to the loss of connectivity depending on the propagation distance between Tx and Rx.

4.3   Blockage Attenuation Over Reflected Paths

Now, we proceed to characterize the blockage attenuation over the reflected paths. To this aim, we subtract the value of the received signal with no blockage impairments from the average value in the blocked state. Table 2 presents the mean blockage attenuation and the mean values of the signal strength propagated directly to the Tx and reflected from the considered materials in blocked and non-blocked cases.

By analyzing the reported data, one may observe that, while the antenna polarization plane orientation greatly impacts on the received signal strength, the impact on the mean blockage value is rather limited. All the material types lead to almost constant blockage attenuation of approximately 9–10 dB for the VP orientation. By comparing the obtained results with those reported in Fig. 1 for non-reflected blockage, one may deduce that the presence of a reflector neither quantitatively nor qualitatively changes the mean attenuation induced by a human blockage phenomenon. The HP orientation is characterized by larger differences between mean blockage attenuations varying in the range of 7–10 dB. However, these changes can be potentially attributed to smaller signal strengths accompanied by reduction in measurement accuracy for drywall and glass.

To provide additional information regarding the attenuation, Fig. 5 reports cumulative distribution functions (CDFs) of blockage attenuations for concrete and drywall materials. The illustration highlights the difference between blockages for different orientations of the antenna polarization plane. Notably, the range of attenuation can be quite large varying between approximately 6 and 10 dB. These differences can be attributed to slight changes in the trajectory of a person crossing the LoS path.

4.4   Fall, Rise and Blockage Times

In addition to attenuation statistics, time-related parameters of blockage are of importance for the design of sub-THz communications systems. In this section, we report blockage, fall and rise times including their mean values and CDFs. Note that the rise time is of special importance in the context of blockage

Table 1. Reflection and blockage losses

Type	Reflection loss, dB	Blockage loss, dB
Concrete, HP	25.85	33.84
Drywall, HP	27.97	37.96
Glass, HP	41.59	48.08
Concrete, VP	18.93	28.76
Drywall, VP	16.87	25.94
Glass, VP	14.79	23.60

Table 2. The mean blockage attenuation

Type	Signal, dBm	Blocked signal, dBm	Blockage attenuation, dB
Direct, LoS	−8.88	−16.21	7.32
Concrete, HP	−34.74	−42.73	7.98
Drywall, HP	−36.86	−46.85	9.99
Glass, HP	−50.48	−56.97	6.48
Concrete, VP	−27.82	−37.65	9.83
Drywall, VP	−25.76	−34.83	9.07
Glass, VP	−23.68	−32.48	8.80

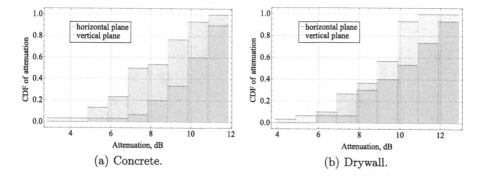

(a) Concrete.   (b) Drywall.

Fig. 5. CDFs of blockage attenuation for concrete and drywall

detection [12,13], while the blockage time is critical for designing algorithms to improve service reliability.

The mean values of the considered metrics are summarized in Table 3. By analyzing the reported values, one may observe that there is a 8% difference between blockage times for concrete and drywall for the VP antenna orientation. Mean rise time for drywall is 30% smaller than for concrete, and their fall times are almost identical. Note that for the HP orientation, all these values differ by 5–8% as evident from Table 3. Still, recalling the results for blockage remedies, these

Table 3. Mean fall, rise and blockage times

Type	Blockage time, ms	Rise time, ms	Fall time, ms
Concrete, HP	316	101	61
Drywall, HP	333	94	66
Concrete, VP	308	92	90
Drywall, VP	285	71	89

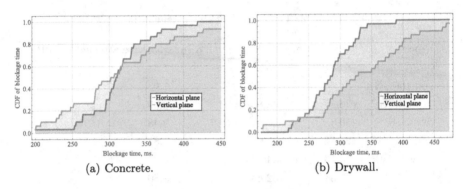

(a) Concrete.   (b) Drywall.

Fig. 6. CDF of blockage times

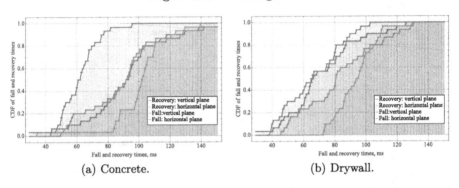

(a) Concrete.   (b) Drywall.

Fig. 7. CDFs of fall and rise times

deviations are expected to not affect the design of blockage detection algorithms [12, 13].

Complementing the mean values, we present the CDFs of blockage, fall and rise times in Fig. 6 and 7. The blockage duration CDFs indicate that CDF for concrete is steeper as compared to drywall and also has a smaller range of approximately 200 ms. Also, the VP antenna orientation leads to a noticeably smoother increase in CDF behavior. In practice, it means that deviation in blockage duration for the HP antenna orientation is much more clustered around its mean-making.

By analyzing CDFs of fall and rise times demonstrated in Fig. 7, one may observe that drywall is characterized by a slightly wider range of values as compared to concrete. The difference in these times for different orientations of the antenna polarization plane is also noticeable. This generally means that the time budget for the detection of blockage events is higher for drywall.

5   Conclusions

As most of the traffic in cellular systems originates indoors, where the link distances are generally shorter while communications over reflected paths are more common as compared to outdoor deployments, in this paper, we performed a measurements campaign at 156 GHz characterizing the reflection and blockage losses of reflected paths. We considered different types of reflection materials typical for the indoor environment including concrete, drywall and glass.

Our main findings are: (i) the orientation of the antenna polarization plane has a profound impact on the reflection losses with horizontally oriented polarization plane losses being at least 7 dB higher as compared to vertically orientated polarization plane, (ii) the reflection material types do not affect the mean blockage attenuation over the reflected paths, (iii) the presence of reflector neither quantitatively nor qualitative changes the mean attenuation induced by a blockage phenomenon, (iv) blockage, fall and rise times for drywall are characterized by slightly smaller mean values as compared to concrete for vertical polarization planes. In general, blockage statistics over reflected paths are similar to that for LoS paths and do not require special communications algorithms design.

Acknowledgements. This study was conducted as a part of strategic project "Digital Transformation: Technologies, Effectiveness, Efficiency" of Higher School of Economics development programme granted by Ministry of science and higher education of Russia "Priority-2030" grant as a part of "Science and Universities" national project. Support from the Basic Research Program of the National Research University Higher School of Economics is gratefully acknowledged.

References

1. Eckhardt, J.M., Petrov, V., Moltchanov, D., Koucheryavy, Y., Kürner, T.: Channel measurements and modeling for low-terahertz band vehicular communications. IEEE J. Sel. Areas Commun. **39**(6), 1590–1603 (2021)
2. Gaspard, I.: Co-and crosspolar scattering measurements at slightly rough walls for indoor propagation channels at mmwaves. In: 2019 IEEE-APS Topical Conference on Antennas and Propagation in Wireless Communications (APWC), pp. 038–041 (2019). https://doi.org/10.1109/APWC.2019.8870411
3. Kokkoniemi, J., Petrov, V., Moltchanov, D., Lehtomäki, J., Koucheryavy, Y., Juntti, M.: Wideband terahertz band reflection and diffuse scattering measurements for beyond 5G indoor wireless networks. In: European Wireless 2016; 22th European Wireless Conference, pp. 1–6. VDE (2016)
4. MacCartney, G.R., Rappaport, T.S., Rangan, S.: Rapid fading due to human blockage in pedestrian crowds at 5g millimeter-wave frequencies. In: GLOBECOM 2017–2017 IEEE Global Communications Conference, pp. 1–7. IEEE (2017)
5. Matthaiou, M., Yurduseven, O., Ngo, H.Q., Morales-Jimenez, D., Cotton, S.L., Fusco, V.F.: The road to 6G: ten physical layer challenges for communications engineers. IEEE Commun. Mag. **59**(1), 64–69 (2021)
6. Moltchanov, D., Sopin, E., Begishev, V., Samuylov, A., Koucheryavy, Y., Samouylov, K.: A tutorial on mathematical modeling of 5G/6G millimeter wave and terahertz cellular systems. IEEE Commun. Surveys Tutor. **24**, 1072–1116 (2022)
7. Nie, S., MacCartney, G.R., Sun, S., Rappaport, T.S.: 72 GHz millimeter wave indoor measurements for wireless and backhaul communications. In: 2013 IEEE

24th Annual International Symposium on Personal, Indoor, and Mobile Radio Communications (PIMRC), pp. 2429–2433. IEEE (2013)

8. Petrov, V., Eckhardt, J.M., Moltchanov, D., Koucheryavy, Y., Kurner, T.: Measurements of reflection and penetration losses in low terahertz band vehicular communications. In: 2020 14th European Conference on Antennas and Propagation (EuCAP), pp. 1–5. IEEE (2020)

9. Petrov, V., Pyattaev, A., Moltchanov, D., Koucheryavy, Y.: Terahertz band communications: applications, research challenges, and standardization activities. In: 2016 8th International Congress on Ultra Modern Telecommunications and Control Systems and Workshops (ICUMT), pp. 183–190. IEEE (2016)

10. Roh, W., et al.: Millimeter-wave beamforming as an enabling technology for 5g cellular communications: theoretical feasibility and prototype results. IEEE Commun. Mag. **52**(2), 106–113 (2014)

11. Shafi, M., et al.: 5g: a tutorial overview of standards, trials, challenges, deployment, and practice. IEEE J. Sel. Areas Commun. **35**(6), 1201–1221 (2017)

12. Shurakov, A., et al.: Empirical blockage characterization and detection in indoor sub-thz communications. Comput. Commun. **201**, 48–58 (2023)

13. Wu, S., Alrabeiah, M., Hredzak, A., Chakrabarti, C., Alkhateeb, A.: Deep learning for moving blockage prediction using real mmwave measurements. In: ICC 2022-IEEE International Conference on Communications, pp. 3753–3758. IEEE (2022)

14. Zaidi, A., Athley, F., Medbo, J., Gustavsson, U., Durisi, G., Chen, X.: 5G Physical Layer: Principles, Models and Technology Components. Academic Press, Cambridge (2018)

On the Automated Text Report Generation in Open Transport Data Analysis Platform

Mark Bulygin$^{(\boxtimes)}$ and Dmitry Namiot

Moscow State University, Leninskiye Gory 1, Moscow 119234, Russian Federation
messimm@yandex.ru

Abstract. According to UN studies, more than 60% of the population will live in cities by 2030. It is important to receive reports on changes in the transport behavior of city residents to change existing and build new transport systems. In this article, we talk about the automated report generation module for an open platform for transport data analysis. We give the classification of transport changes in the city. The work of the module is demonstrated for various types of traffic changes in the example of the analysis of changes in traffic flows in Moscow during the celebrations in early May and the closure of the Smolenskaya metro station for long-term reconstruction. Possibilities of the analysis on time intervals of various extents and lobbies are shown. Proposed ways to further automate the management of transport systems using data collected by the platform. Possible directions for further development of an open platform for transport data analysis are proposed.

Keywords: Data Analysis · Transport Data · Data Processing Platforms · Big Data · Digital Urbanism

1 Introduction

The ever-increasing urbanization leading to rising passenger traffic flows has posed significant challenges in managing and understanding their dynamics. Monitoring and comprehending the behavior patterns of passengers is crucial for optimizing operational efficiency, improving passenger experience, and ensuring their safety. Old conventional reporting methods such as census and surveys often fall short of providing timely, accurate, and detailed insights due to resource constraints and limitations in data analysis. However, the advent of new transport data sources has presented an opportunity to overcome these obstacles and change the way we generate reports on metro passengers' activity.

This scientific article aims to introduce a novel approach to automating text report generation by leveraging open transport data analysis platform. The analytical capabilities of this platform enable us to generate almost real-time, comprehensive reports that can offer actionable insights, inform decision-making processes, and enhance the overall management of public transport systems.

Supported by the Interdisciplinary Scientific and Educational School of Moscow University "Brain, Cognitive Systems, Artificial Intelligence".

The potential benefits of this automated text report generation are vast. Metro operators can gain real-time visibility into passenger flow, identify congestion hotspots, and proactively allocate resources to minimize disruptions, thus enhancing operational efficiency and passenger satisfaction. Regulatory bodies can utilize these reports to evaluate service quality, identify key performance indicators, and implement evidence-based policies. Additionally, researchers can leverage the insights provided by these reports to conduct in-depth studies on various aspects of metro passenger behavior and preferences, leading to further advancements in urban transportation planning and design.

2   On the Modern Data Sources

Previously, censuses, surveys, questionnaires, and manual data collection (with the help of enumerators) were the data sources for solving problems in the transport planning field. Data from such sources was expensive to obtain, contained a large number of inaccuracies, quickly became obsolete, and was collected quite infrequently. Thanks to the development of Internet of Things technologies, cellular communications, and big data storage, many new data sources for research have appeared. The data of modern sources are easier to obtain, their accuracy is higher. They are available with a delay of several minutes. This allows the city authorities to quickly make decisions on changes in the city's transport infrastructure.

2.1   Validators Data

In modern public transport systems in metropolitan areas, smart cards are the primary means of payment. RFID technology is the basis for the production of these cards. Users bring their cards to the validators to pay for the fare. Validators are usually a PoS (Point of Sale) payment terminal. Payment data is captured and stored. Such data most often contains the place identifier and the exact validation time, card identifier, and may also contain some additional data, for example, the type of card or the number of remaining trips.

2.2   Cellular Operators Data

Currently, cellular operators are installing base stations in the metro to improve the level of service. During their operation, cell phones communicate with these stations about signal strength and delay. From this data, cellular operators can determine at which station the cell phone owner began his trip and the endpoint station of this trip.

2.3   Data Types

Post-collection data contains records of each trip. Such individual data is suitable for analyzing the trajectories of individual passengers and allows researchers to make more accurate conclusions. However, such data are too voluminous and inconvenient for storage and analysis. There are approaches that personalize the data of cellular operators. For avoiding these shortcomings, data is aggregated by time (for example, by half-hour intervals) and by space, for example, by metro stations or hoods.

3   Literature Review

F. Calabrese was one of the first to suggest using the data of cellular operators to measure traffic flows in the articles [1,2]. He and his co-authors presented a system for solving several problems of digital urbanism according to the platform developed by Telecom Italia. Of great interest are the works of S. Sobolevsky and his colleagues. Recent articles by these authors are devoted to mobility networks as a predictor of socioeconomic status [3], prediction of urban population-facilities interactions [4], urban zoning using intraday mobile phone-based commuter patterns [5]. Researchers in their work widely use graph neural networks, anomaly detection algorithms, and natural language processing [6].

Currently, many authors support the idea of transforming urban infrastructure systems into smart city systems using data analysis platforms of telecom operators, public transport system validators, and IoT system sensors. [7–10]. Currently, there are several such platform solutions [11–13] that exist and are being developed at once.

4   On the Open Platform Architecture

Currently, there are many platforms for solving digital urbanism problems based on analyzing individual and aggregated data. The architectures of these platforms, according to our review, are fixed. They do not provide the ability to dynamically extend functionality. At the same time, the storage of individual or aggregated transport data, their preliminary semantic check, and visualization of the results problems are solved in all of these platforms.

As part of our research, we are developing a modular platform [14], the architecture of which is shown in the Fig. 1. In this architecture, the general tasks of storing and processing data, as well as visualizing the results, are solved by dedicated modules. New models within a given platform can be defined using an API module. This makes the system architecture open and allows platform users to create new solutions without re-writing data modules.

5   Reports on Transport Usage

For timely response to events taking place in the city, transport services need to receive reports about them. Events according to their impact on the city transport network are divided by us into three types: short-term, medium-term, and long-term.

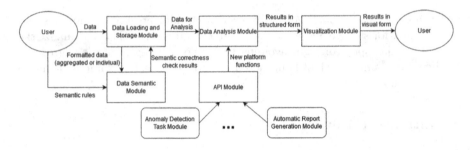

Fig. 1. Platform architecture

5.1 Short-Term Changes

Short-term changes in the city traffic flow may be associated with incidents in the transport network. Examples of such incidents can be a person falling on the rails in the subway, smoke in the subway tunnels, or major accidents on the route of city buses. Such changes cannot be predicted in advance. City services need to understand the number of passengers affected by such an incident to take action, for example, allocating compensation buses. Such changes do not affect the overall volume of the traffic flow, but may only cause delays. To provide reports on such events, a digital twin of the existing transport system can be created. During an incident, the operator can view the workload of the affected areas and assess the scope of necessary measures.

Another reason for short-term changes in the city's transport network may be important social events (concerts, fireworks, football matches). Information about such events is usually known in advance, so there is no need to predict them either. To provide additional transport, it is important to understand the amount of traffic flow generated by such an event. To generate reports on similar events, it is necessary to extract information about the change in traffic flows during similar events in the past.

For short-term changes in the event of incidents, a report in the system should be generated at the request of a transport service employee. For important social events, reports can be generated automatically for all stations and time intervals at the time the anomaly detection algorithm [15] is triggered.

5.2 Mid-Term Changes

Medium-term changes include periodic changes in traffic flow. Such changes may be associated with long holidays (New Year and May holidays in Russia, Chinese New Year and Mid-Autumn Festival in China), as well as vacations for schoolchildren and students. The start and end dates of such changes are usually known in advance, but it is necessary to adjust the transport system to changes in the transport requests of city residents. Changes in the traffic flow are continuous, but then there is a return to normal.

For medium-term changes, the report is generated once during the first manifestation of changes upon triggering the anomaly detection algorithm. Further, these values form a new normality until the end of the period of the medium-term anomaly. This will avoid unnecessary signals in the initial period of change. After the end of the change period, the system returns to the old normal values. Also, a report can be generated at the command of a transport service employee on the last day before the change period. The materials of this report can be used to analyze the return of the transport system to normal operation.

5.3   Long-Term Changes

Long-term changes are associated with major changes in the city. These changes are permanent and then become the new normal. Major changes may take place over a long period. An example of a continuous change in traffic flow would be a new residential complex opening. The traffic flow will increase as the complex is occupied and then after the end of the settlement, it will stabilize.

Changes can occur quite quickly in the case of the opening of new transport facilities, for example, the launch of a new metro station, the introduction of new bus routes, or the construction of new interchange hubs.

In the case of rapid changes, a report is generated at the time the anomaly detection algorithm is triggered for all stations where anomalous values were observed. The observed values are taken as the new normal. In the case of long-term changes, the process of reporting and changing normal values may occur several times as traffic behavior changes.

All described types of changes are presented in Fig. 2

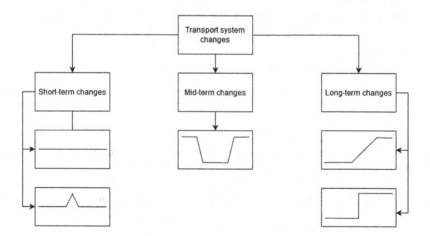

Fig. 2. Type of changes in transport systems

6 Example of Platform Usage

6.1 On the Source Data

The work solution we proposed for generating transport reports was tested on a dataset containing individual data on the validation of transport cards of the Moscow Metro for February 2020 and May 2021. The set contains data on validation time up to a second, validation lobby, card type, and some other additional technical information used for paying for the fare. Note that in the Moscow Metro, the card is validated only once, at the entrance to the station. For this reason, accurate data is only available on the outgoing flow for each station. If it is necessary to obtain information about the incoming stream aggregated over time intervals, data from cellular operators can be used.

6.2 Mid-Term Changes

In Russia, the Spring and Labor Day (May 1) and Victory Day (May 9) are celebrated annually at the beginning of May. Each of these holidays corresponds to several days off in the production calendar. Working people often take vacations on intermediate working days to get a long holiday. The traffic flow of people moving to their workplaces in the morning and to their place of residence from work in the evening is significantly reduced these days. With the end of the holidays, traffic flows return to normal. According to our classification, this event belongs to the medium-term ones. We applied the developed module to the individual transport data of metro users for May 2021. An example of a text report compiled based on the analysis of the traffic flow of the Elektrozavodskaya metro station for May 4 is shown in the Fig. 3.

Station: **Elektrozavodskaya**
Date: 04.05.2021
Event: May Celebrations
Day type: **Weekday**

Report: The total outgoing traffic flow per day was reduced by 8177 people. Typical traffic flow volume value: 18532. Observed value: 10355

Fig. 3. Example of text report on mid-term change

6.3 Short-Term Changes

Traditionally, fireworks are held every year on Victory Day in Moscow. One of the most popular viewing platforms for fireworks is the site on Vorobyovy Gory. Thousands of people become spectators of fireworks. For the city's transport system to withstand the additional load created, it is necessary to take measures following the additional number of people using public transport. The reports generated by the proposed system can help with this. Figure 4 and Fig. 5 respectively present text reports for Lomonosovskiy Prospekt and Universitet metro stations, which are the closest ones for the viewing platform under consideration.

Station: **Lomonosovskiy Prospekt**
Date: 09.05.2021
Event: Fireworks on Victory Day
Day type: **Holiday**

Report: The total outgoing traffic flow per hour interval from 22:00 to 23:00 was increased by 1234 people. Typical traffic flow volume value: 289. Observed value: 1523

Fig. 4. Report on the outgoing traffic flow from Lomonosovskiy Prospekt metro station during Victory Day fireworks

Station: **Universitet**
Date: 09.05.2021
Event: Fireworks on Victory Day
Day type: **Holiday**

Report: The total outgoing traffic flow per hour interval from 22:00 to 23:00 was increased by 2581 people. Typical traffic flow volume value: 952. Observed value: 3533

Fig. 5. Report on the outgoing traffic flow from Universitet metro station during Victory Day fireworks

It is important to note that people who leave the venue using the subway are very likely to arrive at the event also by subway, and at the same station. This information must be taken into account when managing the incoming traffic of these stations in the time intervals before 22:00.

One of the main means of controlling the traffic flow of the metro is to change the interval of trains. Reducing this interval for a long period requires large expenditures of both labor resources and trains.

To more accurately adjust the intervals of metro trains, information on the distribution of the traffic flow over time may be necessary. For these purposes, a load histogram can be created for smaller time intervals. An example of such a histogram for the Lomonosovsky Prospekt station is shown in Figs. 6 and 7.

As we can see in Fig. 6, the main part of the additional load on the Lomonosovskiy Prospekt station fell on four ten-minute time intervals from 22:30 to 23:10. The peak load fell on the period from 22:40 to 22:50. Before 22:30 and after 23:10 the traffic flow slightly exceeded the usual values.

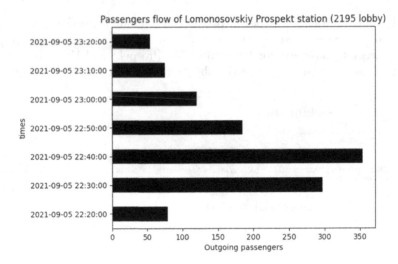

Fig. 6. Load histogram for Lomonosovskiy Prospekt metro station during Victory Day fireworks

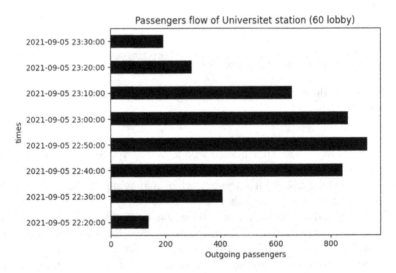

Fig. 7. Load histogram for Universitet metro station during Victory Day fireworks

Figure 7 shows the load distribution of the Universitet metro station. A significant excess of typical values was observed in six ten-minute intervals from 22:30 to 23:30, while, unlike Lomonosovsky Prospekt station, the load peak occurred not in one interval, but in three at once from 22:40 to 23:10.

If necessary for transport services, such reports can be built for intervals of any duration. It should be noted that the delays between the end of the event and the increase in the load on the transport system are due to the need to overcome the distance from the metro station to the observation deck. In the incoming transport stream, on the contrary, the load is observed ahead of the event time for the same reason.

In the Moscow Metro, the platforms are located underground. Escalators located in the lobbies are used to descend to the platforms. Depending on the venue of the event, the load on the lobbies may increase to varying degrees. To set up the movement of escalators, there is a mode for constructing text reports for individual lobbies. During the analyzed event, the main part of the increase in load fell on specific station vestibules. Text reports with analysis by lobbies are presented in Figs. 8 and 9.

Station: **Universitet**
Date: 09.05.2021
Event: Fireworks on Victory Day
Lobbies: 60, 61
Day type: **Holiday**

Lobby 60 report: The total outgoing traffic flow per hour interval from 22:00 to 23:00 was increased by 1976 people. Typical traffic flow volume value: 645. Observed value: 2621, typical value exceeded by 306%

Lobby 61 report: The total outgoing traffic flow per hour interval from 22:00 to 23:00 was increased by 605 people. Typical traffic flow volume value: 912. Observed value: 307, typical value exceeded by 197%

Fig. 8. Report on the outgoing traffic flow from Lomonosovskiy Prospekt metro station with analysis by lobbies

Station: **Universitet**
Date: 09.05.2021
Event: Fireworks on Victory Day
Lobbies: 60, 61
Day type: **Holiday**

Lobby 60 report: The total outgoing traffic flow per hour interval from 22:00 to 23:00 was increased by 1976 people. Typical traffic flow volume value: 645. Observed value: 2621, typical value exceeded by 306%

Lobby 61 report: The total outgoing traffic flow per hour interval from 22:00 to 23:00 was increased by 605 people. Typical traffic flow volume value: 307. Observed value: 912, typical value exceeded by 197%

Fig. 9. Report on the outgoing traffic flow from Universitet metro station with analysis by lobbies

6.4   Long-Term Changes

Long-term changes have a significant impact on the city's transport system and become the new normal. On February 22, 2020, the Smolenskaya station of the Arbatsko-Pokrovskaya line of the Moscow Metro was closed for long-term repairs

(2 years). At the same time, the traffic flow from this station was redistributed between other stations and modes of transport, creating an additional load on the transport system, which needs to be assessed. Information about the total traffic flow passing through this station could be obtained using a text report on the operation of the station on the last day of its operation, presented in Fig. 10.

Station: **Smolenskaya (AP)**
Date: 21.02.2020
Event: Closing Smolenskaya Station
Day type: **Weekday**

Report: The total outgoing traffic flow per day is 39201. Value doesn't exceed typical values

Fig. 10. Report on the outgoing traffic flow of Smolenskaya station (Arbatsko-Pokrovskaya line)

On the first weekday, the system signals a change in traffic flow at several stations. The biggest change occurred at the Smolenskaya station on the Filevskaya Line, which is the closest station to the closed station. Text report on Smolenskaya metro Station is presented in Fig. 11.

Station: **Smolenskaya (F)**
Date: 25.02.2020
Event: Closing Smolenskaya Station
Day type: **Weekday**

Report: The total outgoing traffic flow per day was increased by 22869 people.
Typical traffic flow volume value: 10940. Observed value: 33809. Value exceeded typical values by 209 %

Fig. 11. Report on the outgoing traffic flow of Smolenskaya station (Filevskaya line)

As we can see, from the incoming traffic flow with a typical value of 39201, more than 10,000 passengers moved to the Filevskaya metro station. The rest of the passengers were distributed among other metro stations and modes of transport.

Note that at the neighboring stations of the Kievskaya and Arbatskaya Filevskaya lines, there is a transfer to the Arbatsko-Pokrovskaya line. The distribution of passengers to this station means an increase in passenger traffic only on two sections of the Filevskaya line and a decrease in the flow on one section of the Arbatsko-Pokrovskaya line.

A similar change in the traffic flow was observed throughout all working days of the period under review. Report generation is also available for individual time intervals, which allows transport services to identify time intervals with the greatest changes in the traffic flow.

7   Discussion

The text reports and charts generated by this solution can already be used by city authorities to manage traffic flows. It is important to note that during large social events and activities, changes can occur in the transport network of the entire city. This leads to the formation of a large number of text reports and diagrams. It can be difficult for transport professionals to deal with such large volumes of information. Since the management of transport systems usually occurs not at the level of individual stops/stations, but at the level of routes and lines, in the future the functionality of the system can be extended by methods of report aggregation. The aggregated report will contain less redundant information and reflect the state of the entire control object as a whole, which will make this process faster and easier for a person.

Important for the implementation of the platform is also the issue of interaction between information collection devices (transport card validators, base stations) and the server on which the platform is running. If the data from these devices is available with low latency, then a prompt response to short-term changes and their description is available. In the case of a rare synchronization, for example, once a month, only historical data will be available to prepare for new events.

The information received by the text reporting platform can also be stored and accumulated. Based on it, new algorithms can be proposed for automating traffic flow management. In the intended use case, decisions on the scope and nature of transport measures (number of allocated buses, correction of intervals between trains) are made by the transport officer. The recommendation of measures, as well as the prediction of the success of these measures, can be assigned to new methods of machine learning. The data collected by the platform can be a valuable resource for generating them.

8   Future Development

In the future, we plan to expand the set of tasks solved using the platform. In modern public transport systems, there are concession tickets for various social groups such as schoolchildren, students, pensioners, and the disabled. The fare for these concessionary tickets is significantly lower than the fare for a regular ticket. Social cards can be used by fraudsters to save money on travel. Each of the social groups has its pattern of transport behavior. For example, for schoolchildren in the Moscow region, it is typical to make one trip to the place of study in the interval from 07:30 to 08:00 in the morning, as well as a return trip before 15:00 to the place of residence. The constant use of a social card for trips between 09:00 and 10:00 and return trips between 18:00 and 19:00 may be a signal of fraudulent activities.

We also plan to expand the set of supported data. Sets of both individual and aggregated data can be collected from car-sharing or scooter/bike rental systems. Data on the place and time of the beginning and end of trips can be used to identify areas with low and high demand for sharing facilities. This information can be used to maximize profits.

9   Conclusion

In conclusion, our research has resulted in the development of a highly efficient module that automates the generation of reports on the state of the city's transport system. This module, incorporated into an open transport data analysis platform, enables transport services to create diverse and comprehensive reports based on the proposed classification of changes in the transport network. The reports generated using this module were used to study the changes in traffic in the Moscow metro during the celebrations in early May and the closure of the Smolenskaya metro station for long-term reconstruction. Overall, our module represents a significant advancement in the field of transport data analysis, offering a practical and efficient solution for generating insightful reports on the city's transport system state.

References

1. Calabrese, F., Colonna, M., Lovisolo, P., Parata, D., Ratti, C.: Real-time urban monitoring using cell phones: a case study in Rome. IEEE Trans. Intell. Transp. Syst. **12**(1), 141–151 (2010)
2. Calabrese, F., Di Lorenzo, G., Liu, L., Ratti, C.: Estimating origin-destination flows using opportunistically collected mobile phone location data from one million users in Boston Metropolitan Area. IEEE Pervasive Comput. **10**, 36–44 (2011)
3. Khulbe, D., Belyi, A., Mikeš, O., Sobolevsky, S.: Mobility networks as a predictor of socioeconomic status in urban systems. In: Gervasi, O., et al. (eds.) Computational Science and Its Applications, ICCSA 2023. LNCS, vol. 13957, pp. 453–461. Springer, Cham (2023). https://doi.org/10.1007/978-3-031-36808-0_32
4. Mishina, M., et al.: Prediction of urban population-facilities interactions with graph neural network. In: Gervasi, O., et al. (eds.) Computational Science and Its Applications, ICCSA 2023. LNCS, vol. 13956, pp. 334–348. Springer, Cham (2023). https://doi.org/10.1007/978-3-031-36805-9_23
5. Bogomolov, Y., Belyi, A., Mikeš, O., Sobolevsky, S.: Urban zoning using intraday mobile phone-based commuter patterns in the City of Brno. In: Gervasi, O., et al. (eds.) Computational Science and Its Applications, ICCSA 2023. LNCS, vol. 13957. Springer, Cham (2023). https://doi.org/10.1007/978-3-031-36808-0_35
6. Tikhonova, O., Antonov, A., Bogomolov, Y., Khulbe, D., Sobolevsky, S.L.: Detecting a citizens' activity profile of an urban territory through natural language processing of social media data. Procedia Comput. Sci. **212**, 11–22 (2022)
7. Young, M., Farber, S.: The who, why, and when of Uber and other ride-hailing trips: an examination of a large sample household travel survey. Transp. Res. Part A Policy Pract. **119**, 383–392 (2019)
8. Becker, H., Balac, M., Ciari, F., Axhausen, K.W.: Assessing the welfare impacts of Shared Mobility and Mobility as a Service (MaaS). Transp. Res. Part A Policy Pract. **131**, 228–243 (2020)

9. Holguín-Veras, J., Leal, J.A., Sánchez-Diaz, I., Browne, M., Wojtowicz, J.: State of the art and practice of urban freight management: part i: infrastructure, vehicle-related, and traffic operations. Transp. Res. Part A Policy Pract. **137**, 360–382 (2020)

10. Xu, Z.: UAV surveying and mapping information collection method based on Internet of Things. IoT Cyber-Phys. Syst. **2**, 138–144 (2022)

11. Golubev, A., Chechetkin, I., Solnushkin, K.S., Sadovnikova, N., Parygin, D., Shcherbakov, M.: Strategway: web solutions for building public transportation routes using big geodata analysis. In: Proceedings of the 17th International Conference on Information Integration and Web-Based Applications and Services, December 2015, pp. 1–4 (2015)

12. Feng, H., Lv, H., Lv, Z.: Resilience towarded Digital Twins to improve the adaptability of transportation systems. Transp. Res. Part A Policy Pract. **173**, 103686 (2023)

13. Bogomolov, Y., Sobolevsky, S.: A scalable spatio-temporal analytics framework for urban networks. In: Antonyuk, A., Basov, N. (eds.) Networks in the Global World VI, NetGloW 2022. LNNS, vol. 663, pp. 68–78. Springer, Cham (2023). https://doi.org/10.1007/978-3-031-29408-2_5

14. Bulygin, M.V., Namiot, D.E., Pokusaev, O.N.: On the analysis of individual data on transport usage. Proc. ISA RAN **73**(1), 24–33 (2023)

15. Bulygin, M., Namiot, D.: Anomaly detection method for aggregated cellular operator data. In: 2021 28th Conference of Open Innovations Association (FRUCT), January 2021, pp. 42–48. IEEE (2021)

Developing a Traffic Analysis Suite for Modified Packet Capture File

O. P. Morozova$^{(\boxtimes)}$ ⓘ, M. A. Orlova$^{(\boxtimes)}$ ⓘ, N. A. Naumov ⓘ,
and L. I. Abrosimov ⓘ

National Research University Moscow Power Engineering Institute,
Krasnokazarmennaya 14, Building 1, Moscow 111250, Russia
{MorozovaOP,OrlovaMA,NaumovNA,AbrosimovLI}@mpei.ru
https://mpei.ru

Abstract. The purpose of this work is to develop an application for traffic analysis based on a method of obtaining datasets of network traffic with ground truth already defined. This method allows us to get datasets with accurate ground truth while not violating data privacy since the critical data is stripped and replaced by traffic meta description, which makes it useful for a wide range of traffic analysis methods.

Keywords: Traffic Analysis · ground truth · labeled datasets · a traffic analysis suite

1 Introduction

The need to classify network traffic arose with the proliferation of the internet itself. At first it was brought on by the necessity to identify malicious traffic and attack patterns to ensure the security of networks. Later on, with the development of quality of service (QoS) systems, the need for more granular traffic classification became apparent. The first classifiers were port- and payload-based methods, that used ports and payload patterns, respectively, to identify applications. However, since traffic encryption and dynamic port assignment gained wide recognition the efficiency of those methods has been steadily declining. In their place, a variety of statistics-based methods have been developed. Most of them use different machine learning techniques to build classifiers. Nowadays, both network security monitoring and QoS systems require fast and precise classifiers that are able to meet the challenge of analyzing the ever-changing network environment in real time. However, the development of such classifiers is seriously stunted by unresolved issues in the field of traffic analysis. One of the most significant issue is the collection of big and representative datasets with reliable ground truth that can be used for training and evaluating classifiers.

Authors [1] report a lack of such datasets in multiple domains of traffic analysis: network analytics, intrusion detection and network functions in middleboxes. All those fields require correctly labeled and illustrative datasets, containing different types of network traces. Authors [2] discuss the same issue in the analysis

V. M. Vishnevskiy et al. (Eds.): DCCN 2023, LNCS 14123, pp. 448–461, 2024.
https://doi.org/10.1007/978-3-031-50482-2_35

of the internet of things (IoT) and [3,4] both point out the importance of datasets suitable for deep learning algorithms. Researchers in [5] use their database to analyze the current state of the field of traffic classification. They conclude that most used datasets are either outdated and do not represent the network environment accurately or private and therefore unavailable for verifying and point out that this issue stems from concern for data privacy and enterprise security. This systematic problem is also thoroughly discussed in [6,7]. Even in research dedicated to the development of new classifiers [8], it is pointed out that there is a lack of suitable datasets and the necessity to collect it from scratch.

The works [9–16] tried to solve the issue of the lack of traffic ground truth. There have been several studies dedicated to developing a reliable way to obtain big datasets with dependable ground truth. G'eza szab'o et. Al [9] proposed an active measurement method that allowed validating other classifiers based on their performance on emulated traffic. The ground truth obtained using this method was absolutely accurate since their algorithm recorded an application for every packet and wrote that information into the packet's header. However, this came with a limitation on packet size - it had to be 4 bytes shorter than the maximum allowed packet size to place the application tag in it. Additionally, emulated traffic cannot replicate the variety of real network flows; therefore, classifiers may show worse results in action than during the validation process since their ability to analyze traffic in all its complexity was not tested to the fullest. Similar systems based on emulated network traffic have been deployed in [10,11]. Their absolutely accurate ground truth allowed researchers to evaluate the most common ways to obtain labels for datasets to this day: port- and dpi-based. Multiple sources found out that their accuracy even on emulated flows, without any complications that happen in real networks, has been lacking [10,12].

Another approach to obtaining ground truth was introduced in [13,14]. Those systems are heuristics-based which increases analysis speed but decreases ground truth accuracy since agglomerating methods tend to be less exact in their classification.

A different tool was introduced by the authors of [15]. Tstat can capture traffic in real life or analyze previously recorded traces, creating a log file of flow-level measurements and statistical histograms which is quite useful for network monitoring. However, it does not save any packet-level information that could be used for traffic classification. Additionally, it obtains ground truth for unencrypted packets using DPI-based methods which have been proven ineffective. As for encrypted packets, Tstat implements a SPI-based classifier which also does not provide completely accurate results. Therefore, ground truth obtained using Tstat cannot be called absolutely reliable.

Another tool for network monitoring was proposed in [16]. VBS is a traffic analysis system installed on end users' machines that listens to all connections and collects packets into flows. The application tag for the given flow is determined via socket monitor which grants absolutely correct ground truth for every flow. However, for its usage on Windows a root certificate installation is required and its compatibility with the latest versions in unknown.

Another way to self-collect datasets is by using widely known sniffer applications like Wireshark or TCPdump. However, these methods do not solve the problem of ground truth definition. Often researches have to do it manually or resort to DPI-based classifiers which have multiple shortcomings discussed above. Additionally, the size of datasets becomes disproportionally large as traces contain packets with full payload, that is nowadays rarely used in traffic analysis systems. Therefore, a great part of data in those traces is unimportant.

Many researchers, faced with multiple shortcomings of self-collection of training data, turn to publicly available datasets. However, this method has its own issues. One of the most prominent is the lack of necessary variety of traffic traces. Classifiers built for different purposes require different training datasets and public databases, even though they are growing in number, and are yet to satisfy this demand for diversity. Another point of concern is the definition of ground truth in those datasets. Many of them do not contain any correlation to applications and types of network traffic, and those that do tend to use port- and DPI-based methods to obtain it, which lack accuracy.

As our survey of related works shows, the problem of dataset collection with reliable ground truth is yet to be fully resolved dispite the efforts of researchers. This paper is organized in the following way. Section 1, contains an overview of related work. Section 2. is dedicated to the description of suggested traffic meta description modification of the current most used Packet Capture (PCAP) file format. Section 3 presents an example of traffic analysis using proposed format of file and developed programs. Section 4. describes the work in progress developing of the traffic analysis suite to process modified PCAP files and combine analysis methods for different captured files.

2   Traffic Meta Description

Analysis of the current methods of obtaining training datasets showes that this direction needs further exploration in order to optimize this process. The current analysis approaches are built on the traffic obtained using tools which do not capture application information or when the sniffer is placed in the middle of the traffic path. This makes it more complicated to analyze captured files and to mark up the applications in the traffic.

In light of this, we propose a special formatting for captured packets to be stored in traffic meta description. Each captured packet is to be stored as shown in Fig. 1(b), as opposed to the standard packet storing described in [17] and shown in Fig. 1(a). In the traffic meta description, each packet trace consists of three main parts: header, payload descriptor and an application tag. The header contains all the protocol layers the original packet had. The payload descriptor is what replaces the payload itself since it is removed from the trace. The descriptor contains the payload's size and some statistical measurements that can be used for traffic classification. Finally, the application tag is what solves the problem of ground truth definition. This tag is added directly at the moment of packet collection using a connection tracker; therefore there is no risk of misidentifying the application.

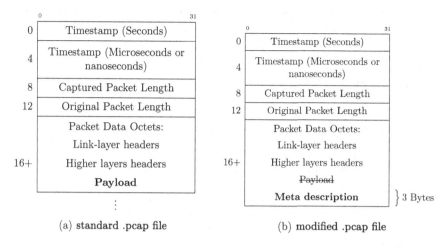

Fig. 1. Standard and modified packet storing in .pcap files

To achieve such packet formatting we developed Python and eBPF programs. The simplified algorithm shown in 1 promises a size reduction of trace files due to the replacement of lengthy payloads with short traffic meta descriptions. To verify this supposition, we conducted an experiment: we repeatedly ran two sniffers on the same PC simultaneously for the same amount of time and compared the size of obtained traces. For sniffing we used one developed program based on the algorithm described in 1 and one widely known tool Wireshark. After a hundred sniffing sessions we calculated the mean growth of file size over time for both our sniffer and Wireshark and the compared obtained dependencies.

As shown in Fig. 2,the application of our algorithm for packet sniffing with traffic meta description does diminish the size growth of the trace file and overall reduces file size. Additionally, since time does not describe flow rate, we calculated the mean size growth over the amount of packets captured that is shown in Fig. 3.

This approach allows us to deal with the multiple problems of dataset collection described above. Since most state-of-the-art classifiers rely on the statistical features of packets and flows, extraction of the payload does not affect the quality of classification, but it does decrease the size of one packet and the whole dataset significantly. For instance, it was observed that an hour of traffic capturing can produce a file up to a few gigabytes in size in .pcap format, if we use standard sniffing software. Using the method depicted in Fig. 1b has shown promising file size reduction - the same amount of sniffing produces only a few megabytes of data in .pcap format. Certainly, removing the payload of every packet and replacing it with much smaller meta description plays the greatest part in size decrement.

Algorithm 1. Packet sniffing algorithm

```
while program running do do
    output_file ← open;
    listen to interfaces;
    if packet captured then
        app_name ← "";
        packet_load_size ← 0;
        get process_id for the packet;
        if process_id exists then
            app_name ← process_name_for_given_id;
            packet_load_size ← load_length_from_the_packet;
            packet_load ← 0;
            packet_meta_layer ← [app_name, packet_load_size];
            packet ← packet + packet_meta_layer;
            output_file ← output_file + packet;
        end if
    end if
end while
```

Comparison of size of standard and modufied .pcap files

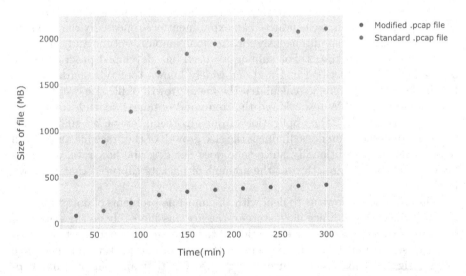

Fig. 2. Comparison of file size of traffic traces over time

This approach makes it possible to collect longer traces using the same amount of memory storage. Additionally, the application tag eliminates the issue of defining ground truth - with traffic meta description obtained dataset already contains ground truth for each packet right after the collection finishes.

Comparison of size of standard and modufied .pcap files

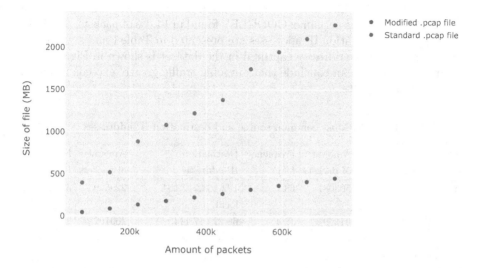

Fig. 3. Comparison of file size of traffic traces over amount of packets captured

Since traffic meta descriptions proved to be an advantageous method of storing packet traces, we used it to develop a traffic analysis suite that allows us to inspect collected traces, sorting packets by generating application. Further description, alongside with test results, will be provided in the presentation. However, here we detail an algorithm for creating trace file with meta description for each packet stored in .pcap format.

3   Traffic Analysis with Traffic Meta Descriptions

To verify the applicability of traffic meta descriptions for network traffic analysis, we used our sniffer program, based on Algorithm 1, to capture traffic on a student PC in the University network. The trace contains network traffic from May 10, 2023. The sniffer ran for 5 h, after which the experiment was deemed concluded successfully. The dataset consisted of 738 829 packets with an oveerall size of 55,9 MB. Here we describe the statistics that were retrieved directly from the trace using the developed programs.

The most commonly used source IP address was 192.168.212.104, which was the student's PC IP address (University LAN), found with a frequency of 68,1%. Next were IP addresses 66.22.217.114 (whois netname: discord-rumow2-1) and 93.158.134.119 (whois netname: YANDEX-93-158-134), found with a frequency of 12,4% and 2,9%, respectively. Table 1 shows all source IP addresses that were found in more than 1% of the packets and Fig. 4 depicts a word map of all source IP addresses found.

Among the most common destinations were also PC IP address 192.168.212.104 and 66.22.217.114, with frequencies of 31,1% and 28,3%, respectively. However, the third most used destination IP address was different - it was 173.194.182.28 (whois netname: GOOGLE), found in 14,0% of packets. Similarly, the most used destination IP addresses are presented in Table 1 and a word map of all destination IP addresses captured in the dataset is shown in Fig. 5.

From Table 1 we can conclude that outgoing traffic greatly exceeds incoming traffic in this dataset.

Table 1. Most common source and destination IP addresses

Source IP address	Amount of packets	Frequency (%)	Destination IP address	Amount of packets	Frequency (%)
192.168.212.104 local	503081	68,1	192.168.212.194 local	229529	31,1
66.22.217.114 whois netname: discord-rumow2-1	91299	12,4	66.22.217.114 whois netname: discord-rumow2-1	209102	28,3
93.158.134.119 whois netname: YANDEX-93-158-134	21529	2,9	173.194.182.28 whois netname: GOOGLE	103136	14,0
139.45.224.73 whois netname: RETN-RU	9750	1,3	173.194.179.36 whois netname: GOOGLE	53261	7,2
139.45.228.1 whois netname: RETN-RU	8240	1,1	93.158.134.119 whois netname: YANDEX-93-158-134	30980	4,2

Regarding source and destination port information, the most used ports are presented in Table 2. Overall, the dataset contains 2960 distinct source ports and 2088 distinct destination ports. As we will show later, port 443 (HTTPS) was the most used and corresponds to a browser (*browser.exe*) which was one of the most active applications during collection of this dataset.

Despite the removal of the payload, we were able to inspect the packet size. Since header length did not change and the payload's size is saved in the traffic meta description, it is possible to easily calculate the total size of every packet. Some statistical values of packet size are presented in Table 3 and packet distribution over size is shown in Fig. 6.

The characteristics described above can be extracted without using the traffic meta description. However, our descriptor allows us to inspect them with regard to the applications that generated each packet. Our trace contains 40 distinct applications. The distribution of the packets between the most active of them is shown in Table 4.

Fig. 4. Word map of all used source IP address

Fig. 5. Word map of all used destination IP addresses

Table 2. Most common source and destination ports

Source port	Amount of packets	Frequency (%)	Destination port	Amount of packets	Frequency (%)
443	129797	17,6	443	282708	38,3
54534	102177	13,8	50001	209102	28,3
50001	91304	12,4	65138	20764	2,8
61619	40957	5,5	55381	9599	1,3
65138	30340	4,1	55384	8238	1,1
57076	10562	1,4			

Table 3. Quantile and descriptive statistics of packet size

Quantile statistics		Descriptive statistics	
Minimum	55	Standard deviation	434,9
5-th percentile	66	Coefficient of variation (CV)	1,418
Median	124	Kurtosis	3,303
95-th percentile	1527	Mean	306,8
Maximum	1538	Median Absolute Deviation (MAD)	52
Range	1483	Skewness	2,227

Additionally, the application tag allows us to conduct session analysis to identify the application responsible for opening the session without needing to inspect the packet itself, as shown in Table 5. This allows us to easily investigate connections between source and destination IP addresses, ports and applications.

Table 4. Most active applications in the trace

Application name	Amount of packets	Frequency (%)
browser.exe	320616	43,4
Discord.exe	245639	33,2
AnyDesk.exe	92140	12,5
steam.exe	38118	5,2
Telegram.exe	10732	1,5
steamwebhelper.exe	10071	1,4

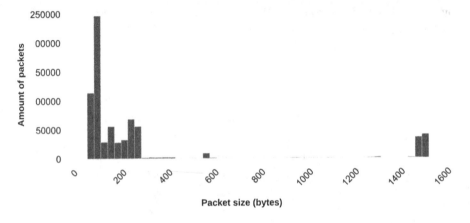

Fig. 6. Distribution of packet size

Table 5. Some sessions opened by the most active applications

Application name	Protocol Protocol	Source IP	Source port	Destination IP	Destination port	Amount of packets
browser.exe	UDP	192.168.212.104	54534	173.194.182.28	443	102177
browser.exe	UDP	192.168.212.104	61619	173.194.182.36	443	40957
browser.exe	UDP	192.168.212.104	57076	173.194.182.36	443	10562
browser.exe	TCP	192.168.212.104	65138	93.158.134.119	443	30337
browser.exe	TCP	93.158.134.119	443	192.168.212.104	65138	20764
steam.exe	TCP	139.45.224.73	443	192.168.212.104	55381	9599
steam.exe	TCP	139.45.228.1	443	192.168.212.104	55384	8238
Discord.exe	TCP	192.168.212.104	55429	162.159.138.234	443	5691
Telegram.exe	TCP	192.168.212.104	57263	149.154.167.50	443	3473

This analysis of an experimental trace allowed us to explore the possible implementations of traffic meta description and demonstrate its benefits for network traffic analysis. Additionally, this examination of packet traces served as a starting point to develop a traffic analysis suite - an application to collect traffic traces in a modified .pcap format and visualize the analysis process.

4   Architecture of the Traffic Analysis Suite

We are developing a web application to perform packet capturing, statistical analysis and visualization of obtained results. Here we provide a general scheme of part of the traffic analysis suite and outline its capabilities. The proposed structure of the application is shown in Fig. 7.

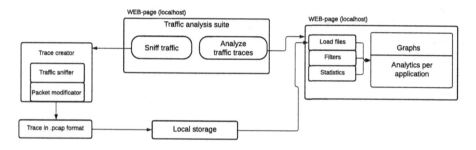

Fig. 7. Architecture of traffic analysis suite

4.1   Main Page

Web-page format was chosen the for the suite application, because this solution removes the problem of compatibility with different operating systems and associated complications in application development. The main page allows access to a traffic sniffer that stores packet traces in modified .pcap format with traffic meta description or to an analyzer of existing traces and performs authentication and authorization processes for suite users.

4.2   Packet Sniffer

The Trace Creator page allows users to configure their sniffing parameters, such as sniff duration, interface or particular applications to listen to and additional options to write in traffic meta description. We intend to cover some statistical values of packet's payload in the descriptor in the future. After activating the sniffing process, packet capturing and modification is described in Algorithm 1. The obtained trace is stored on the local machine and/or in the suite's database (after performing anonymization measures) if the user so chooses.

4.3   Analysis Interface

The analytics page allows users to inspect previously collected packet traces and get statistics per application. We intend to provide tools to configure an interactive plots to show various relations between packet parameters, filter those relations per other parameters, as we did to get Table 5, and compare different traces with each other. Since we plan to include a public database of packet traces with traffic meta descriptions, a tool to import those traces and analyze them alongside self-collected ones will be added. Additionally, we plan to include some traffic analysis methods to classify traffic directly in the application or allow researchers to add their own methods and verify their accuracy on different traces.

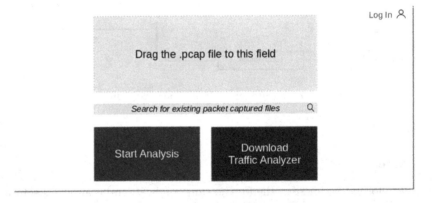

Fig. 8. The main page of the developing suite

Figure 8 shows the main page of the developing suite. Figure 9 shows the customized analytical page with example of generated histogram. The histogram generated automaticaly using Plotly. The user can select the field in dataset and the plot type and obtain the visulized data.

Since the traffic analysis suite is an application primarily focused on implementing a new algorithm of packet capturing and modification of standard .pcap format to include traffic meta descriptions, we provided a description of integral parts of the application to achieve that goal. However, architecture may change and be expanded in the future according to current needs of traffic analysis field.

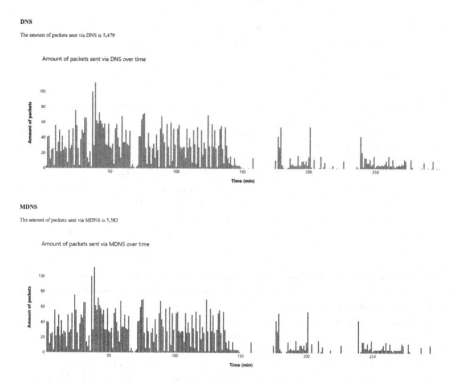

Fig. 9. The analytics page to visualize captured traces

5   Conclusion

In this paper we present a modified formatting for datasets in traffic analysis. Packets are stored without payload, but have a payload descriptor that ensures that no meaningful data is lost from the payload itself and an application tag that serves as ground truth for the packet. We first demonstrate that there is an unresolved issue with the collection of datasets and obtaining ground truth for them. Then we describe our modification for format of datasets where each packet is stored with traffic meta description instead of raw payload and provide an algorithm that achieves that format. We further demonstrate the results of using our algorithm to collect network traffic and provide some analytics of captured traces. Finally, we describe an application that allows to collect datasets in said format and gathering some basic statistics from them in real time - traffic analysis suite.

We conclude that this application is a promising development in the field of traffic classification that allows solving the problem of obtaining reliable ground truth for datasets. However, we see some directions for future improvement of this application, such as exploring of data anonymization methods, expanding payload descriptor so it would give a more comprehensive view of the payload itself, optimizing memory usage and broadening the list of transport protocols that our suite is able to detect.

References

1. Papadogiannaki, E., Ioannidis, S.: 2021 A survey on encrypted network traffic analysis applications, techniques, and countermeasures. ACM Comput. Surv. **54**(6), Article 123, 1–35 (2021). https://doi.org/10.1145/3457904 (Jan 2022)
2. Tahaei, H., Afifi, F., Asemi, A., Zaki, F., Anuar, N.B.: The rise of traffic classification in IoT networks: A survey. J. Netw. Comput. ppli. **154**, 102538 (2020). https://doi.org/10.1016/j.jnca.2020.102538. ISSN 1084–8045
3. Rezaei, S., Liu, X.: Deep learning for encrypted traffic classification: an overview. IEEE Commun. Mag. **57**(5), 76–81 (2019). https://doi.org/10.1109/MCOM.2019.1800819
4. Aceto, G., Ciuonzo, D., Montieri, A., Pescapé, A.: Mobile encrypted traffic classification using deep learning: experimental evaluation, lessons learned, and challenges. IEEE Trans. Netw. Serv. Manage. **16**(2), 445–458 (2019). https://doi.org/10.1109/TNSM.2019.2899085
5. Iglesias, F., Ferreira, D.C., Vormayr, G., Bachl, M., Zseby, T.: NTARC: a data model for the systematic review of network traffic analysis research. Appli. Sci. **10**(12), 4307 (2020). https://doi.org/10.3390/app10124307
6. Getman, A.I., Ikonnikova, M.K.: A survey of network traffic classification methods using machine learning. Program Comput. Soft **48**, 413–423 (2022). https://doi.org/10.1134/S0361768822070052
7. Deart V.Yu., Mankov V.A., Krasnova I.A. Analysis of promising approaches and research on traffic flow classification for maintaining QoS by ML methods in SDN networks. Herald Siberian State Univ. Telecommun. Inform. Sci. (1), 3–23 (2021). (In Russ.) https://doi.org/10.55648/1998-6920-2021-15-1-03-22
8. Dias, K.L., Pongelupe, M.A., Caminhas, W.M., de Errico, L.: An innovative approach for real-time network traffic classification. Comput. Netw. **158**, 143–157 (2019). https://doi.org/10.1016/j.comnet.2019.04.004, ISSN 1389–1286
9. Szabó, G., Orincsay, D., Malomsoky, S., Szabó, I.: On the validation of traffic classification algorithms. In: Claypool, M., Uhlig, S. (eds.) PAM 2008. LNCS, vol. 4979, pp. 72–81. Springer, Heidelberg (2008). https://doi.org/10.1007/978-3-540-79232-1_8
10. Gringoli, F., Salgarelli, L., Dusi, M., Cascarano, N., Risso, F., Claffy, K.C.: GT: picking up the truth from the ground for internet traffic. SIGCOMM Comput. Commun. Rev. **39**(5), 12–18 (2009). https://doi.org/10.1145/1629607.1629610
11. Lizhi, P., Hongli, Z., Bo, Y., Yuehui, C., Tong, W.: Traffic labeller: collecting internet traffic samples with accurate application information. China Commun. **11**(1), 69–78 (2014). https://doi.org/10.1109/CC.2014.6821309
12. Dusi, Maurizio, Gringoli, Francesco, Salgarelli, Luca: Quantifying the accuracy of the ground truth associated with Internet traffic traces. Comput. Netw. **55**(5), 1158–1167 (2011). https://doi.org/10.1016/j.comnet.2010.11.006, ISSN 1389–1286
13. Canini, M., Li, W., Moore, A.W., Bolla, R.: GTVS: boosting the collection of application traffic ground truth. In: Papadopouli, M., Owezarski, P., Pras, A. (eds.) TMA 2009. LNCS, vol. 5537, pp. 54–63. Springer, Heidelberg (2009). https://doi.org/10.1007/978-3-642-01645-5_7

14. Baer, A., et al.: DBStream: a holistic approach to large-scale network traffic monitoring and analysis. Comput. Netw. **107**(Part 1), 5–19 (2016). https://doi.org/10.1016/j.comnet.2016.04.020, ISSN 1389–1286

15. Finamore, A., Mellia, M., Meo, M., Munafo, M.M., Torino, P.D., Rossi, D.: Experiences of Internet traffic monitoring with tstat. IEEE Netw. **25**(3), 8–14 (2011). https://doi.org/10.1109/MNET.2011.5772055

16. Bujlow, T., Balachandran, K., Riaz, T., Pedersen, J.M.: Volunteer-based system for classification of traffic in computer networks. In: 2011 19thTelecommunications Forum (TELFOR) Proceedings of Papers, Belgrade, Serbia, pp. 210–213 (2011). https://doi.org/10.1109/TELFOR.2011.6143528

17. PCAP Capture File Format. https://datatracker.ietf.org/doc/id/draft-gharris-opsawg-pcap-00.html

On Heuristic Algorithm with Greedy Strategy for the Correlation Clustering Problem Solution

Aleksandr Soldatenko⬥, Daria Semenova⁽✉⁾⬥, and Ellada Ibragimova⬥

Siberian Federal University, Krasnoyarsk, Russian Federation
{ASoldatenko,DVSemenova}@sfu-kras.ru

Abstract. The Correlation Clustering (CC) problem is traditionally defined as a problem of partitioning a signed graph without specifying the number of clusters in advance. In this paper, CC problem is considered for undirected and unweighted signed graphs without multiple edges and loops, where error functional is linear combination of intercluster and intracluster errors. In this formulation, the CC problem is NP-complete. Exact algorithms for this problem are time-consuming. Approximate algorithms for solving CC problem often lead to unsatisfactory results, and heuristic algorithms are often non-deterministic in the number of steps leading to a solution. We propose a new heuristic algorithm $SGClust_\alpha$ for the CC problem solving. The main idea of this algorithm is in intracluster error minimizing and optimization of error functional according to the greedy strategy. It was proved that this algorithm takes polynomial time. Numerical experiments were carried out on randomly generated signed graphs.

Keywords: Correlation Clustering · Signed graph · Heuristic Greedy algorithm · Structural balance · Graph partition

1 Introduction

In 1953, in the paper [13], Harary formalized Heider's structural balance theory [12], which describes social relations, and introduced the notion of a sign graph. One of the main properties of the sign graph is the property of balance, which is satisfied if and only if all cycles in a graph are positive. Initially the relations between two groups were considered and it is proved that a graph is balanced if the set of its vertices can be divided into two non-overlapping subsets in such a way that positive edges are inside the clusters and negative edges are between them. In the [8] J. A. Davis elaborated on the property of k-balance when graph vertex set is divided into k groups and called it clustering. He also proved that a graph has a clustering if and only if it does not contain a cycle with exactly one negative edge.

This work is supported by the Krasnoyarsk Mathematical Center and financed by the Ministry of Science and Higher Education of the Russian Federation (Agreement No. 075-02-2022-876).

V. M. Vishnevskiy et al. (Eds.): DCCN 2023, LNCS 14123, pp. 462–477, 2024.
https://doi.org/10.1007/978-3-031-50482-2_36

In the [9] P. Doreian and A. Mrvar were the first to formulate the clustering problem as an optimization model and analyze the changing relationships on the Sampson monastery data. Later, in the [4] Bansal formulated correlation clustering problem, which is a special case of the sign graph partitioning problem.

Nowadays, Correlation Clustering (CC) problem is a well-known unsupervised learning problem aimed at finding a vertices partition in a signed graph with a disagreements minimum number or the maximum number of consistent edges [4]. A disagreement occurs when a positive edge connects vertices from different clusters or a negative edge connects vertices from the same cluster. In such cases, the edge is called inconsistent, and in all others it is called consistent. The peculiarity of this problem is that the number of clusters does not predetermined. Currently, various formulations of the CC problem can be found in the literature. They are classified by following parameters: the number of clusters (set initially, limited by any number or unlimited) [16], the size of clusters (fixed, limited or unlimited) and the type of error functional [9,17,21].

Our attention is focused on the CC problem for undirected and unweighted signed graphs without multiple edges and loops with an error functional in the form of a linear combination of the intercluster and intracluster errors. This formulation was proposed by Doreian et al. in the work [9]. Bansal et al. in the [4] have proved that this problem is NP-hard. Let's note the CC problem may have more than a single solution. Due to the difficulty of the problem under consideration, research to develop heuristic algorithms that find a solution in an acceptable time are being conducted [2,5,7,9,18]. To solve the CC problem, we propose a new heuristic algorithm $SGClust_\alpha$ based on a greedy strategy.

The CC problem arises in various scientific fields such as Community Identification in Social Network, Image Processing, Data Mining, Computational Biology, Telecommunication and Control Systems.

Abdelnasser et al. [1] have proposed a clustering, sub-channel and power allocation framework to be implemented in a semi-distributed fashion in a two-tier OFDMA cellular network. They have proposed correlation clustering approach, which considers the trade-off between the bandwidth and interference, offers a performance that is very close to that of the optimal clustering; however, with much reduced complexity.

Maatouk et al. [19] addressed the problem of clustering and user scheduling with massive MIMO technology, which is regarded as a key factor in the development of 5G networks.

2   Related Works

Figure 1 shows the different variations of the signed graph partitioning problem that can be found in the literature.

This problem is considered for directed, undirected, weighted, and unweighted graphs. The computational complexity for CCP in these cases and works in what they are considered are given in the Table 1. Also, multiplex graphs are used when solving practical problems, for example, when detecting the voting behavior patterns.

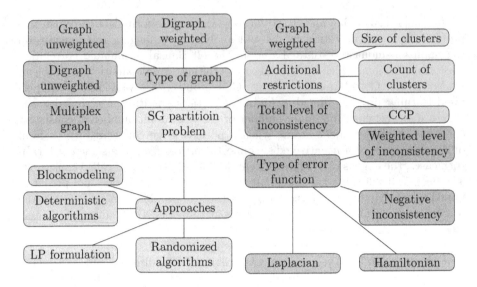

Fig. 1. Taxonomy of Correlation Clustering problem

Among the approaches to solving the problem, the following can be distinguished: LP formulations [6,11,20], randomized algorithms [2,5,7,18], deterministic algorithms [15], blockmodeling [10].

The most popular error functional is the total level of inconsistency, but researchers also suggest using positive inconsistency, negative inconsistency, weighted level of inconsistency [9], Hamiltonian [21], and Laplacian [17] as error functional.

Additional constraints are the size of clusters and the number of clusters. In the case when the number of clusters does not prespecified is singled out as a correlation clustering problem.

The most popular CC problem solving algorithms are described below.

Relocation heuristic (*RH*) [9] is one of the most common heuristics. The initial partition is constructed randomly. Then all possible moves and exchanges of vertices from the current cluster to any other cluster are evaluated. The ones that improve the value of the error functional are applied.

To improve the solution obtained by *RH* M. Brusco and P. Doreian applied *tabu search* metaheuristic [7]. For the partition obtained by the *RH* algorithm, the moves of each vertex to another cluster are evaluated. The ones that most improve or least deteriorate the error are performed. A moved vertex cannot be relocated within a given number of iterations.

An adaptation of the variable neighborhood search algorithm for CC problem is proposed in [7]. For a partition obtained by the *RH* algorithm, vertices with a given probability are moved to a random neighboring cluster, with the probability and radius of move increasing until the solution is improved.

Table 1. Related works

	Complexity	Authors
CCP for unweighted undirected graphs	NP-complete [4]	P. Doreian, A. Mrvar (1996) [9]; N. Bansal, A. Blum, S. Chawla (2002) [4]; N. Ailon, M. Charikar, A. Newman (2008) [2]; V. Il'ev, S. Il'eva, A. Kononov (2016) [16]; N. Arinik, R. Figueiredo, V. Labatut (2019) [3]
CCP for weighted undirected graphs	NP-hard [4]	N. Bansal, A. Blum, S. Chawla (2002) [4]; M. Levorato, R. Figueiredo, Yu. Frota, L. Drummond (2015) [18]
CCP for unweighted directed graphs	NP-hard [22]	P. Doreian, A. Mrvar (1996) [9]; V.A. Traag, J. Bruggeman (2009) [21]; R. Figueiredo, G. Moura (2013) [11]; V. Traag, P. Doreian, A. Mrvar (2018) [22]
CCP for weighted directed graphs	NP-hard [22]	V.A. Traag, J. Bruggeman (2009) [21]; V. Traag, P. Doreian, A. Mrvar (2018) [22]

KwikCluster [2] is 3-approximation algorithm (for complete graphs). For a randomly chosen vertex, all vertices connected to it by a positive edge are placed in the same cluster with it.

Iterated local search heuristic is proposed by Levorato et al. in [18]. First the initial greedy randomized solution is constructed. The Variable Neighborhood Descent (VND) stage constructs a family of r-neighborhood partitions by moving r vertices from one cluster to another. If a partition that improves the solution is found, it is applied, otherwise r is increased. Then, at the perturbation stage, random moves of vertices between clusters are made and VND is repeated.

3   Problem Formulation

3.1   Signed Graph

In the paper the type $\Sigma = (G, \sigma)$ of signed graphs are considered, where $G = (V, E)$ is an undirected, unweighted graph without multiple edges and loops with set of vertices V, $|V| = n \geq 2$ and set of edges E, $|E| = m \geq 1$. In the graph G, each edge is uniquely represented by an unordered pair $e = (u, v)$, where $e \in E$, $u, v \in V$. In this case, it is said that the edge e is incident to the vertices u or v, and the vertices u and v are adjacent. The set of vertices adjacent with v is denoted by $\Gamma(v) = \{u \colon (v, u) \in E\}$. Under the degree of a vertex v the number of edges incident to it is traditionally understood. Obviously, the degree of the vertex v can be represented as $\delta(v) = |\Gamma(v)|$. Then $\Delta = \max\limits_{v \in V} \delta(v)$ is the graph degree. On the edges $(u, v) \in E$ of the graph G, the sign function $\sigma : E \rightarrow \{+, -\}$ is given, it generates a graph edges set partition $E = E^+ \cup E^-$, where E^+ is a set of positive edges, E^- is a set of negative edges.

A signed graph is called k-balanced if the set of its vertices can be divided into k pairwise disjoint nonempty clusters so that all positive edges are inside the clusters and negative edges are between the clusters [14].

3.2 Correlation Clustering Problem

Let's denote the sets system forming a partition of the vertices set V into k subsets by

$$\mathcal{C} = \left\{ C_i \subseteq V : \bigcup_{i=1}^{k} C_i = V, C_i \cap C_j = \emptyset, i \neq j; \ i = \overline{1,k} \right\}. \tag{1}$$

It is known that for an arbitrary signed graph, the k-balance property may have no place. In this case, it is interesting to search for such a partition of the graph set vertices for which it is possible to obtain a k-balanced graph by changing the sign of the edges minimum number. This problem is considered as a graph clustering problem with a special kind of the error functional. The elements of the $C_i \in \mathcal{C}$ partition will be called clusters.

Under a positive error $P(\mathcal{C})$ of the partition (1) a number of positive edges between subsets of $C_1, \ldots, C_k$ will be understood. Let's note that $P(\mathcal{C})$ is the intercluster error calculated by the formula

$$P(\mathcal{C}) = \sum_{i=1}^{k} \sum_{u \in C_i} \sum_{v \in V \setminus C_i} [(u,v) \in E^+], \tag{2}$$

where hereinafter $[\cdot]$ is Iverson's Convention.

Under a negative error $N(\mathcal{C})$ a number of negative edges inside subsets for the partition (1) will be understood. The negative error is the intracluster error calculated by the formula

$$N(\mathcal{C}) = \sum_{i=1}^{k} \sum_{\{u,v\} \subseteq C_i} [(u,v) \in E^-]. \tag{3}$$

In [9] was proposed to represent the total error as a convex combination of the positive and negative errors, this total error depends on the parameter $\alpha \in [0,1]$:

$$Q_\alpha(\mathcal{C}) = \alpha N(\mathcal{C}) + (1 - \alpha) P(\mathcal{C}). \tag{4}$$

An error of the form (4) will be called an α-error of the partition $\mathcal{C}$.

In this paper, the clustering problem of a signed graph is considered in the following formulation [9].

CORRELATION CLUSTERING (CC) PROBLEM

Condition: a signed graph $\Sigma = (G, \sigma)$ is given, where $G = (V, E)$ is an undirected graph, $n = |V| \geq 2, m = |E| \geq 1$

Problem: for a given $\alpha \in [0, 1]$, it is required to find the partition $\mathcal{C}$ of the vertices set V of the signed graph Σ with the minimum total error $Q_\alpha(\mathcal{C})$

In the work [4] it was shown that Correlation Clustering with an error functional in the form (4) at $\alpha = 0.5$ is NP-complete.

Let Φ_k be the set of partitions into k subsets, and $\Phi = \bigcup\limits_{k=1}^{n} \Phi_k$ be the set of all possible partitions of V [9]. Then the cardinality of the solution space Φ equals to the Bell number B_n. The CC problem solution is a set of clusters $\mathcal{C}^*$, which gives the minimum value of the error function (4):

$$\mathcal{C}^* = \arg\min_{\mathcal{C} \in \Phi}\left[\alpha N(\mathcal{C}) + (1 - \alpha)P(\mathcal{C})\right] \tag{5}$$

and $k = |\mathcal{C}^*|$ is a number of clusters. It should be noted that the solution (5) may not be the only one.

4   Main Results

The paper proposes the heuristic algorithm $SGClust_\alpha$ for solving the Correlation Clustering problem. As input, the algorithm takes a signed graph $\Sigma = (G, \sigma)$ and some initial partition $\mathcal{C}_0$. The algorithm result is a partition $\mathcal{C}$ with an error not exceeding the error of the original partition: $Q_\alpha(\mathcal{C}) \leq Q_\alpha(\mathcal{C}_0)$.

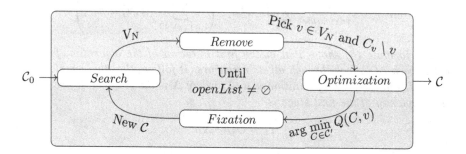

Fig. 2. $SGClust_\alpha$ algorithm scheme

The strategy of the proposed algorithm is presented at Fig. 2 and based on the intracluster error $N(\mathcal{C})$ minimizing without increasing the total error $Q_\alpha(\mathcal{C})$. Each the algorithm iteration consists of the following four stages.

Search:	to find the vertex that makes the greatest contribution to the intracluster error
Removing:	to remove this vertex from its cluster in order to reduce the intracluster error $N(\mathcal{C})$
Optimization:	to find such cluster for this vertex (it can be the initial cluster), joining to which will minimize the total error (4), if such cluster does not exist, then this vertex can be allocated to a new cluster
Fixation:	to fix the given vertex from further movements between clusters

Four stages of the algorithm form a conditional cycle. This cycle guarantees that the algorithm will move each vertex not more than once and then be finished its work by finding some partition $\mathcal{C}$. During the search and optimization stages, the algorithm follows a greedy strategy. And the following theorem is true.

Theorem 1. *The algorithm $SGClust_\alpha$ is correct, i. e. if as input a signed graph $\Sigma = (G, \sigma)$, some initial partition $\mathcal{C}_0$ and a parameter α are given, then this algorithm will find a new partition $\mathcal{C}$ with α-error $Q_\alpha(\mathcal{C}) \leq Q_\alpha(\mathcal{C}_0)$ in time $\mathcal{O}(n^2 \cdot \Delta)$.*

Proof. *The algorithm takes as input a signed graph Σ, error function parameter α and the starting partitioning $\mathcal{C}_0$. From Fig. 2 four main stages of the algorithm can be distinguished $SGClust_\alpha$. Let's look at each of them in sequence.*

Search Stage. At this stage, the vertex with the largest contribution to the intracluster error is searched. For this purpose, the set of vertices V_N is formed according to the following rule:

$$V_N = \arg \max_{v \in openList} \sum_{C_i \in \mathcal{C}} [v \in C_i] \cdot \left(\sum_{u \in \Gamma(v) \cap C_i} [(v, u) \in E^-] \right). \qquad (6)$$

The formula (6) shows that at the first iteration of the algorithm, for each vertex we need to look through all its neighbors. It follows that the complexity of the most time-consuming iteration is $n \cdot \Delta$, where Δ is the graph G degree. Thus, the complexity of the first stage is

$$T_1 = \mathcal{O}(n \cdot \Delta). \qquad (7)$$

Deletion Stage. From the set V_N obtained in the previous step, a vertex v is randomly selected. In this paper, we will assume that the first vertex from the set V_N is always selected, but this does not affect the asymptotic complexity of the algorithm. Note that removing a vertex from a cluster always minimizes the intracluster error while ignoring the intercluster error. The vertex v is removed from the current cluster. This step takes time

$$T_2 = \mathcal{O}(1). \qquad (8)$$

Optimization Stage. This stage evaluates the vertex v joining to all clusters $C_i \in \mathcal{C}$ in order to reduce the error $Q_\alpha(\mathcal{C})$. Note that when moving a vertex v from cluster to cluster, only edges incident to this vertex affect to the error $Q_\alpha(\mathcal{C})$. Consequently, the total error can be reduced only by the vertex error calculated by the following formula:

$$\delta(C_i, v) = \alpha \sum_{u \in \Gamma(v) \cap C_i} [(v, u) \in E^-] + (1 - \alpha) \sum_{u \in \Gamma(v) \setminus C_i} [(v, u) \in E^+]. \quad (9)$$

Note that the formula (9) allows us to compute the vertex contribution to the error $Q_\alpha(\mathcal{C})$ in a time not exceeding $\mathcal{O}(\Delta)$, which in sparse graphs can be much shorter than the direct computation of the error $Q_\alpha(\mathcal{C})$, which can be done in $\mathcal{O}(m)$. To reduce the error value in the optimization step, we need to make sure that the resulting value (9) is smaller than the current vertex contribution to the total error $Q_\alpha(\mathcal{C})$. The current contribution of a vertex to the error is all positive edges incident to it, i.e., $(1 - \alpha) \sum_{u \in \Gamma(v)} [(v, u) \in E^+]$. Thus, if for some cluster C_i the following is true

$$\delta(C_i, v) - (1 - \alpha) \sum_{u \in \Gamma(v)} [(v, u) \in E^+] < 0, \quad (10)$$

then adding a vertex v to cluster C_i reduces the error $Q(\mathcal{C})$. If (10) does not satisfied for each cluster, then vertex v forms a new cluster. In summary, vertex v joins the cluster with the smallest difference $\delta(C_i, v) - (1 - \alpha) \cdot \sum_{u \in \Gamma(v)} [(v, u) \in E^+]$ is realized. This step guarantees the fulfillment of $Q_\alpha(\mathcal{C}) \leq Q_\alpha(\mathcal{C}_0)$. For the proof, we need to consider two cases:

Case 1: *at the stage of removing node v from the cluster, the error $Q_\alpha(\mathcal{C})$ increased, i.e. the value of the node's contribution to the intracluster error was less than the received contribution to the intercluster error;*

Case 2: *at the stage of removing node v from the cluster, the error $Q_\alpha(\mathcal{C})$ decreased, i.e. the value of the node's contribution to the intracluster error was greater than the received contribution to the intercluster error;*

In essence, Case 1 is equivalent to the following inequality

$$\alpha \sum_{u \in \Gamma(v) \cap C_i} [(v, u) \in E^-] < (1 - \alpha) \sum_{u \in \Gamma(v) \cap C_i} [(v, u) \in E^+].$$

Let's consider the inequality (10) for a cluster C_i from which vertex v has been removed:

$$\alpha \sum_{u \in \Gamma(v) \cap C_i} [(v, u) \in E^-] + (1 - \alpha) \sum_{u \in \Gamma(v) \setminus C_i} [(v, u) \in E^+]$$

$$- (1 - \alpha) \sum_{u \in \Gamma(v)} [(v, u) \in E^+] < 0.$$

Let's transform the last sum:

$$\alpha \sum_{u\in\Gamma(v)\cap C_i} [(v,u)\in E^-] + (1-\alpha) \sum_{u\in\Gamma(v)\backslash C_i} [(v,u)\in E^+]$$

$$- (1-\alpha) \left(\sum_{u\in\Gamma(v)\cap C_i} [(v,u)\in E^+] + \sum_{u\in\Gamma(v)\backslash C_i} [(v,u)\in E^+] \right) < 0.$$

By opening the brackets and reducing the similar ones we obtain the following:

$$\alpha \sum_{u\in\Gamma(v)\cap C_i} [(v,u)\in E^-] - (1-\alpha) \sum_{u\in\Gamma(v)\cap C_i} [(v,u)\in E^+] < 0. \qquad (11)$$

The inequality (11) is fully consistent with Case 1 and demonstrates that the inequality (10) will always be satisfied for the cluster from which vertex v has been removed. Thus, v will join this cluster, which does not reduce the error $Q(\mathcal{C})$, or a cluster will be found, for which the smallest value of the difference $\delta(C_i, v) - (1-\alpha) \cdot \sum_{u\in\Gamma(v)} [(v,u)\in E^+]$ is realized, which will reduce the error $Q(\mathcal{C})$.

In turn, Case 2 corresponds to the inequality

$$\alpha \sum_{u\in\Gamma(v)\cap C_i} [(v,u)\in E^-] > (1-\alpha) \sum_{u\in\Gamma(v)\cap C_i} [(v,u)\in E^+].$$

This obviously reduces the error $Q(\mathcal{C})$. Consequently, when (10) is performed, the error $Q(\mathcal{C})$ will decrease. This completes the proof that the algorithm always does not degrade the starting partition, i.e., $Q_\alpha(\mathcal{C}) \le Q_\alpha(\mathcal{C}_0)$.

The worst case for this step is realized if each vertex lies in a separate cluster, then the calculation of the formula (9) will be repeated $n-1$ times. Hence, the complexity of the optimization step is

$$T_3 = \mathcal{O}(n\cdot\Delta). \qquad (12)$$

Fixation Stage. The vertex v is fixed from further moves and removed from the set openList. This operation can be performed in constant time:

$$T_4 = \mathcal{O}(1). \qquad (13)$$

These steps are repeated while the set openList does not be empty, hence the algorithm will be executed exactly n times before completion. Based on the formulas (7), (8), (12), and (13), the algorithm $SGClust_\alpha$ can be completed in time

$$T = n\cdot(T1 + T2 + T3 + T4)$$
$$= n\cdot(\mathcal{O}(n\cdot\Delta) + \mathcal{O}(1) + \mathcal{O}(n\cdot\Delta) + \mathcal{O}(1))$$
$$= \mathcal{O}(n^2\cdot\Delta).$$

Thus the theorem is proved.

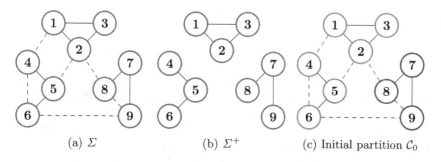

(a) Σ (b) Σ^+ (c) Initial partition $\mathcal{C}_0$

Fig. 3. Construction of the initial partition

Since the strategy proposed above moves each vertex between clusters only once, a successful initial clustering will play a large role in its efficiency. In this paper, the initial partition is chosen as the connected components of the graph $G = (V, E^+)$, i.e. each C_i cluster will contain only vertices from one connected component. Example of construction such initial clustering is shown at Fig. 3. It is easy to prove that vertices belonging to different connected components of a graph built only on positive edges, but lying in the same cluster, will contribute equal to or greater than zero to the total error E_α.

Example 1. *Let's illustrate the work of the SGClustα algorithm by the example of one iteration performing (see Fig. 4). For a given partition $\mathcal{C}_0$ consisting of a single cluster C_1, let $\alpha = 0.5$, then $Q_\alpha(\mathcal{C}_0) = 1$.*

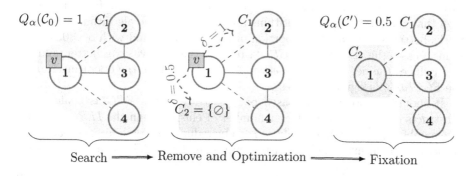

Search $\Longrightarrow$ Remove and Optimization $\Longrightarrow$ Fixation

Fig. 4. An example of one algorithm iteration

Therefore, it is expedient to take the initial partition such that the clusters contain only vertices from one connected component of the graph $G = (V, E^+)$.

5   Computational Experiments

Computational experiments were carried out to evaluate the $SGClust_\alpha$ algorithm effectiveness. All experiments were carried out on a computer with 16GB RAM, an AMD Ryzen 5 3600 6-Core 3.60 GHz processor running the Windows 10 operating system using single-threaded mode. We did multiple comparisons of our algorithm. First comparison was performed with the algorithms *relocation heuristic (RH)* and *Iterated local search (ILS)* on randomly generated graphs by the Waxman method. Next comparison was performed for complete graphs for following algorithms: RH, $KwikCluster$. The purpose of the last comparison was to investigate the $SGClust_\alpha$ and ILS algorithms on real data. Note that for generated graphs the edge signs were generated according to the Bernoulli distribution with different values of the parameter p: 0.25, 0.5 and 0.75. The parameter p reflects the approximate proportion of positive edges in the signed graph during generation, so with $p = 0.25$ in the graph, approximately a fourth of the total number of edges will be positive.

First comparison were carried out for randomly generated connected graphs using the Waxman method with parameters 0.15 and 0.4 [23]. The Waxman method involves generating a vertices set on the unit area plane with further generating edges between vertices. The first parameter determine how strongly the vertices proximity affects the appearance of an edge, the lower the parameter value, the higher the predominance of edges between nearby vertices. The second parameter determines the total edges total proportion in the graph, the higher value than more edges in the graph. Experiments were carried on 1000 of graphs with a vertices fixed number $n \in \{100, 150\}$. Algorithms $SGClust_\alpha$, ILS and RH for α various parameters were compared. The clusters number for the RH algorithm was chosen in the same way as in the first experiments series. The experiments averaged results are given in the Fig. 5. From the Fig. 5 it can be seen that the errors of the algorithms are close, while the algorithm $SGClust_\alpha$ works better with a larger proportion of positive edges and with $\alpha = 0.25$, also in most cases the algorithm $SGClust_\alpha$ reduces the positive error better, and the RH reduces the negative error better. At the same time, the operating time of the $SGClust_\alpha$ is significantly less.

It should be noted that, based on the strategy, the $SGClust_\alpha$ algorithm constructs a solution with a larger number of clusters than the ILS algorithm. The number of clusters in the solution can be reduced by combining non-adjacent vertices representing different clusters into one cluster. This can be performed in polynomial time.

The second series of experiments was performed on randomly generated full sign graphs.

Comparison of $KwikClust$, $SGClust_\alpha$ and RH algorithms on error distribution for different number of vertices $n \in \{50, 100, 200\}$ and fraction of positive edges $p \in \{0.25, 0.5, 0.75\}$ on complete graphs is at Fig. 6. Recall that the $KwikCluster$ algorithm is 3-approximation. Our algorithm showed on average the best solution, among the compared algorithms, along with high time efficiency.

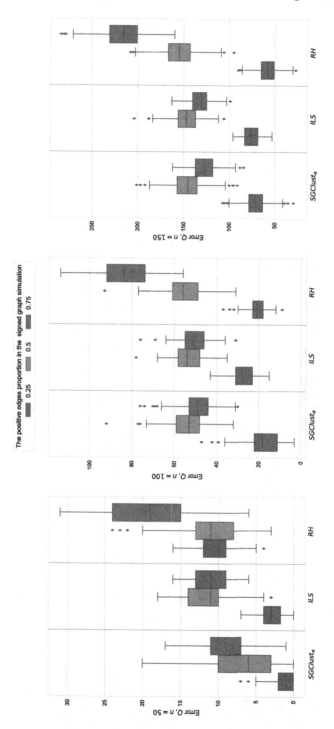

Fig. 5. Comparison of $SGClust_\alpha$, ILS and RH algorithms on error distribution for different number of vertices n and fraction of positive edges p on graphs generated by the Waxman method

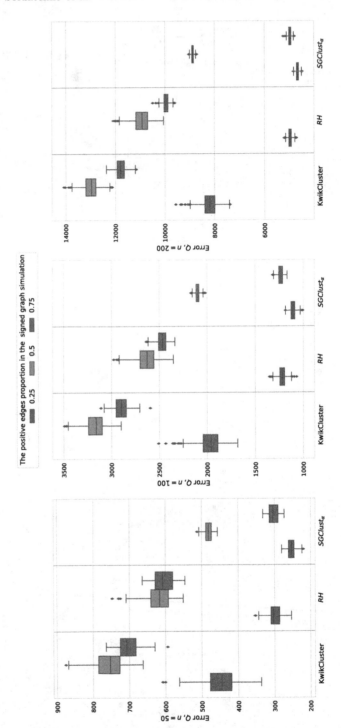

Fig. 6. Comparison of *KwikClust*, *RHSGClust*$_\alpha$ and *SGClust*$_\alpha$*RH* algorithms on error distribution for different number of vertices n and fraction of positive edges p on complete graphs

Table 2. The results of the $SGClust_\alpha$ algorithm on slashdot data

	$SGClust_\alpha$			ILS
n	Intracluster error	Intercluster error	Error	Average error
200	25	22	47	45
300	35	22	57	54
400	32	28	60	57.2
600	62	48	110	109.2
800	101	142	243	240
1000	222	385	607	600
2000	788	1477	2265	2187.2
4000	2603	3723	6326	6213
8000	6801	9352	16153	16073.2
10000	8715	12005	20720	20594.8

The results of the $SGClust_\alpha$ algorithm on slashdot data and comparison with the results of the ILS algorithm are given in the Table 2. Results for ILS algorithm are taken from the paper [18], because due to the random nature of the ILS algorithm, we were unable to obtain the results presented in the authors' paper. Note that the deviation of the $Q(\mathcal{C})$ values presented in Table 2 from the average error value obtained by the ILS algorithm will not exceed 3.7%. Herewith the running time of the proposed algorithm is significantly less.

6   Conclusion

Thus, the developed $SGClust\alpha$ algorithm for the CC problem solving is correct and finds a solution in an acceptable time. Experiments have shown that the optimization quality of positive and negative errors depends on the structure of the original graph. The algorithm $SGClust_\alpha$ wins in the ratio of clustering quality and running time in comparison to the RH, ILS and $KwikCluster$ algorithms. Since practical applications related to the NP-hard CC problem are associated with high-performance systems that require not only acceptable solution quality but also high speed of its computation, the algorithm proposed in this paper can occupy its niche.

References

1. Abdelnasser, A., Hossain, E., Kim, D.I.: Clustering and resource allocation for dense femtocells in a two-tier cellular OFDMA network. IEEE Trans. Wireless Commun. **13**(3), 1628–1641 (2014)
2. Ailon, N., Charikar, M., Newman, A.: Aggregating inconsistent information: ranking and clustering. J. ACM **55**, 684–693 (2008)

3. Arinik, N., Figueiredo, R., Labatut, V.: Multiple partitioning of multiplex signed networks: application to European parliament votes. Soc. Netw. **60**, 83–102 (2020)
4. Bansal, N., Blum, A., Chawla, S.: Correlation Clustering. Mach. Learn. **56**, 89–113 (2002)
5. Bressan, M., Cesa-Bianchi, N., Paudice, A., Vitale, F.: Correlation clustering with adaptive similarity queries. Mach. Learn., 1–25 (2020)
6. Brusco, M.J.: K-balance partitioning: an exact method with applications to generalized structural balance and other psychological contexts. Psychol. Methods **15**, 145–157 (2010)
7. Brusco, M.J., Doreian, P.: Partitioning signed networks using relocation heuristics, tabu search, and variable neighborhood search. Soc. Netw. **56**, 70–80 (2019)
8. Davis, J.A.: Clustering and structural balance in graphs. Hum. Relations **20**, 181–187 (1967)
9. Doreian, P., Mrvar, A.: A partitioning approach to structural balance. Soc. Netw. **18**, 149–168 (1996)
10. Doreian, P., Mrvar, A.: Partitioning signed social networks. Soc. Netw. **31**(1), 1–11 (2009)
11. Figueiredo, R., Moura, G.: Mixed integer programming formulations for clustering problems related to structural balance. Soc. Netw. **35**, 639–651 (2013)
12. Heider, F.: Attitudes and cognitive organization. J. Psychol. **21**, 107–112 (1946)
13. Harary, F.: On the notion of balance of a signed graph. Mich. Math. J. **2**, 143–146 (1953)
14. Harary, F.: Structural balance: a generalization of heider's theory. Psychol. Rev. **63**(5), 227–293 (1956)
15. Ibragimova, E. I., Semenova, D. V. Soldatenko, A.A.: On the study of the properties of the heuristic clustering algorithm of signed graphs. In: 21th International Conference named after A. F. Terpugov Information technologies and mathematical modelling (ITMM-2022), Russia, Tomsk, pp. 259–264 (2022) (in russian)
16. Il'ev, V., Il'eva, S., Kononov, A.: Short survey on graph correlation clustering with minimization criteria. In: Kochetov, Y., Khachay, M., Beresnev, V., Nurminski, E., Pardalos, P. (eds.) DOOR 2016. LNCS, vol. 9869, pp. 25–36. Springer, Cham (2016). https://doi.org/10.1007/978-3-319-44914-2_3
17. Kunegis, J., Schmidt, S., Lommatzsch, A., Lerner, J., Luca, E. W., Albayrak, S.: Spectral Analysis of signed graphs for clustering, prediction and visualization. In: Proceedings of the 2010 SIAM International Conference on Data Mining (SDM), Columbus, Ohio, USA, pp. 559–570 (2010)
18. Levorato, M., Figueiredo, R., Frota, Yu., Drummond, L.: Evaluating balancing on social networks through the efficient solution of correlation clustering problems. EURO J. Comput. Optimiz. **5**(4), 467–498 (2017)
19. Maatouk, A., Hajri, S.E., Assaad, M., Sari, H.: On optimal scheduling for joint spatial division and multiplexing approach in fdd massive mimo. IEEE Trans. Signal Process. **67**(4), 1006–1021 (2018)
20. Queiroga, E., Subramanian, A., Figueiredo, R., Frota, Y.T.: Integer programming formulations and efficient local search for relaxed correlation clustering. J. Global Optim. **81**, 919–966 (2021)

21. Traag, V.A., Bruggeman, J.: Community detection in networks with positive and negative links. Phys. Rev. E **80**, 1–7 (2009)
22. Traag, V., Doreian, P., Mrvar, A.: Partitioning signed networks. In: book: Advances in Network Clustering and Blockmodeling, pp. 225–249. Wiley, New York, United States (2019)
23. Waxman, B.M.: Routing of multipoint connections. IEEE J. Sel. Areas Commun. **6**(9), 1617–1622 (1988)

Decoding of Product Codes in Discrete and Semi-continuous Channels with Memory

Anna Fominykh[1]([✉])(iD) and Andrei Ovchinnikov[2](iD)

[1] Skolkovo Institute of Science and Technology, Moscow, Russia
anna.fominykh@skoltech.ru
[2] HSE University, Saint-Petersburg, Russia
a.ovchinnikov@hse.ru

Abstract. Product code construction is a powerful error-correcting tool for both channels with and without memory. The common approach to decoding product codes is to apply consequent decoders in a sequential manner. The paper examines the influence of memory in the channel on iterative decoding for hard decision, soft decision, and trellis-based decoding algorithms. Also, the attainable performance of iterative schemes with and without knowledge of channel state information is presented.

Keywords: Iterative codes · channels with memory · MAP decoding

1 Introduction

Research in coding theory began with the work of C. Shannon in 1948 [1]. Coding theory develops methods for information processing in order to protect it against errors appearing during data transmission, storage, and processing. Transmission, storage, and processing procedures are often described by mathematical channel models for the convenience of the analysis. The mathematical models of the channel are divided into channels with memory and channels without memory (memoryless channels). In memoryless channels, the errors appearing are independent, while when transmitted over a channel with memory, the errors are not independent and appear in groups, composing error bursts. With the development of information and coding theory, most of the research efforts were aimed at studying discrete channels without memory, while the direction of channels with memory turned out to be a less investigated area, despite the fact that the presence of the memory effect in the channel leads to an "increase" in channel capacity [2].

Several coding techniques could be applied to channels with memory. One of the common techniques is a product (iterative) code design that allows to build longer codes from shorter codes, most often Bose-Chaudhuri-Hocquenghem

The article was prepared within the framework of the Basic Research Program at HSE University.

(BCH) codes or Hamming codes. Some research has been done on the application of low-density parity-check and polar codes for iterative design in channels with memory [3]. Iterative codes, as opposed to other types of codes, have an appropriate structure for burst error correction without the need for additional interleaving. A widely used method for iterative code decoding is the Chase-Pyndiah algorithm. At the same time, the use of short component codes in an iterative design allows for optimal trellis decoding methods. A Berlekamp-Massey (BM) algorithm is another well-known approach to decoding up to half the minimum code distance [4].

The purpose of this paper is to compare the error probability provided by the BM algorithm, the Chase-Pyndiah algorithm, and the optimal trellis-based maximum a posteriori probability (MAP) algorithm in channels with memory, as well as to evaluate the influence of the presence of memory in the channel on the degradation of channel parameters.

2 Model Overview

2.1 Channels with Memory

Most real communication channels possess the effect of error grouping, which is also called "memory". This effect may be caused by the physical transmission environment's features such as reflections, diffraction, and scattering, or the properties of storage systems. When considering channels without memory, in other words, channels with independent errors, the transmission conditions are characterized by the fraction of errors. In channels with memory, the conditions are characterized by the properties of the error bursts (burst length), which are the sections of the transmitted sequences in which errors are possible and beyond which they are absent or extremely unlikely.

Due to the properties of the channels with memory, their conditions may change during the data transmission. Thus, it can be considered that the channel during transmission can be in different states and, from time to time, go from state to state. One of the common ways to represent channels with memory is by using Markov chains, which describe the channel that transits between states [5]. Most frequently, it is assumed that the number of states is finite. Important examples of channels described by Markov chains with two states are the Gilbert and Gilbert-Elliott models [6]. The first state is called a good state (denoted as G) and the second state is called a bad state (denoted as B). Each state is described by its own error probability. Gilbert's model supposes that the error probability in the good state is zero. In 1963, the Gilbert model was generalized by E. Elliott, which suggested the error probability of a good state be non-zero. Being one of the first memory channel models, the Gilbert-Elliott model is still relevant and frequently used to describe actual systems. These models state that the previous state of the channel dictates its current state. Gilbert and Gilbert-Elliot channels are described by a set of probabilities $(P_B, P_G, P_{BG}, P_{GB})$. The probabilities P_B and P_G represent the bit error probabilities in bad and good states, respectively. The pair of probabilities P_{BG} and P_{GB} is called transition probabilities and represent the probability of transitioning from a bad state to a good state and from a good state to a bad state, respectively. The graphical representation of the Gilbert-Elliott channel is presented in Fig. 1.

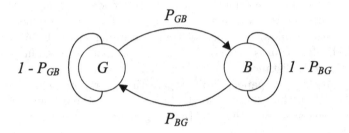

Fig. 1. Gilbert-Elliott channel model.

The unconditional probabilities of being in the bad and good states are calculated as

$$P_B = \frac{P_{GB}}{P_{GB} + P_{BG}}, \quad P_G = \frac{P_{BG}}{P_{GB} + P_{BG}}.$$

Another approach to describing channels with memory is to use a correlated Rayleigh fading model. Fading and scattering phenomena are traditionally described on the example of wireless communication channels, taking into account the features of propagation of a radio signal (multipath): if the transmitting antenna sends a signal pulse to the radio channel, the radio signal, when interacting with obstacles, is reflected in the form of several distorted copies, with the sum of signals representing multiple responses to the same transmitted pulse. The amplitude of the received signal can then be modeled by the Rayleigh distribution, whose probability density function is

$$p(r, \sigma) = \frac{r}{\sigma^2} \exp\left(-\frac{r^2}{2\sigma^2}\right), \quad r \geqslant 0,$$

where σ is the scale parameter of the distribution.

The Rayleigh random variable μ may be written as

$$\mu = \sqrt{x^2 + y^2},$$

where $x, y \sim \mathcal{N}(0, \frac{1}{2})$ are independent Gaussian random variables with zero mean and variance $\frac{1}{2}$.

The fading of the signal amplitude can be expressed using the following model of the Rayleigh communication channel:

$$y = \mu x + \eta,$$

where x is a transmitted signal, μ is a channel gain, which is a Rayleigh random variable with $\mathbb{E}[\mu^2] = 1$, and $\eta \sim \mathcal{N}(0, \sigma^2)$ is an additive white Gaussian noise with zero mean and variance σ^2.

The model of the Rayleigh communication channel is very simple and is a generalization of the model with additive white Gaussian noise by introducing, in addition to the additive component, a multiplicative component of noise called the channel gain. Today, for the modeling of communication systems, much more complex and computationally time-consuming models are used, but the Rayleigh model remains the basic mathematical model, with the help of which, when constructing code-modulation circuits, the channel fading effect can be taken into account.

For the generation of Rayleigh random variables, it was previously assumed that the Gaussian components (and therefore the resulting Rayleigh values) are independent; however, they are often dependent if real communication channels are under consideration. This effect can be taken into account by introducing the correlation coefficient ρ, then the values of μ_i obtained at the i-th moment of time

$$\mu_i = \sqrt{x_i^2 + y_i^2}$$

are generated as

$$x_i = \rho x_{i-1} + \sqrt{1 - \rho^2}\eta_x,$$
$$y_i = \rho y_{i-1} + \sqrt{1 - \rho^2}\eta_y,$$

where $\eta_x, \eta_y \sim \mathcal{N}(0, \frac{1}{2})$.

Several coding techniques could be applied to channels with memory. It was proposed by A. Eckford to apply the channel estimation procedure before the decoding, which adjusts the channel output according to the estimated channel conditions and provides the decoder with updated information [7]. Another way is to optimize a code structure to the existing memory. Some research has been done on the optimization of low-density parity-check and polar codes [8]. Also, one of the common techniques to adapt the code construction for error bursts is an iterative design that allows to build longer codes from shorter codes, most often Bose-Chaudhuri-Hocquenghem (BCH) codes or Hamming codes. Iterative codes, as opposed to other types of codes, have an appropriate structure for burst error correction without the need for additional interleaving.

2.2 Iterative Codes

Iterative or product codes were introduced by P. Elias in [9]. The iterative coding technique is an effective approach to composing long and powerful codes by using short component codes. The simplest case of iterative code with two component codes can be visualized as a matrix where each row corresponds to a codeword of one component code and each column corresponds to a codeword of another code. Generally, more than two component codes may be used to construct iterative code, but each extra code lowers the overall code rate. Further, we will consider the iterative scheme with two BCH component codes.

For the construction of the product or iterative code, two component codes, C_1 with parameters (n_1, k_1), and C_2 with parameters (n_2, k_2), are used. Suppose both codes are in systematic form. At first, the k_2 blocks of k_1 information bits are placed in the k_2 rows of the $k_2 \times k_1$ array as depicted in Fig. 2. Each row is encoded using code C_1 yielding k_2 rows of length n_1. Each column of the obtained array serves as an information block for the second code, i.e., each of the n_1 blocks of length k_2 is encoded using code C_2. The resultant codeword has a length of $n = n_1 n_2$. Commonly, product codes use Hamming codes, BCH codes, and parity check codes as component codes.

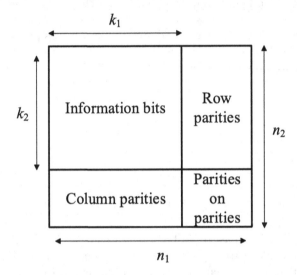

Fig. 2. Example of iterative code.

2.3   Decoding Algorithms

The common approach to decoding iterative codes is to perform subsequent decoding iterations of the component codes. During the decoding process, the received codeword is initially passed to a decoder of a first component code, which produces estimates of the transmitted bits. These estimates are used to update the information about the received bits. In an iterative decoding scheme, the updated information is fed to the decoder of the next component code for the next decoding iteration. This process is repeated multiple times until a stopping criterion is met, such as reaching a maximum number of iterations or satisfying the syndrome checks. During the decoding process, the iterative nature of iterative decoding allows the information from one code to be exchanged and combined with the information from the other code, effectively mitigating the impact of error bursts. Lower error probabilities may be achieved if subsequent decoders exchange information about the reliability of decisions made (the so-called soft decisions) in addition to hard decoding results.

In the paper, the influence of memory in the channel on iterative decoding for hard decision, soft decision, and trellis-based decoding algorithms is investigated. The investigation is being carried out for the BCH codes. As an example of soft decision decoding, the well-known method of iterative code decoding called the Chase-Pyndiah algorithm is considered [10,11]. For each received vector, the Chase algorithm searches the codewords within a sphere of radius $d - 1$ instead of reviewing all the codewords, where d is the minimum distance of the code. All possible error patterns of binary weight not greater than $d - 1$ are considered to compose a candidate codeword list. The resultant codeword list is used to generate the hard decision and the soft output of the coded bits. At the same time, for sufficiently short block component codes, optimal decoding algorithms are known that use a trellis, in particular the Bahl-Cocke-Jelinek-Raviv (BCJR) algorithm, which guarantees optimal symbol-by-symbol MAP decoding [12]. For BCH codes, there exist several hard decoding algorithms that may be categorized as algebraic techniques utilizing the Euclidean or Berlekamp-Massey (BM) algorithm [13].

3 Experiments and Results

This section presents the simulation results for several scenarios in BSC, Gilbert, Gilbert-Elliott, and correlated Rayleigh fading channels. The BSC channel is parameterized by the transition error probability, Gilbert and Gilbert-Elliott channels are described by a set of parameters $(P_B, P_G, P_{BG}, P_{GB})$, and the correlated Rayleigh fading channel depends on the correlation coefficient ρ and signal-to-noise ratio E_s/N_0 with E_s being an energy per symbol and N_0 is a power spectral density of noise.

The simulation scheme is presented in Fig. 3. The source generates $k_1 \times k_2$ information bits. The channel encoder encodes information bits using component BCH codes with parameters n_1 and n_2 equal to 31, and k_1 and k_2 equal to 21 for BSC, Gilbert, and Gilbert-Elliott channels, and n_1 and n_2 equal to 15, and k_1 and k_2 equal to 11 for correlated Rayleigh channel. For BSC, Gilbert, and Gilbert-Elliott channels, the modulator block is omitted. For correlated Rayleigh fading, binary phase-shift keying is used. The data is passed through the channel. Conventionally, for Gilbert and Gilbert-Elliott channels, it is assumed that channel parameters estimation should be carried out. In the experiments, we consider two cases: no channel state information (no CSI) and perfect knowledge about channel state information (CSI). On the receiver, demodulation is performed, and the demodulated data is fed to the decoder. After decoding, the error probability is estimated.

To summarize, the considered scenarios are

- Error probability comparison for 1 and 3 decoding iterations for Chase-Pyndiah, BCJR, and BM algorithms in binary-symmetric channel (BSC).
- Error probability comparison for 3 decoding iterations for Chase-Pyndiah, BCJR, and BM algorithms in Gilbert and Gilbert-Elliott channels with perfect knowledge about channel states.

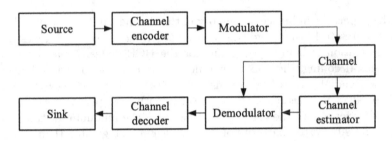

Fig. 3. Simulation scheme.

- Error probability comparison for 3 decoding iterations for Chase-Pyndiah, BCJR, and BM algorithms in Gilbert and Gilbert-Elliott channels without knowledge about channel states.
- Error probability comparison for 1 and 3 decoding iterations for Chase-Pyndiah, BCJR, and BM algorithms in the correlated Rayleigh fading channel with different correlation coefficients ρ.

In the simulation figures, the horizontal axis is plotted in terms of channel bit error probability (CBEP), which is calculated as CBEP $= (P_{GB}P_B + P_{BG}P_G)/(P_{GB} + P_{BG})$. The simulation parameters are: $P_B = 0.5$, $P_{GB} = 0.01$, P_{BG} varies, and $P_G = 0$ in the Gilbert channel and $P_G = 0.01$ in the Gilbert-Elliott channel.

The error probability comparison for 1 and 3 decoding iterations for decoding algorithms in the BSC is given in Fig. 4. The simulation results show that the performance of the BM algorithm with iterations improves not as much as the performance of BCJR and Chase-Pyndiah.

The error probability comparison for 3 decoding iterations for decoding algorithms in Gilbert and Gilbert-Elliott channels without knowledge about channel states is presented in Figs. 5 and 6, respectively. The performance of the decoders is slightly worse in the Gilbert channel compared to the Gilbert-Elliott channel.

The error probability comparison for 3 decoding iterations for decoding algorithms in Gilbert and Gilbert-Elliott channels with perfect knowledge about channel states is presented in Figs. 7 and 8, respectively. Knowledge about channel state information is more beneficial for the BCJR algorithm in the Gilbert channel compared to the Gilbert-Elliott channel. The opposite behavior is observed for the Chase-Pyndiah algorithm.

The knowledge of the channel state information is modeled by a corresponding log-likelihood ratio (LLR) magnitude calculation. For the no channel knowledge case, all LLR values have the same magnitude computed as $l_i = (2y_i - 1)\log\frac{1-\text{CBEP}}{\text{CBEP}}$ for some ith bit, so all the bits have the same reliability depending only on the channel parameters. For the case with the knowledge about the channel states, the bits that fall in the good state are initialized with high reliabilities as $l_i = (2y_i - 1)\log\frac{1-P_G}{P_G}$, with clipping if necessary, so the maximum value of l_i is fixed to some huge constant to avoid decoder instability.

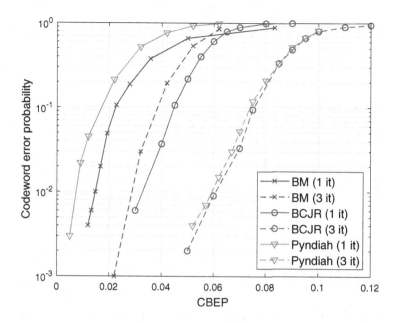

Fig. 4. Error probability comparison for 1 and 3 decoding iterations in BSC.

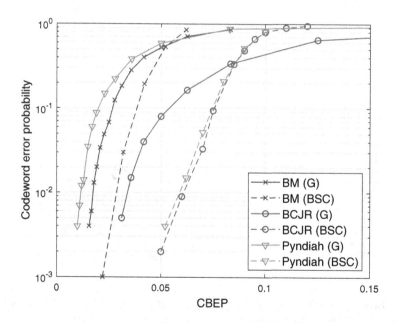

Fig. 5. Error probability comparison for 3 decoding iterations in Gilbert and BSC channels.

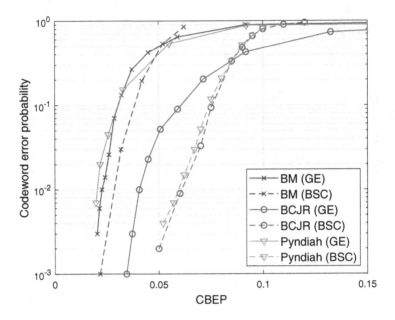

Fig. 6. Error probability comparison for 3 decoding iterations in Gilbert-Elliott and BSC channels.

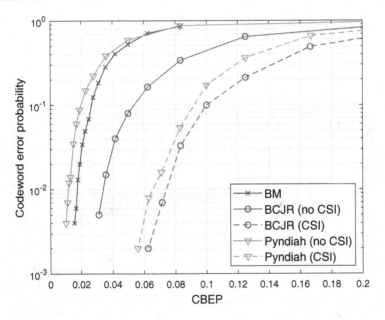

Fig. 7. Error probability comparison for unknown CSI and known CSI cases for 3 decoding iterations in Gilbert channel.

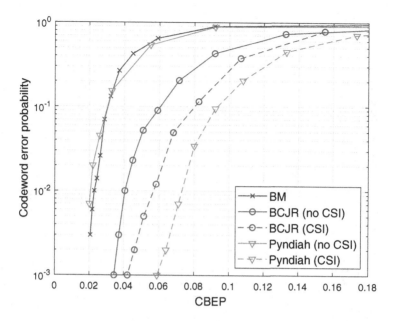

Fig. 8. Error probability comparison for unknown CSI and known CSI cases for 3 decoding iterations in Gilbert-Elliott channel.

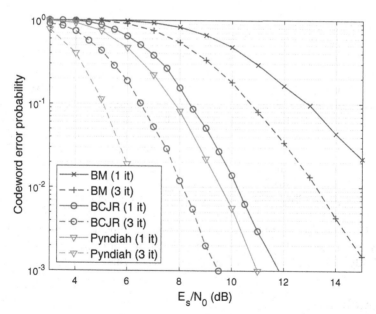

Fig. 9. Error probability comparison for 1 and 3 decoding iterations in correlated Rayleigh fading channel with $\rho = 0$.

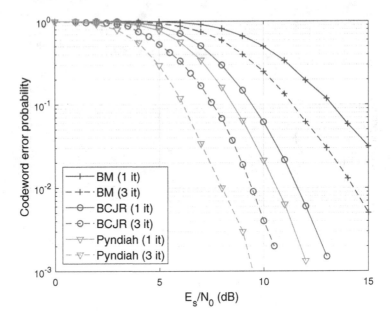

Fig. 10. Error probability comparison for 1 and 3 decoding iterations in correlated Rayleigh fading channel with $\rho = 0.9$.

For bits in a bad state, the LLRs are computed as $l_i = (2y_i - 1) \log \frac{1-P_B}{P_B}$ to facilitate the uncertainty about the channel in a bad state.

The error probability comparison for 1 and 3 decoding iterations for decoding algorithms in the correlated Rayleigh fading channel with $\rho = 0$ is given in Fig. 9. The simulation results show that the error correcting performance of the BM algorithm is still degraded in comparison to the performance of BCJR and Chase-Pyndiah. The performance of the BCJR algorithm in the correlated Rayleigh channel differs from the Gilbert and Gilbert-Elliott channels. The BCJR codeword error probability in the correlated Rayleigh channel is worse compared to the Chase-Pyndiah algorithm with the same number of iterations.

The error probability comparison for 1 and 3 decoding iterations for decoding algorithms in the correlated Rayleigh fading channel with $\rho = 0.9$ is given in Fig. 10. The mutual arrangement of the curves remains as in the case with $\rho = 0$. The BCJR codeword error probability is still worse compared to the Chase-Pyndiah algorithm with the same number of iterations.

4   Conclusion

In this work, the error probabilities of the BM, Chase-Pyndiah, and BCJR algorithms were compared in channels with and without memory. An investigation of the influence of channel parameters on error probability degradation is provided.

It was observed that the Chase-Pyndiah algorithm and the BCJR decoding algorithm show different behavior in the discrete and semi-continuous channels with memory. Also, the attainable performance of iterative schemes with and without knowledge of channel state information is analyzed.

References

1. Shannon, C.E.: A mathematical theory of communication. Bell Syst. Tech. J. **27**, 379–423 (1948)
2. Gallager R.G.: Information Theory and Reliable Communication, vol. 588. Wiley, New York (1968)
3. Ovchinnikov, A.A., Fominykh, A.A.: Evaluation of error probability of iterative schemes for channels with memory. In: 2023 Wave Electronics and its Application in Information and Telecommunication Systems (WECONF), pp. 1–5. IEEE, St. Petersburg (2023)
4. Lin, S., Li, J.: Fundamentals of Classical and Modern Error-Correcting Codes. Cambridge University Press, Cambridge (2022)
5. Bremaud, P.: Markov Chains. Springer, Cham (2020)
6. Elliott, E.O.: Estimates of error rates for codes on burst-noise channels. Bell Syst. Tech. J. **42**, 1977–1997 (1963)
7. Eckford, A.W., Kschischang, F.R., Pasupathy, S.: Analysis of low-density parity-check codes for the Gilbert-Elliott channel. IEEE Trans. Inf. Theory **51**(11), 3872–3889 (2005)
8. Fominykh, A.A., Ovchinnikov, A.A.: Comparative analysis of polar and LDPC codes in space and satellite communication systems. In: 2023 Wave Electronics and its Application in Information and Telecommunication Systems (WECONF), pp. 1–4. IEEE, St. Petersburg, Russia (2023)
9. Elias, P.: Error-free coding. IRE Trans. Inf. Theory **4**, 29–37 (1954)
10. Chase, D.: Class of algorithms for decoding block codes with channel measurement information. IEEE Trans. Inf. Theory **18**, 170–182 (1972)
11. Pyndiah, R.M.: Near-optimum decoding of product codes: block turbo codes. IEEE Trans. Commun. **46**, 1003–1010 (1998)
12. Bahl, L.R., Cocke, J., Jelinek, F., Raviv, J.: Optimal decoding of linear codes for minimizing symbol error rate. IEEE Trans. Inf. Theory **20**(2), 284–287 (1974)
13. Berlekamp, E.R.: Algebraic Coding Theory, McGraw-Hill Series in Systems Science. McGraw-Hill, New York (1968)

Minimizing the Peak Age of Information in LoRaWAN System Based on the Importance of Information

Dmitriy Kim[1](✉) ⓘ, Andrey Turlikov[2] ⓘ, Natalya Markovskaya[2] ⓘ,
and Kairat Bostanbekov[3] ⓘ

[1] Narxoz University, Zhandossov Street 55, Almaty, Kazakhstan
dmitriy.kim@narxoz.kz
[2] HSE University, Kantemirovskaya Street 3A, St. Petersburgh, Russia
{a.tiurlikov,nmarkovskaya}@hse.ru
[3] KMG Engineering LLP, Z05H9E8 Astana, Kazakhstan

Abstract. The communication network based on LoRaWAN technology consists of N nodes and carries out one-way transmission of monitoring data to the base station (BS). The messages from each node are sent to the BS at random time intervals and independently of each other. If they have the same spreading factor (SF) and airtime of messages from two or more nodes overlap, a collision occurs and none of the messages reaches the BS. We assume that the importance of up-to-date information from each node is different, and the peak age of information (PAoI) is used to measure its freshness. We define the functional as the maximum PAoI among all nodes in the system, and our task is to find the parameters under which its minimization is achieved. We choose various message intensities to account for the different importance of information from each node, formulate and solve the optimal SF allocation problem, find a lower bound for the sought functional, and demonstrate by numerical example that our solution is close to the lower bound.

Keywords: LoRaWAN · Optimization problem · ALOHA · Peak Age of Information · SF allocation · Importance of up-to-date information

1 Introduction

Long Range Wide-Area Networks (LoRaWAN) technologies might be used to monitor the results of measurements at objects located over a large area and enable low-cost data transmission and collection systems. One of the key features of LoRa technology is the ability to exploit the trade-off between message transmission duration and delivery reliability through the use of a parameter such as SF. The value of this parameter determines both the number of bits that can be transmitted in a single signal when using LoRa modulation and the duration of the signal duration. Increasing the SF value by one unit the duration of the signal doubles. Messages that are transmitted with different SF values can be successfully received at the BS. There are many works in which the following scenario is considered. There is some area where the devices and BS are located. Given the distances of the devices from the base station and a path-loss model, it is required to assign for each device such a value of SF, which minimizes or maximizes some functional that characterizes the performance of the system. In most of the works, such a functional is the probability of message delivery (or package delivery rate) to the BS. A review of such works is given in [1,2]. Some works [3] consider the problem of jointly choosing the SF value and the code rate of a forward error correction code. The large number of parameters that describe the scenario under consideration makes it difficult to compare the results of the work among themselves. It follows from the results that the farther the device is located from the base station, the higher the SF value should be assigned to the device (see Fig. 11 from [2]).

There are works [4,5], in which the authors neglect the signal attenuation in the channel and in this case the problem is reduced to the choice of the number of devices, to which a certain value of SF is assigned. In this approach, the number of devices decreases as the SF value increases. There are few works (see, for example, [6]) in which the assignment of SF takes into account such a feature as Age of Information.

We consider the communication network based on LoRaWAN technology which consists of N nodes and carries out one-way transmission of monitoring data to the BS. There are three classes of devices in LoRaWAN [7]: Class A, B, and C . Devices of Class A initiate transmission itself and send packets as they have been generated. We consider a low-cost monitoring system based on LoRaWAN devices of Class A with transmission at random intervals. If during the sending of a message from one node an another node transmits its message, a collision occurs and both messages are lost (ALOHA type of protocol).

We assume that information from different nodes has various "importance". Our goal is to use different message sending intensities depending on the importance of the node and the allocation of the nodes themselves to different spreading factors (SF allocation) to optimize some functional that characterizes the communication system. For example, to monitor voltage fluctuations in the low-voltage (LV) electrical grid, the voltage in the nodes near the transformer changes "more slowly" than in the remote nodes. To assess the state of the LV grid by monitoring the voltage at the nodes, information from the nodes remote (according to the grid topology) from the transformer must be received more frequently than from other nodes. We can take this into account by varying the intensity of sending messages from different nodes: if the voltage in a node behaves more predictably, the intensity of sending messages may be low, and vice versa - nodes with large voltage fluctuations send their messages with a higher intensity.

At the same time, a very high total message sending intensity from all nodes can occupy the transmission medium and thus increase the number of message collisions. At the same time, it is needed that the information received will be "fresh". The same can happen when the total message sending intensity is low, i.e. when all nodes rarely send up-to-date information. To account for the freshness of information, we use the notion of the Age of Information [8].

The age of information is a concept that reflects the freshness of the information [9,10] and was defined as $\Delta(t) = t - u(t)$, i.e. the difference of the current time t instant and the time stamp of the received update $u(t)$. The more general form of updating delay, so called Cost of Update Delay (CoUD) (see [10,11]), was suggested as a stochastic process that increases as a function of time between updates:

$$C(t) = f\left(t - u(t)\right),$$

where $f(\cdot)$ is a non-negative, monotonically increasing function. We use linear function

$$f(t) = \sigma^2 t \tag{1}$$

for some constant $\sigma^2 > 0$, which characterizes the "importance" of information of different nodes. A more tractable age metric as peak age of information (PAoI) (see Fig. 1) was introduced in works [8,12].

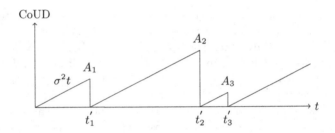

Fig. 1. CoUD trajectory with a linear function (1) , $t_1', t_2', t_3', \ldots$ – time instants of the information update, $A_1, A_2, A_3, \ldots$ – peak values of CoUD.

The peak age-of-information (PAoI) metric (see [13]) is defined as:

$$PAoI = \lim_{M \to \infty} \frac{\sum_{m=1}^M A_m}{M}.$$

It is easy to see that if $A_1, A_2, A_3, \ldots$ are independent and identically distributed then

$$PAoI = \mathbf{E}A_1.$$

In [13], the authors consider close problem of optimizing the PAoI by controlling the arrival rate of update messages and derive properties of the optimal solution for the M/G/1 and M/G/1/1 models.

To take into account the "importance" of information from different nodes, we use different growth rates σ_i^2, $i = 1, 2, \ldots, N$, [14] for CoUD of each node. It means that each node has its own $PAoI_i$, $i = 1, 2, \ldots, N$.

Based on the concept of Age of Information it is possible to construct a lot of different functionals characterizing the communication system. In our work, we consider the functional:

$$MPAoI = \max\{PAoI_1, PAoI_2, \ldots, PAoI_N\},$$

which means maximal average peak value among all nodes in the system.

The idea of our work is to choose such parameters of the communication system at which the desired functional is minimized. We first find the message sending intensities from different nodes (Sect. 2), and then partition the entire set of nodes into spreading factors (Sect. 3) to further reduce the maximum PAoI among all nodes. Section 5 is devoted to another functional for optimization. In Sect. 5, we obtain an estimate of the lower bound for our functional. In Sect. 6 we compare results for equal and different message sending intensities.

2 One SF: Problem Formulation for Minimizing of MPAoF

In this section we assume that all nodes use the same SF with airtime of one message $Q > 0$ and the same frequency [1].

We have N nodes and each of them sends messages with intensities $\lambda_i > 0$, $i = 1, 2, \ldots, N$. It means that intervals between messages of node i have exponential distribution with parameter $\lambda_i > 0$. We suppose that they are independent for each node.

Now let us formulate the optimization problem for one SF:
to find $(\lambda_1^*, \lambda_2^*, \ldots, \lambda_N^*)$ at which

$$MPAoI = \max\{PAoI_1, PAoI_2, \ldots, PAoI_N\} \to \min. \tag{2}$$

If during the time interval $(-Q, Q)$ from the instant one node sends a message, the other nodes do not send messages, then the message is successfully delivered to the BS. Otherwise, a collision occurs and the messages are lost. This assumption is fulfilled in practice when all nodes are approximately the same distance from the BS. According to this distance, the spreading factor is chosen so that in the absence of collisions the messages are delivered successfully.

Based on these assumptions and by performing a series of arguments [15] we consider a simplified model for the total stream of messages

$$\Lambda = \lambda_1 + \lambda_2 + \cdots + \lambda_N$$

and use the properties of the Poisson process. According to this model the stream of successfully delivered messages from ith node might be defined as Poisson with intensity $\lambda_i p$, where $p = e^{-2Q\Lambda}$ and the total intensity of all successfully delivered messages is equal to

$$\Lambda \times p.$$

Put $e_{\lambda_i p}$ as an exponentially distributed random variable with intensity $\lambda_i p$. In this case, we can conclude that

$$PAoI_i = \mathbf{E}\left(\sigma_i^2 e_{\lambda_i p}\right) = \frac{\sigma_i^2}{\lambda_i p}.$$

The optimization problem (2) can be written in the form: to find $(\lambda_1^*, \lambda_2^*, \ldots, \lambda_N^*)$ at which

$$\max\left\{\frac{\sigma_1^2}{\lambda_1 p}, \frac{\sigma_2^2}{\lambda_2 p}, \ldots, \frac{\sigma_N^2}{\lambda_N p}\right\} \to \min. \tag{3}$$

The solution to the optimization problem might be achieved if for some minimal $K > 0$

$$\frac{\sigma_1^2}{\lambda_1 p} = \cdots = \frac{\sigma_i^2}{\lambda_i p} = \cdots = \frac{\sigma_N^2}{\lambda_N p} = K.$$

Then $\sigma_i^2 = K\lambda_i p$, $i = 1, 2, \ldots, N$, and

$$\sum_{i=1}^{N} \sigma_i^2 = K\Lambda p$$

or

$$K = \frac{\sum_{i=1}^{N} \sigma_i^2}{\Lambda p}.$$

It means that we have to find the maximal value of the intensity of the total stream of successfully delivered messages. Consider the optimization problem:

$$\Lambda \times e^{-2Q\Lambda} \to \max.$$

Lemma 1. *The maximal value of the intensity of the total stream of successfully delivered messages*

$$\Lambda^* = \frac{1}{2Q}. \tag{4}$$

Then we obtain that minimal $K > 0$ is equal to

$$K^* = 2eQ \sum_{i=1}^{N} \sigma_i^2 \tag{5}$$

and optimal intensity for ith node

$$\lambda_i^* = \frac{1}{2Q} \frac{\sigma_i^2}{\sum_{i=1}^{N} \sigma_i^2}.$$

Remark 1. If all nodes have the same importance, i.e.

$$\sigma_i^2 = \sigma^2,$$

then optimal values

$$\lambda_i^* = \lambda = \frac{1}{2QN}.$$

3   One SF: Another Optimization Problem Formulation

In this section, we try to understand the physical meaning of the parameter σ_i^2, which is responsible for the importance of information. This require describing what is being measured by the devices. We will assume that the behavior in time of the measured object can be described by a random process. Then, we introduce a new functional, unrelated to $MPAoI$, for which the parameter σ_i^2 has some meaning. As a result, we formulate a new optimization problem whose solution coincides with the solution of the problem based on $MPAoI$. If we obtain the same result on the basis of different assumptions, it indicates the naturalness of the approach used.

Assume that device in each node measures readings of some parameter (e.g. voltages in LV grid) with stochastic dynamics in time, i.e. each node at random intervals measures values of random parameter (voltages, temperature, etc.). We assume that the parameter might be describes as random process $U_i(t)$, $i = 1, 2, \ldots, N$, where $t > 0$ – time.

Let $e_{\lambda_i p}(1), e_{\lambda_i p}(2), \ldots, e_{\lambda_i p}(n), \ldots$ be independent copies of random variable $e_{\lambda_i p}$ for every $i = 1, 2, \ldots, N$. Define

$$\tau_i(1) = e_{\lambda_i p}(1),$$

$$\tau_i(2) = e_{\lambda_i p}(1) + e_{\lambda_i p}(2),$$

$$\tau_i(3) = e_{\lambda_i p}(1) + e_{\lambda_i p}(2) + e_{\lambda_i p}(3), \ldots.$$

Let $U_i(t)$, $i = 1, 2, \ldots, N$, be a Lévy process, i.e., a homogeneous random process with independent increments and $\mathbf{E}U_i(1)^2 < \infty$. Then changes in the measured parameter during the time between successfully delivered messages (measurements)

$$U_i(\tau_1) - U_i(0), \ U_i(\tau_1 + \tau_2) - U_i(\tau_1), \ U_i(\tau_1 + \tau_2 + \tau_3) - U_i(\tau_1 + \tau_2), \ldots$$

are independent random variables distributed like $U_i(e_{\lambda_i p})$.

Consider the uncertainty of the information from node i as the variation of the measured parameter for the time between measurements

$$V_i = D\left(U_i\left(e_{\lambda_i p}\right)\right).$$

The more uncertainty in the time between measurements is associated with a particular node, the more important the information from it is. As an another to $MPAoI$ risk functional characterizing the system might be studied maximal uncertainty among all nodes.

Now we formulate the optimization problem: to find $(\lambda_1^*, \lambda_2^*, \ldots, \lambda_N^*)$ at which risk functional is minimal:

$$\max\{V_1, V_2, \ldots, V_N\} \to \min.$$

If we suppose that variation

$$DU_i(1) = \sigma_i^2$$

then

$$V_i = D\left(U_i\left(e_{\lambda_i p}\right)\right) = \frac{\sigma_i^2}{\lambda_i p}$$

and we obtain the same problem formulation as (3) with PAoI.

We obtained that σ_i^2, which is responsible for the importance of information, can be interpreted as the variance of the measured parameter at node i. If we consider the voltage fluctuations in the nodes of an LV electrical grid, the more distant (in the sense of grid topology) they are from the transformer, the larger their fluctuations (or variance) are, so they have higher information importance for the grid operator compared to the nodes near the transformer, where the voltage fluctuations are small.

4   SF Allocation for MPAoI Minimizing

The LoRa technology uses distributed spectrum modulation with 6 orthogonal spreading factors (SF $= 7, 8, \ldots, 12$). Messages with different SF can be transmitted simultaneously. A smaller SF provides a higher data rate. A larger SF increases the receiver's sensitivity and therefore the range of the system [1–3].

There are many works (see, for example, [3,16]), where the task of assigning spreading factors is considered without taking into account the specifics of the transmitted data.

We can reduce MPAoI by allocating nodes to different SF. We need to divide indices of all nodes $G = \{1, 2, \ldots, N\}$ into $1 \leq k \leq 6$ sets $G_{13-k}, G_{13-k+1}, \ldots G_{12}$, such that $G = G_{13-k} \cup \cdots \cup G_{12}$, $G_i \cap G_j = \emptyset$, $i \neq j$, where G_i – node indices with SF$= i$ and G_j – node indices with SF$= j$. We assume that message transmission in each of the SF occurs independently of the nodes with other SF. The optimal allocation of nodes allows us to reduce the MPAoI value.

Let $Q_7, Q_8, \ldots, Q_{12}$ be airtime for our SFs and we need to split the set G into k subsets so that

$$\max\{MPAoI_{13-k}, MPAoI_{13-k+1}, \ldots, MPAoI_{12}\} \to \min,$$

where $MPAoI_{13-k}$, $k = 1, 2, \ldots, 6$, – maximal average peak value among nodes with $SF = k$. The last expression might be rewritten as

$$\max\left\{2eQ_{13-k}\sum_{i \in G_{13-k}}\sigma_i^2, \ldots, 2eQ_{12}\sum_{i \in G_{12}}\sigma_i^2\right\} \to \min.$$

If we define

$$\nu = \max_{13-k \leq j \leq 12} Q_j \sum_{i \in G_j}\sigma_i^2,$$

then

$$\nu \geq Q_j \sum_{i \in G_j}\sigma_i^2, \quad j = 13 - k, 13 - k + 1, \ldots, 12,$$

and we can perform our task as well known a linear integer programming problem.

Let x be a vector from $N \times k$ elements

$$x = (x_1, x_2, \ldots, x_{kN}), \ x_i \in \{0, 1\},$$

where

$$x_i + x_{N+i} + \cdots + x_{(k-2)N+i} + x_{(k-1)N+i} = 1, \ i = 1, 2, \ldots, N.$$

Put

$$s_{13-k} = (\sigma_1^2, \sigma_2^2, \ldots, \sigma_N^2, 0, \ldots, 0),$$

$$s_{13-k+1} = (0, \ldots, 0, \sigma_1^2, \sigma_2^2, \ldots, \sigma_N^2, 0, \ldots, 0),$$

$$\ldots,$$

$$s_{12} = (0, \ldots, 0, \sigma_1^2, \sigma_2^2, \ldots, \sigma_N^2).$$

For brevity, we can rewrite $\sum_{i \in G_j} \sigma_i^2 = s_j x^T$, $j = 13 - k, 13 - k + 1, \ldots, 12$, where x^T is a transposed vector x. It is to formulate the optimal SF allocation problem:

$$\nu \to \min, \tag{6}$$

$$\nu - Q_j s_j x^T \geq 0, \ j = 13 - k, 13 - k + 1, \ldots, 12.$$

$$x_i \in \{0, 1\}, \ i = 1, 2, \ldots, kN,$$

$$x_i + x_{N+i} + \cdots + x_{(k-2)N+i} + x_{(k-1)N+i} = 1, \ i = 1, 2, \ldots, N.$$

Example 1. Let $N = 30$. Assume that the distance from the base station is such that spreading factor 10 messages are successfully delivered in the absence of collisions. Then spreading factors 11 and 12 can also be used for transmission. It means that $k = 3$ and we have to allocate 30 nodes among SF= 10, SF= 11 and SF= 12. Put $Q_{10} = 0.3707$ sec., $Q_{11} = 0.8233$ sec. and $Q_{12} = 1.4828$ sec. Let

$$\sigma_i^2 = \frac{i}{10}, \quad i = 1, 2, \ldots, N.$$

Then

$$G_{10} = \{2 - 8, 14 - 20, 22 - 26\},$$

$$G_{11} = \{9, 27 - 30\},$$

$$G_{12} = \{1, 10 - 13, 21\},$$

$$2e \max \left\{ Q_{10} \sum_{i \in G_{10}} \sigma_i^2, Q_{11} \sum_{i \in G_{11}} \sigma_i^2, Q_{12} \sum_{i \in G_{12}} \sigma_i^2 \right\} =$$

$$2e \max \{10.157, 10.126, 10.083\} = 55.22.$$

Remark that integer programming is a quite time consuming procedure. Therefore, it is important to find a lower bound for our functional. In other words, if we obtain a solution satisfying the constraints of the optimization problem (6) and close enough to the lower bound, we can stop the computational process of searching for the optimal solution and, thus, significantly reduce the number of calculations.

5   Lower Bound for MPAoI

Define $S = \sum_{i=1}^{N} \sigma_i^2$. Let

$$S_j = \sum_{i \in G_j} \sigma_i^2, \ j = 13 - k, 13 - k + 1, \ldots, 12.$$

It is easy to find the lower bound if we consider the system of equations:

$$\begin{cases} S_{13-k} + S_{13-k+1} + \cdots + S_{12} = S, \\ Q_i S_i = Q_j S_j, \ i, j = 13 - k, 13 - k + 1, \ldots, 12. \end{cases}$$

It is easy to prove the following lemma:

Lemma 2.

$$\max \left\{ 2eQ_{13-k} \sum_{i \in G_{13-k}} \sigma_i^2, \ldots, 2eQ_{12} \sum_{i \in G_{12}} \sigma_i^2 \right\} \geq 2e \times \frac{\prod_{j=13-k}^{12} Q_j}{\sum_{i=13-k}^{12} \frac{\prod_{j=13-k}^{12} Q_j}{Q_i}} \times S.$$

Let us illustrate the lemma with an

Example 2. Under the conditions of the previous example for three SF:

$$2e \times \frac{Q_{10}Q_{11}Q_{12}}{Q_{10}Q_{11} + Q_{11}Q_{12} + Q_{10}Q_{12}} \times S = 2e \times 9.466 = 51.463.$$

6   MPAoI Comparison for Cases of Equal and Different Message Sending Intensities

If all N nodes have the same SF and the same message sending intensities $\lambda_i^* = \lambda^* = \frac{1}{2QN}$ (see Remark 1). We can find that

$$\lambda^* = \frac{1}{2QN}.$$

Then $\lambda_i^* p = \frac{1}{2QeN}$ and the functional $MPAoI$ is equal to

$$2QeN \max \left\{ \sigma_1^2, \sigma_2^2, \ldots, \sigma_N^2 \right\} = 2eQN \max_i \sigma_i^2.$$

By comparing with (5), we can see that the more "unequal" the importance of information from different nodes, the greater the difference will be between $MPAoI$ for the case of different and equal message sending intensities.

Next, consider the case with multiple SFs. Suppose we have a partition of all nodes with indices $G = \{1, 2, \ldots, N\}$ into $1 \leq k \leq 6$ non-intersecting sets $G_{13-k}, G_{13-k+1}, \ldots G_{12}$. It is interesting to compare the maximum value of MPAoI for the case of different message sending intensities and the case of identical intensities.

Let

$$N_{13-k} = \sum_{i=1}^{N} I\left(\sigma_i^2 \in G_{13-k}\right), \quad k = 1, 2, \ldots, 6,$$

where $I(\cdot)$ – indicator function. It is easy to see that for equal message sending intensities

$$\max\{MPAoI_{13-k}, MPAoI_{13-k+1}, \ldots, MPAoI_{12}\}$$

$$= 2e \max\left\{Q_{13-k} N_{13-k} \max_{i \in G_{13-k}} \sigma_i^2, \ldots, Q_{12} N_{12} \max_{i \in G_{12}} \sigma_i^2\right\} \quad (7)$$

and it is a new problem how to split all nodes by spreading factors in order to minimize the sought value. We will not do that task, but instead study a few numerical examples for comparison.

Example 3. Under the partition $G = G_{10} \cup G_{11} \cup G_{12}$ from the Example 1

$$2e \max\left\{Q_{10} N_{10} \max_{i \in G_{10}} \sigma_i^2, Q_{10} N_{11} \max_{i \in G_{11}} \sigma_i^2, Q_{12} N_{12} \max_{i \in G_{12}} \sigma_i^2\right\} = 101.5758.$$

Clearly, the solution (6) is not an optimal partition of G on SF for (7). This is therefore the reason why such a high value for the sought functional was obtained in Example 3. Let us consider another partition

$$G = \tilde{G}_{10} \cup \tilde{G}_{11} \cup \tilde{G}_{12},$$

where $\tilde{G}_{10} = \{19 - 30\}$, $\tilde{G}_{11} = \{10 - 18\}$, $\tilde{G}_{12} = \{1 - 9\}$.

Example 4. Under the conditions of Example 1 with the partitioning

$$G = \tilde{G}_{10} \cup \tilde{G}_{11} \cup \tilde{G}_{12}$$

$$2e \max\left\{Q_{10} N_{10} \max_{i \in G_{10}} \sigma_i^2, Q_{10} N_{11} \max_{i \in G_{11}} \sigma_i^2, Q_{12} N_{12} \max_{i \in G_{12}} \sigma_i^2\right\}$$

$$= 2e \max\{13.34520, 13.33746, 12.01068\} = 72.55203.$$

7 Conclusion

We study communication system, which works based on LoRaWAN technology. The system consists of N nodes with devices of Class A. Every node initiates transmission with some information and sends messages to the BS at random time intervals independently of each other. Every node belongs to one out of k SF, which has a different airtime. If nodes have the same spreading factor and airtime of messages from two or more nodes overlap, a collision occurs and none of the messages reaches the BS. In this work, we used a simplified model that does not take into account the distance from the BS and path-loss model, as in the works [3,16].

According to our model, every node has different importance for the system. Such metric as the peak age of information (PAoI) with various growth rates is used to measure the freshness and importance of information from the nodes. We choose the functional describing the entire system as the maximal PAoI among all nodes and investigate how to minimize it.

At first, we choose various message intensities to account for different freshness of information from each node and then formulate the problem of optimal SF node allocation. At the end, we give an example and also find the lower bound for our functional, which might be used to reduce calculations for optimization problem.

Acknowledgments. This research of Dmitriy Kim and Kairat Bostanbekov is funded by the Science Committee of the Ministry of Science and Higher Education of the Republic of Kazakhstan (Grant No. AP19680230) and was prepared by Andrey Turlikov and Natalya Markovskaya within the framework of the Basic Research Program at HSE University.

Author Contribution. Dmitriy Kim and Kairat Bostanbekov developed the mathematical model, Andrey Turlikov formulated the problem, Natalya Markovskaya obtained numerical results of the examples.

References

1. Silva, F.S.D., et al.: A survey on long-range wide-area network technology optimizations. IEEE Access **9**, 106079–106106 (2021)
2. Jouhari, M., Saeed, N., Alouini, M.S., Amhoud, E.M.: A survey on scalable LoRaWAN for massive IoT: Recent advances, potentials, and challenges. IEEE Communications Surveys & Tutorials, IEEE (2023)
3. Bankov, D., Khorov, E., Lyakhov, A.: LoRaWAN modeling and MCS allocation to satisfy heterogeneous QoS requirements. Sensors **19**(19), 4204 (2019)
4. Tiurlikova, A., Stepanov, N., Mikhaylov, K.: Method of assigning spreading factor to improve the scalability of the LoRaWAN wide area network. In: 2018 10th International Congress on Ultra Modern Telecommunications and Control Systems and Workshops (ICUMT), pp. 1–4. IEEE, (2018)

5. Gusev, O., Turlikov, A., Kuzmichev, S., Stepanov, N.: Data delivery efficient spreading factor allocation in dense lorawan deployments. In: 2019 XVI International Symposium Problems of Redundancy in Information and Control Systems (REDUNDANCY), pp. 199–204. IEEE, (2019)
6. Wang, Z., Xu, X., Zhaom J.: Spreading factor allocation and rate adaption for minimizing age of information in LoRaWAN. In: 2022 IEEE 24th Int Conf on High Performance Computing & Communications; 8th Int Conf on Data Science & Systems; 20th Int Conf on Smart City; 8th Int Conf on Dependability in Sensor, Cloud & Big Data Systems & Application (HPCC/DSS/SmartCity/DependSys), pp. 482–489. IEEE, (2022)
7. "What is LoRaWAN,". www.lora-alliance.org/about-lorawan/ (2022)
8. Yates, R.D., Sun, Y., Brown, D.R., Kaul, S.K., Modiano, E., Ulukus, S.: Age of information: an introduction and survey. IEEE J. Sel. Areas Commun. **39**(5), 1183–1210 (2021)
9. Chen, X., Gatsis, K., Hassani, H., Bidokhti, S.S.: Age of information in random access channels. IEEE Trans. Inf. Theory **68**(10), 6548–6568 (2022)
10. Yates, S.K.R., Gruteser, M.: Real-time status: how often should one update. In: IEEE International Conference Computer Communication (INFOCOM), pp. 2731–2735 (2012)
11. Sun, Y., Uysal-Biyikoglu, E., Yates, R., Koksal, C.E., Shroff, N.B.: Update or wait: how to keep your data fresh. In: 35th Annual IEEE International Conference Computer Communication (INFOCOM), pp. 1–9 (2016)
12. Costa, M., Codreanu, M., Ephremides A.: Age of information with packet management. In: Proceedings of the IEEE International Symposium Information Theory (ISIT), pp. 1583–1587 (2014)
13. Huang, L., Modiano, E.: Optimizing age-of-information in a multi-class queueing system. In: Proceedings of the IEEE International Symposium Information Theory (ISIT), pp. 1681–1685 (2015)
14. Kosta, A., Pappas, N., Ephremides, A. Angelakis, V.: Age and value of information: non-linear age case. In: 2017 IEEE International Symposium on Information Theory (ISIT), pp. 326–330. Aachen (2017)
15. Kim, D., Georgiev, G., Markovskaya, N.: A model of random multiple access in unlicensed spectrum systems. In: 2022 Wave Electronics and its Application in Information and Telecommunication Systems, WECONF (2022)
16. Garlisi, D., Tinnirello, I., Bianchi, G., Cuomo, F.: Capture aware sequential waterfilling for LoRaWAN adaptive data rate. IEEE Trans. Wireless Commun. **20**(3), 2019–2033 (2021)

Precoder for Proportional Fair Resource Allocation in Downlink NOMA-MIMO Systems

Ilya Levitsky⬚, Sergey Tutelian⁽✉⁾⬚, Aleksey Kureev⬚, and Evgeny Khorov⬚

Institute for Information Transmission Problems of the Russian Academy of Sciences, Moscow, Russia
{levitsky,tutelian,kureev,khorov}@wireless.iitp.ru

Abstract. Non-Orthogonal Multiple Access (NOMA) allows the base station (BS) to simultaneously transmit data to such a number of users that exceeds the number of transmit antennas. For that, the users are grouped according to their channel conditions and the user with a stronger channel uses a method called Successive Interference Cancellation (SIC) to reduce interference from the signals for other users in a group with a weaker channel. In downlink NOMA-MIMO (Multiple-Input Multiple-Output) systems the BS uses specific precoders to serve multiple users. Typically, such precoders form beams focused on users with higher channel gains of each group. Such an approach leads to a high interference at users with lower channel gain and, as a result, their throughput degrades and fairness in the network decreases. To solve this problem, the paper proposes a novel semi-orthogonal precoder that forms non-orthogonal beams for users within each group while maintaining orthogonality between the groups. The obtained simulation results show that the proposed precoder significantly improves geometric mean user throughput and outperforms the state-of-the-art approach in the area of moderate signal-to-interference-plus-noise ratio and can be used jointly with existing precoders for downlink NOMA-MIMO systems.

Keywords: Fairness · MIMO · NOMA · precoder construction · Zero Forcing

1 Introduction

Downlink Non-Orthogonal Multiple Access (NOMA) is a promising multiple access technology that is able to increase the spectral efficiency of future wireless systems. The core idea of NOMA is to allow a base station (BS) to transmit data to several users, using the same time and frequency resources. It is achieved by superposing the messages intended for different users in the power domain.

The research has been carried out at IITP RAS and supported by the Russian Science Foundation (Grant No. 21-19-00846, https://rscf.ru/en/project/21-19-00846/).

The users with higher channel gains employ the Successive Interference Cancellation (SIC) technique to effectively mitigate inter-user interference [1].

The principles of downlink NOMA can be applied in Multiple-Input Multiple-Output (MIMO) systems. In NOMA-MIMO systems [2–4], the users are split into several groups, each of which contains users with highly-correlated channels. The number of groups may equal the number of transmitter antennas, which corresponds to the number of transmit beams. The transmitter allocates a specific beam for each group, and within a group, users apply the NOMA principles described above. Many studies [4–6] consider a case with two users per group, namely the cell-edge user and the cell-center user, as a practical case. Generally, the BS builds a precoder that forms beams focused on the cell-center user. Existing research [7] proves that this approach is a close-to-optimal solution for maximizing aggregated throughput performance of the downlink NOMA-MIMO system. However, this approach leads to significant inter-group interference affecting the cell-edge user in the same group. On top of that, the cell-edge user suffers because of a misaligned precoder for the beam. Consequently, their throughput degrades, and resource allocation becomes unfair.

The goal of this study is to enhance the performance of cell-edge users in the downlink NOMA-MIMO system without compromising the experience of cell-center users. The paper proposes a novel approach to construct a semi-orthogonal precoder that forms non-orthogonal beams for users within each group while maintaining orthogonality between the groups. The proposed design ensures that the precoder orients the transmit beam in such a way that a beam intended for each user does not cause any interference to users in separate groups. We demonstrate that the developed precoder is particularly effective in the area of moderate signal-to-interference-plus-noise ratio (SINR), providing more than 5% gain compared to the state-of-the-art approach where beamforming is done on the cell-center user. Moreover, the proposed precoder demonstrates its effectiveness in joint usage with existing precoders, increasing the area of applicability of NOMA-MIMO systems.

In the remainder of this paper, we use the following notations:

- bold lowercase symbols are used to denote vectors;
- calligraphic symbols denote sets;
- A^T is the regular transposition of matrix A;
- A^H is the Hermitian transposition of matrix A;
- $|\mathcal{S}| = S$ is the cardinality of set $\mathcal{S}$;
- a_k is the k-th element in vector $\mathbf{a}$;
- $\mathbb{E}(A)$ is the expectation of matrix A;
- $\mathbb{I}[.]$ is the indicator function;
- $\mathrm{tr}(A)$ is trace of matrix A;
- I is the identity matrix;
- $\mathbb{C}$ is a set of complex numbers.

The rest of the paper is organized as follows. Section 3 describes the system model. Section 4 describes the semi-orthogonal precoder. In Sect. 5, we explain the simulation scenario and provide numerical results. Section 6 concludes the paper.

2   Related Works

The problem of Precoder construction for NOMA-MIMO systems has attracted massive attention in the literature. Consider an approach where the signals transmitted by each user are coded via an exclusive precoding vector. Using this idea, various studies [8–14] investigate rate maximization, total power minimization, and max-min fairness for NOMA systems. One study approaches the sum-rate maximization problem by using a minorization-maximization algorithm [8]. The paper [9] extends previous research and adds quality-of-service (QoS) constraints, also considering various decoding orders on the receiver. A further exploration to maximize the sum rate decreases the precoder construction complexity by taking the singular value decomposition of channel state information (CSI) [10]. Two more studies individually examine total power minimization in terms of QoS requirements and under target interference level constraints [11,12]. Lastly, the max-min fair (MMF) precoding design problem is addressed for a multi-antenna base station, and power allocation (PA) problems to achieve MMF in single antenna transmitter NOMA systems were also evaluated [11,13,14].

Consider an approach, where one precoding vector is shared among a group of users [2,15–20]. Using this concept, the paper [15] aims to optimize the weighted sum rate with total power constraints for two-user groups. For clustered downlink NOMA systems, the authors of the paper [2,16] propose a sub-optimal user clustering algorithm, and the optimal power allocation policy to maximize the weighted sum rate. Authors of the paper [17] design a method to maximize the sum rate of cell-center users while ensuring QoS constraints on the rates of weaker users through a joint power allocation and precoder design using successive convex approximation and semi-definite relaxation. This problem was further extended to multi-cell networks [18]. Lastly, papers [19,20] address the issue of minimizing total transmission power for downlink clustered NOMA.

However, all these papers do not aim to improve the performance of cell-edge users. In literature [21–23], various techniques are proposed to tackle this problem. These techniques depend on incorporating extra antennas or additional functionality into the users or completely restructuring the system model, both posing implementation challenges. For example, paper [21] considers a cloud radio access network with coordinated multipoint transmissions, and the paper [22] uses a cooperative relaying scheme, considering a cell-center user as a relay. Meanwhile, the authors of the paper [23] overcome the low channel gain of cell-edge users by equipping them with multiple antennas.

This research aims to improve the performance of cell-edge users in the downlink NOMA-MIMO system, without the performance loss for cell-center users. To do that, we propose a semi-orthogonal precoder, that consists of non-orthogonal beams for users within each group, while preserving orthogonality between different groups.

3   System Model

Consider a downlink multiuser NOMA-MIMO system, where a BS communicates with a set of users $\hat{\mathcal{K}}$ with cardinality $|\hat{\mathcal{K}}| = \hat{K}$. The BS has M antennas, whereas each user has a single antenna. The BS intends to send data to all users, but it may choose a subset of users to service at a time. Thus, BS selects a *configuration*, which is a tuple $(\mathcal{K}, f)$ that includes a subset $\mathcal{K} \subset \hat{\mathcal{K}}$ of $|\mathcal{K}| = K$ users, and a grouping function f that subdivides $\mathcal{K}$ into G mutually non-overlapping subsets $\mathcal{K}_g$. The grouping function f takes an index of a user k and returns the group number g and the user's index i inside this group[1]:

$$f : \mathcal{K} \rightarrow (\mathcal{G}, \mathcal{K}_g), \ \mathcal{G} = \{1, ..., G\}, \ \mathcal{K}_g = \{1, ..., K_g\}. \tag{1}$$

For simplicity, let $K_g \leq 2$. $K_g > 2$ rarely notably improves the performance [24]. If $K_g = 2$, we refer to the users as cell-edge and cell-center ones. The system model can be easily adapted to include an arbitrary maximum number of users per group.

The BS sends data to users in $\mathcal{K}$, encoded into the symbol s_k for the k-th user. Symbols s_k are independent and normalized.

Thus, the BS forms a vector $\mathbf{s} = [s_1, ... s_K]^T \in \mathbb{C}^{K \times 1}$ such that $\mathbb{E}[\mathbf{s}\mathbf{s}^H] = \mathrm{I}$.

To send data to the user k, the BS uses precoder $\mathbf{p}_k \in \mathbb{C}^{M \times 1}$. Hence, for all simultaneously serviced users, the full precoder matrix is $\mathrm{P} = [\mathbf{p}_1, ..., \mathbf{p}_K] \in \mathbb{C}^{M \times K}$. We remind here that each user k has a distinct precoder vector $\mathbf{p}_k$. The resulting transmitted signal $\mathbf{x} \in \mathbb{C}^{M \times 1}$ equals $\mathbf{x} = \mathrm{P}\mathbf{s}$. The average total power is constrained by E_{max}:

$$\mathbb{E}[\mathbf{x}\mathbf{x}^H] = \mathrm{tr}(\mathrm{P}\mathrm{P}^H) \leq E_{max} \tag{2}$$

The received signal for each k-th user is y_k, modeled as

$$y_k = \mathbf{h}_k^H \sum_{l=0}^{K} \mathbf{p}_k s_k + n_k. \tag{3}$$

The $\mathbf{h}_k \in \mathbb{C}^{M \times 1}$ represents a channel coefficient vector for user k. The noise component n_k is a circularly symmetric complex Gaussian random variable with zero mean and variance σ^2, and they are i.i.d. for all k. Without loss of generality, grouping f indexes users so the cell-edge user comes first: $||\mathbf{h}_{g,1}|| < ||\mathbf{h}_{g,2}||$.

To receive symbol s_k, the k-th user either performs SIC or treats the interference as noise. We assume an ideal SIC, where cell-center users can always decode data for cell-edge users. If a user i in a group g decodes symbol $s_{g,j}$, the signal-to-interference-plus-noise ratio (SINR) equals

$$\gamma_{g,(j \rightarrow i)} = \frac{|\mathbf{h}_{g,i}^H \mathbf{p}_{g,j}|^2}{|\mathbf{h}_{g,i}^H \mathbf{p}_{g,i}|^2 \cdot \mathbb{I}[j = 1] + \sum_{k \notin \mathcal{K}_g} |\mathbf{h}_{g,i}^H \mathbf{p}_k|^2 + \sigma^2}, \tag{4}$$

---

[1] Thanks to f, we can switch between global user indexing k and in-group indexing $f(k) = (g, i)$: $a_k \equiv a_{f(k)} \equiv a_{g,i}$, where a is any user property.

where the first term of the denominator represents the intra-group interference (from the signal for the cell-center user), the second one is the inter-group interference, and the third one is the Gaussian noise variance. The value of $\gamma_{g,(j \to i)}$ shall be high enough to allow the cell-center user to decode the message for the cell-edge user and perform SIC successfully.

The corresponding achievable rate for user i when decoding its data is

$$R_{g,i} = \log_2 \left(1 + \gamma_{g,(i \to i)}\right). \tag{5}$$

4   Precoder Design

Let K users be selected, and groups $\mathcal{G} = \{1, ..., G\}$ be formed. Given the channel state information $\mathbf{h}_k$ for each k-th user, the precoder matrix is $\tilde{P} = [\tilde{\mathbf{p}}_1, ..., \tilde{\mathbf{p}}_K]$, where $\tilde{\mathbf{p}}_k$ is the precoder vector for user k. To reallocate the power among the users, the precoder can be multiplied $\tilde{P}$ by a diagonal matrix D: $P = \tilde{P}D$. For simplicity, we allocate power to a group g proportional to K_g. Inside each group g, we further search for the best power allocation coefficients $\alpha_{g,i}$, as described in Sect. 5.

4.1   Zero Forcing Precoder

For non-NOMA systems, a widely used precoder is zero-forcing (ZF). Given the channel matrix $H = [\mathbf{h}_1, ..., \mathbf{h}_K]$, the precoder equals:

$$\tilde{P} = [\tilde{\mathbf{p}}_1, ..., \tilde{\mathbf{p}}_K] = ZF(H) \equiv H(H^H H)^{-1}, \tag{6}$$

Here we highlight the main property of the ZF precoder.

$$ZF([\mathbf{h}_1, ..., \mathbf{h}_i]) = [\mathbf{p}_1, ..., \mathbf{p}_i] : \quad \mathbf{h}_i^H \mathbf{p}_j = 0 \ \forall i \neq j. \tag{7}$$

The downside of the ZF precoder is that the effective channel gain for some users is low if they have correlated channel vectors. Moreover, K cannot exceed the number of antennas M at the BS.

4.2   State-of-the-Art Precoder for NOMA-MIMO Systems

The widely-used state-of-the-art (SOTA) NOMA-MIMO precoder forms the precoder matrix so that the precoder vectors for each group are mutually orthogonal. The precoder matrix is based on the channel state information of the cell-center user in each two-user group. Let $\hat{H} = [\mathbf{h}_{1,K_1}, ..., \mathbf{h}_{G,K_G}]$. Then, the precoder equals:

$$\hat{P} = [\hat{\mathbf{p}}_1, ..., \hat{\mathbf{p}}_g, ..., \hat{\mathbf{p}}_G] = ZF(\hat{H}) = \hat{H}(\hat{H}^H \hat{H})^{-1}, \tag{8}$$

where $\hat{\mathbf{p}}_g$ is the unweighted precoder vector for the group g. As all users in the group get the same unweighted precoder vector, respective columns in $\hat{P}_{M \times G}$ are duplicated to get $\tilde{P}_{M \times K}$.

We note that inside two-user groups, the SOTA precoder cancels inter-beam interference on cell-center users only. The signal quality on cell-edge users deteriorates due to not only misaligned precoding but also inter-beam interference.

4.3   Proposed Precoder

We propose to use a semi-orthogonal precoder that is created as follows. Consider a user k from a group $\mathcal{K}_g$ and all users from other groups. We build a precoding vector for this user k:

$$\tilde{\mathbf{p}}_k = \mathrm{ZF}([\mathbf{h}_k, \mathbf{h}_{l_1}, ..., \mathbf{h}_{l_{K-K_g}}])[1, 0, ..., 0]^T_{1 \times K - K_g + 1} \ \forall l_j \notin \mathcal{K}_g : k \in \mathcal{K}_g. \quad (9)$$

Performing this procedure for each user, we get $[\tilde{\mathbf{p}}_1, ..., \tilde{\mathbf{p}}_K] = \tilde{\mathrm{P}}$.

This precoder allows us to eliminate inter-group interference. Also, the cell-edge user receives a better signal than with the SOTA precoder. To construct this precoder, M shall not exceed $K - K_g + 1$.

We point out that ZF is a special case of both SOTA and proposed precoding when all groups have only one user. Hence, both of these NOMA-based precoders are expected to show superior performance compared to ZF.

5   Numerical Results

To compare the performance of the proposed precoder and SOTA against ZF, we use simulation. For the scenario from the system model in Sect. 3, we generate channel matrices using QuaDRiGa simulation environment with the "Winner Indoor A1 LOS Scenario" [25]. Channel noise variance $\sigma^2 = -94$ dBm. The system contains a BS with $M = 3$ or $M = 4$ antennas and $\hat{K} = 4$ single-antenna users.

We consider two scenarios for user locations. Figure 1 demonstrates Scenario 1 for simulation: two pairs of users are located on two straight lines that intersect at the BS with angle $\theta = 45°$ between lines. Two cell-center users have pathloss PL_{center} and two cell-edge users have pathloss PL_{edge}. The pathloss difference between cell-edge and cell-center users in dB is $PL_{diff} = PL_{center} - PL_{edge}$.

Figure 2 illustrates a more general scenario. There are also two cell-center and two cell-edge users with pathlosses PL_{center} and PL_{edge} respectively. The difference is that users are not positioned in straight lines, but can take random positions in corresponding areas.

We consider a slotted system. In every slot $t \in T$, the BS chooses the configuration $(\mathcal{K}, f)$ and the precoder P. There are several strategies for the precoder selection: ZF, SOTA, and semi-orthogonal precoder. Also, we consider the fourth strategy, where the BS can select any precoder considered. The latter is denoted as SOTA+Semi-orthogonal.

The set of users selected for transmission in each slot is chosen based on the solution of the following optimization problems. The first considered problem is maximizing the geometric mean of all users rates:

$$\max_{(\mathcal{K}, f), \mathrm{P}} \prod_{k \in \hat{\mathcal{K}}} R_k \quad (10a)$$

$$\mathrm{s.t.} \ \mathrm{tr}(\mathrm{PP}^H) \leq E_{max}, \quad (10b)$$

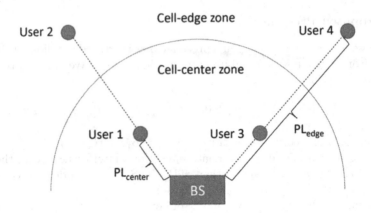

Fig. 1. Scenario 1.

where R_k is the total rate of user k in all slots: is a sum of its rates in each slot:

$$R_k = \sum_{t \in T} R_{f(k,t)}, \tag{11}$$

$R_{f(k,t)}$ is the rate of user k in slot t, $f(k,t)$ is grouping function f for user k in this slot. We also consider geometric mean maximization of the cell-edge users rates subject to a minimum rate constraint for cell-center users:

$$\max_{(\mathcal{K},f),P} \prod_{k \in \hat{\mathcal{K}}_{edge}} R_k \tag{12a}$$

$$\text{s.t. } R_k \geq R_k^{th}, \ \forall k \in \hat{\mathcal{K}}_{center}, \tag{12b}$$

$$\text{tr}(PP^H) \leq E_{max}, \tag{12c}$$

where $\hat{\mathcal{K}}_{edge}$ is set of cell-edge users, $\hat{\mathcal{K}}_{center} = \hat{\mathcal{K}} \setminus \hat{\mathcal{K}}_{edge}$, R_k^{th} is a minimum rate constraint for the user k.

For the given objective function (10) or (12) the BS chooses the best configurations for every slot in the recurring slot chain using an exhaustive search. To limit the search options we set the number of slots in a sequence $|T| = 4$.

In the series of experiments, we fix $PL_{center} = 80$ dB and change the maximum transmit power E_{max} and PL_{diff}. This is equivalent to fixing the transmit power and changing the path loss PL_{center} and PL_{edge}. We first consider the results obtained in Scenario 1. Figures 3 and 4 shows the relative gain of various strategies versus ZF for various values of $E_{RX}^{edge} = E_{max} - PL_{center} - PL_{diff}$. Figure 3 shows the results for the optimization problem (10) and Fig. 4 shows the results for the optimization problem (12) with $R_k^{th} = 0.5R_k^*$, where R_k^* is the maximum achievable rate for user k. The user k gets the maximum rate when it is the only one served, $\mathcal{K} = \{k\}$ for every slot, and R_k^* can be defined from such a case.

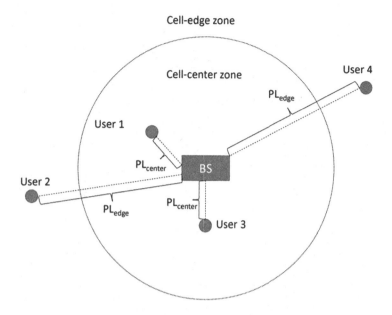

Fig. 2. Scenario 2.

It can be seen that a higher path loss difference between cell-edge and cell-center users gives a greater increase in throughput for precoders with NOMA (SOTA or semi-orthogonal), which is consistent with the prior general observation of NOMA behavior. However, if users are close to the BS, i.e., for high $E_{RX}^{edge} \gtrsim -60$ dBm, NOMA precoders give minimal gain compared to ZF. The reason is that power gains using ZF with perfect channel knowledge for users are high enough to outweigh performance drops associated with high channel correlation. However, with NOMA-based precoders, the interference from the cell-center signal is much higher than the Gaussian noise at cell-edge users, so if PL_{diff} is fixed, SINR stays the same even when channel gains increase. So it is more advantageous to use only MIMO at close distances. Moreover, for $M = 4$, ZF works better and the maximum gain of NOMA-based precoders is also smaller.

For small $E_{RX}^{edge} < -80$ dBm, SOTA achieves a greater throughput gain compared to a semi-orthogonal precoder. For $M = 3$ there is a 10–15% difference in the gain against the ZF precoder. It happens because cell-edge users have rather low channel gains, and interference reduction for cell-edge users with the semi-orthogonal precoder has almost no effect on the system performance in terms of the geometric mean of throughput. At the same time, SOTA obtains higher SINR values on cell-center users. As a result, it is easier for BS to allocate more power to cell-edge users. Note that for $M = 4$ difference in gain drops from 10–15% to about 7%.

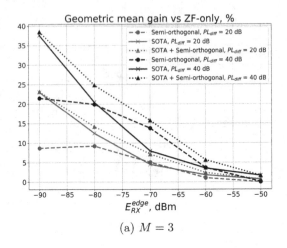

(a) $M = 3$

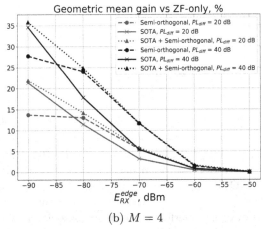

(b) $M = 4$

Fig. 3. The gains in the geometric mean of throughput of all users with NOMA-MIMO precoders compared with the ZF precoder, Scenario 1.

For medium values -80 dBm $\lesssim E_{RX}^{edge} \lesssim -60$ dBm, the semi-orthogonal precoder shows better results compared to both ZF and SOTA ones. In this case, the interference nulling for cell-edge users plays an essential role in the considered problem. Channel conditions of cell-edge users are insufficient for ZF to overcome correlation issues. However, they are sufficient to have a noticeable contribution to system performance. For example, the gain in the geometric mean of throughput for cell-edge users from the proposed precoder is 15%, which is 7% higher than that from SOTA for $M = 3$ and $E_{RX}^{edge} = -70$ dBm.

Note that the join usage of various precoders shows better results than using them separately. This strategy is able to use the advantages of both NOMA-based precoders depending on the channel conditions. The effectiveness of this strategy is observable for $E_{RX}^{edge} = -80$ dBm, where joint usage of both NOMA-

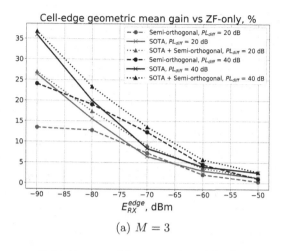

(a) $M = 3$

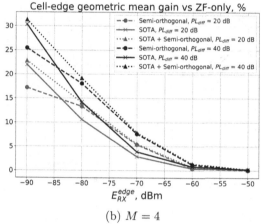

(b) $M = 4$

Fig. 4. The gains in the geometric mean of throughput of cell-edge users with NOMA-MIMO precoders compared with the ZF precoder, Scenario 1.

based precoders achieves up to 24% gain in the geometric mean of throughput of cell-edge users and up to 25% gain in the geometric mean of throughput of all users.

Let us now consider the results for Scenario 2, shown in Fig. 5 and 6. As in Scenario 1, Fig. 5 shows the results for the problem (10) and Fig. 6 shows results for the problem (12) with $R_k^{th} = 0.5R_k^*$.

It can be seen that most of the conclusions drawn for Scenario 1 apply to Scenario 2 as well. However, it is worth noting the following differences. First, all the gains of NOMA precoders are approximately twice lower because the users are not located on the same straight line in Scenario 2, resulting in a lower average channel correlation within cell-edge/cell-center pairs of users. Therefore, using ZF results in higher average SINR values on average.

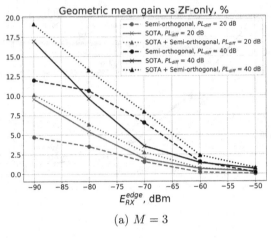

(a) $M = 3$

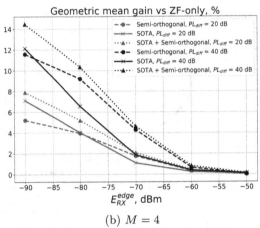

(b) $M = 4$

Fig. 5. The gains in the geometric mean of throughput of all users with NOMA-MIMO precoders compared with the ZF precoder, Scenario 2.

Second, there is a larger relative difference for medium values of E_{RX}^{edge} between the semi-orthogonal precoder and SOTA. This is due to the fact that when using SOTA, there is only one spatial beam per user pair using NOMA. As a result, when the users are no longer in the same straight line, the cell-edge users obtain lower SINR values.

To summarize, we demonstrate application areas of NOMA-based precoders for various pathloss values. If all users are close enough, then the BS can use just ZF precoding without NOMA. If cell-edge users have insufficient channel quality, the BS shall use the SOTA solution to satisfy cell-center users and provide the remaining resources to cell-edge users. Finally, if there are cell-edge users with channel gains that exceed channel noise, then the usage of the proposed precoding obtains the best performance results. The proposed precoder demonstrates

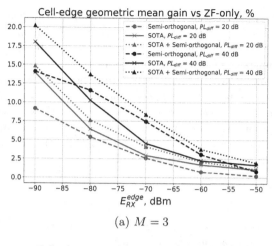

(a) $M = 3$

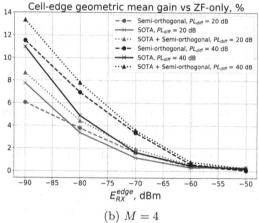

(b) $M = 4$

Fig. 6. The gains in the geometric mean of throughput of cell-edge users with NOMA-MIMO precoders compared with the ZF precoder, Scenario 2.

superior gain in the geometric mean of throughput moderate SINR range, and, hence, is preferable for implementation together with the existing precoders for NOMA-MIMO systems.

6   Conclusion

In this paper, we have proposed a novel approach to enhance the performance of cell-edge users in downlink NOMA-MIMO systems without negatively impacting the user experience of cell-center users. The proposed precoder design, which independently constructs precoders for each user within a group, effectively eliminates inter-group interference. The developed precoder is particularly effective in the area of moderate SINRs, outperforming the state-of-the-art approach where beamforming is done on the cell-center user. To examine the developed precoder, we perform extensive simulation using the QuaDRiGa channel generator and demonstrate that the joint usage of the proposed and state-of-the-art precoders improves the geometric mean of throughput in time-slotted downlink NOMA-MIMO systems, taking into account QoS requirements of cell-center users. The proposed precoder demonstrates its effectiveness in joint usage with existing precoders, increasing the area of applicability of NOMA-MIMO systems. However, the proposed approach assumes an increase in the number of antennas on the base station, which warrants further investigation into the cost-benefit analysis of this assumption.

References

1. Khorov, E., Kureev, A., Levitsky, I., Akyildiz, I.F.: Prototyping and experimental study of non-orthogonal multiple access in Wi-Fi networks. IEEE Network **34**(4), 210–217 (2020)
2. Ali, S., Hossain, E., Kim, D.I.: Non-orthogonal multiple access (NOMA) for downlink multiuser MIMO systems: user clustering, beamforming, and power allocation. IEEE Access **5**, 565–577 (2016)
3. Zeng, M., Yadav, A., Dobre, O.A., Poor, H.V.: A fair individual rate comparison between MIMO-NOMA and MIMO-OMA. In: IEEE Globecom Workshops (GC Wkshps), pp. 1–5. IEEE (2017)
4. Kimy, B., et al.: Non-orthogonal multiple access in a downlink multiuser beamforming system. In: 2013 IEEE Military Communications Conference (MILCOM 2013), pp. 1278–1283. IEEE (2013)
5. Zeng, M., Yadav, A., Dobre, O.A., Tsiropoulos, G.I., Poor, H.V.: On the sum rate of MIMO-NOMA and MIMO-OMA systems. IEEE Wirel. Commun. Lett. **6**(4), 534–537 (2017)
6. Ding, Z., Adachi, F., Poor, H.V.: The application of MIMO to non-orthogonal multiple access. IEEE Trans. Wireless Commun. **15**(1), 537–552 (2015)
7. Nguyen, V.-D., Tuan, H.D., Duong, T.Q., Poor, H.V., Shin, O.-S.: Precoder design for signal superposition in MIMO-NOMA multicell networks. IEEE J. Sel. Areas Commun. **35**(12), 2681–2695 (2017)
8. Hanif, M.F., Ding, Z., Ratnarajah, T., Karagiannidis, G.K.: A minorization-maximization method for optimizing sum rate in the downlink of non-orthogonal multiple access systems. IEEE Trans. Signal Process. **64**(1), 76–88 (2015)
9. Zhu, F., Lu, Z., Zhu, J., Wang, J., Huang, Y.: Beamforming design for downlink non-orthogonal multiple access systems. IEEE Access **6**, 10956–10965 (2018)
10. Chen, C., Cai, W., Cheng, X., Yang, L., Jin, Y.: Low complexity beamforming and user selection schemes for 5G MIMO-NOMA systems. IEEE J. Sel. Areas Commun. **35**(12), 2708–2722 (2017)
11. Alavi, F., Cumanan, K., Ding, Z., Burr, A.G.: Beamforming techniques for nonorthogonal multiple access in 5G cellular networks. IEEE Trans. Veh. Technol. **67**(10), 9474–9487 (2018)
12. Chen, Z., Ding, Z., Xu, P., Dai, X.: Optimal precoding for a QoS optimization problem in two-user MISO-NOMA downlink. IEEE Commun. Lett. **20**(6), 1263–1266 (2016)
13. Timotheou, S., Krikidis, I.: Fairness for non-orthogonal multiple access in 5G systems. IEEE Signal Process. Lett. **22**(10), 1647–1651 (2015)

14. Choi, J.: Power allocation for max-sum rate and max-min rate proportional fairness in NOMA. IEEE Commun. Lett. **20**(10), 2055–2058 (2016)
15. Sun, X., Duran-Herrmann, D., Zhong, Z., Yang, Y.: Non-orthogonal multiple access with weighted sum-rate optimization for downlink broadcast channel. In: 2015 IEEE Military Communications Conference (MILCOM 2015), pp. 1176–1181. IEEE (2015)
16. Ali, M.S., Tabassum, H., Hossain, E.: Dynamic user clustering and power allocation for uplink and downlink non-orthogonal multiple access (NOMA) systems. IEEE Access **4**, 6325–6343 (2016)
17. Sun, X., et al.: Joint beamforming and power allocation design in downlink non-orthogonal multiple access systems. In: IEEE Globecom Workshops (GC Wkshps), pp. 1–6. IEEE (2016)
18. Sun, X., et al.: Joint beamforming and power allocation in downlink noma multiuser mimo networks. IEEE Trans. Wireless Commun. **17**(8), 5367–5381 (2018)
19. Choi, J.: Minimum power multicast beamforming with superposition coding for multiresolution broadcast and application to noma systems. IEEE Trans. Commun. **63**(3), 791–800 (2015)
20. Liu, Z., Lei, L., Zhang, N., Kang, G., Chatzinotas, S.: Joint beamforming and power optimization with iterative user clustering for MISO-NOMA systems. IEEE Access **5**, 6872–6884 (2017)
21. Georgakopoulos, P., Akhtar, T., Mavrokefalidis, C., Politis, I., Berberidis, K., Koulouridis, S.: Coalition formation games for improved cell-edge user service in downlink NOMA and MU-MIMO small cell systems. IEEE Access **9**, 118484–118501 (2021)
22. Do, T.N., da Costa, D.B., Duong, T.Q., An, B.: Improving the performance of cell-edge users in NOMA systems using cooperative relaying. IEEE Trans. Commun. **66**(5), 1883–1901 (2018)
23. Le, C.-B., Do, D.-T., Voznak, M.: Wireless-powered cooperative MIMO NOMA networks: design and performance improvement for cell-edge users. Electronics **8**(3), 328 (2019)
24. Liu, F., Petrova, M.: Performance of proportional fair scheduling for downlink PD-NOMA networks. IEEE Trans. Wireless Commun. **17**(10), 7027–7039 (2018)
25. Jaeckel, S., Raschkowski, L., Börner, K., Thiele, L.: QuaDRiGa: a 3-D multi-cell channel model with time evolution for enabling virtual field trials. IEEE Trans. Antennas Propag. **62**(6), 3242–3256 (2014)

Surrogate Models for the Compressibility Factor of Natural Gas

Olga Kochueva[✉][iD] and Ruslan Akhmetzianov[iD]

National University of Oil and Gas "Gubkin University", 65 Leninsky Prospekt,
Moscow 119991, Russia
kochueva.o@gubkin.ru

Abstract. The paper presents an example of the so-called surrogate modeling. This is a computer modeling technique where machine learning methods are used to build a fast (surrogate) model that allows you to get a result with acceptable accuracy on data from a complex (physically proven, built on a solution of systems of nonlinear algebraic equations or partial differential equations) and resource-intensive model of an object or process. We develop and analyze the models for calculating the compressibility factor trained on a large amount of data calculated with AGA-8 equation of state. The presented models can be applied to replace the original model when analyzing the development of risk situations and searching for optimal gas transportation modes, in software designed for staff training on computer simulators. The feature and the novelty of the proposed study is not only an analysis of the accuracy of the obtained models for the complex multicomponent composition of natural gas, but also an analysis of the derivatives of the compressibility factor with respect to pressure and temperature, which is important for conducting calculations of nonstationary modes of gas transmission and calculating the sound speed in the gas.

Keywords: Machine learning · Surrogate modelling · Compressibility factor

1 Introduction and Motivation

Hydraulic calculations are the primary tool for rational technical decisions related to the design and operation of pipeline systems. The compressibility factor is introduced into the gas equation of state to account for its real properties and depends on the pressure, temperature, and gas component composition. Most explicit and implicit approximations have been developed based on the accumulated significant volume of experimental data. The calculation of the compressibility factor with regard to gas fraction composition can be performed according to the equation of state (in recent years, the work of most researchers is based on GERG-2008, AGA8 [1,2], AGA10), where the problem is reduced to an iterative method of solving a nonlinear equation. The time of calculations plays a significant role in non-stationary modeling of gas transfer modes in main pipelines as well as determining optimal mode of gas transmission and parameters identification of a gas transportation system. The work aimed to build approximations to calculate the compressibility factor using machine learning (ML) algorithms and to analyze the quality of the resulting models.

So-called surrogate modeling has gained popularity relatively recently with the development and wide application of machine learning and has become an alternative to traditional numerical algorithms. An example of application of this approach for thermodynamic calculations of multicomponent phase equilibria is given in [3]. The authors developed Artificial Neural Network (ANN) based on data generated using a semi-empirical gamma-phi model framework (eNRTL for the liquid phase and Peng Robison for the vapor phase). The accuracy of the surrogate model predictions of the vapour-liquid equilibrium for the proposed model is found to be satisfactory as the results provide an average absolute relative difference of 0.5% compared to the estimates obtained with a rigorous thermodynamic model. It is shown that the speed of ML based surrogate models can be about 10 times faster than interpolation methods and about 1000 times faster than rigorous vapour-liquid equilibrium calculations.

In [4], as an alternative to time-consuming finite element analysis for simulation the thermal behavior of multi-plate clutches, the authors propose surrogate models using various machine learning methods (polynomial regression, decision tree, support vector regressor, Gaussian process and neural networks). They evaluated all models with respect to their ability to predict the maximum clutch temperature based on the loads of a slip cycle and concluded that ML models fundamentally achieve good results, at that ANN provided the best results.

It should be noted that the studies [3,4] are quite near in subject area to the problem under consideration, so performing calculations involving thermodynamic models is a promising area for the application of surrogate modeling. Particularly interesting is the tradeoff between the accuracy of the calculations and the computational effort required by models.

In recent decades, finding the best computational method for the compressibility factor has been a hot research issue. In [5] the relationship in explicit form, constructed on the basis of a set of experimental data (3038 records) is presented. The correlation is built as a fraction, the numerator and denominator of which are functions containing the reduced pressures and temperature in various degrees and their logarithms, there are 20 coefficients in the formula. The paper [6] presents the formula that is a modification of the implicit relation [7]. The resulting model contains 19 coefficients, and the polynomial functions of reduced temperature and pressure and exponential functions of temperature are used in the calculation. A feature of [8] is the division of the range of reduced pressure values into 2 subsets, a separate model is built for each range. Due to this it became feasible to reach a higher accuracy of the model, the group method of data handling (GMDH) was used in the work. The authors note that the models built using GMDH do not include logarithmic or exponential functions and can be easily used in programming of flow-meters. The paper [9] introduces a model based on a multidimensional nonlinear relationship, built on a sample of 6988 values obtained from the digitized Standing-Katz diagram. The authors give a formula that is the quotient of two polynomials, the formula has 20 coefficients, a mean absolute percentage error (MAPE) about 1.5% is indicated. To calculate the compressibility factor in [10], an artificial neural network with two hidden layers is proposed, it was trained on 4158 experimental data entries. The papers [11,12] presented correlations to calculate the Z-factor explicitly based on genetic programming or Symbolic Regression (SR) method. These two separate studies considered different ranges of pressure and temperature, presented different correlations and reported accuracy of the models (MAPE) of 2.5% and 0.03%, respectively). In [13], two different models based on statistical regression and multi-layer feed-forward neural network (MLFN) were developed to predict Z-factor of natural gas based on the experimental data of 1079 samples. The authors report correlation coefficient of 0.967 and 0.979, respectively, which corresponds to coefficients of determination R^2 0.935 and 0.958. In [14] the authors presented artificial network based on fuzzy inference system Tagaki-Sugeno to predict Z-factor. They analyzed around 6500 published and unpublished data points with a wide range of Z-factor from several oil fields in the Middle East. The obtained model performs with MAPE of 0.13%.

Therefore, it is difficult to compare the results of all mentioned models since the testing was carried out on a disparate data. In many works, authors usually present cross-plots determining the ratio between the obtained and experimental values of the compressibility factor, they also provide the determination coefficient values and the average absolute error, but the majority of them do not conduct a complete analysis of the models including the study of the derivatives of the compressibility factor with respect to pressure and temperature. Our study fills this gap.

2   Generating a Data Set

The aim of the work was to obtain and analyze models for the compressibility factor, which could replace the iterative procedure proposed in [2], for the ranges of pressure, temperature and gas compositions, typical for the main gas pipelines modes. Such approximations, provided that the deviation of the compressibility factor calculated by them will not exceed 0.05–0.1% of the result obtained by the procedure [2], can be used when carrying out calculations of non-stationary modes of gas transfer, when searching for the optimum conditions, in simulators - everywhere where the speed of calculation is important.

Natural gas is a mixture of methane, hydrocarbons from C_2 to C_6, it also may contain nitrogen, carbon dioxide, and helium, Fig. 1 presents the components and their respective contents in the natural gas transported through the long-distance pipelines.

Since many components are contained in natural gas in small amounts, the study planned to analyze the possibility of reducing the number of input variables without compromising the model quality, therefore the molar mass of the gas

Component	MIN, %	MAX, %
Methane	92.2	98.87
Nitrogen	0.5	0.9
Carbon dioxide	0.025	0.435
Ethane	0.33	4.7
Propane	0.0051	0.95
Isobutane	0.0066	0.26
n-Butane	0.0025	0.24
Isopentane	0.0025	0.045
n-Pentane	0.0025	0.03
n-Hexane	0.0025	0.0124
Hydrogen	0.001	0.0023
Oxygen	0.0046	0.0085
Helium	0.0086	0.0147

Fig. 1. Composition of natural gas transported through main pipelines.

was added to the input parameters. The authors didn't pursue the goal to obtain universally applicable models, thus we specified a temperature in the range from 273 to 333 K and two pressure ranges: $P_1 \in [3.5 - 5.6]$ MPa and $P_2 \in [5.5 - 7.5]$ MPa for main pipelines of different capacities. For different gas compositions, pressure values were set with a step of 0.02 MPa, temperature with a step of 2K, and the values of the compressibility factor were calculated with the method [2]. The algorithm for generating the initial datasets is presented in Fig. 2.

1. Set methane fraction ω_1, $\omega_1 \in [92.2, 98.8]$ %, increment $\delta_{\omega_1} = 0.2$ %
  2. Set the temperature $T = T_0$, $T \in [273, 323]$ K, $\delta_T = 2$ K
    3. Set the pressure $P = P_0$, $P_1 \in [3500, 5600]$, $P_2 \in [5500, 7500]kPa$, $\delta_P = 20$ kPa
      4. Randomly select a list of unique indexes $\Psi = \{\xi_i \mid i = 1 \dots N - 1, \xi \sim U(1, N - 1)\}$
      5. Randomly generate the components of the vector $\Omega = \Omega(\omega_i)$: $\omega_i \sim U(a_i, b_i), i \in \Psi$; $\omega_{\Psi[-1]} = 1 - \sum \omega_i$
      6. Calculate the compressibility coefficient $Z(P, T)$ according to [2]
     7. $P = P + \delta_P$
  8. $T = T + \delta_T$
9. $\omega_1 = \omega_1 + \delta_{\omega_1}$

Fig. 2. Generation of initial datasets.

Thus, initial data for model training and testing were generated with the number of entries 24722 for the range P_1 and 23028 for the range P_2. It should be noted that the specified ranges in both pressure and temperature are not a limitation for the proposed technique.

3   Results and Discussion

We considered 3 sets of input parameters that differ in the completeness of information about the gas composition, they are presented in Fig. 3.

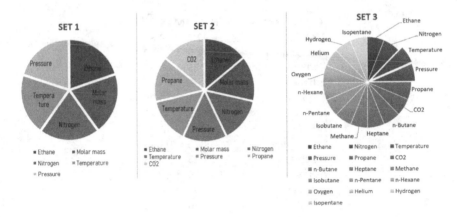

Fig. 3. 3 sets of input parameters.

We generated 4 models, based on Random Forest (RF) algorithm [15], Group Method Data Handling (GMDH) [16], Symbolic Regression (SR) [17], artificial neural network (ANN). Two of them - SR and GMDH present the approximation in explicit form. Symbolic Regression is not as widely used machine learning algorithm as RF or ANN, its advantage is computational efficiency, the possibility to control the complexity of the obtained correlations, the fact that dependence can be obtained in explicit form. A number of publications [18–22] emphasize the effectiveness of its use to solve a number of practical problems. The detailed description of models built with SR method is presented in [12].

The developed models for calculating the compressibility factor have $R^2 = 0.99$, MAE $= 0.0002$–0.001, MAPE $= 0.02$–0.1%. In Table 1, a mean absolute percentage error (MAPE) for pressure range P_2 for different methods is given.

Table 1. MAPE, (%) for test sample for pressure range P_2 for different machine learning methods.

	Linear Regression (MLR)	Random Forest (RF)	Symbolic Regression (SR)	Neural Network (ANN)	Group Method Data Handling (GMDH)
Set 1	0.37	0.06	0.05	0.04	0.03
Set 2	0.29	0.04	0.03	0.03	0.02
Set 3	0.11	0.05	0.03	0.02	0.02

As can be seen from Table 1, the molar mass and two components (ethane and nitrogen), can be successfully used for calculating the compressibility factor instead of the full gas composition for the specified pressure range. The additional error when using as input data Set 1 instead of Set 3 does not exceed 0.02% for all models except MLR. Table 1 shows that the model based on GMDH algorithm has the best metrics, the models obtained by other machine learning methods lose slightly in accuracy, although their MAPEs satisfy the requirement to be less than 0.1%.

The hypotheses of constancy of the mathematical expectation, the normality of the distribution, the absence of autocorrelation, and the constancy of the variance were tested for the residuals of each model. We used ANOVA for testing the hypotheses of constancy of the mathematical expectation, Goldfeld-Quandt test for checking homoscedasticity, Pearson χ^2 test for the hypothesis about normal distribution of the residuals and plotting autocorrelation function (ACF). Testing normality and autocorrelation of the residuals for ANN model is shown in Fig. 4.

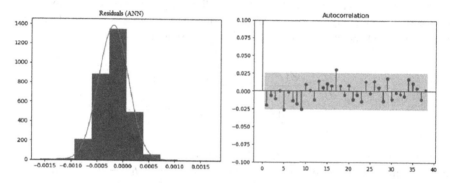

Fig. 4. Testing normality and autocorrelation of the residuals for ANN compressibility factor model.

For all models, conclusions were drawn about the constancy of the mathematical expectation of the residuals, the normality of the distribution, the absence of autocorrelation, and the constancy of the variance.

The main technique currently used in modeling fluid flow in long pipelines is continuum mechanics. To analyze unsteady nonisothermal flow, three equations are considered for the three mode parameters: pressure $P(x,t)$, flow rate $q(x,t)$, and temperature $T(x,t)$, where x is spatial variable and t is time, the equations of unsteady nonisothermal flow in a complete notation can be found in [23]. In some cases it is justified to use simplified models: a) nonstationary isothermal flow - two equations regarding $P(x,t), q(x,t)$, b) stationary nonisothermal flow - two equations regarding $P(x), T(x)$. If one does not simplify the system a priori, the equations will include partial derivatives of the compressibility coefficient $\partial z/\partial P$ and $\partial z/\partial T$. In [24,25] a few questions are posed related to the legitimacy of some of the assumptions made in the derivation and further transformations of system of partial differential equations. The partial derivative of the compressibility factor with respect to pressure is also used to calculate the speed of sound in gas. The speed of sound is a quantity used in various applications such as leak detection. When a pipe ruptures, detecting the leak as quickly as possible helps to reduce the loss of transported product and prevent situations that are dangerous for people and the environment. Small leaks are more difficult to detect, and it is also difficult to determine the location of the leak, which is estimated from the moments of signal detection by remote sensors. Sound propagates through the gas pipeline (through the transported product) at a velocity of about 450 m/s. For such fast processes, small (relative) errors in calculating the speed of sound in gas can lead to significant errors in the identification of the leak location, which will make it difficult to localize the failure location on the route. All this indicates that the accuracy of determining the derivative $\partial z/\partial P$ has a certain applied importance.

We compared the derivatives of the compressibility factor with respect to pressure and temperature calculated by each model with the derivative numerically calculated with method [2]. For explicit approximations the derivatives were calculated analytically, for implicit ones the estimates of the derivatives were calculated using a central finite-difference scheme:

$$\frac{\partial Z}{\partial P} = \frac{Z(P + h_p, T) - Z(P - h_p, T)}{2h_p}; \tag{1}$$

$$\frac{\partial Z}{\partial T} = \frac{Z(P, T + h_t) - Z(P, T - h_t)}{2h_t}. \tag{2}$$

For numerical calculating $\partial Z/\partial P$ the increment $h_p = 0.001\,\text{MPa}$ was used, and for numerical calculating $\partial Z/\partial T$ the increment $h_t = 0.1\text{K}$ was taken. Figure 5 presents some examples of derivatives comparison.

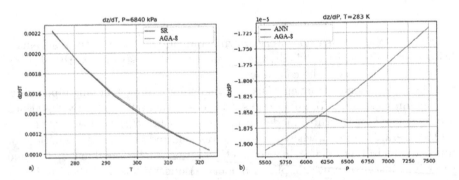

Fig. 5. Derivatives of the compressibility factor with respect to pressure for SR method (on the left) and with respect to temperature for ANN (on the right).

Models built with SR and GMDH methods demonstrate good correspondence of the derivatives of the compressibility factor both with respect to pressure and to temperature with the derivatives of compressibility factor calculated with method [2]. Figure 5 on the left presents graph of $\partial Z/\partial T$ for $P = 6.84\,\text{MPa}$ for

SR model. The model built with GMDH method showed good performance, its mean absolute percentage error for $\partial Z/\partial P$ when $T = 283$ K is about 1.4%, and for $\partial Z/\partial T$ when $P = 6.48$ MPa is less 1%. Models built with RF and ANN methods demonstrate poor agreement with the derivatives of compressibility factor calculated with method [2]. Figure 5 on the right shows a graph of $\partial Z/\partial P$ for $T = 283$K for ANN model.

We have tested the models developed for the interval P_2 for pressure $P <$ 5.5 MPa and $P > 7.5$ MPa. The predictive performance of the GMDH model and SR model for pressure values beyond the intervals where the models were trained, turned out to be of acceptable quality. ANN and RF models had poor performance for the interval $P > 7.5$ MPa (see Fig. 6). In Fig. 6, graphs on the left show cross-plots for predicted vs actual values for pressure in the interval P_2, graphs on the right show cross-plots for predicted vs actual values for pressure $P > 7.5$ MPa.

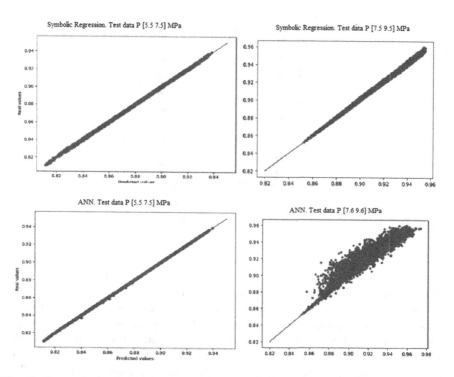

Fig. 6. Cross-plots predicted vs actual compressibility factor for SR model (upper graphs) and ANN model (lower graphs).

Unsteady flows in industrial gas pipelines are a topic of in-depth study by many researchers in recent years. This is due to the continuous increase in the share of renewable energy sources in the energy balance of many countries. Solar and wind power, which are very much dependent on weather conditions, are developing the most successfully. In this situation, gas becomes an alternative source of energy, the need for which varies dramatically with changes in the weather and time of day. The structure of the gas supply system also changes in the direction of increasing reserve capacities and bringing them closer to the consumer. In methodological terms, this leads to the need to solve optimization problems of joint management of the modes of hybrid power and gas supply systems. That is why, one of the main objectives of our study was to develop the computationally effective approximation of the compressibility coefficient, which can be used in models of unsteady gas flow, in solving optimization problems, etc. MAPE and computation time for different machine learning algorithms are shown in Fig. 7. Calculation time is given in seconds to calculate 100 values of the compressibility coefficient for the samples from the test set. The last bar corresponds to the method [2], since it is the source of the dataset, its error is assumed to be 0.

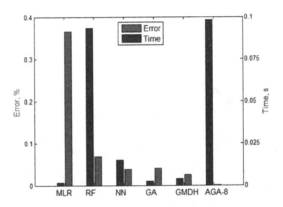

Fig. 7. MAPE and computation time for different machine learning algorithms.

From Fig. 7, we can see that the best computational time belongs to multiple linear regression (MLR), symbolic regression (genetic algorithm (GA)) and GMDH. Random Forest algorithm gives approximately the same computational time as method [2], so it does not make sense to use it as a surrogate model. Since the computation time for the SR model is reduced by up to 50 times compared to solving the AGA-8 equation of state, the same gain will be achieved in solving optimization problems. The GMDH model runs about 25 times faster than the numerical solution of the equation of state, which is also a good value. The ratio between error and speed of computation is the best among the considered methods for GMDH and symbolic regression.

4   Conclusion

The developed surrogate models for calculating the compressibility factor have $R^2 = 0.99$, MAE = 0.0002–0.001, MAPE = 0.02–0.1%. The error of the calculation methods used to obtain the initial data for model building does not exceed 0.2%. The results indicate that the goal of the work has been achieved, and approximations have been obtained that can successfully replace computationally demanding iterative procedures without loss of accuracy. The best results including the shape of the derivatives of the compressibility factor with respect to pressure and temperature, showed models based on GMDH method and symbolic regression. The resulting models are easy to integrate into existing software and in higher-level models. When developing specialized software products, the choice of mathematical and algorithmic models should be determined in accordance with the purpose of the software product and limitations on the operating range. For models for the compressibility factor of natural gas is essential to take into account the accuracy of determining the derivatives $\partial z/\partial P$ and $\partial z/\partial T$. The study has shown that despite the high accuracy, the application of ANN has essential disadvantages when it is necessary to use derivatives of the obtained functions and when the input data is outside the intervals on which the model was trained.

References

1. ISO 12213-2:2006. Natural Gas - Calculation of Compression Factor Switzerland. ISO, Geneva (2006)
2. Repository for the Supplementary Files to AGA 8 NIST USA. www.pages.nist.gov/AGA8/. Accessed 10 Feb 2022
3. Carranza-Abaid, A., Svendsen, H., Jakobsen, J.: Surrogate modelling of VLE: integrating machine learning with thermodynamic constraints. Chem. Eng. Sci. X **8**, 100080 (2020). https://doi.org/10.1016/j.cesx.2020.100080
4. Schneider, T., Bedrikow, A.B., Dietsch, M., Voelkel, K., Pflaum, H., Stahl, K.: Machine learning based surrogate models for the thermal behavior of multi-plate clutches. Appl. Syst. Innov. **5**, 97 (2022) https://doi.org/10.3390/asi5050097
5. Azizi, N., Behbahani, R., Isazadeh, M.A.: An efficient correlation for calculating compressibility factor of natural gases. J. Nat. Gas Chem. **19**, 642–645 (2010). https://doi.org/10.1016/S1003-9953(09)60081-5
6. Kareem, L.A., Iwalewa, T.M., Al-Marhoun, M.: New explicit correlation for the compressibility factor of natural gas: linearized z-factor isotherms. J. Petrol. Explor. Prod. Technol. **6**, 481–492 (2016). https://doi.org/10.1007/s13202-015-0209-3
7. Hall, K.R., Yarborough, L.: A new equation-of-state for Z-factor calculations. Oil Gas J. **71**, 82–92 (1973)
8. Lin, L., Li, S., Sun, S., Yuan, Y., Yang, M.: A novel efficient model for gas compressibility factor based on GMDH network. Flow Measur. Instrument. **71**, 101677 (2020). https://doi.org/10.1016/j.flowmeasinst.2019.101677
9. Wang, Y., Ye, J, Wu, Sh.: An accurate correlation for calculating natural gas compressibility factors under a wide range of pressure conditions. Energy Rep. **8**(2), 130–137 (2022). https://doi.org/10.1016/j.egyr.2021.11.029
10. Azizi, N., Rezakazemi, M., Zarei, M.M.: An intelligent approach to predict gas compressibility factor using neural network model. Neural Comput. Appl. **31**(1), 55–64 (2019)
11. Towfighi, S.: An empirical equation for the gas compressibility factor. Z. Pet. Sci. Technol. **38**, 24–27 (2020)

12. Kochueva, O., Zadorozhnyy V.: Analysis of approximations of the gas compressibility factor derived from genetic algorithms. E3S Web Conf. **97**, 01005 (2023). https://doi.org/10.1051/e3sconf/202339701005. Mathematical Models and Methods of the Analysis and Optimal Synthesis of the Developing Pipeline and Hydraulic Systems 2022

13. Ghanem, A., Gouda, M.F., Alharthy, R.D., Desouky, S.M.: Predicting the compressibility factor of natural gas by using statistical modeling and neural network. Energies **15**, 1807 (2022). https://doi.org/10.3390/en15051807

14. Al-Gathe, A., Baarimah, S., Al-Khudafi, A.: Modelling gas compressibility factor using different fuzzy methods. AIP Conf. Proc. **2443**, 030031 (2022). https://doi.org/10.1063/5.0092029

15. Breiman, L.: Random forests. Mach. Learn. **45**, 5–32 (2001). https://doi.org/10.1023/A:1010933404324

16. Madala, H.R., Ivakhnenko, O.G.: Inductive Learning Algorithms for Complex Systems Modeling. CRC Press, Boca Raton (1994)

17. Koza, J.: Genetic programming: on the programming of computers by means of natural selection. The MIT Press, Cambridge (1992)

18. Saghafi, H., Arabloo, M.: Development of genetic programming (GP) models for gas condensate compressibility factor determination below dew point pressure. J. Petrol. Sci. Eng. **171**, 890–904 (2018) https://doi.org/10.1016/j.petrol.2018.08.020

19. Kochueva, O., Nikolskii, K.: Data analysis and symbolic regression models for predicting CO and NOx emissions from gas turbines. Computation **9**, 139 (2021). https://doi.org/10.3390/computation9120139

20. Kochueva O.: Razrabotka modelej prognozirovaniya vybrosov oksidov ugleroda i azota gazovyh turbin na osnove geneticheskih algoritmov. Delovoj zhurnal Neftegaz.RU (in Russian) **5–6**(125–126), 14–20 (2022)

21. Praks, P., Lampart, M., Praksová, R.; Brkić, D., Kozubek, T., Najser, J.: Selection of appropriate symbolic regression models using statistical and dynamic system criteria: example of waste gasification. Axioms **11**, 463 (2022). https://doi.org/10.3390/axioms11090463

22. Angelis, D., Sofos, F., Karakasidis, T.: Artificial intelligence in physical sciences: symbolic regression trends and perspectives. Archiv. Comput. Methods Eng. **30**, 3845–3865 (2023). https://doi.org/10.1007/s11831-023-09922-z

23. Helgaker, J., Oosterkamp, A., Langelandsvik, L., Ytrehus, T.: Validation of 1D flow model for high pressure natural gas pipelines. J. Nat. Gas Sci. Eng. **16**, 44–56 (2014). https://doi.org/10.1016/j.jngse.2013.11.001

24. Sukharev, M., Kochueva, O., Zhaglova, A.: Experimental study of wave processes in main gas pipelines under normal operating conditions. Fluids **8**(2), 45 (2023). https://doi.org/10.3390/fluids8020045

25. Sukharev, M., Kochueva, O.: Phenomenological study of the dynamics of pressure distribution in a gas flow in a long-distance pipeline. E3S Web Conf. **102**, 01006 (2019). https://doi.org/10.1051/e3sconf/201910201006. Mathematical Models and Methods of the Analysis and Optimal Synthesis of the Developing Pipeline and Hydraulic Systems 2019.

26. Sukharev M.G., Samoilov R.V., Kritinina A.S.: Natural gas compressibility factor: approximations comparison and selection criteria. Automat. Inform. Fuel Energy Complex (in Russian) **1**(594), 42–54 (2023). https://doi.org/10.33285/2782-604X-2023-1(594)-42-54

Numerical Evaluation of the Optimal Precoder Design with Delayed CSI

Alexander Kalachikov$^{(\boxtimes)}$ (iD)

Siberian State University of Telecommunications and Information Sciences,
Kirova Street 86, Novosibirsk 630102, Russia
330rts@gmail.com

Abstract. In paper the numerical investigation of the impact of delayed channel state information (CSI) due to user movement and caused channel aging on the performance of multiuser precoder in downlink MISO system. The time variant CSI is obtained by the minimum-meansquare-error (MMSE) channel estimation. We consider Zero Forcing (ZF) algorithm and numerical optimization based solution of calculating precoder vectors maximizimg sum rate of multiuser system. For numerical simulation the QUADRIGA channel model reflecting the real propagation conditions for moving users is used. The obtained performance of multiuser Zero Forcing and optimization based beamforming in spatially correlated channel are compared based on the empirical cumulative density function of the sum rate of multiple users.

Keywords: Multiuser precoding · QUADRIGA 3GPP channel model · ZF precoding · optimal precoding design MMSE channel estimation

1 Introduction

Massive multiple-input multiple-output (MIMO) systems enchance the capacity of multi-user MIMO systems by using beamforming over a transmit antenna array on the base station achieving spatial multiplexing gain [1].

Most of the performance gain of massive MIMO depends heavily on the accurate channel state information at the base station (BSs). The actual channel estimation is of great importance for precoding in downlink (DL) and combining in uplink (UL). In OFDM systems with moving user the rapidly time-varying channel may change over several consecutive symbols. For mobile users the channel impulse response is time varying and the coherence time is reduced. The signal processing at the base comprising channel estimation, scheduling, resource allocation, computing the precoding vectors causes the delays that can exceed the coherence time, the time duration after which CSI is considered as outdated.

The mismatch between the channel coefficients obtained by the channel estimation and used for precoding and the actual channel coefficients refers to channel aging [2].

The publication has been prepared according to the state order of Mintsifry Rossii No. 071-03-2023-001.

It was observed that the moderate-mobility scenario at 30 km/h leads to as much as 50 of the performance reduction compared to low-mobility scenario [3].

Therefore, the study of channel aging effects is crucial for complex system simulation in scenarios with moving users.

An aging channel model was considered in [3] under assumption that the temporal autocorrelation is described by the Jakes-Clarke model and equal Doppler shift assumption, resulting in an autocorrelation function (ACF) given by the zeroth-order Bessel function of the first kind. But real channels have limited angular spread of multipath components and the practical temporal correlation can differ from Jakes-Clarke model. For the numerical simulation of the channel estimation under channel aging the realistic channel model should be used reflecting a non-isotropic scattering with spatiall correlation.

In this paper, we investigate the effect of channel aging on a precoder performance in multiuser downlink MISO system with vehicle users in a more realistic scattering scenario with spatially correlated channels.

2   System Model

2.1   System Model

In the typical scenario [4] the Base Station with massive MIMO equipped N_T transmit and receive antennas serves K users UE each having one antenna. The N_T antennas at the BS are defined as two-dimensional (2D) uniform rectangular array (URA).

It is assumed that the system operates in TDD mode and the fast fading channel coefficient between BS and the k-th user moving user (UE) is represented by $\mathbf{h}_k(t)$. The uplink (UL) channel is estimated by sending pilots from each UE to BS and due to reciprocity is used as the downlink DL channel estimation.

During the channel estimation perion users transmit pilot sequences of length τ symbols. The channel remains constant during this period. In data transmission period $(T - \tau)$ OFDM symbols the channel varies from symbol to symbol.

In matrix form the MU-MIMO channel matrix for time index n and subcarrier index s is composed as $\mathbf{H}_{n,s} = [\mathbf{h}_{1,n,s} \ldots \mathbf{h}_{K,n,s}]^T$ and the received vector is defined as

$$\mathbf{y}_{n,s} = \mathbf{H}_{n,s}^T \mathbf{x}_{n,s} + n_{n,s} \tag{1}$$

By using precoding (beamforming) the received signal for user k is defined as

$$y_k = \mathbf{h}_k^T \mathbf{w}_k s_k + \sum_{j \in S, j \neq k} \mathbf{h}_k^T \mathbf{w}_j s_j + \mathbf{n}_k, \, for \, k = 1, \ldots, K \tag{2}$$

where the sum term corresponds to the interference from other users.

2.2   Channel Aging

Due to movement of the UEs the temporal variations in the propagation environment arise which affect the channel coefficient in a resource slots. The estimated channel response becomes outdated over time which degrades the system performance. System performance evaluation and design require an channel model that correctly describes the spatial and temporal correlations of the time varying channel coefficients. the time-variant channel vector for the k-th user at time n $\mathbf{h}_k[n]$ can be modeled as a function of its initial state $\mathbf{h}_k[0]$ and an innovation component [5] as

$$\mathbf{h}_k[n] = \rho_k[n]\mathbf{h}_k[0] + \overline{\rho}_k[n]\mathbf{g}_k[n]$$

where time 0 corresponds to the last symbol transmitted in the channel estimation period, $\mathbf{g}_k[n]$ represents the independent innovation component at the time instant n, $\rho_k[n]$ represents the temporal correlation coefficient of channel vector between the channel realizations at time 0 and n. For isotropic scattering propagation environment $\rho_k[n] = J_0(2\pi f_D T_s n)$, where $J_0(\cdot)$ is the zero order first kind Bessel function, T_s is the sampling interval (OFDM symbol duration), f_D is the maximum Doppler frequency shift for the user with velocity v and carrier frequency f_c. The difference between the actual channel $\mathbf{h}_k[n]$ and the part $\rho_k[n]\mathbf{h}_k[0]$ can be considered as the channel approximation error affecting on the performance of precoding for the interference reduce.

2.3   Channel Estimation

The precoder calculation require channel state information (CSI) on the BS side, which is obtained during an uplink training phase by exploiting channel reciprocity in a time-division duplex (TDD) system.

First the Least Square (LS) estimation is aplied to the received signal for each user on the pilot positions p of SRS sequence to obtain the esimate of the channel coefficients $\tilde{\mathbf{h}}_k[n]$. The LS estimate is computed by division of received symbols on corresponding values of pilot sequence. The estimate obtained at pilot symbol period is used as the initial state $\mathbf{h}_k[0]$ to compute estimates of the channels $\hat{\mathbf{h}}_k[n]$ at other symbols of this resource slot and period of pilot insertion in resourse slot depend on the channel aging effect.

The minimum mean square error (MMSE) estimation of channel coefficiens is computed as

$$\hat{\mathbf{h}}_k[n] = \mathbf{w}_k \tilde{\mathbf{h}}_k[n]$$

where $\mathbf{w}_k$ are wieghts of MMSE estimator [nguen]. The MMSE weights are computed as $\mathbf{w}_k = \boldsymbol{\Phi}_k^{-1}\mathbf{r}_k$ where $\boldsymbol{\Phi}_k = E\{\tilde{\mathbf{h}}_k\tilde{\mathbf{h}}_k^H\}$ - channel autocovariance matrix, $\mathbf{r}_k = E\{h_k^*\mathbf{y}_k\}$ is cross-covariance vector [6].

2.4   ZF Precoding

The ZF precoding vector $\mathbf{w}_k$ of the k-th UE is orthogonal to the transpose conjugate channel vectors of all other UEs $\mathbf{h}_k^H\mathbf{w}_j = 0$ for $j \neq k$. The ZF precoding matrix W is composed of all precoding vectors $\mathbf{w}_j$ and is computed as the pseudoinverse of the channel matrix of the selected users as $\mathbf{W} = \mathbf{H}(\mathbf{H}^H\mathbf{H})^{-1}$.

The sum rate is the sum of the rates achieved by the UEs on all subcarriers and is depend on the signal to interference noise ratio (SINR) of each UE. The SINR at the $k - th$ UE on one subcarrier is

$$SINR_k = \frac{|\mathbf{h}_k^T \mathbf{w}_k|^2}{\sum_{j \neq k} |\mathbf{h}_k^T \mathbf{w}_j|^2 + K\sigma^2/P} \tag{3}$$

The achieved multiuser sum rate is determined as

$$R_{BF} = \sum_{k=1}^{K} (log_2(1 + SINR_k)), \tag{4}$$

This metric is used to evaluate the spectral efficiency of considered precoding algorithms under channel aging effect [7].

2.5   Optimization Based Precoding Design

The performance of the ZF precoding in the case of fast moving user and delayed CSI is far from optimal. The task of finding optimal precoding vectors can be formulated and solved as optimization problem of maximizing the expected sum-rate over the realizations of the realistic channel model. The Second-Order Cone Programming (SOCP) formulation is used to find the stationary points of this optimization problems.

The objective is to design precoding vectors $\mathbf{w}_1, ..., \mathbf{w}_K$ that maximize the multiuser sum rate R_{BF} depending of individual user $SINR_k$ under the BS tramsmit power constraint.

$$\max_{w_1, ..., w_K \in C^{N_t}} R_{BF} \tag{5}$$

$$s.t. \sum_{k=1}^{K} \|\mathbf{w}_k\|^2 < P \tag{6}$$

The cost function R_{BF} depends on the SINRs which are non-convex functions of the precoding vectors $\mathbf{w}_1, ..., \mathbf{w}_K$. The design of transmit precoder vectors can be transformed as the problem of minimizing the total transmit power subject to SINR constraints γ_k at each of the K receivers. The values of γ_k are the minimal acceptable SINR for the kth user. The corresponding minimization problem is formulated as follows

$$\min_{w_1, ..., w_K \in C^{N_t}} \sum_{k=1}^{K} \|w_k\|^2 \tag{7}$$

$$s.t. \ SINR_k \geq \gamma_k \tag{8}$$

The solution of this problem gives the precoding vectors that achieves the required SINR using the minimum of power. The problem can be reformulated as a convex problem. The constraints $SINR_k \geq \gamma_k$ can be rewritten as

$$|\mathbf{h}_k^T \mathbf{w}_k|^2 \geq \gamma_k \sum_{j \neq k} |\mathbf{h}_k^T \mathbf{w}_j|^2 + \gamma_k \sigma^2.$$

The inner prodct $\mathbf{h}_k^T \mathbf{w}_k$ must be real valued and positive and the consaraint can be written as

$$\mathbf{h}_k^T \mathbf{w}_k \geq \sqrt{\gamma_k \sum_{j \neq k} |\mathbf{h}_k^T \mathbf{w}_j|^2 + \gamma_k \sigma^2}$$

$$Im(\mathbf{h}_k^T \mathbf{w}_k) = 0.$$

The reformulated SINR constraint is a second order cone constraint [8,9].

The optimization problem 8 can be efficiently solved using the software implementations of convex optimization metods, such as CVX [12].

3   Numerical Results

In this section the performance of the precoding algorithms are presented by computer simulation. The performances of ZF precoding and optimization based design are evaluated by a system level simulation in terms of the average spectral efficiency based on the QUADRIGA channel model [8,9]. Simulation parameters are provided in Table 1.

Table 1. Simulation parameters

Parameter	Value
Channel model	QuaDRiGa version 2.2
Scenario	BERLIN UMi NLOS
Center frequency	4.5 GHz
Number of subcarriers	512
Utilized bandwidth	20 MHz
Number of BS transmit antennas	16
Number of selected users	4
Speed of users	30, 60 kmh
Number of multi-path clusters	15

The simulation scenario is BERLIN UMi NLOS. According to the UMi scenario the users are uniformly distributed around the BS in the area of 200 m

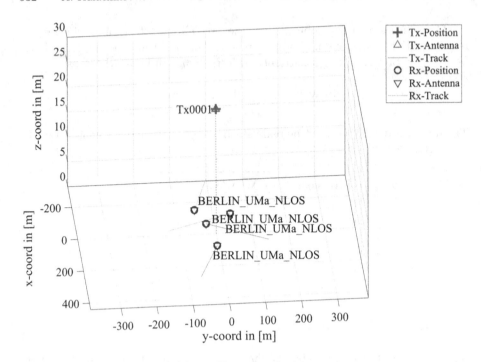

Fig. 1. Users distribution

from the transmitter. In Fig. 1 the random users distribution with corresponding linear tracks foe each user is presented.

The path loss model and shadow fading are disabled and the narrowband channel vectors for selected subcarrier are normalized to unit power. The BS is equipped with 16 antennas. The number of randomly distributed users is set as K = 4 and each user is equipped with single antenna. The center frequency is set to 4.5 GHz. For each of users the linear movement track with specified user speed is defined. The lengths of all the simulated tracks are equals 200 m which gives approximately 6000 snapshots of the channel impulse responses at speed 30 kmph and 20000 snapshots of the channel impulse responses at speed 60 kmph. The simulation scenario is set to BERLIN UMi NLOS. In Fig. 2 the resulting Doppler spectrum for channel of selected user at speed 60 kmph is presented. For the channel between BS and user 1 the Doppler shift is 250 Hz.

The generated channel for given user and transmit antenna at BS consist of multipath components. The frequency response is calculated for bandwidth 20 MHz according to 512 subcarriers of transmit signals. For each subcarrier the beamforming vector is computed according to algorithm ZF and optimal solution.

3.1 Selection of Interval of Pilot Symbols in Slot

Channel aging effect has to taken into account for the appropriate selection interval of pilot symbols insertion in the transmission slot. The CSI obtained in

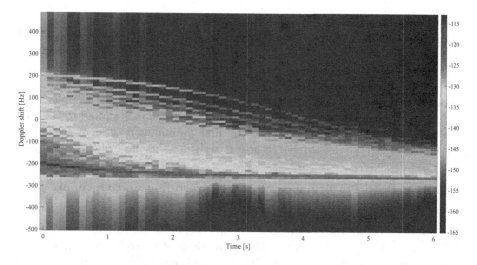

Fig. 2. Doppler spectrum of the channel of User 1 at speed 60 kmph

pilot position is used for precoding calculation in subsequent data symbols of
the slot. The delayed CSI affect on multi-user interference and degrades the user
SINR and the sum spectral efficiency. For selection of interval of the pilot symbols
in slot the user SINR for different speed of user movement were calculated uzing
ZF precoding (3) and the optimization based precoding.

In Fig. 3 the downlink SINR values are presented for two value of speed
of moving users at first 60 symbol index. The first symbol correspond to pilot

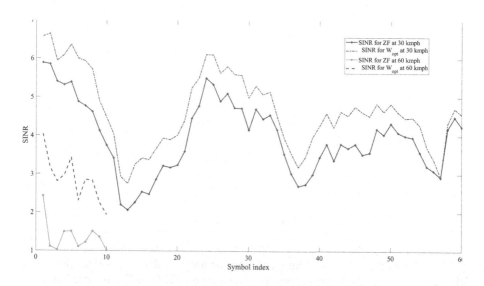

Fig. 3. Downlink SINR with symbol index

symbol and the other are the data symbols using the obtained channel estimate at the pilot position for precoder vector calculation. The corresponding MISO channel realizations were obtained in Quadriga channel model. The fluctuating values of SINR corresponds to temporal channel correlation coefficient. The update rate of the channel realization and the corresponding time instant of the model $T_s = 0.001$ s. When the speed of users and the normalised Doppler shift $f_D T_s$ are increased the position of the first SINR minimum is shifted to the left. The value of indice corresponding to the first minimum is used to select the reasonable length of pilot symbol insertion to reduce the effect of channel aging.

3.2   Sum Rate Performance of of ZF Precoding and Optimal Precoder Design

This section provides the numerical results to observe the impact of channel aging on the sum rate of MISO system. The comparison of precoder performances is presented using the cumulative distribution functions (CDFs) of sum rate. The SOCP optimization problem is solved numerically using the convex optimization software CVX [10].

The impact of imperfect CSI is presented for the speed of the users 30 and 60 km/h for SNR value equals 18 dB. Figure 4 provides an comparison of precoder performances of 4 users and for the speed of the users 30 kmph and 60 kmph.

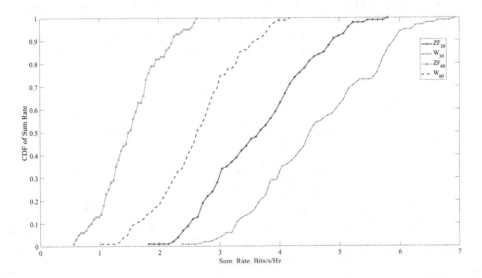

Fig. 4. CDF of Sum Rate

At low mobility of v = 30 kmph the performance difference is remarkable for ZF and optimization based precoders. The median SE for ZF precoder comprisesn 3.8 bps in comparison with the median SE for optimization based precoder

4.5 bps. The SE of the optimization based precoder outperforms the ZF precoder. The sum-rate of optimization based precoder is about 1.8 times of that of the ZF precoder at SNR = 18 dB for the 30 kmph.

With the increase of users speed the performance loss of ZF precoder becomes more notable. At medium mobility of v = 60 kmph the performance loss is remarkable for ZF and optimization based precoders. The median SE for ZF precoder reduces to 1.3 bps. The median SE for optimization based precoder reduces to 2.6 bps in comparison with user speed 30 kmph. The SE of the optimization based precoder as well outperforms the ZF precoder. The sum-rate of optimization based precoder is about 2 times of that of the ZF precoder at SNR = 8 dB for the 60 kmph.

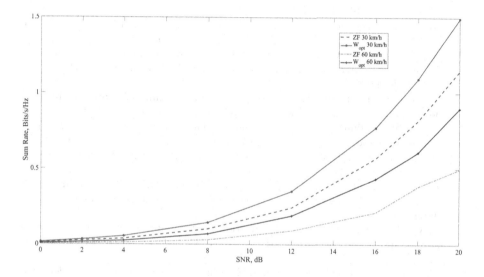

Fig. 5. Sum Rate vs SNR

Simulation results provided in Fig. 5 are the sum-rate performance for two speed values obtained by averaging of the sum-rate of 4 users for different SNR values. The performance gain of the optimization based precoder precoders compared to ZF become more significant when SNR increases for both of speed values.

4   Conclusion

Using the obtained SINR performance we can select appropriate interval between pilot symbols in the transmission slot not greater than the interval of the first minimum of SINR to reduce the effect of channel aging. The results show that the performance gain of precoder based on SOCP is more significant than ZF precoder in scenario with larger speed of users. The performance gain of the optimization based precoder precoders compared to ZF become more significant when SNR increases for both of speed values.

References

1. Castaneda, E., Silva, A., Gameiro, A., Kountouris, M.: An overview on resource allocation techniques for multi-user MIMO systems. IEEE Commun. Surv. Tutor. **19**(1), 239–284 (2017)
2. Truong, K.T., Heath, R.W.: Effects of channel aging in massive MIMO systems. J. Commun. Netw. **15**, 338–351 (2013)
3. Yin, H., Wang, H., Liu, Y., Gesbert, D.: Addressing the curse of mobility in massive MIMOWith Prony-based angular-delay domain channel predictions. IEEE J. Sel. Areas Commun. **38**, 2903–2917 (2020)
4. 3GPP, NR; Physical channels and modulation, 3rd Generation Partnership Project (3GPP), Technical Specification (TS) 38.211, 10, version 16.3.0
5. Chopra, R., Murthy, C.R., Suraweera, H.A., Larsson, E.G.: Performance analysis of FDD massive MIMO systems under channel aging. IEEE Trans. Wireless Commun. **17**(2), 1094–1108 (2018)
6. Nguyen, L.H., Rheinschmitt, R., Wild, T., Brink, S.: Limits of channel estimation and signal combining for multipoint cellular radio. In: Proceedings of the 8th International Symposium on Wireless Communication Systems, pp. 176–180 (2011)
7. Zheng, J., Zhang, J., Bjornson, E., Ai, B.: Impact of channel aging on cell-free massive MIMO over spatially correlated channels. IEEE Trans. Wireless Commun. **20**(10), 6451–6466 (2021)
8. Bengtsson, M., Ottersten, B.: Optimal and suboptimal transmit beamforming. In: Godara, L.C. (ed.) Handbook of Antennas in Wireless Communications. CRC Press (2001)
9. Yu, W., Lan, T.: Transmitter optimization for the multi-antenna downlink with per-antenna power constraints. IEEE Trans. Signal Process. **55**(6), 2646–2660 (2007)
10. Jaeckel, S., Raschkowski, L., Boerner, K., Thiele, L., Burkhardt, F., Eberlein, E.: QuaDRiGa - Quasi Deterministic Radio Channel Generator. User Manual and Documentation. Tech. Rep. v2.2.0, Fraunhofer Heinrich Hertz Institute (2019)
11. Jaeckel, S., Raschkowski, L., Boerner, K., Thiele, L.: QuaDRiGa: a 3-D multicell channel model with time evolution for enabling virtual field trials. IEEE Trans. Antennas Propag. (2014)
12. Grant, M., Boyd, S.: CVX: Matlab software for disciplined convex programming, version 2.1. www.cvxr.com/cvx

Correction to: Analysis of Queuing Systems under *N* Policy with Different Server Activation Strategies

Greeshma Joseph, Varghese Jacob(iD), and Achyutha Krishnamoorthy(iD)

Correction to:
Chapter 14 in: V. M. Vishnevskiy et al. (Eds.): *Distributed Computer and Communication Networks: Control, Computation, Communications*, **LNCS 14123,**
https://doi.org/10.1007/978-3-031-50482-2_14

In the originally published version of chapter 14, the affiliation of author Greeshma Joseph and Achyutha Krishnamoorthy has been updated to CMS College, Kottayam, India; and the affiliation of author Varghese Jacob has been updated to Government Arts and Science College, Nadapuram, India.

The updated version of this chapter can be found at
https://doi.org/10.1007/978-3-031-50482-2_14

© The Author(s), under exclusive license to Springer Nature Switzerland AG 2024
V. M. Vishnevskiy et al. (Eds.): DCCN 2023, LNCS 14123, p. C1, 2024.
https://doi.org/10.1007/978-3-031-50482-2_42

Author Index